Springer Texts in Business and Economics

More information about this series at http://www.springer.com/series/10099

Volker Böhm

Macroeconomic Theory

Springer

Volker Böhm
Faculty of Business Administration and Economics
Bielefeld University
Bielefeld, Germany

ISSN 2192-4333 ISSN 2192-4341 (electronic)
Springer Texts in Business and Economics
ISBN 978-3-319-60148-9 ISBN 978-3-319-60149-6 (eBook)
DOI 10.1007/978-3-319-60149-6

Library of Congress Control Number: 2017954931

Printed on acid-free paper

This Springer imprint is published by Springer Nature
The registered company is Springer International Publishing AG
The registered company address is: Gewerbestrasse 11, 6330 Cham, Switzerland

"The composition of this book has been for the author a long struggle of escape, and so must the reading of it be for most readers if the author's assault upon them is to be successful, – a struggle of escape from the habitual modes of thought and expression. The ideas which are here expressed so laboriously are extremely simple and should be obvious. The difficulty lies, not in the new ideas, but in escaping from the old ones, which ramify, for those brought up as most of us have been, into every corner of our minds."

John Maynard Keynes '*The General Theory of Employment, Interest, and Money*'

December 13, 1935

Preface

The main chapters of this book developed out of the core material of a graduate course in *Macroeconomic Theory* which was given at Bielefeld University over several years. At first the objective was to present the microeconomic foundations of macroeconomic analysis for an economy in the short run deriving prices, wages, output, and employment for a monetary economy within the framework of temporary general equilibrium theory emphasizing Keynesian features and contrasting the results under disequilibrium with those of Classical Macroeconomics. As new results resolving dynamic issues were obtained for both the disequilibrium as well as for the equilibrium modeling, comparisons between the two approaches were possible within the same general model. Emphasis of the course material shifted gradually from only a comparative statics analysis to a joint presentation of a static and a dynamic analysis side by side. In the end a synthesis was obtained of extending the traditional intertemporal sequential modeling to a full dynamic analysis for the equilibrium *and* the disequilibrium description of temporary allocations. This allows an analysis of macroeconomic issues with and without noise, in particular the consequences of stationary fiscal policies on allocations, on the quantity of money, as well as on the stability of the evolution of the economy.

The book offers a unified general equilibrium approach starting from the competitive temporary equilibrium model in the spirit of Hicks, Keynes, and others. The material contrasts and compares the two main competing approaches in macroeconomics within the same intertemporal structure of a closed monetary economy: the one postulating full price flexibility to guarantee equilibrium in all markets at all times under perfect foresight or rational expectations, versus the so-called disequilibrium approach where trading occurs at non-market-clearing prices and wages while these adjust sluggishly from period to period in response to market disequilibrium signals. In summary, the book presents a synthesis of equilibrium and disequilibrium analysis within a prototype monetary macroeconomic model unifying the features of temporary equilibrium and Keynesian models. Toward the end limits of time and space imposed the exclusion of material extending the model to include inventory and government debt. The available draft sections which were partly presented in class still have to be completed and brought to a level of presentation comparable to the current one.

A book with material of this scope is the outcome of several attempts initiated over many years without a systematic planning at the beginning. However, as the first ideas developed more than twenty years ago it became clear that I was led to write and circulate my own notes rather than following the presentation of chapters from macroeconomic textbooks or from published articles. Their specificity often did not fulfill my criteria to teach macroeconomics from a unifying perspective while using known tools from microeconomics, general equilibrium theory, and dynamic analysis to obtain the disequilibrium extension for a unifying macroeconomic theory.

Many colleagues, coworkers, and former students have contributed to the completion through remarks and suggestions to successive drafts. I am particularly grateful to Costas Azariadis, Thorsten Pampel, George Vachadze, and Gerd Weinrich who have read versions of chapters at the later stages and helped to make the final presentation more readable, containing fewer imprecisions, minimizing misunderstandings, and hopefully avoiding errors. While teaching the material not only in Bielefeld but also at international departments the contact with advanced students has been one of the major sources to understand the need for intuitive presentations of material when using new methods. Among the many students who have been involved in the presentations of the course material I am especially indebted to Oliver Claas who accompanied my teaching of the material during several rounds, tutoring, designing exercises and solutions, as well as substituting in class occasionally. Our research on labor market models with bargaining has led to joint results which are included partly in the book. His dissertation points at new and challenging findings to resolve issues in labor economics within the macroeconomic context. Over many years Oliver has accompanied the development of the manuscript with questions, critical remarks, and an invaluable support in the use of LaTeX suggesting software for figures, diagrams, and numerical results, and for programming support with MACRODYN. Hans-Walter Lorenz introduced me to the extended graphical world of pstricks within LaTeX providing graphical additions to describe qualitative properties geometrically which often require long verbal explanations. I was privileged to complete my endeavor during the last five years in a prospering scientific environment with student researchers, programmers, and colleagues. I am grateful for the support by the Department of Business Administration and Economics, the Center for Mathematical Economics (IMW), and the Research Center for Mathematical Modelling (RCM^2) at Bielefeld University providing generous assistance in my continued struggle uphill to reach a level of completion for the results which now make up the main message of the book.

Bielefeld, July 10, 2017 *Volker Böhm*

Contents

List of Figures

Chapter 1
Introduction

The descriptive framework of an economy with a decentralized and disaggregated market structure – as observed in most Western monetary economies – is characterized by dominant private ownership of individual economic agents. They interact in markets together with central organizations of significant size such as governments or central banks which represent the institutional elements of empirical economies with markets. They serve as the explanatory framework for the corresponding abstractions used in the theoretical building blocks of a macroeconomic model of an economy as laid out by general equilibrium theory of private ownership economies in the sense of Arrow and Debreu (see Arrow, 1951; Debreu, 1959; Arrow & Hahn, 1971) and extensions thereof (following Keynes, 1936; Hicks, 1939; Patinkin, 1965, and others).

From a historical perspective macroeconomic theory is a relatively young discipline within economics. R. G. D. Allen attributed the coining of the term *macroeconomics* to Ragnar Frisch in 1933 (see Allen, 1967, Chapter 1) most likely referring to Frisch's contribution in the famous Festschrift in Honour of Gustav Cassel 1933 as the source. There, however, he explicitly introduces the term *macrodynamics* and not macroeconomics. The first explicit mentioning of the term is yet unclear[1]. With the introduction of the term the division of economic analysis into the two subfields microeconomics and macroeconomics in the years was established to distinguish also between the two classes of models. Over the last eighty years both terms have undergone changes in their connotation and in the usage of criteria to distinguish between them. The recognition of the necessity to design an aggregative model for an economy with complete recording of intermarket flows was most forcefully presented in *The General Theory of Employment, Interest, and Money* by Keynes (1936). He seems to have offered one of the first complete macroeconomic analysis of the determination of national output, employment, and money holdings for a closed system of real and monetary flows arising from trade in interdependent markets.

[1] Others have been using and popularizing the term later on, but its first introduction in a publication seems still unknown even for active researchers at the time (see for example Samuelson, 1974)

Modern macroeconomic theory analyzes foremost the structure and development over time of an aggregative closed economy from a positive or descriptive point of view. Macroeconomic analysis also addresses principle normative questions of the evaluation of outcomes seen as being induced by regulatory measures enforced by the central agencies. In other words, macroeconomic theory is designed to serve as a toolbox for the control and the evaluation of the performance of economic policies over time based on a diversity of performance criteria among them the welfare of consumers in the economy as a whole.

1.1 A Unified Approach in Macroeconomic Modeling

Most modules used in macroeconomics are justified on the basis of an underlying microeconomic decentralized economy. One of the decisions to be made when designing a macroeconomic model is the choice of generality of the underlying microeconomic structure to be portrayed. The degree of heterogeneity of consumers, producers, commodities, and markets has to be chosen versus a reduced form of sectors describing prototypical microeconomic characteristics to represent the aggregate properties of the respective sector in the macro model. Independent of the choice of heterogeneity of agents and markets, three criteria seem to be of importance in evaluating the quality of a macroeconomic model to be used as an underlying structure for an empirically oriented dynamic analysis.

First: Current economic understanding supports the view that one of the essential features of a *macroeconomic* model is to guarantee aggregate consistency of monetary flows within the model between the interacting agents and markets (in contrast to a microeconomic one typically with limited consistency only).

Second: Since one of the major debates concerning macroeconomic configurations addresses issues of the possibility of unemployment – or potential disequilibrium in more general terms – it seems necessary to set the frame of the unifying model in such a way allowing equilibrium in markets as a special case of a more general theory of allocations under non-market clearing conditions.

Third: To guarantee a dynamic perspective for the analysis which allows comparisons between empirical measurable data and experimental/numerical time series a formulation is to be used which fulfills the requirements used in other disciplines treating dynamic phenomena allowing the application of methods from the theory of dynamical systems. The methods and modeling structures chosen in the sequel are an attempt to follow these three major principles to guarantee a unified approach to be maintained throughout the presentation and to be applicable for most issues to be discussed.

1.2 Closed Monetary Systems with Intertemporal Consistency

Building Blocks of Macroeconomics

From a methodological point of view the *macroeconomic* perspective of an economic analysis may be defined along different criteria depending on the objective of the investigation, on the essentials of the specific analysis, or of different degrees of generality. It is undisputed among theorists, however, that a reasonably interesting macroeconomic model should have a microeconomic underpinning from first principles and a consistent intertemporal embedding. In other words, an integration of the Arrow-Debreu model has to be found into the sequential framework with an infinite horizon setting as laid out, for example, in Hicks (1939). The theory of economic growth is the area where aspects of intertemporal allocations and sustainability play the decisive role (as, for example, in Azariadis, 1993). However, most contributions address primarily issues of intertemporal allocations within barter economies (often with one commodity/market only), a treatment of intertemporal allocative questions within monetary economies is excluded.

In order to go beyond barter economies (and to avoid some of the pitfalls when trying to analyze monetary aspects within real economies only; see Hahn, 1982) it is desirable to extend such one-commodity models to a general and sufficiently rich structure of the Arrow-Debreu-Hicksian type augmented by introducing money and other paper assets as the means to store consumption value intertemporally. The goal must be to describe a prototype macroeconomic model as a *complete and closed monetary system* of economic flows between agents at any one time, i.e. of all transactions between agents and institutions on markets, real and monetary, and of all transfers between agents satisfying *intertemporal consistency* with respect to money and assets. These requirements follow from the usage of the models of general equilibrium theory. They define the essential difference as compared to microeconomic models which typically exhibit an inherent partial, open, and incomplete recording of flows, transactions, and transfers. Their partial consistency implies a limited validity only of their properties for an economy as a whole. The completeness of recorded flows and intertemporal consistency has to be judged according to the principles of an integrated national income accounting system.

Standard terminology in economics uses the term *general equilibrium* modeling to mean complete recording (or consistency) of flows as opposed to the term *partial equilibrium* analysis for the situation of incomplete recordings of interagent flows or intermarket consistency. Using the word equilibrium is misleading, unfortunately, since neither the partial nor the complete recording of all interagent flows stipulates necessarily the notion of equilibrium in the sense of equality or consistency of desired transactions between agents.

Types of Models

There is an almost unlimited range of differentiating features to design economic systems with closed-flow properties. The macroeconomic literature reveals a rich range of the heterogeneity of markets, commodities, and agents:

- real models versus monetary models, markets with real or nominal assets,
- models with certainty and perfect foresight or uncertainty,
- deterministic or stochastic/random models,
- models with a finite or an infinite horizon,
- markets with time differential trading: spot, forward, and futures markets.

One of the fundamental results of the theory of economic closed-flow systems shows that the settlement of consistent intertemporal competitive allocations of an economy with a complete system of spot and futures markets and sufficient information can be sustained by a system of contracts decided on once and for all at the beginning of all times. The declared trades need to be executed only at all later dates. As a consequence, equilibrium future trading requires no money, no assets, no revision of plans, no reopening of trading, no spot markets in the future after settling prices and trades at one date only.

If in addition, the time horizon of such models is finite, nominal assets do not possess a positive terminal value. Under rationality and by induction nominal assets do not have a positive value at any earlier trading period. As a consequence, paper assets are redundant in such economies and have zero value. Therefore, models with a finite time horizon do not qualify as useful models for the description of dynamic issues in monetary economies relating to empirical macroeconomic phenomena. The argument for the redundancy of assets hinges on the assumption that the notion of equilibrium ignores intertemporal consistency with respect to the past. In other words, such equilibria have no past. The initial state has no predecessor in time, a condition which is difficult to be fulfilled in empirical macroeconomics. Therefore, a sufficiently rich macroeconomic analysis describing monetary and empirically observable dynamic features – built on market exchanges and valuations with agents optimizing intertemporally – requires a dynamic model

- with money, bonds, or other paper assets for intertemporal storage of wealth,
- with heterogeneous agents and markets, i.e. with more than one real commodity,
- with infinite horizon, to allow for positive valuation of assets in equilibrium,
- with uncertainty and an incomplete system of contingent and futures markets.

1.3 Temporary Disequilibrium versus Equilibrium

A theory of allocations in intertemporal economies with multiple markets and agents in closed-flow systems defining *equilibrium configurations* only cannot describe and analyze simultaneously *disequilibrium configurations* and the process of search for a price system *in any period* which is supposed to guarantee temporary equilibrium[2]. When prices do not induce market clearing the equilibrium paradigm does not define trades, incomes, and budgets independently of whether there is no equilibrium or whether the search process to equilibrate demand and supply has not reached a fixed point. Nothing can be inferred about economic activities from the model in such disequilibrium situations about the existence or multiplicity of a temporary equilibrium since it is simply assumed. If the search process for equilibrium prices is perceived of as a tâtonnement procedure (as in the Walrasian tradition for given demand and supply functions of agents), trade is not allowed while prices adjust. Therefore, the 'price dynamics of tâtonnement' remains a virtual mechanism without trading. No trade levels or incomes are determined during the adjustment. Thus, to describe the price dynamics in a market as a tâtonnement process has little relevance as a descriptive mechanism for dynamic characteristics of an economy. From a macroeconomic perspective this means, in particular, that all aggregate measures of income and expenditures, of trade volume or levels of activity, even levels of employment/unemployment are not defined as long as prices change. No observable data would be generated from such a virtual experiment.

To overcome this deficiency in designing a macroeconomic model it is necessary to model temporary market situations so that trades, payments, and incomes are well defined at non-equilibrium prices. This should be done in such a way that equilibrium configurations arise as special cases of disequilibrium. The so-called non-Walrasian approach, developed in the mid-1970's and presenting an extension of general equilibrium theory originating from Keynesian ideas, was designed to describe disequilibrium situations in markets (see for example Benassy, 1975b; Barro & Grossman, 1976; Malinvaud, 1977). Its main principles can be used to describe microfounded disequilibrium allocations for all markets for any temporary situation at non-Walrasian prices. This defines trades, incomes, and budgets in an arbitrary period. However, it begs the immediate question of how prices will be adjusted for the next period. The temporary disequilibrium configurations should not be considered 'an equilibrium' in any sense, and the rationing occurring should be used to define associated price adjustments, making the disequilibrium configuration a truly short-run event.

Therefore, in order to design a dynamic macroeconomic theory allowing disequilibrium configurations within the Hicksian intertemporal framework one is to develop an intertemporal theory of disequilibria *together* with an associated mechanism of prices changing between periods. These have to be defined jointly as a non-tâtonnement process since they must be related conceptually to the rationing configuration. The 'equilibrium over time' (as suggested by Hicks, 1939) should be

[2] Hicks (1939) was well aware of the fact that finding the temporary equilibrium in every period required another process of iteration which he neglected by assumption. "Since we shall not pay much attention to the process of equilibration which must precede the formation of the equilibrium prices, our method seems to imply that we conceive of the economic system as being always in equilibrium." Hicks (1939), p. 131. By assuming existence of a unique temporary equilibrium his position remained totally classical in spite of the fact that Keynes (1936) three years before published his fundamental critique of the classical theory of unemployment (as exemplified by Pigou (1933); see Keynes (1936), Chapter 19).

extended to more general situations of sequences of prices *and* of disequilibria to overcome the dichotomy of a virtual price adjustment as tâtonnement on the one hand (to determine the equilibrium) and the determination of trades, transfers, and budgets on the other.

The goal is to present the equilibrium and the disequilibrium configurations within the same microeconomic model for an identical aggregative monetary environment. When the economy is simple – and specific enough to guarantee a unique temporary equilibrium price vector at each date and arbitrary levels of expectations and all state variables – sequences of prices, expectations, and allocations *in equilibrium* serve as a benchmark for the dynamics under market clearing for a monetary economy. For an operative macroeconomic theory, however, the modeling framework should be of the disequilibrium type in any period at given prices, expectations, and other state variables. In this way, a tâtonnement search for equilibrium prices is avoided. The joint adjustment of prices, expectations, and all other state variables should be the object of the intertemporal analysis. The stationary states[3] of these disequilibrium dynamics must be the objects to be examined in the long run. The convergence in the sense of dynamic stability of the disequilibrium sequences should be the criteria for their observability and predictability, see Section 1.4. In the end this leads to extending the static Arrow-Debreu Model with equilibria not to another static variant of *equilibria at non-Walrasian prices*, but to an intertemporal Arrow-Debreu-Hicks-Keynes Model under price dynamics and disequilibrium determination of trades, transfers, and budgets. This modeling procedure also provides the possibility to examine and integrate the existing partial disequilibrium models of the Keynesian literature (specifically those of the AS-AD type as well as of those using the IS-LM framework) as special cases of a temporary disequilibrium configuration.

1.4 Recursive Intertemporal Economies as Dynamical Systems

From Intertemporal Equilibrium to Dynamics

A description of the dynamic evolution of an economy as an *intertemporal equilibrium process under consistent expectations* (as an 'equilibrium over time' in the sense of Hicks, 1939, p. 132) excludes two essential features from the modeling: the possibility to describe and analyze *disequilibrium configurations* as well as *non-perfect predictions* of a macroeconomic process over time. Hicks himself had seri-

[3] It is an open issue to find conditions under which steady states in a monetary environment exist, in particular to describe the properties of price adjustment rules whose fixed points induce disequilibrium trades in the intertemporal environment. It is also unclear in general whether these correspond to the equilibria at fixed prices in the sense of the literature (as used by Drèze, 1975, for example). It seems that, in a competitive environment, there can be no steady states with disequilibrium if prices always respond in a sign preserving way to disequilibrium signals.

ous doubts that an economy with temporary equilibrium in each period could ever evolve with perfect foresight, p. 132:

> "An economy in perfect equilibrium over time is like the sun in *Faust*:
>
> ihre vorgeschrieb'ne Reise
> vollendet sie mit Donnergang.
>
> The degree of disequilibrium marks the extent to which expectations are cheated, and plans go astray. No economic system ever does exhibit perfect equilibrium over time; nevertheless the ideal is approached more nearly at some time than others."

He uses the term disequilibrium to describe intertemporal imperfections and discusses a list of reasons why these occur naturally, primarily because plans cannot be sufficiently coordinated intertemporally except possibly in a so-called 'Futures Economy'.

The restrictive implications of the principle of market clearing for a macroeconomic allocations in a given period as the only descriptive device was discussed in detail in the previous Section 1.3. It will be argued below that these appear even more forcefully in the intertemporal setting to abandon the concept of intertemporal or sequential allocations in favor of a strictly dynamic macroeconomic model. Assuming in addition equality of prices with previous predictions to prevail along the complete time path of an economy excludes in addition the possibility to characterize intertemporal inconsistencies or *non-perfect predictions* in a macroeconomic recursive environment. Both phenomena are commonly observed features of macroeconomic time series, a fact which seems worth a consideration to include these as explanatory factors within a theoretical model as well. From a time series point of view it is desirable to replace the intertemporal modeling of the sequential equilibrium structure by a forward recursive dynamical system for two fundamental reasons.

First: Since out-of-equilibrium trading is not part of the description of the model, intertemporal equilibria are assumed to be reached via a perceived equilibrating process as in a tâtonnement world. This is assumed to be uniquely stable, occurring outside the model and, therefore, an unobservable construct for the modeler. As a consequence, an equilibrium of the economy is a solution of an *infinite system* of equations, i.e. a fixed point, and *not* the stationary rest point of a dynamical system describing a disequilibrium adjustment of a non-tâtonnement environment. In this situation no properties of such a process are derived from economic characteristics of the model. In other words, the properties of equilibrium states are totally disconnected from the dynamic description of the perceived adjustment process whose rest points are the equilibria.

For the same formal reason, the permanent consistency of forecasts/expectations and future events becomes a fixed point in the space of sequences defined as a simultaneous equality of an infinite list of forecasts and realizations independent of any rule which may link past observations to predictions for future events. Thus, equilibria with perfect foresight or rational expectations are not the result of a sequential (or recursive) process of adaptation or learning where inequalities are allowed and whose limiting states exhibit the postulated rationality. Such an expectations process is not modeled and conceptually detached from the equilibrium rationality postulated for the solution. It neglects essentially the interaction between the forward recursive expectation formation and its impact on the outcome for future temporary equilibria, i.e. denying the existence of an expectations feedback. Thus, the characterization of allocations and expectations over time as an equilibrium in a sequential but *non-recursive* way (in the strict sense of the theory of dynamical systems) makes the macroeconomic approach essentially static and not accessible to the methods of dynamical systems theory or econometric time series analysis. To cite Grandmont:

> "The fact that time appears explicitly as an index in this formulation should not deceive the reader. The above equilibrium concept is inherently static: *all elements of the sequences* (x_t) *and* $(x^e_{t,t+1})$ *are determined simultaneously* by an outside observer, hence emphasis on the word 'intertemporal' in the definition. All markets, past, present, and future are equilibrated at the same time.", Grandmont (1988), p. XIII–XXIV.

Second: To investigate dynamic properties of such equilibria becomes very difficult or even impossible, since the mathematical techniques for a structural analysis of stability, of dependence on parameters, and on initial conditions are usable for dynamical systems only. They are not applicable to systems of implicitly defined solutions. For many sufficiently complex macroeconomic systems intertemporal equilibria with consistent expectations may not exist at all, so that a dynamic equilibrium analysis under consistent expectations is impossible per se. For many macroeconomic equilibrium systems with expectational consistency it is also not known whether a forward recursive form exists. Often a recursive solution may exist, but it cannot be found or characterized analytically, or it exists only locally and cannot be extended to forward invariant sets.

Regarding equilibria in stochastic environments research of the past years has shown that intertemporal equilibria in models under rational expectations often cannot be shown to exist, which may be due to inconsistent policy variables, bankruptcies, or other parametric restrictions in the model. If they do the solutions may turn out to be non-stationary or dynamically unstable. In this case, the equilibrium theory of rational expectations equilibria becomes vacuous and inapplicable. Instability implies essentially that the economic development of any such model could not be observed in an experimental environment since the time series would be diverging. Thus, such models cannot serve in a meaningful way as a measuring rod to compare the theoretical findings with empirical data. These are often assumed to be drawn from a stationary environment. Since equilibria are defined as simultaneous solutions of *a complete set of infinitely many conditions* to hold at all times, no 'out of equilibrium' allocations can be accounted for, so that the same critique as in the deterministic case applies[4]:

[4] This debate (see for example Chapter 1 Lindahl, 1939) is as old as equilibrium theory itself after Walras. He defines equilibrium as a zero of excess demand, a necessary condition of a stationary state of a dynamic price process which is not made explicit. The *perceived* adjustment of prices assumed to operate in the background leading to equilibrium and referred to as a tâtonnement process remains virtual. No trading occurs *before* equilibrium is reached, a fact which limits the

- adjustments of prices and trading out of equilibrium are not part of the descriptive framework of the model and cannot be analyzed,
- the role of non-rational expectations cannot be assessed excluding an analysis of the effects of the correction of beliefs and of learning,
- the stability of intertemporal adjustments cannot be examined within the model since it lacks a *forward* recursive structure.

This identifies equilibrium macroeconomics as an extension of static Walrasian equilibrium theory under rational expectations and tâtonnement without descriptive elements for a theory of price and stock adjustments, of expectations adjustment, and without an operative concept of stability.

In summary, these arguments imply three good reasons to try to develop a strictly forward recursive theory in macroeconomics by following the principles laid out above.

(1) In order to escape the tâtonnement trap, it is necessary to formulate *explicitly* the laws of motion of all variables describing a state of the economy as part of the conceptual primitives. This means, in particular, that the *adjustment of prices* needs has to be defined (together with the main state variables) as a consequence of trading at non-market-clearing prices and the *adjustment of expectations* has to be given as a result of correcting predictions. Under special circumstances and specific conditions recursive rules may be found and specified under which the orbits of the economic dynamical system describing economic activities at each date are in fact equilibria with rational expectations. To identify these special macroeconomic environments and analyze the properties of their mechanisms is one of the first objectives of such a dynamic macroeconomic theory.

(2) To ward off against the general cases when equilibria may not exist, a theory of trading at non-Walrasian prices in any one period has to be defined. It provides a formal basis independent of whether equilibrium prices and allocations exist, of whether they are unique or not, or whether they are analytically computable. The disequilibrium trading rules should be described in a forward recursive format since they correspond essentially to allocation mechanisms generating outcomes from signals. Most expectation formation principles correspond to statistical methods based on data from time series. Their computation again guarantees a forward recursive mapping.
The complexity of interacting markets with disequilibrium trading calls for high dimensional nonlinear models typically without analytical or even computable solutions for temporary equilibria or fixed points. Thus, with an explicit recursive structure for trading, there is no conceptual restriction to the degree of heterogeneity, as long as the size of the model does not outgrow the computing facilities of available machines. A priori, there is no need to consider only specific analytical forms of outcomes in any one period.

---

usefulness of such models for markets with non-converging price dynamics: agents never trade out of equilibrium!

(3) Finally, once the model is formulated as a dynamical system the tools of mathematical theories become available, alongside with powerful computational facilities for extensive and highly informative qualitative experimentation. Both of these have been improved greatly during the last decades to deal with non linearities *and* random perturbations simultaneously. Such *random dynamical systems* are an appropriate mathematical framework as time series generators for economic models. Their properties can be analyzed using numerical procedures to obtain data to examine stability, characteristics of stationary orbits, and to carry out a bifurcation analysis. This also provides a reliable tool to investigate *endogenous* causes of instability and of business cycles and to carry out reliable comparisons of numerical orbits with data from empirical time series.

1.5 The Principle Features of the Book

The motivation to develop a monetary macroeconomic model of an Arrow-Debreu-Hicks-Keynesian economy described as an explicit dynamical system derives its justification from the general result presented in Chapter 3.

> *Under a set of acceptable standard neoclassical assumptions there exists a sufficiently rich class of low dimensional macroeconomic models within a recursive extension in the Keynesian spirit possessing a consistent embedding of the equilibrium paradigm both with respect to the issue of equilibrium versus disequilibrium as to adaptive versus rational expectations formation.*

This resolves the issue of nonexistence with the intertemporal equilibrium approach. Rational expectations equilibria are generated as orbits of dynamical systems under specific structural neoclassical assumptions. The descriptive features of Keynesian theory supply the modules to be used to extend the equilibrium model to a consistent disequilibrium theory for all markets (using the rules of rationing as in Benassy, 1975b; Barro & Grossman, 1976; Malinvaud, 1977, and others) for a short run analysis in a prototype model. In other words, there exists a unifying dynamic monetary macroeconomic model with a well specified microfounded intertemporal structure inducing orbits which can be tested against the standard criteria of equilibria and rationality. There seem to be no conceptual restrictions to economic situations inducing higher dimensional models but maintaining the explicit dynamical systems structure. Thus, extensions to more general workable models with inventories, government debt, bonds or assets, and other more extended generalizations are feasible.

Embedding the disequilibrium features in an overlapping generations model of consumers, the same natural microeconomic basis as in competitive business cycle theory is used with a disequilibrium trading rule. Standard price and wage adjustment principles which respond to disequilibrium situations in each market are invoked to describe explicit dynamics of prices and wages. The process of expectations formation is described by a class of recursive forecasting rules, which include as special cases those inducing rational expectations or perfect foresight, if they exist. In this way, an explicit forward recursive dynamical system is obtained. Moreover, the conditions are provided under which recursive perfect forecasting rules exist under market clearing implying the possibility of dynamic orbits along which market clearing and rational expectations prevail. Taking a strict *dynamic* perspective creates the two important methodological advantages of making the results (1) accessible to numerical analysis in deterministic and in stochastic environments and (2) comparable to and testable against empirical data.

The book does not treat any aspects of economic growth. It presents a dynamical monetary model of a stationary private ownership economy with overlapping generations of consumers of the Keynesian type.

Chapter 2 begins with a presentation of the microeconomic foundations of an intertemporal theory of trading among overlapping generations of consumers along the lines of the literature on intertemporal closed-flow economies (as suggested by Hicks, 1939; Patinkin, 1965; Grandmont, 1974, 1983). Sections 2.1 to 2.3 present the general case of exchange economies discussing the role of expectations for the existence of temporary equilibrium and providing sufficient conditions for two basic existence results of temporary equilibria.

Chapter 3 presents the generic microeconomic elements of the prototype macroeconomic model with atemporal production. Section 3.2 provides a complete description of allocative properties of the competitive temporary general equilibrium of a monetary economy. Sections 3.3 to 3.7 analyze extensions to noncompetitive equilibria with monopolistic competition on the commodity and on the labor market, with bargaining, and efficiency wages. Situations under noncompetitive commodity pricing are analyzed in Section 3.3 which containing a discussion of the so-called New Keynesian approach. Sections 3.4 to 3.6 present a detailed analysis of alternative forms of noncompetitive wage determination with subsections presenting producer and union monopolies (3.4.1 and 3.4.2), efficiency wages (3.7.3), and efficient wage bargaining (3.5.2). The allocative properties of the different cases are compared in Section 3.6 for an economy with isoelastic characteristics of consumers and producers. Section 3.7 presents a specific example of endogenous wage rigidities under competitive conditions.

Chapter 4 provides an analysis of the dynamics of the benchmark *equilibrium model* with *perfect foresight* in the deterministic case and under *rational expectations* with stochastic production. It contains a detailed presentation of the tools and concepts necessary to obtain a strictly forward recursive model whose orbits generate equilibria with the required properties. Sufficient conditions for existence and stability are provided, indicating the limitations of a dynamic description of rational expectations equilibria. **Chapter 5** applies the analytical methods of Chapter 4 to establish the dynamic features of basic cases of stationary fiscal and monetary policies.

Chapters 6 and 7 extend the temporary equilibrium setup to a short-run Keynesian disequilibrium model following the non-Walrasian tradition. The allocative properties are presented in detail in Chapter 6 with a complete analysis of comparative statics results usually associated with the effectiveness of short-run policy measures. Section 6.3 shows how the IS-LM model is to be integrated into the disequilibrium framework. The dynamic analysis of the disequilibrium extension with adaptive/recursive adjustments of prices and price expectations follows in Chapter 7. It shows that the disequilibrium adjustment induces endogenous business cycles through the impact of specific nonlinearities of the model *and* of the presence of inherent regime switching between different disequilibrium regimes. A multitude of attracting business cycle phenomena occur depending on the ranges of the parameters. The analysis describes the relevant ranges of these using bifurcation theory and numerical analysis identifying causal relationships between policy parameters, productivity parameters, and their impact on business cycle characteristics.

Random perturbations are introduced in the disequilibrium model in **Chapter 8** which presents a detailed numerical analysis of different experimental scenarios using i.i.d. perturbations of selected parameters. The analysis exhibits and reveals first and foremost the power of a numerical analysis available for explicit recursive nonlinear random dynamical systems to provide experimental qualitative and quantitative results of statistical relationships for attracting stationary time series.

Second, the experiments provide a contrast to standard usage of numerical methods in stochastic general equilibrium models. They indicate the sources for differences in the results between the nonlinear dynamic approach here and results for the same class of models if linearizing techniques were to be used for *nonrecursive* models to approximate the stationary solution of a nonlinear *recursive system.*

Third, the analysis describes numerous scenarios of business cycles in stochastic environments using the principles of bifurcation theory, i.e. examining the role of the non-stochastic parameters on the long-run development of the endogenous variables. These are applicable for any well specified parametric model of the class of economies presented here. Their properties can be determined experimentally and numerically. The numerical results show a wide range of economic and statistical features, characterizing, for example, properties of a long-run Phillips curve, of the real-wage-employment tradeoff, or of the role of fiscal policy.

Chapter 2
Microeconomic Foundations

2.1 The Basic Intertemporal Model with Money

The description of the intertemporal trading structure of commodities, agents, and markets is the one originally proposed by Hicks (1939) and used later by Patinkin (1965). The mathematical presentation here follows Grandmont (1983) which corresponds to an extension of the general finite-dimensional Arrow-Debreu economy to an infinite repetition (sequence) of trading periods of finite-dimensional economies. Let time evolve in a discrete way, $(\ldots, t-1, t, t+1, \ldots)$; $\quad t \in \mathbb{Z}$, where a trading period t is interpreted as the shortest conceptual time length during which markets are opened and trading at equilibrium prices is determined[1]. Let $l \geq 1$ denote the number of commodities traded in each period, with one market for each commodity. Money serves as a unit of account. At first and for simplicity, it is also the only store of value between periods, no credit is allowed, no other assets are available. These can be introduced in a straightforward way. Prices of commodities are expressed in terms of money and trading in markets is carried out in exchange for money.

At any date/period t the consumption sector of the economy consists of finitely many consumers each with a finite lifetime remaining in the economy. Let $A \neq \emptyset$, $|A| < \infty$, denote the set of consumers in the market in period t with the typical consumer denoted $a \in A$. For each $a \in A$, the integer $n_a \geq 0$ denotes the number of periods of economic life of a remaining after t. In other words, $1 + n_a$ describes the total number of periods of economic activity of agent a.

Given these general data, consumers are characterized in the usual way by their consumption sets, their preferences, and their endowments. For the purposes here, assume that consumers can consume nonnegative amounts of all commodities in any period, so that the individual life time consumption sets are equal to the nonnegative orthant $\mathbb{R}_+^{l(1+n^a)}$. Therefore, the consumption bundles are given by vectors of the form $c^a = (c, c_\tau) := (c, c_1, \ldots, c_{n^a})$ with $c_\tau \in \mathbb{R}_+^l$ and $\tau = 1, \ldots, n^a$.

[1] Hicks refers to a trading week where markets open on Monday and equilibrium prices are determined by Friday. Then trades are carried out and goods and services are exchanged against money at equilibrium prices.

Let $e^a_\tau \in \mathbb{R}^l_+$; $\tau = 0, \dots, n^a$ denote consumer a's endowment of commodities in period τ which is assumed to be given and known with certainty. Define $e^a = (e_0, e_1, \dots, e_{n^a}) \in \mathbb{R}^{l(1+n^a)}_+$ to be his life time endowment of commodities which is a feasible consumption plan under the assumptions made. In addition to his lifetime endowment of commodities consumer a has an initial nonnegative quantity of money $m^a_0 \geq 0$ which may be the result of savings in the previous period or of a transfer. Intertemporal preferences of each consumer a are described by a continuous utility function $u^a : \mathbb{R}^{l\cdot(1+n^a)}_+ \to \mathbb{R}$; $c^a \mapsto u^a(c^a)$ defined on the consumption set $\mathbb{R}^{l\cdot(1+n^a)}_+$.

2.2 Intertemporal Decisions of Consumers

Let $(p, p^a) := (p, p^a_1, \dots, p^a_{n^a}) \in \mathbb{R}^{l(1+n^a)}_+$ denote the list of *intertemporal* prices for consumer a, where $p \in \mathbb{R}^l_+$ are current market prices known and common to all consumers but $p^a := (p^a_1, \dots, p^a_{n^a})$ are the specific expected/predicted prices for future periods by each consumer a which can differ across consumers. Let $c^a := (c, c_1, \dots, c_{n^a}) \geq 0$ denote a consumption plan and let $m^a = (m_1, \dots, m_{n^a}) \geq 0$ a savings plan of a. These are called *feasible* for prices $(p, p^a) \geq 0$ if

$$pc + m_1 = pe_0 + m^a_0$$

$$p^a_\tau c_\tau + m_{\tau+1} = p^a_\tau e_\tau + m_\tau, \qquad \tau = 1, \dots, n^a$$

A feasible concumption plan $\tilde{c}^a$ is called an *optimal consumption plan* for prices (p, p^a), if it is a solution to the optimization problem

$$\tilde{c}^a = \Xi(m^a_0, p, p^a) := \arg\max_{c^a} \left\{ u^a(c^a) \;\middle|\; \begin{aligned} pc + m_1 &= pe_0 + m^a_0 \\ p^a_\tau c_\tau + m_{\tau+1} &= p^a_\tau e_\tau + m_\tau, \\ \tau &= 1, \dots, n^a \end{aligned} \right\}. \tag{2.2.1}$$

The mapping $\Xi(m^a_0, p, p^a)$ is consumer a's demand function (correspondence). Consumer theory yields the following result for the optimization problem (2.2.1).

Proposition 2.2.1.
Let u^a be continuous, strictly concave, and strictly monotonically increasing. For every $(p, p^a) \gg 0$, $(e^a, m^a_0) \gneqq 0$, (2.2.1) has a unique solution and $\Xi(m^a_0, p, p^a)$ is a continuous function.

For the analysis of allocations and equilibria in the current period it is apparent that trading in the current period affects the pair (c, m_1) while the remaining components of the optimal consumption plan $\tilde{c}^a$ and of m_τ are intended to be carried out in the future. Thus, it is useful to consider the list of current intended trades (c^a, m^a_1) of each consumer. Let $\varphi^a(m^a_0, p, p^a)$ denote the first l components of $\Xi^a(m^a_0, p, p^a)$, describing the consumption demand in the current period. In addition, define the desired money holding at the end of the current period as

$$\mu^a(m_0^a, p, p^a) := m_0^a + pe_0 - p\varphi^a(m_0^a, p, p^a). \qquad (2.2.2)$$

Equivalently, this yields $\mu^a(m_0^a, p, p^a) - m_0^a$ as consumer a's net savings. The two functions φ^a and μ^a are the relevant functions to determine behavior in the current period. All other variables are optimal plans to be carried out in the future. They can be adjusted in future periods (after reoptimization under new conditions), unless price expectations and endowments for future periods have been foreseen correctly. Therefore they are irrelevant for current demand and supply behavior. The following properties of the demand for consumption commodities and for money hold.

Proposition 2.2.2.
Under the assumptions of Proposition 2.2.1:

$$\Xi(m_0^a, p, p^a) \text{ is homogeneous of degree zero in } (m_0^a, p, p^a) \qquad (2.2.3)$$
$$\varphi^a(m_0^a, p, p^a) \text{ is homogeneous of degree zero in } (m_0^a, p, p^a) \qquad (2.2.4)$$
$$\mu^a(m_0^a, p, p^a) \text{ is homogeneous of degree one in } (m_0^a, p, p^a) \qquad (2.2.5)$$

satisfying

$$p\,(\varphi^a(m_0^a, p, p^a) - e_0) + \mu^a(m_0^a, p, p^a) - m_0^a = 0 \qquad (2.2.6)$$

for all $(m_0^a, p, p^a) \in \mathbb{R}_+ \times \mathbb{R}_+^{l(1+n^a)}$.

Rewriting these properties in terms of the excess demand function one obtains

$$z^a(m_0^a, p, p^a) := \varphi^a(m_0^a, p, p^a) - e_0.$$

This implies $pz^a(m_0^a, p^a) + \mu^a(m_0^a, p, p^a) - m_0^a = 0$, i.e. the value of individual excess demand in the goods market and the money market is equal to zero for all current prices. In other words, in temporary analysis nonsatiation and binding budget constraints imply that Walras' Law holds for every agent a individually *for all current prices p and* any list of *future expected prices* p^a. However, the demand functions are no longer homogeneous of degree zero in current prices alone, since changes of the latter at *given expected prices* for the future clearly influence relative intertemporal prices. This implies changes of intertemporal rates of substitution and therefore of optimal consumption plans.

2.2.1 The Geometry of the Budget Set and of Demand

It is useful to examine whether there are qualitative differences in consumption demand as compared to standard static one-period optimization. As an example, consider a typical consumer with a two-period optimization decision, i.e. with $n^a = 1$, $l = 1$, $m_0 > 0$, $(e_0, e_1) \gg 0$. His intertemporal consumption demand is defined by

$$\max \{u(c, c_1) \,|\, pc + m = pe_0 + m_0,\ p_1 c_1 = m + p_1 e_1,\ m \geq 0\}.$$

For smooth and concave utility functions $u(c, c_1)$ standard consumer theory implies typical representations of price effects and of income/wealth effects in a two-dimensional representation $(c, c_1) \in \mathbb{R}^2_+$.

Wealth Effects

The slope of the budget line in consumption space $\mathbb{R}^2_+$ is given by the expected rate of inflation p/p_1 and the restriction of no credit implies a kink in the intertemporal budget set of each consumer, see Figure 2.1 **a**. Therefore, changes of initial money

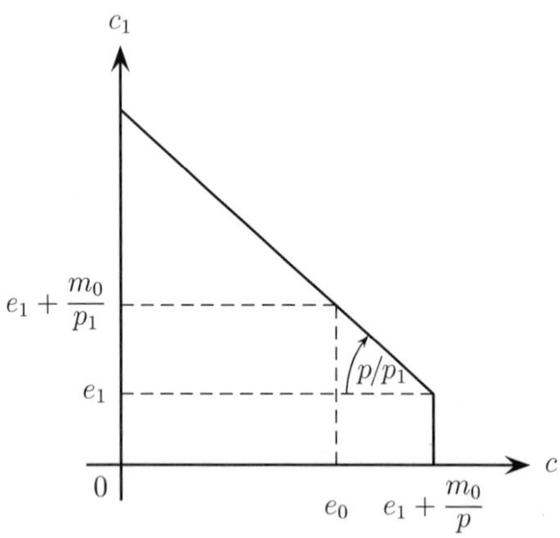

a Budget set

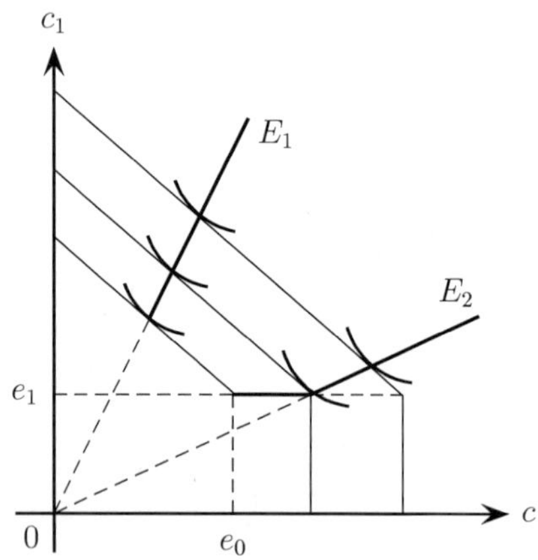

b Two expansion paths: $p/p_1 \gtrless MRS(e_0, e_1)$

Fig. 2.1 The role of money balances; p, p_1 constant

balances $m_0 \geq 0$ induce regular wealth-expansion paths as long as the marginal rate of substitution at the endowment point $(e_0 + m_0/p, e_1)$ is greater than or equal to the (inverse of) the expected rate of inflation p/p_1. In particular, preferences satisfying

$$MRS(e_0, e_1) < \frac{p}{p_1}$$

and normality of consumption in both periods implies interior consumption plans for all $m_0 \geq$ and with $c < e_0$ for $m_0 = 0$. This induces a strictly monotonically increasing expansion path as E_1 in Figure 2.1 **b**. For $MRS(e_0, e_1) > p/p_1$, the optimal consumption plan at $m_0 = 0$ is the corner solution $(c, c_1) = (e_0, e_1))$ and the expansion path is horizontal with $c_1 = e_1$ up to a critical positive value of m_0, as shown by E_2. For larger values the optimal consumption plan does not have a binding credit constraint.

Price Effects

Changes of current or future expected prices induce intertemporal substitution in consumption. Under smooth preferences they imply a regular offer curve if the respective no-trade position is interior. For changes of prices p in the current period this is $(e_0, e_1 + m_0/p_1)$ which is independent of the current price p. If the demand point is regular and interior changes of the current price induce a rotation of the budget line around the endowment point (as in Figure 2.2 **a**) implying an associated offer curve (as in Figure 2.2 **b**) and inducing higher indirect utility than at the no-trade position for all price changes. Then, the Slutzky decomposition implies that the current price effect is negative if consumption in both periods is a normal good. The consumer changes from the demand side to the supply side of the current

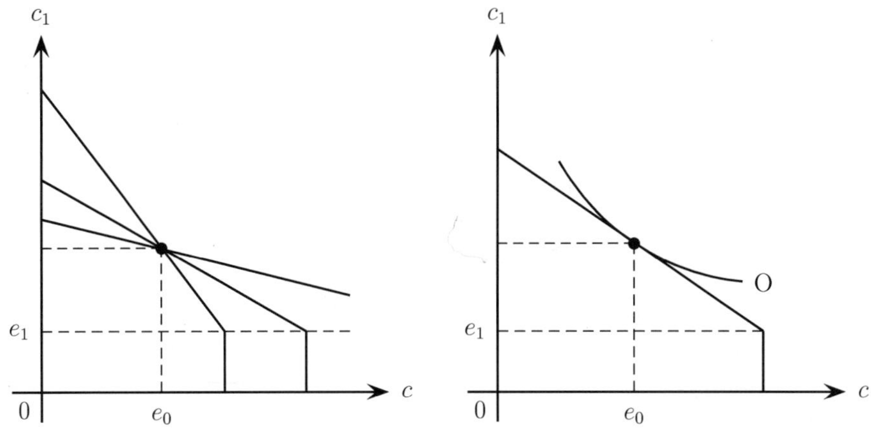

a Price effects **b** Offer curve

Fig. 2.2 Changes of the current price; (m_0, p_1) given

market induced by intertemporal substitution in consumption if and only if savings change from positive to negative, i.e.

$$MRS\left(e_0, e_1 + \frac{m_0}{p_1}\right) \lesseqgtr \frac{p}{p_1} \iff \text{savings} \gtreqless 0 \iff \text{supply} \gtreqless 0, \qquad (2.2.7)$$

which can be seen from Figure 2.2 **a**. In other words, he is a net supplier of the consumption good whenever he is a net saver.

Changes of the expected price p_1 for future consumption induce a rotation of the budget line around the point of maximal current consumption $(e_0 + m_0/p, e_1)$ which is independent of the future expected price p_1. This is the optimal consumption plan on the boundary of the budget set for prices $p/p_1 < MRS(e_0 + m_0/p, e_1)$. Higher relative prices induce a regular offer curve as shown in Figure 2.3 **b** with a negative own-price effect under normality. However, the offer curve is not always globally monotonic and may become backward-bending implying a negative cross-price effect for current consumption demand.

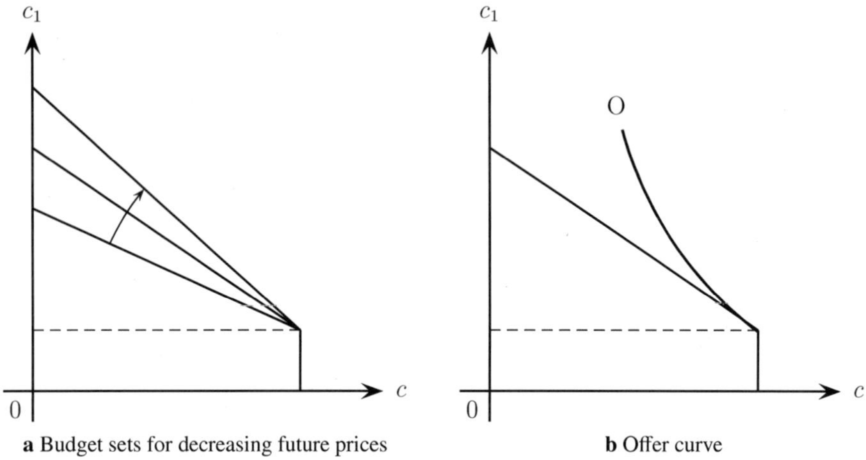

Fig. 2.3 Decreasing the expected future price p_1 for (m_0, p) given

In situations without credit restrictions and a uniform interest rate $r > 0$ on savings or on credit, the consumption decision without a sign restriction on final money holdings m becomes

$$\max\{u(c, c_1) \mid pc + m = pe_0 + m_0,\ p_1c_1 = (1 + r)m + p_1e_1\}$$

implying a slope of the intertemporal budget line equal to $(1 + r)\frac{p}{p_1}$. Demand for credit is positive if and only if

$$MRS\left(e_0 + \frac{m_0}{p}, e_1\right) > (1 + r)\frac{p}{p_1} \tag{2.2.8}$$

with an optimal consumption decision $c^* > e_0 + m_0/p$. Thus, the amount saved $s(r) = m_0 + p(e_0 - c^*) < m_0$, see Figure 2.4. Under normality the Slutzky decomposition implies that the demand for credit is a decreasing function of the interest rate r. Conversely, positive savings is an increasing function of the interest rate implying that the desired final money balances is an increasing function of the interest rate.

Demand for Consumption and Money

In some situations it is useful to analyze and portray geometrically the tradeoff between the two decisions to be made in the current period, i.e. the level of current consumption and the demand for money as functions of the current parameters of

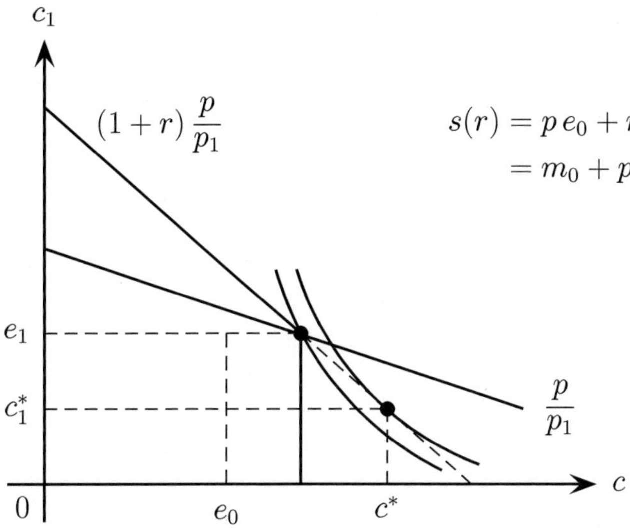

Fig. 2.4 The demand for credit $s(r) < 0$ at interest rate $r > 0$

the decision (m_0, p) for fixed expected prices p_1. For such a purpose the consumer

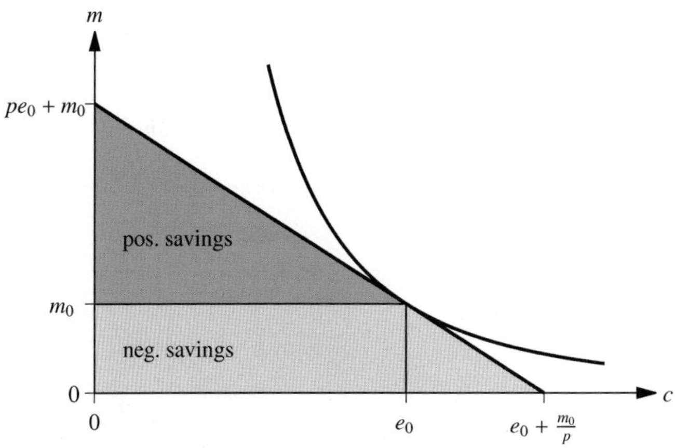

Fig. 2.5 Offer curve of consumption and demand for money: p_1 constant

decision problem can be rewritten as

$$\max\{u(c, c_1) \,|\, pc + m = pe_0 + m_0,\ p_1 c_1 = m + p_1 e_1, m \geq 0\}$$

$$= \max\left\{u\left(c, \frac{m + p_1 e_1}{p_1}\right) \,\middle|\, pc + m = pe_0 + m_0, m \geq 0\right\},$$

defining convex indifference curves of the function $u(c, m/p_1)$ in $\mathbb{R}^2_+$. These depend on future prices p_1. The associated budget line has slope p and a no-trade point (e_0, m_0). In this case the optimal decision $(\varphi(m_0, p, p_1), \mu(m_0, p, p_1))$ can be portrayed as an offer curve in two-dimensional space for positive current prices, see Figure 2.5.

As a final example consider the situation of a consumer-worker with $n^a = 1$, $l = 1$ who offers labor $0 \leq \ell \leq \ell_{\max}$ in the first period of his life. He consumes in both periods, but has no endowment, i.e. $e_0 = e_1 = 0$. However, let his initial money balances be positive $m_0 > 0$. His intertemporal preferences are given by a concave utility function $u : [0, \ell_{\max}] \times \mathbb{R}^2_+ \to \mathbb{R}$, $(\ell, c, c_1) \mapsto u(\ell, c, c_1)$ which is monotonically increasing in (c, c_1) and decreasing in l. For a given wage $w > 0$ and prices (p, p_1), his labor-consumption decision is defined by

$$\max_{\ell,c,c_1} \{u(\ell, c, c_1) \mid pc + m = m_0 + w\ell, \quad p_1 c_1 = m \geq 0\}$$

inducing an equivalent maximization to determine current optimal decisions (l, c, m) given by the solution of

$$\max_{\ell,c,m} \left\{ u\left(\ell, c, \frac{m}{p_1}\right) \middle| m \geq 0 \right\}.$$

This maximization implies indifference contours in $\mathbb{R}^3$ and associated indifference curves of $u(\ell, c, m/p_1)$ intersecting the budget set and which depend on future prices p_1. Under concavity these form concentric convex upper contour sets with an interior optimal decision at the center, Figure 2.6 **b**. As conclusion, the optimal labor-

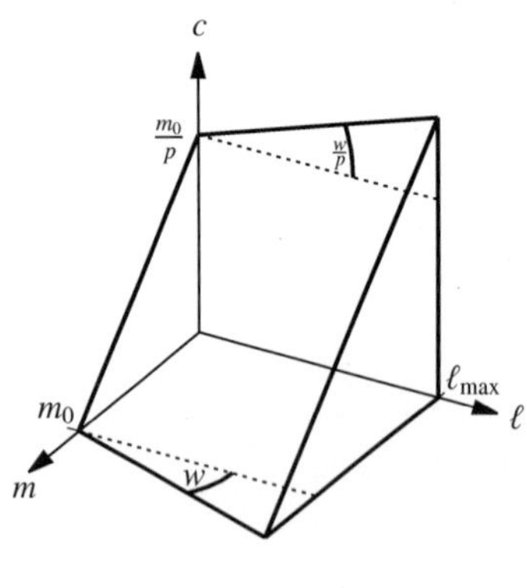

a Budget set of consumer-worker

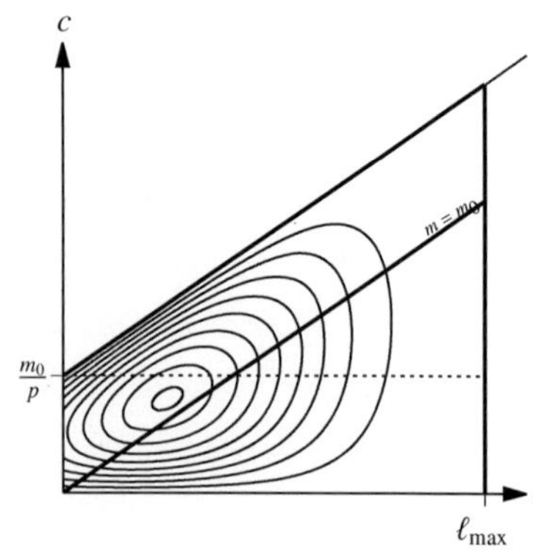

b Labor supply and consumption

Fig. 2.6 Consumption demand, labor supply, and savings of consumer-worker: $m_0 > 0$, $(e_0, e_1) = 0$

consumption plan inherits all the standard properties from static household optimization: current consumption demand and labor supply are functions homogeneous of degree zero in (w, p, p_1, m_0) while the demand for final money balances is homogeneous of degree one. In other words, the functions describing *current* behavior of the typical consumer in an intertemporal monetary setting inherit the standard homogeneity properties from static consumer theory with respect to *all* monetary variables which includes his subjective price expectations. Therefore, when future forecasts are fixed current demand and supply cannot be homogeneous in current prices and wealth. This may occur only under special assumptions concerning expectations formation.

2.2.2 The Intertemporal Utility Index: An Alternative Approach

Given *expected prices* $p^a = (p_1, \dots, p_{n^a})$ by consumer a, one may consider his optimal consumption and savings decision of the current period in a two step procedure, using the Bellman principle. This separates conceptually the list of intertemporal decisions of a consumer into those elements which he *has to decide on* in the current period, i.e. the vector of current consumption and his final money holdings and into the list of those decisions which he *plans* to decide on in the future. Since the set of all future decisions depend in a unique way on each given decision (c, m_1) in the current period, one may define for any triple $(c, m_1, p^a) \in \mathbb{R}^l_+ \times \mathbb{R}_+ \times \mathbb{R}^{l \cdot n^a}$ the intertemporal utility index

$$v(c, m_1, p^a) := \max_{c^a} \{u^a(c, c^a) \mid p^a_\tau c_\tau + m_{\tau+1} = p^a_\tau e_\tau + m_\tau, \quad \tau = 1, \dots, n^a\}$$

where $c^a = (c_1, \dots, c_{n^a})$ is the vector of future consumption plans and (m_τ) is the vector of planned money holdings. $v(c, m_1, p^a)$ is the maximum obtainable utility in the future, i.e. the maximal expected utility under the price expectations prevailing in the current period and it is homogeneous of degree zero in (m_1, p^a). Then, the fundamental result of intertemporal optimization shown by Bellman (1957) states that under very general assumptions the following two equalities hold.

$$\begin{aligned} &\max_{c,m_1} \{v(c, m_1, p^a) \mid pc + m_1 \le pe_0 + m_0, m_1 \ge 0\} \\ &= \max_{c,c^a} \left\{ u(c, c^a) \;\middle|\; \begin{array}{lll} pc_0 + m_1 = pe_0 + m^a_0 & m_1 \ge 0 & \\ p^a_\tau c_\tau + m_{\tau+1} = p^a_\tau e_\tau + m_\tau & m_\tau \ge 0, & \tau = 1, \dots, n^a \end{array} \right\} \end{aligned} \tag{2.2.9}$$

and

$$\begin{pmatrix} \varphi(m_0, p, p^a) \\ \mu(m_0, p, p^a) \end{pmatrix} = \arg\max_{c,m_1} \{v(c, m_1, p^a) \mid pc + m_1 \le pe_0 + m_0, m_1 \ge 0\}. \tag{2.2.10}$$

Here $\varphi(m_0, p, p^a)$ and $\mu(m_0, p, p^a)$ are the demand functions of current consumption and of final money holdings respectively as introduced in equation (2.2.2). In other words, the two step maximization leads to the same maximum and the same set of maximizers in the current period.

Clearly, v is a derived notion of utility function with "money and expectations in the utility function" to be maximized on a single budget set

$$\max_{c,m} \{v(c, m, p^a) \mid pc + m \leq pe_0 + m_0, m \geq 0\} .$$

In other words, desired money holdings and forecasts appear as if they influence intertemporal preferences between consumption and money holdings while the maximization is performed with respect to a single budget restriction as in static houshold optimization independent of forecasts. Obviously the statement is consistent with the fact that the true preferences described by the function u do not change because of monetary effects, but that the implications for optimal intertemporal decisions for future periods appear as dependent on monetary effects through the sequence of budget sets changing the level sets of the indirect intertemporal utility function v. This implies that the analytical or geometric features of optimization of consumers in an intertemporal macroeconomic context can be portrayed with the same typical two-dimensional apparatus as in static consumer theory noting that current price and income effects determine changes of the budget set while changes of expectations induce changes of the contours of the intertemporal utility index via different forecasts.

In summary, the results of the last sections revealed that the optimal consumption behavior of consumers in an intertemporal setting with money as a store of value can be analyzed with the methods of static consumer theory when the number of commodities and the planning horizon are finite. Nonsatiation in consumption implies binding budget constraints under optimization. Therefore, excess demand functions for current commodities and the demand for money satisfy Walras' Law in current prices for all values of future expected prices. Moreover, the application of the standard microeconomic apparatus shows that the consumer's response to changes of wealth, current and future expected prices implies global homogeneity properties with respect to all monetary parameters. In the language of traditional microeconomic theory this means that money illusion prevails at the individual level implying that an equiproportionate change of all prices, current and expected, and of wealth leaves real consumption decisions invariant. However, when future expected prices are fixed, the homogeneity does not hold for equiproportionate changes of wealth and current prices.

2.2.3 Formation of Price Expectations

In a competitive environment when agents take the prices on spot markets as given the information for prices by an individual can be split into two parts. The first one consisting of current market prices which are *certain*, *objective*, and *identical* for all agents in any given period. In contrast, expected prices for the future are *un-*

certain and *subjective*. As a consequence, forecasts differ across agents. In order to decide on his own forecasts, it is assumed that each agent uses a forecasting rule or function which summarizes his subjective beliefs of relevant aspects of the economic evolution. Such a rule may be defined or based on a statistical or econometric model, adopted from an advisory agency, the government, or his own calculations from market data, past prices, other observable variables, or possibly unobservable shocks. For a rational economic agent this implies that he uses some form of an economic model for which he calculates his individual forecast.

In the general case it is reasonable to assume that agents form expectations on the basis of current and past data, for example a forecasting function would be of the form $\psi : \mathbb{R}^{l\cdot(T+1)} \to \mathbb{R}^{l\cdot n_a}$, where T is the finite memory of the forecasting system. In this case, future expected prices $p^a = \psi(p, p_{-1}, p_{-2}, \ldots, p_{-T})$ become a function of current and of past data. More generally, the function ψ itself could depend on the particular the period under consideration when agents adjust their forecasting model in each period due to some learning process. From a recursive point of view, however, the values of past prices $(p_{-1}, p_{-2}, \ldots, p_{-T})$ are given, describing the particular history of the economy up to this point. They are a given datum and unalterable ("Not heaven itself upon the past hath power", Hicks, 1939, p. 130). Under this aspect, the functional relationship with respect to past prices may be an important analytical property of a forecasting method, but its importance for making the actual forecast for the future development in any given period can be considered as secondary ("What would be my forecast if the history had been different?"). Thus, for an analysis of *current demand behavior* with respect to *current prices* the influence of the history would have to be incorporated into the functional form and the dependence on past prices can be ignored safely for analytical reasons. Thus, for temporary equilibrium analysis in a forward looking model forecasts can be considered typically as continuous functions of current prices alone, i.e. a forecasting rule is a mapping

$$\psi : \mathbb{R}^l_+ \to \mathbb{R}^{l\cdot n_a}, \qquad p \mapsto \psi(p) = (\psi_1(p), \ldots, \psi_{n_a}(p)) = (p_1, \ldots, p_{n_a}). \tag{2.2.11}$$

In the literature different properties of expectations functions are discussed and used either in a microeconomic or a macroeconomic context[2]. Prominent ones are so-called *static expectations* describing the situation where the agent assumes that all prices in the future are identical to current prices (assumed for example by Patinkin, 1965), in other words, for all p, $\psi_i(p) = p, i = 1, \ldots, n_a$. In this case, subjective expected inflation rates between any successive periods are zero for all commodities and expected relative prices in future periods are the same as in the current period. As a consequence the consumer behaves along offer curves with intertemporal marginal rates of substitution equal to unity.

Homogeneous functions with respect to current prices form a more general class of expectation functions. Most empirical formulas derived from statistical indices satisfy a linearity or a constant scaling property which makes them homogeneous

[2] Hicks (1939), Chapter XVI contains an early discussion of the role of different properties of forecasting rules.

functions. An expectation function ψ is called homogeneous of degree $r > 0$ when the mapping is homogeneous of degree $r > 0$ in every component $i = 1, \ldots, n_a$ for future periods , i.e. when

$$\psi(\lambda p) \equiv \left(\psi_i(\lambda p)\right)_{i=1}^{n_a} = \left(\lambda \psi_i(p)\right)_{i=1}^{n_a} \equiv \lambda^r \psi(p), \quad \text{for all} \quad \lambda > 0, \quad p \in \mathbb{R}_+^l.$$

For $r > 1$, ($r < 1$), an increase of the current price level implies that an equiproportionate parametric change of all current prices induces equal and successively increasing (decreasing) rates of expected inflation for all commodities in each future period leaving relative expected prices unchanged. In other words, for any $h = 1, \ldots, l$ and $i = 1, \ldots, n_a$, the expected inflation factor is $p_{h,i}/p_h = (1 + \pi)^r$ for any percentage price change $\pi > -1$ in the current period. For that reason, such

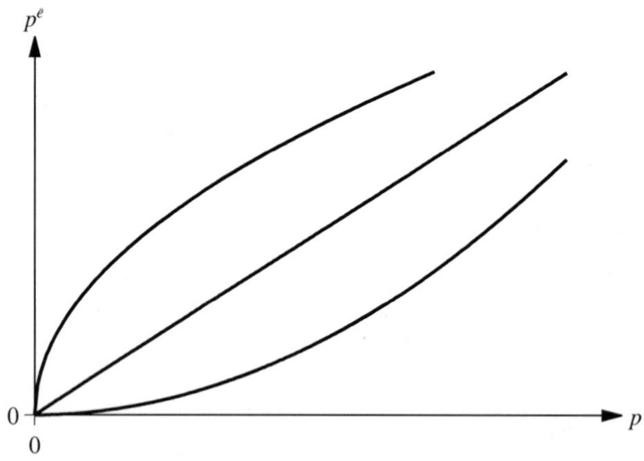

Fig. 2.7 Forms of price expectations: deflationary, unit elastic, inflationary

expectations are referred to as *inflationary* for $r > 1$ or *deflationary* for $r < 1$, see Figure 2.7. For $r = 1$ they are called *unit elastic* (see Hicks, 1939, p. 205). This implies in particular that expected price levels are linearly dependent on the current one.

Demand with Expectations Formation

It is evident that the demand behavior with respect to current prices of an agent in a competitive environment changes decisively when he believes that future prices change in a systematic way with current prices according to his subjective expectations function. In this case all current price effects will differ substantially from those derived from static consumer theory. In order to examine this influence, it is useful to substitute the expectations function into the intertemporal utility index

$$V_\psi(c,m,p) := v(c,m,\psi(p))$$

and define the associated demand function

$$\varphi_\psi(m_0,p) := \varphi(m_0,p,\psi(p)) = \arg\max_{c,m}\left\{V_\psi(c,m,p)\,|\,pc+m=m_0+pe_0\right\}. \quad (2.2.12)$$

With the expectations formation ψ, the demand function and the utility index are functions of current data alone. However, the functional form of the expectations function must have a decisive influence on intertemporal expected maximal utility and on consumer demand implying that changes of current prices induce two effects on current demand behavior: the pure substitution effect from the change of current prices plus the additional effect induced through the perceived change of expected future prices. Together they are responsible for strong or weak intertemporal substitution effects in current demand. Therefore, in general, the demand function φ_ψ may exhibit quite different price effects than those in static demand theory. In particular, it cannot be expected in general that demand is homogeneous of degree zero in money balances and current prices. Unit elastic expectations functions provide the unique case under which the homogeneity prevails, as stated in the next lemma.

Lemma 2.2.1.
If ψ is homogeneous of degree one, then

- $\tilde{V}_\psi(c,m,p)$ *is homogenous of degree zero in* (m,p),
- $\varphi_\psi(m_0,p)$ *is homogeneous of degree zero in* (m_0,p).

Under these conditions, consumption demand in an intertemporal model exhibits most of the common properties known from static demand theory.

In order to understand the effects of changes of current prices from market data in the general case with l commodities it is useful to decompose a price effect on current demand into a so-called *real balance effect* and an *intertemporal substitution effect* (see Grandmont, 1983). Consider a λ-fold increase of the current price level induced by an equiproportionate increase of all current prices, i.e. $p \mapsto \lambda p$. This induces an associated $1/\lambda$ change of purchasing power of real balances from m_0 to m_0/λ. In other words, the real balance effect corresponds to the income effect of static consumer theory measuring the expenditure induced compensating variation of a price/wealth loss. Then, if the induced demand effect is more or less than the one implied by the pure change of real balances, then one may say that there is a *positive* or a *negative* intertemporal substitution effect induced through an expectations function ψ. Specifically, consider the excess demand function

$$Z_\psi(p,m_0) := \varphi_\psi(p,m_0) - e_0.$$

Then, one may decompose the total effect of a proportional change of all price by λ into a *substitution effect* and a *real balance effect* written as

$$\begin{aligned}\Delta Z(\lambda) :&= Z_\psi(\lambda p, m_0) - Z_\psi(p, m_0)\\ &= \underbrace{Z_\psi(\lambda p, m_0) - Z_\psi\left(p, \frac{m_0}{\lambda}\right)}_{\text{intertemporal substitution effect}} + \underbrace{Z_\psi\left(p, \frac{m_0}{\lambda}\right) - Z_\psi(p, m_0)}_{\text{real balance effect}}.\end{aligned} \tag{2.2.13}$$

Thus, the existence of specific expectations function with a direct link from current

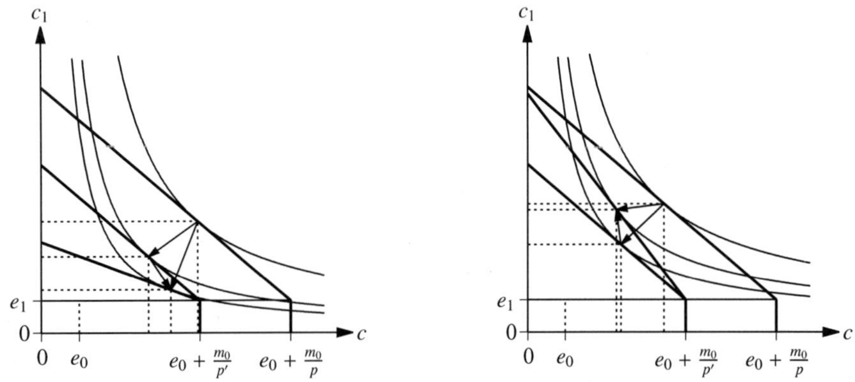

a A negative substitution effect **b** A small positive substitution effect

Fig. 2.8 Intertemporal substitution vs. real balance effects

prices to expected future prices may cause a positive substitution into (an increase of) current demand when its price rises. In static demand theory such an effect would correspond to the existence of a Giffen good requiring a commodity to be inferior induced by nonconcave preferences. In contrast, under intertemporal optimization such effects may occur under concavity (without inferiority) due to expectations effects.

2.3 Temporary Equilibrium in a Monetary Economy

In most models with intertemporal production inventory holding by a producer becomes a major issue and challenge to describe intertemporal production possibilities as well as the intertemporal objective function of a producer. The storage of commodities becomes an additional form of real intertemporal transfer of value for producers when outside money remains the only other possibility for the transfer of value between periods, an option consumers do not have access to. In most models the optimization of producers induces temporary demand and supply functions in competitive markets which have the same form as those of consumers. They depend on current prices, on future price expectations, and on endowments which now consist of money balances and inventory. They often satisfy the same intertemporal homogeneity properties as those of consumers when objectives are defined by some real entity or a linear homogeneous function in nominal prices and money balances. As a consequence, demand and supply functions by an individual producer are structurally identical to those of consumers which satisfy Walras's Law. Therefore, aggregate demand and supply functions are of the same form as in exchange economies. Thus, when searching for the conditions of temporary equilibria in an economy with money, with private agents satisfying individual budget restrictions, temporary equilibrium prices are those which guarantee a zero of aggregate excess demand across all agents. It turns out that most issues of existence, multiplicity, and properties of equilibrium prices can be studied using consumer excess demand functions only.

Definition 2.3.1. Given a set A of consumers with the characteristics as described in Section 2.2 in an arbitrary period t, a temporary equilibrium is a vector of strictly positive current prices $p \in \mathbb{R}^l_{++}$ such that the commodity market and the money market clear, i.e.

$$
\begin{aligned}
&\sum_{a \in A} z^a(\bar{m}^a, p, p_1^a, \ldots, p_{n^a}^a) = 0 \\
&\sum_{a \in A} \mu^a(\bar{m}^a, p, p_1^a, \ldots, p_{n^a}^a) = \sum_{a \in A} \bar{m}^a
\end{aligned}
\tag{2.3.1}
$$

.

It suffices to find a vector $p \in \mathbb{R}_{++}$ which clears the l commodity markets. Walras' Law then implies that the money market is cleared as well.

An equilibrium price vector p consists of *nominal or absolute* prices in terms of money. If p is an equilibrium price vector, then λp with $\lambda > 0$ is typically not an equilibrium price vector since the excess demand functions z^a are not homogeneous of degree zero in p at given expectations and money holdings $\left(\bar{m}^a, p_1^a, \ldots, p_{n^a}^a\right)_{a \in A}$. In addition, one cannot expect that equilibrium prices imply money illusion at given expectations in the following sense:

$$
\sum_{a \in A} z^a(\bar{m}^a, p, p_1^a, \ldots, p_{n^a}^a) = 0 \implies \sum_{a \in A} z^a(\lambda \bar{m}^a, \lambda p, p_1^a, \ldots, p_{n^a}^a) = 0, \text{ some } \lambda > 0.
$$

Finally, if all agents have the same characteristics (preferences, endowments, price expectations) markets will typically not clear with zero trade in all markets, i.e. a trivial no-trade equilibrium typically does not exist at arbitrary identical expectations even when endowments are strictly interior in all successive periods $\tau \in \{1, \ldots, n^a\}$ for all $a \in A$. Therefore, temporary equilibria, if they exist at given characteristics, induce non-zero trades with or without diversity/heterogeneity of agents characteristics.

Proposition 2.3.1.
Let endowments and expectations $(e^a, \bar{m}^a, p^a)$ *be given with* $e_0^a \gg 0$, $p^a \gg 0$ *for* $a \in A$ *such that* $\sum \bar{m}^a > 0$. *There exists a temporary equilibrium price* $p \gg 0$

clearing commodity markets and the money market if the assumption in Proposition 2.2.1 hold.

The assumptions of Proposition 2.2.1 guarantee that excess demand functions are well behaved on the boundary since the intertemporal utility index is concave and monotonic. Price changes in the current period guarantee sufficiently strong intertemporal substitution effect since expected prices (p^a) are fixed and strictly positive. Thus, existence for given expected prices is guaranteed under the same assumptions as in static Arrow-Debreu economies.

Strictly positive equilibrium prices $p \gg 0$ imply that money has positive value, i.e. money has positive purchasing power for any commodity. Such equilibria are referred to as *monetary* temporary equilibria. If an equilibrium were to require that the price of some good $p_h \rightarrow \infty$, an associated equilibrium would be called non-monetary implying zero real money balances.

Definition 2.3.2. Let

$$\mathcal{P}(\{\bar{m}^a, p^a\}_{a\in A}) := \left\{ p \in \mathbb{R}^l_+ \;\middle|\; \sum_{a\in A} z^a(\bar{m}^a, p, p^a) = 0 \right\} \tag{2.3.2}$$

denote the set of equilibrium prices.

The mapping $\mathcal{P} : \mathbb{R}^{n_a} \times \mathbb{R}^{l \cdot n_a} \rightarrow \mathbb{R}^l$ is called the equilibrium **price law**. Since individual excess demand functions are homogeneous of degree zero in $(\bar{m}^a, p, p^a)$, the price law is homogeneous of degree one in $(\{\bar{m}^a, p^a\}_{a\in A})$. In other words, the implications of the quantity theory of money hold in temporary equilibrium: a simultaneous positive rescaling of the quantity of money held by each individual agent *and* of all individually expected future prices by the same positive multiplicative factor induces a pure nominal effect of the same size on all equilibrium prices without changing the equilibrium allocation. Only in this sense money is neutral when the rescaling does not change the distribution of money holdings among agents *and* expected prices in all future periods by all agents move in the same proportion.

Surely, $\mathcal{P}$ may not be a function. Notice that it stipulates that *current equilibrium prices p* depend on *expected prices* for the future, i.e. equilibrium prices in any period are a function of expectations for future periods. Thus, the sequential dating of expectations *lead* current equilibrium prices a property referred to as an *expectational lead* in the price law. In other words, the current equilibrium price vector p_t depends on the expectations formed in t for future periods $t + n$, $n = 1, \ldots,$. Other things being equal this means that constant expectations over time imply constant equilibrium prices over time modulo changes in money balances. Therefore, the evolution of temporary equilibrium prices depends in an essential way on the evolution of expectations and on the evolution of money balances.

2.3.1 Temporary Equilibrium with Consistent Expectations

For a temporary equilibrium analysis it may be of interest to analyse those situations when markets clear *and* expected prices for each agent coincide with those which he predicts in the current period using his personal expectation function as given in Section 2.2.3. While this may seem to be a reasonable attempt to obtain consistency between prices, predictions for the future, and demand the rational for the concept remains an individualistic one[3]. If predictions differ across agents there will be generic widespread disagreement at an equilibrium prices. In fact, the consistency of supply and demand reflected in the existence of *current* prices may actually require the heterogeneity in predictions for the future, a fact often described as a necessary condition for nontrivial equilibria in asset markets. Moreover, if they exist, supporting expected *future* prices does not reveal much information on the validity or correctness of the forecasting rule. Consistency in this sense only confirms the pointwise coincidence of predictions for the chosen functional form of the forecasting rule. It does not imply a correct prediction for the future in the sense of a vanishing forecasting error since future prices are not part of the consistency requirement. Thus, the consistency does not imply perfect foresight or rational expectations in any sense. If the latter were the objective of consistency the price in temporary equilibrium would have to be compared with *past* predictions which are assumed to be fixed and not part of the definition of consistency.

Let the forecasting rules of all agents be given as in equation (2.2.11) by a family of mappings $\{\psi^a\}_{a\in A}$ describing individually expected future prices as functions of the current price vector p. This implies in a natural way the following notion of an equilibrium with consistent expectations.

Definition 2.3.3. A price system $p^* \gg 0$ is called a temporary equilibrium with consistent expectations relative to $\{\psi^a\}_{a\in A}$ if

$$\sum_{a\in A} z^a(\bar{m}^a, p^*, p^a) = 0 \quad \text{and} \quad p^a = \psi^a(p^*) \quad \text{for all} \quad a \in A. \tag{2.3.3}$$

Sufficient conditions for the existence of equilibria are given in the next theorem.

Theorem 2.3.1 (Grandmont 1983).
Let the characteristics $\{(u^a, e^a, \bar{m}^a, \psi^a), a \in A\}$ of an exchange economy be given. If for all $a \in A$:

(a) $e^a \gg 0$,

(b) u^a *is continuous, strictly quasi-concave, strictly increasing,*

(c) $\sum_{a\in A} \bar{m}^a > 0$,

(d) ψ^a *is bounded above and below by strictly positive constants* $0 \ll \underline{q} \leq \psi^a \ll \overline{q}$,

(e) ψ^a *is continuous,*

there exists a monetary temporary equilibrium with consistent expectations.

[3] Grandmont (1983) assumes that all agents use the same forecasting rule.

A sketch of the proof (see Grandmont, 1974, 1983; Schulz, 1985, for details) reveals the importance of the boundedness assumption imposed on the forecasting rule. Given the list $\{z^a(\bar{m}^a, p, p^a)\}_{a\in A}$ of excess demand functions, let $\mathcal{P}(\{\bar{m}^a, p^a\}_{a\in A}) := \left\{p \in \mathbb{R}^l_+ \mid \sum_{a\in A} z^a(\bar{m}^a, p, p^a) = 0\right\}$ denote the equilibrium price law as defined in (2.3.2). A temporary equilibrium with consistent expectations is a fixed point of the mapping $F : \mathbb{R}^l \to \mathbb{R}^l$ defined by $F := \mathcal{P} \circ \psi$, $p \mapsto F(p) = \mathcal{P}(\psi(p))$. The function F is continuous if both $\mathcal{P}$ and ψ are continuous. Let $K \subset \mathbb{R}^{l\cdot n_a}_{++}$ denote a nonempty compact cube with $\Psi(p) \in K$ and let $C \subset \mathbb{R}^l_{++}$ denote a compact convex set such that $\mathcal{P}(K) \subset C$. Then, the mapping $F : C \to C$ satisfies the conditions of Brouwer's fixed point theorem. Hence, there exists $p^* \in \mathbb{R}^l_{++}$ such that $\sum_{a\in A} z^a(\bar{m}^a, p^*, \psi^a(p^*)) = 0$.

It was one of the new insights exhibited by Grandmont (1974, 1983) that consistency with expectations could not be obtained in general in a temporary monetary equilibrium with unbounded forecasting rules. Most notably, homogeneous rules, in particular unit elastic ones, combined with standard neoclassical assumptions on preferences and demand would not guarantee the existence of finite commodity prices to enable market clearing situations in a monetary economy. Thus, equilibria supported by prices inducing a positive finite value of money as the intertemporal medium of storage could not be guaranteed, a property which was typically assumed within the neoclassical synthesis of monetary theories. As an implication the result also shows that the so-called classical dichotomy of money fails to hold under non-homogeneous expectations even when equilibria exist, an observation which contradicts the neutrality hypothesis of monetary policy.

From a methodological point of view the reasoning of neoclassical monetary theory contains essentially two incorrect arguments.

First: *The validity of Walras Law for aggregate excess demand functions generated from optimizing behavior of individual agents under the usual assumptions of interiority, continuity, and convexity necessarily implies the existence of positive equilibrium prices when forecasts for future prices depend on current prices.*

Second: *Positive equilibrium prices, if they exist, are homogeneous of degree one in nominal units of prices and money, for a given distribution of money balances. This property is usually referred to as the* hypothesis of the neutrality of money *which asserts that the price law is homogeneous of degree one in nominal money balances alone.*

Grandmont (1974, 1983) exhibits first that with endogenous expectations the short run demand functions for goods are not homogeneous of degree zero and the short run demand function for money is not homogeneous of degree one. He then shows that the assertion of existence with endogenous expectations is not necessarily valid under the usual assumptions of interiority, continuity, and convexity for consumers unless bounded forecasting rules are used by agents. He indicates that the error consists of an oversight of the necessity of intertemporal substitution effects generated via the forecasting mechanisms to be sufficiently strong in order to compensate the potentially feasible real balance effects which by themselves cannot induce excess demand changes in each real market. In particular, these effects are and insufficient if agents use unit elastic forecasting rules which are those commonly assumed to prevail by both classical as well as neoclassical monetary theorists. Such rules, however, are a necessary condition for the hypothesis of neutrality[4].

2.3.2 *Failure of Existence of Consistent Monetary Equilibria*

Consider a single consumer with separable concave utility for two periods $U(c, c_1) = u(c) + v(c_1)$. Given the budget constraints

$$pc + m = pe + \bar{m}$$
$$p_1^e c_1 = p_1^e e_1 + m,$$

his inverse demand at his initial bundle $(e, e_1, \bar{m})$ is given by

$$\frac{p}{p_1^e} = \frac{u'(e)}{v'(e_1 + \bar{m}/p_1^e)}$$

implying the explicit form of the price law (zero excess demand for one agent!)

$$p = \mathcal{P}(\bar{m}, p_1^e) := u'(e)\frac{p_1^e}{v'(e_1 + \bar{m}/p_1^e)}. \tag{2.3.4}$$

If v is strictly concave, $\mathcal{P}$ is monotonically increasing in p_1^e while $\mathcal{P}(\bar{m}, p_1^e)/p_1^e$ is strictly decreasing satisfying

$$\lim_{p_1^e \to 0} \frac{\mathcal{P}(\bar{m}, p_1^e)}{p_1^e} > 0 \quad \text{and} \quad \lim_{p_1^e \to \infty} \frac{\mathcal{P}(\bar{m}, p_1^e)}{p_1^e} < \infty. \tag{2.3.5}$$

If $\mathcal{P}(\bar{m}, 0) = 0$ the graph of $\mathcal{P}$ lies in a strictly positive cone of $\mathbb{R}_+^2$ of feasible expected rates of inflation induced by the price law. Hence, for any continuous forecasting function ψ whose graph does not intersect this cone no consistent temporary equilibrium exists, see Figure 2.9 **a**. For all positive prices $p > 0$ the value of excess demand $Z_\psi(\bar{m}, p)$ with consistent expectations will be positive (for example for ψ) or negative (for φ). Therefore, all increasing sequences of pairs of prices and consistent expectations $(p_t, p_{t+1}^e = \psi(p_t))_{t=0}^\infty$ diverge to infinity in the case of positive excess demand, or all decreasing sequences will converge to zero under φ in the case of excess supply. This property implies in particular, that stationary equilibria may fail to exist if the cone of feasible rates of inflation does not contain the identity, i.e. the line $p_1^e = p$ of zero inflation.

For the special case when $v(c_1) = \ln c_1$ zero excess demand implies the affine price law

---

[4] For additional comments and a discussion of the debate between Classical versus Neoclassical theories of money see (Grandmont, 1983, Chapter 1), also Friedman (1965); Patinkin (1965); Hahn (1965).

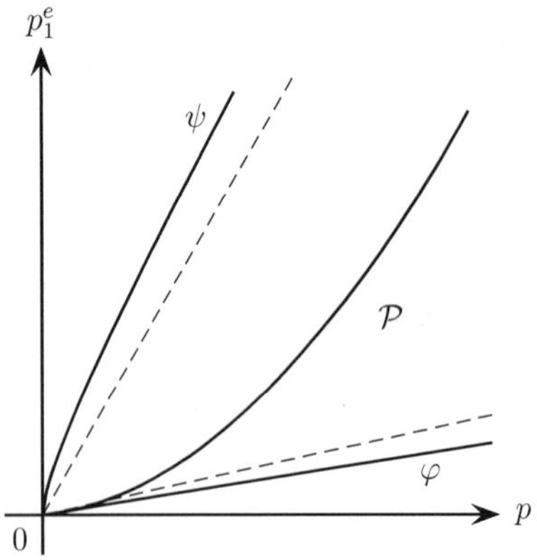

a Two inconsistent forecasting functions φ, ψ

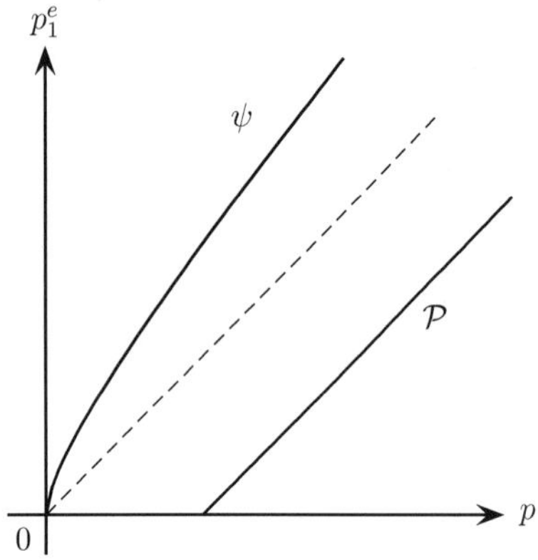

b Failure with an affine price law

Fig. 2.9 Failure of existence of monetary equilibria consistent with expectations

$$p = \mathcal{P}(\bar{m}, p_1^e) = u'(e)\left(p_1^e e_1 + \bar{m}\right) \tag{2.3.6}$$

with $\mathcal{P}(\bar{m}, 0) > 0$. In this case, any unit elastic expectations function $p_1^e = \alpha p$ is inconsistent whenever $\alpha \geq 1/(e_1 u'(e))$, see Figure 2.9 **b**. Hence, for forecasting functions such as ψ or φ there will be either *excess supply* or *excess demand* at **all prices**. The result does not use any specific properties of preferences or endowments which determine $\mathcal{P}$. Therefore, this feature is generic in the space of economies with smooth intertemporal preferences, positive endowments, and continuous forecasting functions. By separating the parameters which guarantee existence of equilibria at given forecasts, i.e. a nonempty price law, from the properties of the forecasting rules it is evident that the failure originates from the interplay of the forecasting rules with the price law and *not* from the structural intertemporal features of agents characteristics.

2.3.3 *On the Validity of the Neoclassical Hypothesis*

Given the standard assumptions on preferences, endowments, (and technology) for a dynamic economy in an intertemporal setting with money (or other nominal assets to allow intertemporal transfer of wealth) full flexibility of prices is said to guarantee strong real balance effects affecting excess demand in all markets sufficiently and inducing a monetary equilibrium at finite prices in all markets at any time. Therefore, fully flexible prices will always guarantee existence of a temporary equilibrium with a positive value of money. A tâtonnement process on excess demands will be able to find the monetary equilibrium. The argument holds true for any value of fixed forecasts of future prices given the normal assumptions of interiority, continuity, and convexity for a temporary Arrow-Debreu economy with outside money. This is no longer true when forecasts are assumed to depend on current prices via a forecasting rule. In such a situation, strong real balance effects for a given forecast are not sufficient to guarantee existence since they may be offset by an induced expectations effect. In particular, expectations can be *too flexible* with respect to current prices eliminating a substitution effect altogether to prevent existence. Most importantly, unit elastic forecasting rules which are necessary for the neutrality of money in equilibrium may induce no intertemporal substitution effect, eliminating intertemporal substitution required for a change of sign of excess demand in the current market.

The economic reasons for the failure of existence and of the neutrality postulate of money of the neoclassical hypothesis are therefore twofold. Price flexibility in a given period guarantees only global control of marginal rates of substitution in that period, but not of intertemporal subjective marginal rates unless expected future prices are constrained to a strictly positive interval. With price dependent forecasts arbitrarily small or large current prices imply global control of the purchasing power of money balances *in the current period only* and not necessarily in future periods. Global price effects in current periods do not necessarily induce global price effects in expectations. There may be a lack of willingness to intertemporal substitution, when price expectations 'move too much' with current prices. The purchasing power effect on real balances from current prices may not be sufficient, contrary to neoclassical perception. In essence, *the interaction* between individual characteristics and the impact of the forecasting rule on expectations are responsible for existence or nonexistence. If there is no temporary equilibrium for a given forecasting rule the neutrality postulate is vacuous and not relevant. If, however, equilibria exist, the neutrality of money *under the given forecasting rule* requires that the rule is locally unit elastic.

Chapter 3
Models of Monetary Equilibrium

The objective of this chapter is to formulate a basic prototype *monetary macroeconomic model* built along the microeconomic principles of Chapter 2, which will serve as the benchmark for a consistent macroeconomic analysis of the *short-run allocations*, i.e. of the determination of output and employment with associated prices and wages in each given period, as well as for the *long-run behavior* of the economy under alternative dynamic scenarios: i.e. with adaptive or rational expectations, under random perturbations, and under different government policies. The modeling structure will be as simple as possible at the beginning and as detailed as necessary in each subsequent section to capture the main intertemporal ingredients of a monetary economy with production in order to discuss the specific economic questions posed in each section.

3.1 Markets and Agents

Economic activity in each period occurs on two real markets, one for a single homogeneous produced good, purchased and sold at a nominal price $p > 0$, and one for a homogeneous input labor, traded at a nominal wage rate $w > 0$. Labor is the only input to be used to produce the commodity. Unless otherwise stated the two markets operate in a competitive fashion implying that all agents act as price takers. Capital equipment is assumed to be constant and non-depreciating. There will be no inventory holding in the basic prototype model. Such a possibility can be introduced easily by extending the scope of the production function in order to investigate its additional implications for temporary allocations and prices as well as for the dynamics of the economy.

In the basic version of the model there are three sectors in the economy composed of consumers, producers, and of a government combined with a central bank.

The Government Sector

In its combined role the government and central bank issues fiat money which serves as a unit of account for all trades. In this situation, money is the only store of value available for consumers. In later chapters government debt, bonds, and other assets will be introduced implying that the role of a central bank will have to be described separately from the fiscal activities of the government which are described next. In each period the government buys/consumes a constant quantity $g \geq 0$ of the produced commodity inducing a constant stream of real public benefits to the private sector which are assumed to be constant in each period. To finance its expenditures pg the government collects taxes from consumers at proportional rates $0 \leq \tau_w \leq 1$ on wages and $0 \leq \tau_\pi \leq 1$ on profit income respectively. A deficit between public expenditure pg and tax revenue in any period will be financed by an *increase* of the amount of money held by consumers (if positive) or by a *decrease* (if negative).

The Production Sector

There exists a finite number $n_f \geq 1$ of producers (firms). All firms $j \in \left\{1, \ldots, n_f\right\}$ are owned by consumers and produce in each period the commodity from labor using the same production function $F : \mathbb{R}_+ \to \mathbb{R}_+$. Since there is no possibility to hold inventory the decisions of each firm involves no intertemporal considerations. Therefore, under competitive conditions each firm attempts to maximize profit at given prices and wages in each period, all of which is distributed to the owners/shareholders. Formally this implies that for each pair of prices and wages $(p, w) \gg 0$ labor demand by each producer is defined as

$$z^\star = h\left(\frac{w}{p}\right) := \arg\max_z pF(z) - wz \tag{3.1.1}$$

inducing a commodity supply of $y^\star = F\left(h\left(\frac{w}{p}\right)\right)$.

The Consumption Sector

The private consumption sector consists of overlapping generations of two-period-lived consumers who can be either *worker consumers* or *shareholder consumers*, who do not supply labor. In the first period of their life shareholders are the owners of the firms and they receive profit during that period only. A generation of shareholders $i \in \{1, \ldots, n_s\}$ consists of n_s consumers[1] whose two period preferences over consumption are convex and homothetic given by a homogeneous concave utility function $U_s : \mathbb{R}^2_+ \to \mathbb{R}$, $(c_1, c_2) \mapsto U_s(c_1, c_2) \equiv c_1 u_s(c_2/c_1)$, where u_s is assumed to be strictly increasing and concave. Given the shareholders' net nominal profit in-

[1] The integer number n_s will be used as the index/name of a shareholder as well as for the cardinality of the set of shareholders $|\{1, \ldots, i, \ldots, n_s\}| \equiv n_s$. No confusion should arise.

come[2] $\pi \geq 0$ when young, the current and the expected future price (p, p^e), and the homotheticity assumption, the consumption demand function ϕ_s of a young shareholder in the first part of his life takes on a multiplicative form

$$\phi_s(p, p^e, (1-\tau_\pi)\pi) := c_s(\theta^e)(1-\tau_\pi)\frac{\pi}{p} := \arg\max_{0\leq c\leq 1}\left\{c\, u_s\left(\frac{1-c}{\theta^e c}\right)\right\}(1-\tau_\pi)\frac{\pi}{p} \tag{3.1.2}$$

where $0 \leq c_s(\theta^e) \leq 1$ is the propensity to consume out of net income π and $\theta^e := p^e/p$ is the expected (gross) rate of inflation. A shareholder's savings function (i.e. his demand for money when young) is given by

$$M_s(\theta^e, \pi) := (1 - c_s(\theta^e))(1-\tau_\pi)\pi. \tag{3.1.3}$$

The individual consumption demand function is homogeneous of degree zero while the demand for money is homogeneous of degree one in (p, p^e, π).

For the basic model it will be assumed that the profit from producers is distributed among the shareholders according to a given ownership structure for given shares[3] $\theta_{ij} \geq 0$ with $\sum_{j=1}^{n_f} \theta_{ij} = 1$ for every $i = 1, \ldots, n_s$ and $j = 1, \ldots, n_f$. In other words, all profits are payed to shareholders. As a consequence, aggregate savings and aggregate consumption demand by shareholders is linear in aggregate net profits.

Each generation of consumer-workers consists of $n_w \geq 1$ consumers whose intertemporal preferences are given by a utility function $U_w : \mathbb{R}^3_+ \to \mathbb{R}$, $(c_1, c_2, \ell) \mapsto U_w(c_1, c_2, \ell)$ where $\ell \geq 0$ is the amount of labor offered in the first period of his life. It is assumed that the preferences are separable between consuming goods (c_1, c_2) in both periods and the disutility of working ℓ and that preferences for consumption are convex and homothetic (as for shareholders). Thus, the utility function of a worker-consumer takes the form

$$U_w(c_1, c_2, \ell) \equiv c_1 u_w\left(\frac{c_2}{c_1}\right) - v(\ell)$$

where $u_w : \mathbb{R}_+ \to \mathbb{R}$ is concave and increasing while $v : [0, \ell_{\max}] \to \mathbb{R}$ is convex and nondecreasing with $v(0) = 0$. Depending on the circumstances $\ell_{\max}$ may be finite or infinite. If the two conditions of monotonicity and concavity/convexity are strict unique interior solutions for a worker-consumer will be guaranteed. This motivates the following assumption.

Assumption 3.1.1 *The function $u_w : \mathbb{R}_+ \to \mathbb{R}_+$ is strictly concave and increasing while $v : [0, \ell_{max}] \to \mathbb{R}_+$ is strictly convex and increasing. They satisfy the respective Inada conditions defined by*

---

[2] Since all variables refer to the same time period, the subscript t will be suppressed whenever possible, in order to reduce notation. Notice that the expected price level $p^e \equiv p^e_{t,t+1}$ is the price expected in period t for period $t+1$, which is assumed to be decided on by the consumer at the beginning of period t.

[3] The same name θ is used here locally to denote the profit shares which should not be confused with the variable θ^e used for the expected inflation rate.

$$\lim_{x \to 0} u'_w(x) = \infty \qquad \lim_{x \to \infty} u'_w(x) = 0 \tag{3.1.4}$$

$$\lim_{\ell \to 0} v'(\ell) = 0 \qquad \lim_{\ell \to \ell_{max}} v'(\ell) = \infty. \tag{3.1.5}$$

The next section discusses the qualitative implications of the assumptions of homogeneity and separability for the labor supply behavior and consumption demand. Lemma B.1.1 states and derives the results formally. The homogeneity of the worker's utility function U_w with respect to (c_1, c_2) implies that his propensity to consume out of wage income is a function of expected inflation alone, which is defined in the same way as for the shareholder by

$$c_w(\theta^e) := \arg\max_{0 \le c \le 1} \left\{ c\, u_w \left(\frac{1-c}{\theta^e c} \right) \right\} \quad \text{with} \quad V(\theta^e) := \max_{0 \le c \le 1} \left\{ c\, u_w \left(\frac{1-c}{\theta^e c} \right) \right\}. \tag{3.1.6}$$

The number $V(\theta^e)$ is the indirect utility for a unit of net real income which is strictly decreasing in θ^e with elasticity E_V satisfying $0 \le -E_V(\theta^e) \le 1$. This implies in turn that the labor supply function for an interior solution is obtained as

$$\phi^\ell \left(\frac{w}{p}, \theta^e \right) := (v')^{-1} \left(\frac{(1-\tau_w)w}{p} V(\theta^e) \right) \tag{3.1.7}$$

which is increasing in the real wage and decreasing in expected inflation under the respective Inada condition of Assumption 3.1.1. This induces an individual commodity demand function as

$$\phi_w \left(\frac{w}{p}, \theta^e \right) := c_w(\theta^e) \frac{(1-\tau_w)w}{p} \phi^\ell \left(\frac{w}{p}, \theta^e \right), \tag{3.1.8}$$

whose elasticity with respect to real wages is larger than one. If there are $n_w \ge 1$ identical workers per period in the economy, aggregate labor supply can be written as a function of the real wage and of the expected (gross) rate of inflation in a multiplicative form

$$N \left(\frac{w}{p} V(\theta^e) \right) := n_w \phi^\ell \left(\frac{w}{p}, \theta^e \right).$$

Different alternative properties of the functions u_w and the disutility function of work v imply a wide range of distinct labor supply functions. First, under strict convexity of v, labor supply is an increasing function of the nominal wage rate. If u_w has a constant elasticity $0 \le 1 - \delta \le 1$ one has (see Lemma B.1.1)

$$c(\theta^e) = \delta \quad \text{and} \quad V(\theta^e) = \left(\frac{1}{\theta^e} \right)^{1-\delta} \delta^\delta (1-\delta)^{1-\delta} \quad \text{with} \quad E_V(\theta^e) = \delta - 1. \tag{3.1.9}$$

In particular, when the worker derives no utility from consuming when young then $\delta = 0$, so that $c_w(\theta^e) = 0$ and $V(\theta^e) = 1/\theta^e$. This makes the labor supply function dependent on the expected real wage w/p^e alone and independent of current prices.

Second, if $v(\ell) = \bar{w}\ell$ is linear, labor supply may become unbounded at positive real wages unless v is restricted to a compact interval $[0, \ell_{\max}]$. In this case, labor supply would become

$$\phi^{\ell}\left(\frac{w}{p}, \theta^e\right) = \begin{cases} \ell_{\max} & (1-\tau_w)\frac{w}{p}V(\theta^e) > \bar{w} \\ [0, \ell_{\max}] & (1-\tau_w)\frac{w}{p}V(\theta^e) = \bar{w} \\ 0 & (1-\tau_w)\frac{w}{p}V(\theta^e) < \bar{w} \end{cases} \tag{3.1.10}$$

which is no longer a function for every w. Note that $\bar{w}$ identifies a net real reservation wage below which the consumer does not supply any labor. At $\bar{w}$ he is indifferent of working any amount between zero and $\ell_{\max}$. If $\bar{w} = 0$, consumers suffer no disutility from work implying that labor supply is constant and equal to $\ell_{\max}$ for all positive net nominal wages w. Thus, the structural assumption of separable preferences between consumption and leisure combined with intertemporal homogeneity in consumption implies a wide range of qualitative features with alternative properties for the aggregate labor supply function.

Figure 3.1 provides a geometric description of the stationary structure of the monetary economy in an arbitrary period with its two markets and three sectors. The circles identify the two markets for labor **L** and the consumption good **X**. The three sectors are marked by square boxes labeled **P** for the producer, **G** for the government, and **Y | O** for the two successive generations of consumers with its young and old members respectively in the current period. Notice that in this economy the government and the producer sector are the same and identical in every period (since they are infinitely lived economic agents) while the members of the consumption-worker sector changes in every period but maintains a stationary composition.

The dashed lines with arrows indicate the real flows between sectors through markets while the solid lines correspond to the associated expenditure/monetary flows. Since the economy is closed feasibility of market transactions imply that the sum of all real inflows and outflows of each market sum to zero. While each of the private sectors (consumers and producers) run a balanced budget in each period, the monetary inflows (taxes from the young) and outflows (purchases of commodities from producers) by the government sector do not balance necessarily. This implies the typical skewed intertemporal structure of markets, inducing the young generation to accept a level of savings M_{t+1} (a promise of purchasing power for the future) which may be different from the previous one M_t and financed by a government deficit or surplus.

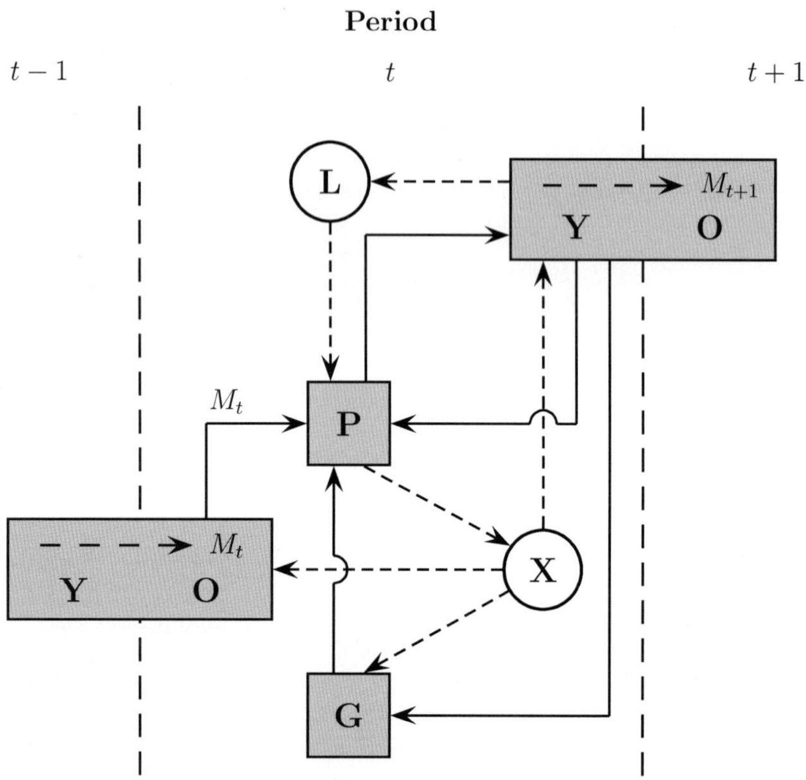

Fig. 3.1 Profiles of intratemporal and intertemporal flows in the economy

3.2 Competitive Temporary Equilibrium

For any given pair (M, p^e) of money balances M held by old consumers in the current period and expected prices p^e held by young consumers, a competitive temporary equilibrium is defined as a pair $(p, w) \gg 0$ of prices and wages which induce zero excess demand in the labor market and in the commodity market such that income consistency holds for all consumers, i.e.

$$
\begin{aligned}
0 &= n_f h\left(\frac{w}{p}\right) - n_w \phi^{\ell}\left(\frac{w}{p}, \frac{p^e}{p}\right) \\
0 &= \frac{M}{p} + g + \sum_{i=1}^{n_s} \phi_s(p, p^e, (1-\tau_\pi)\pi_i) + n_w \phi_w\left(\frac{w}{p}, \frac{p^e}{p}\right) - n_f F\left(h\left(\frac{w}{p}\right)\right) \qquad (3.2.1) \\
\pi_i &:= \left(\sum_{j=1}^{n_f} \theta_{ij}\right)\left(pF\left(h\left(\frac{w}{p}\right)\right) - wh\left(\frac{w}{p}\right)\right), \qquad \forall i \in \{1, \ldots, n_s\},
\end{aligned}
$$

for given shares $\theta_{ij} \geq 0$ with $\sum_{i=1}^{n_s} \theta_{ij} = 1$ for every $j = 1, \ldots, n_f$. The third equation imposes income consistency of aggregate commodity demand which implies that Walras' Law holds for commodity excess demand in the second equation for the monetary economy including the government budget. This corresponds to the aggregate income flow equation in a closed monetary economy, guaranteeing that total expenditure in the economy is equal to total income in any given period.

In general situations, in particular under heterogeneity among consumers, all three equations imply a high degree of interdependencies, since the wage and the price both appear in all three equations. From a structural point of view, however, the first two equations describe the immediate price effects on excess demands in the two markets, while the third equation imposes the role of income effects on the demand side in the economy. In fact the second and third equation together imply that in nominal terms aggregate income consistency of a closed economy (without investment) must hold, namely that total consumption expenditures must be equal to total profit income Π plus total wage income W, i.e. $Y = \Pi + W$ and

$$M + pg + (1 - \tau_\pi)c_s\left(\frac{p^e}{p}\right)\Pi + (1 - \tau_w)c_w\left(\frac{p^e}{p}\right)W \equiv Y = \Pi + W. \tag{3.2.2}$$

With homothetic intertemporal preferences of consumers, as assumed here throughout, the second equation can be rewritten as

$$\underbrace{pn_f F\left(h\left(\frac{w}{p}\right)\right)}_{\text{agg. income}} = M + pg + (1 - \tau_\pi)c_s\left(\frac{p^e}{p}\right)\underbrace{p\left(n_f F\left(h\left(\frac{w}{p}\right)\right) - \frac{w}{p}h\left(\frac{w}{p}\right)\right)}_{\text{agg. profit}}$$
$$+ (1 - \tau_w)c_w\left(\frac{p^e}{p}\right)\underbrace{wn_w\phi^\ell\left(\frac{w}{p}, \frac{p^e}{p}\right)}_{\text{agg. wages}}.$$

Imposing *income consistency of wage income and profits* under competitive factor pricing one obtains a necessary aggregate condition of income-expenditure equivalence

$$pn_f F\left(h\left(\frac{w}{p}\right)\right) = M + pg$$
$$+ pn_f F\left(h\left(\frac{w}{p}\right)\right)\left[(1 - \tau_\pi)c_s\left(\frac{p^e}{p}\right)(1 - E_F) + (1 - \tau_w)c_w\left(\frac{p^e}{p}\right)E_F\right],$$

since the income shares are given by the respective factor elasticities in production

$$E_F\left(h\left(\frac{w}{p}\right)\right) \equiv \frac{h(w/p)F'(h(w/p))}{F(h(w/p))}.$$

Combined with the first equation of (3.2.1) this yields two equilibrium conditions for competitive temporary equilibrium in the two unknowns (p, w)

$$0 = n_f h\left(\frac{w}{p}\right) - n_w \phi^l\left(\frac{w}{p}, \frac{p^e}{p}\right)$$

$$pn_f F\left(h\left(\frac{w}{p}\right)\right) = M + pg \tag{3.2.3}$$

$$+pn_f F\left(h\left(\frac{w}{p}\right)\right)\left[(1-\tau_\pi)c_s\left(\frac{p^e}{p}\right)\left(1 - E_F\left(h\left(\frac{w}{p}\right)\right)\right) + (1-\tau_w)c_w\left(\frac{p^e}{p}\right)E_F\left(h\left(\frac{w}{p}\right)\right)\right]$$

whose solution defines the temporary equilibrium prices and wages $(p, w) \gg 0$ for any given level of aggregate money balances and price expectations by private consumers.

3.2.1 Income Consistent Aggregate Commodity Demand

The equality of the second equation of (3.2.3) is surely weaker than the equality postulated by the second equation of (3.2.1), since it requires income consistency only and does not impose commodity market clearing. However, this relation proves to be a workable and consistent aggregate demand relation, since it transforms the implication of Walras' Law from the microeconomic perspective into a fundamental macroeconomic relationship. From national income accounting rules for a closed economy (without net capital formation), for any feasible consistent trade across markets total nominal national income Y equals total expenditures E of all sectors since there are no effects from depreciation, investment, or storage. Therefore, $Y = E = C + G$ where E consists of total consumption expenditure C of the private sector plus spending by the government sector $G = pg$ with g government real demand.

Under the structural assumption of overlapping generations of consumers, aggregate private consumption expenditure consists of spending from assets plus spending financed by net wage income and net profits. Let W denote total wage income and Π denote total profits, while M is the total quantity of money (asset value) in the current period to be used for consumption by old consumers. Therefore, given the homotheticity in intertemporal consumption and the separability of the consumption-leisure tradeoff for consumer-worker preferences, if $0 < c_w < 1$ and $0 < c_s < 1$ are the propensities to consume out of net income by the two income groups, and if $0 \le \tau_w, \tau_\pi \le 1$ are the proportional tax rates for the two income groups, total spending on private consumption in the current period is given by

$$C = M + c_w(1-\tau_w)W + c_s(1-\tau_\pi)\Pi.$$

Therefore, the identity of total expenditures and aggregate income imposes aggregate income consistency in the form of

$$Y = M + G + c_w(1-\tau_w)W + c_s(1-\tau_\pi)\Pi \tag{3.2.4}$$

which can be rewritten and solved for aggregate income as

$$Y = \frac{M + G}{1 - c_w(1 - \tau_w)\frac{W}{Y} + c_s(1 - \tau_\pi)\frac{\Pi}{Y}}.$$

Thus, one obtains the standard multiplier formula for total income showing that total GDP can always be written as a multiple of aggregate exogenous expenditure.

If p denotes the output price (or equivalently the price index of production) and y output (or real income), then using the homotheticity of preferences plus the property of competitive factor income shares implied by (3.1.1) one can separate real demand from nominal expenditures by

$$\begin{aligned} py &= M + pg + c_w(1 - \tau_w)py\frac{W}{py} + c_s(1 - \tau_\pi)py\frac{\Pi}{py} \\ &= M + pg + (c_w(1 - \tau_w)E_F + c_s(1 - \tau_\pi)(1 - E_F))\,py, \end{aligned}$$

where E_F denotes the elasticity of the production function F while competitive factor pricing implies income shares equal to E_F and $1-E_F$ respectively. This equation implies an income consistent aggregate demand relation

$$\begin{aligned} y^D &= \frac{1}{p}\frac{M + pg}{1 - c_w(1 - \tau_w)E_F - c_s(1 - \tau_\pi)(1 - E_F)} \\ &= \frac{1}{p}\frac{M + pg}{1 - c_s(1 - \tau_\pi) + E_F\left[c_s(1 - \tau_\pi) - c_w(1 - \tau_w)\right]} \end{aligned} \tag{3.2.5}$$

which has to hold for any price p at which y^D is traded. This shows that consumption expenditures out of current income is a weighted sum of the net propensities to consume of the two groups of consumers, which could be written as $c(1 - \tau) \equiv c_w(1-\tau_w)E_F + c_s(1-\tau_\pi)(1-E_F)$. As a consequence from (3.2.5) it is useful to define for the competitive case a general *income-consistent* aggregate demand function as

$$\begin{aligned} y^d &= D\left(\frac{M}{p}, \frac{p^e}{p}, \frac{w}{p}, g, \tau_w, \tau_\pi\right) \\ &:= \frac{M/p + g}{1 - (1 - \tau_w)c_w\left(\frac{p^e}{p}\right)E_F\left(h\left(\frac{w}{p}\right)\right) - (1 - \tau_\pi)c_s\left(\frac{p^e}{p}\right)\left(1 - E_F\left(h\left(\frac{w}{p}\right)\right)\right)} \\ &= \frac{M/p + g}{1 - (1 - \tau_\pi)c_s\left(\frac{p^e}{p}\right) + E_F\left(h(\frac{w}{p})\right)\left[(1 - \tau_w)c_w\left(\frac{p^e}{p}\right) - (1 - \tau_\pi)c_s\left(\frac{p^e}{p}\right)\right]} \end{aligned} \tag{3.2.6}$$

where consumption propensities and income shares are now appropriately evaluated at their respective competitive levels depending on expected inflation and on the real wage. This aggregate income-consistent commodity demand function has the Keynesian multiplier form, stipulating that it is always a linear multiple of exogenous real demand from assets and the government, $M/p + g$. In full generality the multiplier is a relatively complex construction where price expectations and the real wage enter in different ways. The form reveals, however, in which way the different structural properties of the economy influence aggregate demand. Demand from asset holders and from the government are essentially exogenous to the incomes process, while the size of the multiplier varies due to expectations effects, income distribution effects, or the heterogeneity of intertemporal preferences between consumers.

With perfect competition in both markets the real wage induces an impact on aggregate demand only through the income distribution defined by the features of the production technology. Thus, aggregate demand will be independent of the real wage if the functional income distribution is constant, in particular when the production function is isoelastic. Expectations have an impact on the multiplier only through the gross propensities to consume of the two types of consumers. Thus, if intertemporal preferences are isoelastic future price expectations play no role in aggregate demand. Consumer heterogeneity enters in two forms, one is through the propensities to consume $c_w(\theta^e)$ and $c_s(\theta^e)$ coupled with an effect from potential differentiable taxation, $\tau_w \neq \tau_\pi$. Thus, consumer heterogeneity matters when the net propensities to consume are different.

The form of the aggregate demand function (3.2.6) corresponds to the well known multiplier formula of the traditional Keynesian expenditure model, except that the price has been made explicit. Its multiplier reflects the necessary parameters of the underlying structural features of the economy: income distribution and expectations. Moreover, it seems impossible to write any other aggregate demand relationship which captures income distribution effects as well as demand effects from consumer demand for all prices p. Its application in the model with wage bargaining (Section 3.5) shows its usefulness in noncompetitive environments. The fundamental reason is the formal equivalence of this equation with the second equation of the definition of temporary equilibrium in (3.2.3) *after* taking account of the income effects. This provides, ultimately, the justification for choosing an aggregate demand function of the type as suggested in the basic Assumption 3.2.1. In other words, consistency of incomes on the aggregate level of demand is tantamount to guaranteeing that Walras' Law holds. Equation (3.2.5) is an exact and *explicit* demand function if the multiplier does not depend on the commodity price. For example, if the propensities to consume of the two income groups are constant and production is isoelastic, then equation (3.2.5) is the *only* aggregate demand function of this economy which is income-consistent[4].

Most of the subsequent analysis will abstract from distributional issues, making it reasonable to assume that the income distribution is relatively insensitive to price effects or that it does not matter[5]. In addition, however, under Assumption 3.2.1 of homothetic intertemporal preferences for both income groups, the propensities to

[4] Traditional Keynesian analysis uses the income-expenditure consistency as the principle criterion to define the IS-curve without deriving its specific form from microeconomic assumptions. Without the overlapping generations structure, money balances do not enter into the IS-equation. Nevertheless, with the Keynesian assumption of the demand for money, the LM-curve, the definition of the aggregate demand function satisfying consistency of total income and expenditure induces an aggregate demand function depending on money balances and on prices; see Section 6.3.

[5] The income-consistent aggregate demand function proves to be useful in other macroeconomic models as well where distributional questions are examined. One such case occurs under bargaining presented in Section 3.5 (see also Böhm & Claas, 2017; Claas, 2017, for details).

consume could be functions of the expected inflation rate $\theta^e := p^e/p$. In order to take proper account of these two possibly competing issues, the general functional form of the income-consistent *aggregate demand function* chosen will be

$$y = D\left(\frac{M}{p}, \theta^e, g, \tau\right) := \frac{M/p + g}{1 - c(\theta^e)(1 - \tau)}, \tag{3.2.7}$$

which covers the structural assumptions given in Section 3.1. This will be consistent with its general properties and the specific multiplier formula in other cases when more specific results are desirable. Summarizing, under a uniform proportional tax rate on incomes and identical propensities to consume between income groups, the income consistent aggregate demand function must be of the form (3.2.7). It is homogeneous of degree zero in money, price expectations, and current prices with a price elasticity $0 > E_D(p) > -1$ if $c'(p^e/p) \geq 0$. In the sequel, reduced forms of the two versions (3.2.6) and of (3.2.7) will be used. Depending on the context only those variables will be included as arguments of the demand function which enter in the associated analysis. Those which are constant will be suppressed.

3.2.2 Prices and Wages, Output and Employment

Given the microeconomic structure of the prototype economy defined above, it is now possible to address the main macroeconomic issues for the performance of closed, monetary, and dynamic economies to provide answers to the basic, commonly and most frequently asked questions of standard macroeconomics (as for example in Blanchard & Fischer, 1989; Dornbusch & Fischer, 1991; Felderer & Homburg, 1987; Blanchard, 2003; Abel & Bernanke, 2005).

- What are the main determining factors of the level of aggregate economic activity in temporary competitive equilibrium, i.e. the determinants of the level of output and employment, of the level of prices and wages?
- Does the role of these determining factors change in an important way when the behavior of private agents is noncompetitive with strategic interaction or under cooperative behavior?
- What is the impact of different government policies on the macroeconomic levels in temporary equilibrium and in the long run?
- What determines the dynamic evolution of economic activity?
- What is the role of expectations formation, in particular what determines economic activity in the long run under perfect foresight or rational expectations?
- Are there specific structural causes inducing business cycles?
- What role do stochastic perturbations on the technology or on demand play for the evolution?

In order to find qualitative or quantitative answers to these questions more specific assumptions and properties of the underlying structural properties of the economy as described in Section 3.1 need to be introduced. Let $n_f \geq 1$ denote the number of identical competitive (price and wage taking) firms, each with the same neoclassical instantaneous production function $F : \mathbb{R}_+ \to \mathbb{R}_+$, while the consumption sector consists of $n_w \geq 1$ two-period-lived consumer-workers and $n_s \geq 1$ shareholders. Let the following assumption characterize the associated *macroeconomic* functional relationships in its most basic form without an expectations effect and without a distribution effect in demand, which are consistent with the set of microeconomic assumptions laid out in Section 3.1 above .

Assumption 3.2.1

(i) The production function $F : \mathbb{R}_+ \to \mathbb{R}_+$ for each firm is strictly concave, strictly monotonically increasing with $F(0) = 0$ and satisfies the Inada conditions

$$\lim_{z\to 0} F'(z) = \infty \qquad \textit{and} \qquad \lim_{z\to\infty} F'(z) = 0. \tag{3.2.8}$$

(ii) The aggregate labor supply function $N : \mathbb{R}_+ \to [0, n_w \ell_{\max}]$ is strictly monotonically increasing and of the form[6]

$$L^s = N\left(\frac{w}{p} V\left(\frac{p^e}{p}\right)\right) \quad \textit{with} \quad N(0) = 0, \tag{3.2.9}$$

where $V : \mathbb{R}_+ \to \mathbb{R}_+$ is strictly decreasing and bijective.

(iii) Government consumption $g \geq 0$ and tax rates $0 \leq \tau_w = \tau_\pi \equiv \tau \leq 1$ are uniform.

(iv) The aggregate income-consistent commodity demand function is of the form

$$y^D = D\left(\frac{M}{p}, \frac{p^e}{p}, g, \tau\right) = \frac{M/p + g}{1 - c(p^e/p)(1-\tau)} \tag{3.2.10}$$

which is strictly increasing in M and g and assumed to be strictly decreasing in p.

(v) Maximal output exceeds minimal demand

$$n_f F\left(\frac{n_w \ell_{max}}{n_f}\right) > \frac{g}{1 - c(1-\tau)} \tag{3.2.11}$$

All functions are assumed to be twice continuously differentiable.

In item (ii) the tax rate for workers has been suppressed as an argument of aggregate labor supply. Instead of item (ii) one may assume the Inada conditions 3.1.1 on the micro level for consumer-workers. The Inada conditions (3.2.8) for the production function imply that the demand for labor of each firm at given prices and wages is described by a function of the real wage alone

$$h\left(\frac{w}{p}\right) := \arg\max_{z\geq 0} \{pF(z) - wz\}, \tag{3.2.12}$$

[6] $\ell_{\max} = +\infty$ is allowed when the disutility of work is isoelastic as in Section 3.2.7.

which is continuously differentiable, strictly monotonically decreasing, and globally invertible on $\mathbb{R}_{++}$. Thus, labor demand is strictly positive for each positive real wage. As a consequence, aggregate labor demand is a surjective monotonically decreasing function of the real wage. Similarly, the aggregate labor supply function is simply an aggregate image of individual labor supply and strictly increasing in the real wage.

3.2.3 Competitive Temporary Equilibrium

Definition 3.2.1. A temporary equilibrium at (M, p^e, g, τ) with $(M, p^e) \gg 0$ is a pair $(p, w) \gg 0$ such that for each vector (M, p^e, g, τ)

$$\begin{aligned} D\left(\tfrac{M}{p}, g, \tau\right) - n_f F\left(h\left(\frac{w}{p}\right)\right) &= 0 \\ n_f h\left(\frac{w}{p}\right) - N\left(\frac{w}{p} V\left(\frac{p^e}{p}\right)\right) &= 0. \end{aligned} \tag{3.2.13}$$

Under Assumption 3.2.1 a unique temporary competitive equilibrium exists for all $(M, p^e) \gg 0$. This implies that there exist two mappings, one determining nominal equilibrium prices, which is called the *price law* and a second one determining nominal equilibrium wages called the *wage law*, i.e.

$$p = \mathcal{P}(M, p^e, g, \tau) \tag{3.2.14}$$

$$w = \mathcal{W}(M, p^e, g, \tau). \tag{3.2.15}$$

Both mappings are a result of the interaction of supply and demand in the commodity market and in the labor market. The equilibrium price and wage are determined simultaneously and their solution cannot be separated or solved stepwise due to cross market income effects. The equilibrium pair (p, w) in turn induces equilibrium allocations consisting of individual levels of employment, production, consumption, and savings, as well as of government expenditures and tax revenue. In other words, the price law and the wage law determine a level of full employment given by a mapping $L = \mathcal{L}(M, p^e, g, \tau)$

$$\mathcal{L}(M, p^e, g, \tau) := n_f h\left(\frac{\mathcal{W}(M, p^e, g, \tau)}{\mathcal{P}(M, p^e, g, \tau)}\right) \equiv N\left(\frac{\mathcal{W}(M, p^e, g, \tau)}{\mathcal{P}(M, p^e, g, \tau)} V\left(\frac{p^e}{\mathcal{P}(M, p^e, g, \tau)}\right)\right) \tag{3.2.16}$$

and an output level determined by a mapping

$$y = \mathcal{Y}(M, p^e, g, \tau) := n_f F\left(h\left(\frac{\mathcal{W}(M, p^e, g, \tau)}{\mathcal{P}(M, p^e, g, \tau)}\right)\right) \equiv D\left(\frac{M}{\mathcal{P}(M, p^e, g, \tau)}, g, \tau\right). \tag{3.2.17}$$

3.2.4 Properties of the Price Law and the Wage Law

Lemma 3.2.1.
Let the government parameters (g, τ) and Assumption 3.2.1 be given. Then,

(a) The price and the wage law $\mathcal{P} : \mathbb{R}^2_+ \to \mathbb{R}_+$, $\mathcal{W} : \mathbb{R}^2_+ \to \mathbb{R}$ are both functions which are homogeneous of degree one and monotonically increasing in (M, p^e).

(b) Their partial elasticities with respect to expectations and to money balances are less than one and satisfy[7]

$$0 < E_{\mathcal{P}}(p^e) < E_{\mathcal{W}}(p^e) < 1 \tag{3.2.18}$$

$$0 < E_{\mathcal{W}}(M) < E_{\mathcal{P}}(M) < 1. \tag{3.2.19}$$

(c) An increase of expected prices decreases output and employment while an increase of money balances increases output and employment.

(d) If aggregate demand is homogeneous of degree zero in (p, M), then the quantity theory of money holds, i.e. the functions $\mathcal{P}$ and $\mathcal{W}$ are homogeneous of degree one in (M, p^e) with

$$p = M\mathcal{P}\left(1, \frac{p^e}{M}\right), \qquad w = M\mathcal{W}\left(1, \frac{p^e}{M}\right) \tag{3.2.20}$$

and $\mathcal{Y}$ and $\mathcal{L}$ are homogeneous of degree zero in (M, p^e) with

$$L = \mathcal{L}\left(1, \frac{p^e}{M}\right), \qquad y = \mathcal{Y}\left(1, \frac{p^e}{M}\right). \tag{3.2.21}$$

Items (a) to (c) can be shown applying the Implicit Function Theorem to equations (3.2.14) and (3.2.15) while the homogeneity properties (d) follow directly from the definitions and (3.2.13).

The economic implications of these properties are immediate and support the standard intuition. An increase in the nominal variables (M, p^e) induce primarily positive demand effects which imply intuitively expected increases in the two nominal equilibrium prices. Under homogeneity of commodity demand this implies that the price law is homogeneous of degree one and concave. Statement (b) of the lemma implies that the real wage is increasing in expected prices inducing a reduction of employment and output, as stated in item (c). Conversely, an increase in money balances induces a decrease of the real wage and thus an increase in output and employment. In other words, there is a *positive tradeoff* between money balances and expected prices (inflation) on output and employment. Figure 3.2 portrays the graph and the level curves of the price function (subfigures **a** and **b**) and the

[7] The partial elasticities are defined in the usual way, using the short hand notation $E_{\mathcal{P}}(p^e) := \frac{\partial \mathcal{P}(M,p^e)}{\partial p^e} \frac{p^e}{\mathcal{P}(M,p^e)} \equiv E_{\mathcal{P}}(p^e)(M, p^e)$. Notice that an elasticity between zero and one of a partial function does not imply that the partial or the global function is concave. However, concavity and monotonicity are sufficient to guarantee that the partial elasticities are less than one.

level curves of the employment function (subfigure **c**). These confirm the respective qualitative comparative statics effects and implications of the quantity theory.

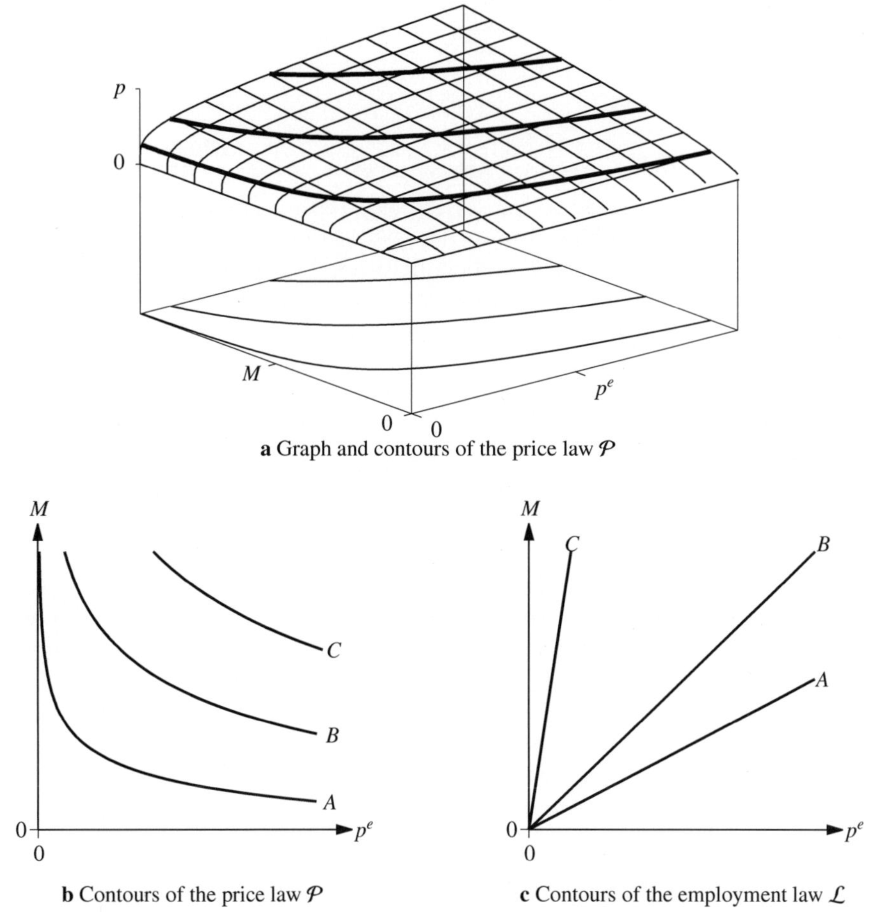

Fig. 3.2 The role of money and expectations on prices and employment for $A < B < C$

Differentiability and the homogeneity of the price law imply further structural properties in equilibrium. First, the sum of the two elasticities is equal to one at all values (M, p^e), i.e.

$$E_{\mathcal{P}}(p^e) + E_{\mathcal{P}}(M) = 1. \tag{3.2.22}$$

Second, each level set of the function $\mathcal{P}(M, p^e)$ contains essentially all equilibrium information between real money balances and expected inflation, in particular its unit contour $p = p\mathcal{P}(M/p, p^e/p) = 1 = \mathcal{P}(m, \theta^e) \equiv m\mathcal{P}(1, \theta^e/m) \equiv m\mathcal{P}(q^e)$ where $q^e := p^e/M$. Then, at temporary equilibrium, the values of expected inflation

$p^e/p =: \theta^e$ and of real money balances $m := M/p$ must satisfy the implicit relation $1 = m\mathcal{P}(1, \theta^e/m)$. This induces a monotonic relation between real money balances and expected inflation. Applying the Implicit Function Theorem, one obtains

$$E_m(\theta^e) := \frac{dm}{d\theta^e}\frac{\theta^e}{m} = -\frac{E_{\mathcal{P}}(\theta^e)}{1 - E_{\mathcal{P}}(\theta^e)}, \tag{3.2.23}$$

where

$$\frac{\partial p}{\partial M} = \mathcal{P}(1, q^e) - q^e \frac{\partial \mathcal{P}(1, q^e)}{\partial q^e} \quad \text{and} \quad \frac{\partial p}{\partial p^e} = \frac{\partial \mathcal{P}(1, q^e)}{\partial q^e}.$$

In other words, at all equilibria, real money balances and expected rates of inflation are always negatively correlated with elasticity $\frac{E_{\mathcal{P}}(q^e)}{1-E_{\mathcal{P}}(q^e)}$. As a consequence, all intertemporal changes of money balances and expectations will imply changes of real money balances and expected rates of inflation along the unit contour of the price law, while parametric changes (changes of government parameters, demand shocks, or productivity shocks) imply displacements of this curve.

It is often assumed that aggregate labor supply is inelastic, i.e. it is given by a constant $N(w/p, p^e/p) \equiv L_{\max} > 0$, a property which can also be deduced from the same set of structural properties when the disutility of work is constant and labor has an upper bound. In this case, existence and uniqueness of equilibria follow under the remaining assumptions, implying special properties of the equilibrium mappings. First of all, expectations from the labor market side play no role. Most importantly, the labor market equilibrium determines a constant equilibrium real wage, depending on the technological features (n_f, F) and $L_{\max}$ alone. Equilibrium output is constant for all (M, p^e). The price level is determined by the commodity market separately and the price law is linear in money, unless there are expectations effects in the aggregate demand function. The implications of the quantity theory hold if the aggregate demand function is homogeneous of degree zero in (M, p^e, p).

3.2.5 Aggregate Supply and Aggregate Demand

In monetary macroeconomic theory it is customary and informative to analyze the existence of temporary equilibria and its properties in a two step procedure, which in effect is an attempt to separate the role and importance of price and income effects from those of expectations between the two markets. This is possible under the basic structural Assumption 3.2.1.

Since the labor market equilibrium condition in (3.2.1) contains no income effects, a labor market clearing real wage can be determined parametrically as a function of prices and price expectations which is independent of income effects from the demand side of the commodity market. Define the real wage $\alpha := w/p$ and the expected rate of inflation by households/workers as $\theta^e := p^e/p$. Then, the full employment condition can be written as

$$n_f h(\alpha) - n_w \phi^\ell(\alpha, \theta^e) \overset{!}{=} 0 \quad \Longleftrightarrow \quad (1-\tau_w)\alpha V(\theta^e) \overset{!}{=} v'\left(\frac{n_f}{n_w} h(\alpha)\right) \tag{3.2.24}$$

whose solution induces a unique equilibrium real wage function

$$\alpha = W(\theta^e, \tau_w) \tag{3.2.25}$$

which is increasing in θ^e and in τ_w. This implies

$$\begin{aligned} 0 < \frac{\partial W}{\partial \theta^e}(\theta^e, \tau_w) &= -\frac{(1-\tau_w)\alpha V'}{(1-\tau_w)V(\theta^e) - \frac{n_f}{n_w} v'' h'} \\ &< -\frac{\alpha \theta^e V'(\theta^e)}{\theta^e V(\theta^e)} = -\frac{\alpha}{\theta^e} E_V(\theta^e) < \frac{\alpha}{\theta^e}, \end{aligned} \tag{3.2.26}$$

showing that W has an elasticity with respect to θ^e which is less than one. Therefore, taking labor market clearing into account, one obtains the equilibrium full employment level as $L = n_f h(W(\theta^e, \tau_w)) \equiv n_w \phi^\ell(W(\theta^e, \tau_w), \theta^e)$. This induces an associated full employment output level as a function of prices and price expectations which is called the *aggregate supply function AS* defined as

$$y^s = AS(\theta^e, \tau_w) := n_f F(h(W(\theta^e, \tau_w))) \equiv n_f F\left(\frac{n_w}{n_f} \phi^\ell(W(\theta^e, \tau_w), \theta^e)\right). \tag{3.2.27}$$

Thus, the aggregate supply function defines the total feasible output/supply which is consistent with competitive labor market clearing[8]. The properties of the production function F and of the real wage function W imply that AS is a *decreasing* function of the expected inflation rate and of the wage tax τ_w. Thus, output y^s is increasing in the commodity price p and decreasing in the wage tax. Figure 3.3 displays the aggregate supply function in two different diagrams for a given tax rate τ_w. Using the assumptions on the microeconomic level one can summarize the results for labor market clearing and aggregate supply as follows. The proof follows from Lemma B.1.1.

Lemma 3.2.2.
Assume that the production function of each firm satisfies the concave Inada conditions (3.2.8) *and let aggregate labor supply satisfy* (3.2.9)*, or alternatively let consumer-worker characteristics satisfy the respective Inada conditions* (3.1.4) *and* (3.1.5) *given in Assumption 3.1.1. Then:*

(a) The real wage function $\alpha = W(\theta^e, \tau_w)$ clearing the labor market is strictly increasing in θ^e such that $0 < E_W(\theta^e) < 1$ with

[8] The concept of the aggregate supply function *AS* is not defined in the same way everywhere in the literature (compare Blanchard, 2003; Lucas, 1973) and economists disagree about its usefulness (see Barro, 1994). However, whatever is its definition in a given model it proves to be a useful analytical device in macroeconomics to reveal the different roles of effects from prices, income, or wealth against those from expectations.

$$\lim_{\theta^e \to 0} W(\theta^e, \tau_w) = F'\left(\frac{n_w \ell_{max}}{n_f}\right) \quad \textit{and} \quad \lim_{\theta^e \to \infty} W(\theta^e, \tau_w) = \infty. \tag{3.2.28}$$

(b) The aggregate supply function $AS(\theta^e, \tau_w)$ is strictly decreasing and globally defined for $0 < \theta^e < \infty$ with

$$\lim_{\theta^e \to 0} AS(\theta^e, \tau_w) = n_f F\left(\frac{n_w \ell_{max}}{n_f}\right) \quad \textit{and} \quad \lim_{\theta^e \to \infty} AS(\theta^e, \tau_w) = 0. \tag{3.2.29}$$

The equilibrium real wage has a positive lower bound and aggregate supply has a positive upper bound if $\ell_{\max} < +\infty$. From (3.2.25) one obtains directly the identity

$$0 > E_{AS}(\theta^e) \equiv E_{F \circ h}(\alpha)\, E_W(\theta^e) = \underbrace{E_F(h(\alpha))\, E_h(\alpha)}_{<0}\, \underbrace{E_W(\theta^e)}_{<1} \tag{3.2.30}$$

between the elasticity of aggregate supply and its multiplicative composition into the production elasticities and the elasticity of the equilibrium real wage function. This yields structural information about the elasticity of the aggregate supply function

$$0 > E_{AS}(\theta^e) > E_F(h(\alpha))\, E_h(\alpha) = \frac{E_F(h(\alpha))}{E_{F'}(h(\alpha))} = \frac{(F'(h(\alpha)))^2}{F(h(\alpha))F''(h(\alpha))}, \tag{3.2.31}$$

since $F'(h(\alpha)) \equiv \alpha$ and $\alpha = W(\theta^e, \tau_w)$. Thus, the elasticity of *aggregate supply*

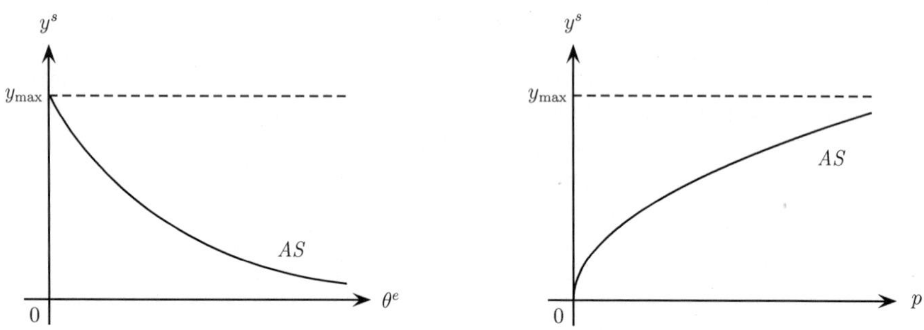

a Aggregate Supply **b** Aggregate Supply given p^e

Fig. 3.3 Aggregate Supply function

under labor market clearing is always smaller (in absolute value) than the elasticity of the *notional supply* function of each producer (at the associated equilibrium real wage), which is itself determined by the curvature

$$\frac{(F'(h(\alpha)))^2}{F(h(\alpha))F''(h(\alpha))}$$

of the production function F alone. In other words, the functional relationship of *aggregate supply to price* described by the aggregate supply function differs systematically from the corresponding partial elasticities of the production sector due to the intermarket equilibrating wage effects. It is not simply the aggregate description of the individual supply behavior of the production sector. This feature implies that empirically observable correlations of supply side data show the properties of the market elasticities only. Inferences from such data to the corresponding properties of the production functions must take into account the elasticities of the supply side or of the equilibrium real wage function as well.

In a second step using the income consistent aggregate demand function (3.2.7), equilibrium in both markets simultaneously is now obtained if $y^s = y^d$, i.e. a price level $p > 0$ solving

$$D\left(\frac{M}{p}, g, \tau\right) = AS\left(\frac{p^e}{p}, \tau_w\right) \tag{3.2.32}$$

defines a competitive temporary equilibrium and the price law as $p = \mathcal{P}(M, p^e, g, \tau)$. Existence of a positive solution requires that $\ell_{\max}$ is sufficiently large (or that g is

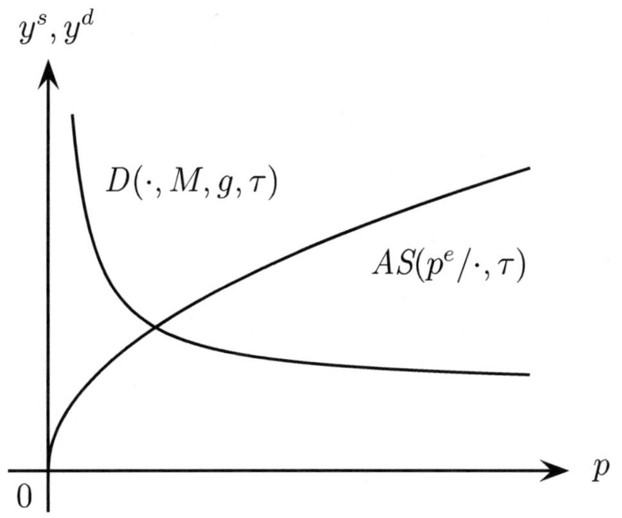

Fig. 3.4 Temporary equilibrium in AS-AD Model; $\tau = \tau_w = \tau_\pi$ and (M, p^e, g) given

sufficiently small), namely that

$$n_f F\left(\frac{n_w \ell_{\max}}{n_f}\right) > \frac{g}{1 - c(1 - \tau)}$$

as stipulated by condition (v) of Assumption 3.2.1. Its other conditions imply that the equilibrium price is unique and positive for all $(p^e, M) \gg 0$ and $(g, \tau = \tau_w = \tau_\pi)$. Surely, the solution of (3.2.32) satisfies the properties listed in Lemma 3.2.1. Figure 3.4 portrays the essential geometric features of the two step procedure. The diagram also reveals qualitatively the comparative statics effects on output, prices, and income with respect to money balances, expectations, and government parameters.

From the general equilibrium perspective of the economy the price law $\mathcal{P}$ becomes the central mapping for the analysis of the macroeconomy inducing all properties of the temporary equilibrium allocation. Therefore, applying the Implicit Function Theorem to (3.2.32) one can derive all comparative statics properties of the price law from which one also obtains the implications on the temporary equilibrium levels of output and employment through (3.2.17) and (3.2.16).

The comparative statics effects with respect to the fiscal parameters (g, τ) on the temporary equilibrium levels are straightforward. A change of government demand induces price and output effects with unique signs given by

$$\begin{aligned} \frac{\partial \mathcal{P}}{\partial g} &= -\frac{\partial D/\partial g}{\partial D/\partial p + \frac{p^e}{p^2}\partial AS/\partial \theta^e} > 0 \\ \frac{\partial \mathcal{Y}}{\partial g} &= -\frac{(\mathcal{P}(M, p^e, g, \tau))^2}{p^e} \cdot \frac{\partial AS}{\partial \theta^e} \cdot \frac{\partial \mathcal{P}}{\partial g} > 0. \end{aligned} \tag{3.2.33}$$

Therefore, for each level of money balances and expectations, higher government demand is associated with higher prices, output, and employment through its effect on aggregate demand. These results can also be seen geometrically from Figure 3.4 since the change of government demand leaves the aggregate supply function unaffected. Thus, when expectations and income distribution effects play no role for aggregate demand, prices, output and the employment level increase with government demand implying in particular that nominal GDP increases with higher government demand. In other words, the real and the nominal government multiplier are both positive. However, this is made possible only through an increase in prices and wages but a decrease of the real wage which induces more employment.

Similarly, one obtains the comparative statics result for a uniform increase of both tax rates. Since the tax rates have a negative effect on aggregate supply and on aggregate demand their net impact on equilibrium prices can be either positive or negative, i.e. one finds that

$$\frac{\partial \mathcal{P}}{\partial \tau_w} = -\frac{\partial D/\partial \tau - \partial AS/\partial \tau_w}{\partial D/\partial p + \frac{p^e}{p^2}\partial AS/\partial \theta^e} \gtrless 0. \tag{3.2.34}$$

However, the output effect can be signed uniquely since

$$\frac{\partial \mathcal{Y}}{\partial \tau_w} = \frac{\partial AS}{\partial \tau_w} - \frac{p^e}{(\mathcal{P}(M, p^e))^2}\frac{\partial AS}{\partial \theta^e}\frac{\partial \mathcal{P}}{\partial \tau} = \frac{\dfrac{\partial AS}{\partial \tau_w}\dfrac{\partial D}{\partial p} + \dfrac{p^e}{p^2}\dfrac{\partial AS}{\partial \theta^e}\dfrac{\partial D}{\partial \tau}}{\dfrac{\partial D}{\partial p} + \dfrac{p^e}{p^2}\dfrac{\partial AS}{\partial \theta^e}} < 0. \tag{3.2.35}$$

Therefore, the traditional result known from Keynesian expenditure theory (under price/wage rigidity and unemployment) generalizes to the competitive temporary equilibrium model as well, i.e. a decrease of a uniform general income tax rate

$\tau = \tau_w = \tau_\pi$ induces an increase of output and employment while the effect on nominal GDP may go either way. Using the same method, an increase of money balances M also induces an upward shift of the aggregate demand function increasing output, prices, and GDP. In contrast, higher price expectations induce a downward shift of the aggregate supply function increasing prices and decreasing output and employment. Notice, however, that nominal income $py = \mathcal{P}(M, p^e)\mathcal{Y}(M, p^e)$ is still an increasing function in p^e, since the elasticity of the aggregate demand function is greater than minus one for fixed (M, g, τ).

Finally, the properties of aggregate demand and of aggregate supply induce important global properties of the price law in addition to homogeneity, concavity, and monotonicity. For every $M > 0$, one obtains two further properties of the limiting behavior of the expected inflation rates with respect to price expectations when AS is strictly decreasing[9]. Since the aggregate supply function is continuous and strictly decreasing, it is invertible on its range with respect to the rate of expected inflation. Therefore, its inverse exists which has an explicit representation with properties given in the following lemma. These will be useful for the dynamic analysis under rational expectations or perfect foresight presented in Chapter 4.

Lemma 3.2.3.
For every $M > 0$ one has:

$$\lim_{p^e \to 0} \frac{p^e}{\mathcal{P}(M, p^e)} = \frac{1}{M} \lim_{p^e \to 0} \frac{p^e}{\mathcal{P}\left(1, \frac{p^e}{M}\right)} = 0$$

$$\lim_{p^e \to \infty} \frac{p^e}{\mathcal{P}(M, p^e)} = AS^{-1}\left(\frac{g}{1 - c(1-\tau)}\right) > 0. \tag{3.2.36}$$

If aggregate supply AS is invertible with respect to θ^e, the price law $\mathcal{P}$ has an inverse with respect to p^e for all $p \geq M\mathcal{P}(1, 0)$, given by

$$p^e = \mathcal{P}^e(p, M, g, \tau) := p\, AS^{-1}\left(D\left(\frac{M}{p}, g, \tau\right)\right). \tag{3.2.37}$$

In addition, $\mathcal{P}^e(p, M)/p$ is a strictly increasing function which satisfies

$$\lim_{p \to M\mathcal{P}(1,0)} \frac{\mathcal{P}^e(p, M)}{p} = \lim_{p \to M\mathcal{P}(1,0)} AS^{-1}\left(D\left(\frac{M}{p}, g, \tau\right)\right) = 0 \tag{3.2.38}$$

$$\lim_{p \to \infty} \frac{\mathcal{P}^e(p, M)}{p} = \lim_{p \to \infty} AS^{-1}\left(D\left(\frac{M}{p}, g, \tau\right)\right) = AS^{-1}\left(\frac{g}{1 - c(1-\tau)}\right). \tag{3.2.39}$$

Therefore, the expected rate of inflation $\theta^e := p^e/p$ is a strictly increasing function of the equilibrium price, since the price law has elasticity less than one with respect

[9] To reduce notation the tax rate τ_w will be suppressed as an argument from now on. Parametric changes of the taxation for workers will not be discussed further in this chapter.

to expectations. In Figure 3.5 the inverse $\mathcal{P}^e(p, M)$ has been drawn as a convex increasing function since the price law is assumed to be concave in p^e.

To complete the analysis of the price law, additional implications for the range of values observable in equilibrium can be derived from its general properties. Let

$$\mathcal{E}_{\mathcal{P}} := \left\{(\theta^e, m) \in \mathbb{R}^2_+ \mid \theta^e = \frac{p^e}{\mathcal{P}(M, p^e)}, m = \frac{M}{\mathcal{P}(M, p^e)}, (M, p^e) \gg 0\right\} \tag{3.2.40}$$

denote the set of possible real states $(\theta^e, m) \in \mathbb{R}^2_+$ in equilibrium for the economy under $\mathcal{P}$, called the *equilibrium set* of $\mathcal{P}$. Using the homogeneity of $\mathcal{P}$ one finds that

$$(\theta^e, m) \in \mathcal{E}_{\mathcal{P}} \iff \mathcal{P}(m, \theta^e) = \mathcal{P}\left(\frac{M}{\mathcal{P}(M, p^e)}, \frac{p^e}{\mathcal{P}(M, p^e)}\right) = \frac{1}{\mathcal{P}(M, p^e)}\mathcal{P}(M, p^e) = 1.$$

In other words, the set $\mathcal{E}_{\mathcal{P}}$ of real balances and expected inflation rates in equilibrium coincides with the unit contour of the price law $\mathcal{P}$ in the space $\mathbb{R}^2_+$ of (M, p^e), see the discussion of Lemma 3.2.1 and Figure 3.2. Therefore, changes of the state variables $(M, p^e) \gg 0$ imply observable movements on the equilibrium set (along the unit contour) while changes of any parameters of the economy induce displacements of the equilibrium set (the unit contour).

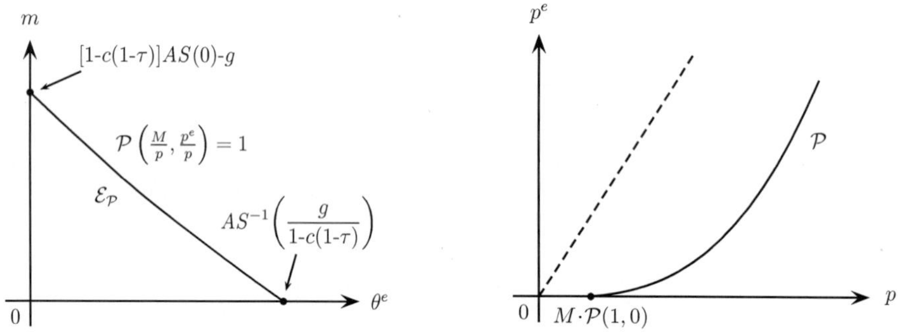

Fig. 3.5 Real money balances and expected inflation in temporary equilibrium

Exploiting the advantages of the AS–AD framework one step further (when expectations effects are separated from the wealth effects) the equilibrium set has an explicit representation given by the equation

$$m = [1 - c(1 - \tau)]AS(\theta^e) - g. \tag{3.2.41}$$

It is the set of points for which the equilibrium condition $D(M/\mathcal{P}(M, p^e)) = AS(p^e/\mathcal{P}(M, p^e))$ is satisfied for any given $(M, p^e) \gg 0$. Under Assumption 3.2.1, $AS(0) = y_{\max} > g/[1-c(1-\tau)]$. With $AS(0) < \infty$ and $g > 0$ equation (3.2.41) yields a downward sloping curve as the equilibrium manifold $\mathcal{E}_{\mathcal{P}}$, as depicted in Figure 3.5

a, which intersects both axes. Thus, in temporary equilibrium with bounded maximal output both real money balances and expected inflation rates have a positive upper bound. Since the graph of this function is the unit contour of the price law this implies that the price law must be a concave function if AS is convex in θ^e. Then, the inverse $\mathcal{P}^e(p, M)$ is a convex increasing function as shown in Figure 3.5 **b**.

3.2.6 Prices and Wages: A Two-Step Geometric Analysis

While the procedure underlying the AS–AD approach reveals the properties of the mechanism determining the equilibrium price level and real GDP in a most transparent way, the allocative implications between the labor market and the output market regarding the real wage adjustment are hidden. To reveal these properties it is useful to analyze the temporary equilibrium equations (3.2.13) applying an alternative two-step procedure. Consider first the labor market clearing condition

$$n_f h\left(\frac{w}{p}\right) - N\left((1-\tau_w)\frac{w}{p}V\left(\frac{p^e}{p}\right)\right) = 0.$$

Together with the Inada conditions it implies the existence of the market clearing real wage derived in (3.2.25) as a function of the expected inflation rate $w/p = W(\theta^e, \tau_w)$. Therefore, for each given expected price p^e, one obtains an explicit wage-price function under labor market clearing, defined by

$$w = LE(p, p^e) := pW\left(\frac{p^e}{p}, \tau\right) \tag{3.2.42}$$

whose elasticity with respect to p satisfies

$$0 < E_{LE}(p) = 1 - E_W(\theta^e) < 1 \tag{3.2.43}$$

(see (3.2.26)). Therefore, LE is an increasing function of p with an elasticity less than one everywhere. In addition, one also finds that changes of expectations have a positive effect on the wage function, i.e.

$$\frac{\partial}{\partial p^e} LE(p, p^e) = \frac{\partial W}{\partial \theta^e}\left(\frac{p^e}{p}\right) > 0,$$

i.e. increases in the expected price level implies an upward shift/rotation of the graph of LE (see Figure 3.6 **a**). LE has been drawn as a concave function which need not be the case, since the elasticity condition implies only that $LE(p, p^e)/p$ is nonincreasing.

Next consider the commodity market equilibrium condition

$$D(p, M, g, \tau) - n_f F\left(h\left(\frac{w}{p}\right)\right) = 0$$

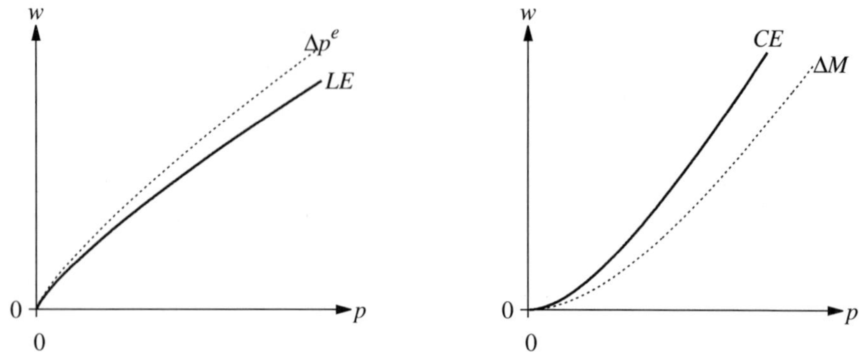

a Labor market: $\Delta p^e > 0$

b Commodity market: $\Delta M > 0$

Fig. 3.6 The role of expectations and of money balances

which is assumed to be independent of price expectations. Under the Inada conditions for the production function F this implies the existence of an associated wage function for each price p clearing the commodity market given by

$$w = CE(p, M) := p\,h^{-1}\left(F^{-1}\left(\frac{D(p, M, g, \tau)}{n_f}\right)\right) \tag{3.2.44}$$

$$= pF'\left(F^{-1}\left(\frac{D(p, M, g, \tau)}{n_f}\right)\right).$$

Since its derivative satisfies

$$\frac{\partial}{\partial p}CE(p, M) = \frac{w}{p} + \frac{p}{n_f h' F'}\frac{\partial D}{\partial p} > \frac{w}{p}, \tag{3.2.45}$$

CE is an increasing (convex) function with elasticity greater than one in p. In addition, an increase in money balances implies a downward shift/rotation of its graph, i.e.

$$\frac{\partial}{\partial M}CE(p, M, g, \tau) = \frac{p}{n_f h' F'}\frac{\partial D}{\partial M} < 0, \tag{3.2.46}$$

(see Figure 3.6 **b**). Finally, a temporary equilibrium is a pair $(p, w) \gg 0$ such that $w = LE(p, p^e) = CE(p, M, g, \tau)$. For (p^e, M), existence and uniqueness is guaranteed by a unique positive zero of the mapping

$$H(p) := LE(p, p^e) - CE(p, M)).$$

Notice, that H is strictly concave in p and $H(p)/p$ is strictly decreasing satisfying

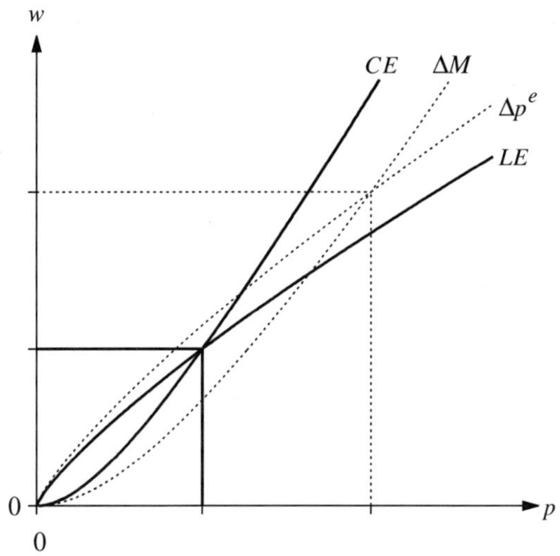

Fig. 3.7 Prices and wages in temporary equilibrium: $\Delta M > 0$ or $\Delta p^e > 0$

$$\lim_{p \to 0} \frac{H(p)}{p} = +\infty \qquad \text{and} \qquad \lim_{p \to \infty} \frac{H(p)}{p} \leq -k \tag{3.2.47}$$

for some $k > 0$. Therefore, a unique zero $p^* > 0$ exists. Moreover, the equilibrium pair satisfies $(p^*, w^*) = (p^*, LE(p^*, p^e)) = (\mathcal{P}(M, p^e), \mathcal{W}(M, p^e))$. Summarizing one obtains the following lemma whose proof is an alternative method to show existence and uniqueness of a temporary equilibrium without using the notion of the aggregate supply function.

Lemma 3.2.4.
Given Assumption 3.2.1, there exists a unique temporary equilibrium $(p, w) \gg 0$ *for every feasible* (M, p^e, g, τ).

Figure 3.7 displays the existence configuration and the possible the comparative statics effects of an increase of money balances or of price expectations on equilibrium prices and wages. In a similar fashion, the changes of all other parameters (g, τ, n_f) on equilibrium values are easily portrayed geometrically using the two functions LE and CE and their respective properties. Figure 3.8 reveals the implications of the homogeneity of the price law and of the wage law when money balances and expectations undergo an equiproportionate increase inducing an identical increase of the equilibrium wage and of the equilibrium price while the equilibrium real wage remains unchanged. In this case the implications of the quantity theory of money hold.

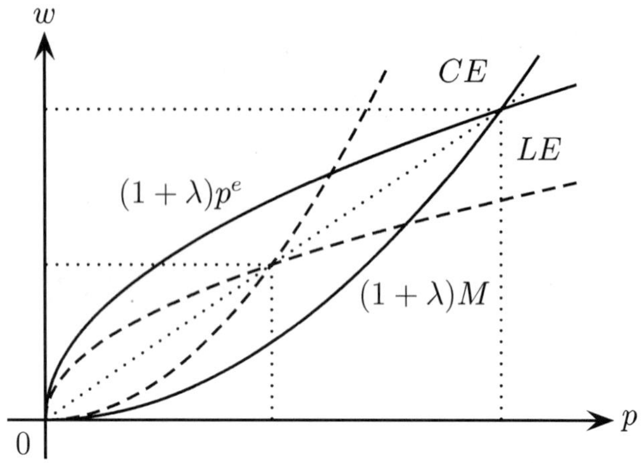

Fig. 3.8 An equiproportionate change of money balances and expectations

3.2.7 Isoelastic Functions: An Example

It is often informative to use specific functional forms for producer and consumer characteristics in order to verify the correctness of the intuition about the qualitative properties of a model, to analyze its characteristics, but also to derive more specific quantitative properties of its features. It turns out that the assumption of isoelastic functions for producers and consumers induces explicit algebraic relationships which yield additional insight into the features of the economy. Models with isoelastic functions are frequently used by many authors in macroeconomics (for example Lucas, 1972; Blanchard & Fischer, 1989, and others).

Consider the production function $F : \mathbb{R}_+ \to \mathbb{R}_+$ given in the isoelastic form

$$F(L) = \frac{A}{B} L^B \qquad \text{with} \qquad A > 0, \quad 0 < B < 1. \tag{3.2.48}$$

If $\alpha := w/p$ denotes the real wage, this yields the individual labor demand function

$$h(\alpha) = \left(\frac{A}{\alpha}\right)^{\frac{1}{1-B}} \tag{3.2.49}$$

implying aggregate supply of the production sector as

$$n_f F(h(\alpha)) = \frac{A_1}{B}(\alpha)^{\frac{B}{B-1}} \qquad \text{where} \qquad A_1 := n_f A^{\frac{1}{1-B}}. \tag{3.2.50}$$

For the worker characteristics assume that the function u_w describing intertemporal preferences has elasticity one (no consumption when young)and that the function v defining the disutility of work has elasticity $1 + C$, $C > 0$. $u_w = id$ implies that $V(\theta^e) \equiv 1/\theta^e$ and $c_w(\theta^e) = 0$. Then, the labor supply function $N : \mathbb{R}_+ \to \mathbb{R}_+$ is given by

$$L^S = N\left(\frac{(1-\tau_w)w}{p^e}\right) = n_w\left(\frac{(1-\tau_w)w}{p^e}\right)^{1/C} \quad \text{with} \quad C > 0. \tag{3.2.51}$$

Labor market equilibrium

$$n_f h(\alpha) = n_f \left(\frac{A}{\alpha}\right)^{\frac{1}{1-B}} \stackrel{!}{=} N((1-\tau_w)\alpha/\theta^e) = n_w(1-\tau_w)^{1/C}\left(\frac{\alpha}{\theta^e}\right)^{1/C}$$

implies an isoelastic equilibrium real wage function given by

$$\alpha = W(\theta^e) = \left(\frac{n_f}{n_w(1-\tau_w)^{1/C}}\right)^{\frac{C(1-B)}{C+1-B}} A^{\frac{C}{C+(1-B)}} (\theta^e)^{\frac{(1-B)}{(C+1-B)}}, \tag{3.2.52}$$

and an aggregate supply function

$$AS(\theta^e) = n_f F(h(W(\theta^e))) = \frac{A_1 A_2}{B}(\theta^e)^{\frac{-B}{C+(1-B)}} \tag{3.2.53}$$

where

$$A_2 := \left(\frac{n_f}{n_w(1-\tau_w)^{1/C}}\right)^{\frac{-CB}{C+1-B}} A^{\frac{CB}{(B-1)(C+(1-B))}},$$

which is isoelastic as well. Thus, aggregate supply is a downward sloping isoelastic function of the expected inflation rate θ^e, which is globally invertible on $\mathbb{R}_{++}$. The elasticity of the aggregate supply function may be greater or less than minus one, depending on the values of B and C, i.e. one has

$$B < (C + (1-B) \quad \Leftrightarrow \quad B < \frac{1+C}{2}. \tag{3.2.54}$$

Thus, $C > 1$ or $B < \frac{1}{2}$ implies that commodity supply as a function of the price (given p^e) has an elasticity

$$E_{AS}(p) = E_{AS}(\theta^e) \cdot (-1) = |E_{AS}(\theta^e)| < 1, \tag{3.2.55}$$

inducing a strictly concave isoelastic aggregate supply function in p. For a relatively small compact set of parameters

$$\left\{(B,C) \in \mathbb{R}^2_+ \,\middle|\, 1 + C \leq 2B\right\}$$

the aggregate supply function is linear or strictly convex in p.

The isoelastic aggregate supply function (3.2.53) in logarithmic form

$$\log y = a + b(\log p^e - \log p), \qquad b < 0, \quad a > 0 \tag{3.2.56}$$

is often referred to as the Lucas supply function (Lucas, 1972), who seems to have been the first who used it extensively to model aggregate supply within the macroeconomy.

Next, consider the demand side, when the utility function u_s of shareholders has elasticity $1-c_s$, $0 < c_s < 1$. Then, his net propensity to consume is $c_s(\theta) = c_s(1-\tau_\pi)$ implying an average propensity to consume $c(1-\tau) = c_s(1-\tau_\pi)(1-B)$. This yields an aggregate demand function of the form

$$D\left(\frac{M}{p}\right) = \frac{M/p + g}{1 - c(1-\tau)} \quad \text{with} \quad 0 < c < 1, \quad 0 \le \tau \le 1, \tag{3.2.57}$$

which has no expectations feed back. The equilibrium condition $AS(\theta^e) = D(M/p)$ implies a unique positive equilibrium price $p = \mathcal{P}(M, p^e)$ for every $(M, p^e) \gg 0$, but also an explicit inverse $\mathcal{P}^e$ of the price law $\mathcal{P}$ with respect to p for every $M \ge 0$ given by

$$p^e = \mathcal{P}^e(p, M) := p\left[\frac{B}{A_1 A_2}\,\frac{M/p + g}{1 - c(1-\tau)}\right]^{\frac{C+(1-B)}{-B}}. \tag{3.2.58}$$

It is a strictly increasing, strictly convex function in p with elasticity greater than one satisfying

$$\lim_{p\to 0} \frac{\mathcal{P}^e(p, M)}{p} = 0 \qquad \text{and} \qquad \lim_{p\to\infty} \frac{\mathcal{P}^e(p, M)}{p} = \begin{cases} \infty & g = 0 \\ < \infty & g > 0 \end{cases}.$$

Therefore, the inverse (3.2.58) as well as the price law itself are monotonically increasing bijective functions for any given (M, g, τ), a property which will prove extremely useful in Chapter 4 for the discussion of properties required to describe the dynamics with perfect foresight. Notice that the global invertibility requires that $\mathcal{P}$ is surjective on $\mathbb{R}_+$, thus equilibrium prices can become arbitrarily small for any given value $M > 0$ and output cannot be bounded. Therefore, $\ell_{\max} = +\infty$, which is an implicit condition when assuming a constant elasticity of the disutility function v of labor[10].

3.2.8 On Rigidities and Nonexistence of Equilibria

The analysis so far has shown that unique interior competitive equilibria exist for all positive values (M, p^e) of expectations and aggregate money balances under the general Assumption 3.2.1. In other words, the equilibrium price law $\mathcal{P}$ is globally defined on $\mathbb{R}^2_+$. Its main key ingredients are the global properties induced by the Inada condition for the production function and an equivalent property on the disutility of labor for workers (the so-called convex Inada condition for v). Together these implied that the price law was also differentiable with strictly positive partial derivatives with respect to both state variables. This section investigates the implications of some weaker conditions which are often considered as economically important. In particular, it will be examined to what extent these generalizations cause degen-

[10] Compare the properties 3.2.38 stated for the bounded case in Lemma 3.2.3.

erate effects of the price law or even failure of existence of temporary competitive equilibria.

Linear Disutility and Bounded Labor Supply

The assumption on the form of labor supply in the previous sections implies essentially the global properties of the aggregate supply function. In many cases it is assumed that individual labor supply has a positive finite upper bound. Consider the case with a linear disutility function $v : [0, \ell_{\max}] \to \mathbb{R}$ with constant $v' \geq 0$. In this case individual labor supply is of the form

$$\phi_\ell(w, p, p^e) = \begin{cases} 0 & \dfrac{w}{p} < \dfrac{v'}{(1-\tau_w)V(\theta^e)} \\[2ex] [0, \ell_{\max}] & \dfrac{w}{p} = \dfrac{v'}{(1-\tau_w)V(\theta^e)} \\[2ex] \ell_{\max} & \dfrac{w}{p} > \dfrac{v'}{(1-\tau_w)V(\theta^e)} \end{cases} \tag{3.2.59}$$

which is an upper-hemicontinuous and convex valued correspondence, but not a function any more. If producers' technologies still satisfy the Inada conditions, for every $\theta^e > 0$, the excess demand for labor is a strictly decreasing function of the real wage except at a unique critical level $\bar{\theta}$ which solves

$$F'\left(\frac{n_w \ell_{\max}}{n_f}\right) = \frac{v'}{(1-\tau_w)V(\bar{\theta})}.$$

Therefore, there exists a unique positive market clearing real wage given by

$$\begin{aligned} W(\theta^e) &= \begin{cases} F'\left(\dfrac{n_w \ell_{\max}}{n_f}\right) & \theta^e < \bar{\theta} \\[2ex] \dfrac{v'}{(1-\tau_w)V(\theta^e)} & \bar{\theta} \leq \theta^e \end{cases} \\ &= \max\left\{F'\left(\frac{n_w \ell_{\max}}{n_f}\right), \frac{v'}{(1-\tau_w)V(\theta^e)}\right\} \end{aligned} \tag{3.2.60}$$

which is piecewise differentiable, homogeneous of degree zero in (p, p^e), and non-decreasing in expected inflation. The formula shows that, on the one hand, the equilibrium real wage can never be lower than $v'/(1-\tau_w)V(\theta^e)$ which is the reservation wage for a worker. On the other hand, because of the maximal labor supply $n_w \ell_{\max}$ an equilibrium real wage cannot be smaller than the marginal product of F at average employment per firm, since this would induce positive excess demand for labor otherwise (see Figure 3.9 **b**. It still satisfies $0 \leq E_W(\theta^e) < 1$. The labor market equilibrium curve in price-wage space is given by

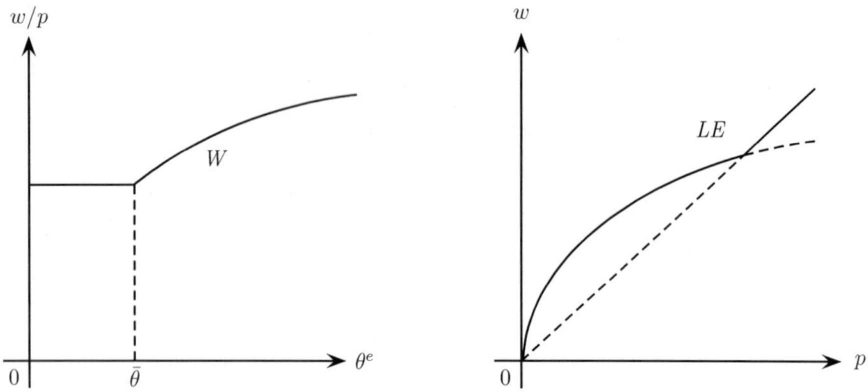

a Equilibrium real wage **b** Labor market clearing for given p^e

Fig. 3.9 Labor market clearing with linear disutility of labor

$$LE(p) := \max\left\{pF'\left(\frac{n_w\ell_{\max}}{n_f}\right), \frac{pv'}{(1-\tau_w)V(\theta^e)}\right\} \tag{3.2.61}$$

which is strictly increasing with elasticity less than or equal to one for each p^e (see Figure 3.9 **b**). Finally, one obtains for the aggregate supply function

$$y^s = AS(\theta^e) = n_f F(h(W(\theta^e))) = n_f F\left(h\left(\max\left\{F'\left(\frac{n_w\ell_{\max}}{n_f}\right), \frac{v'}{(1-\tau_w)V(\theta^e)}\right\}\right)\right) \tag{3.2.62}$$

which is bounded above. Thus, aggregate supply is a nonincreasing function of ex-

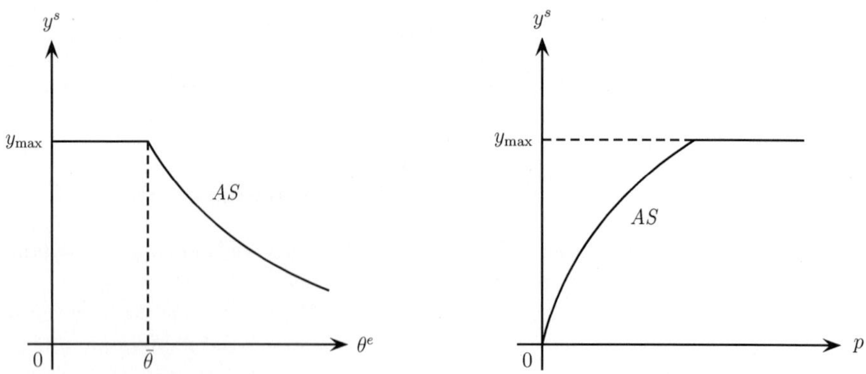

a Aggregate supply **b** Aggregate supply for given p^e

Fig. 3.10 Aggregate supply with linear disutility of labor pected inflation which is constant for low levels of inflation and a nondecreasing function of prices for given expectations p^e which reaches a positive maximum at a finite critical price (see Figure 3.10). Since aggregate demand has a strictly positive level for $p \to \infty$ for each positive level of government demand, the boundedness of aggregate supply requires that government demand is not too large for existence of a temporary equilibrium, i.e. $D(M/p, g, \tau) = AS(p^e/p)$ has a positive solution in p if and only if

$$y^s_{\max} := n_f F\left(\frac{n_w \ell_{\max}}{n_f}\right) > \frac{g}{1 - c(\tau)}. \tag{3.2.63}$$

Figure 3.11 displays two situations of existence and uniqueness under this condition. Therefore, with linear disutility of labor, the price law $p = \mathcal{P}(M, p^e)$ is a well defined

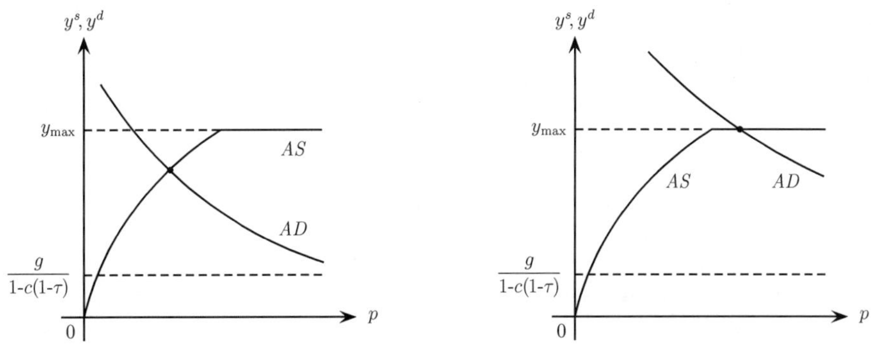

a M small **b** M large

Fig. 3.11 Existence of temporary equilibrium with linear disutility of labor; p^e given

continuous function, strictly increasing in money balances, and homogeneous of degree one in (M, p^e). The homogeneity yields that $p = p^e\mathcal{P}(M/p^e, 1)$. Together with the monotonicity this implies that there exists a unique level of expected real balances $\overline{(M/p^e)}$ such that equilibrium inflation rates p^e/p are less than the critical level $\bar{\theta}$ is and only if expected real money balances are larger than the critical level $\overline{(M/p^e)}$, formally one has

$$\bar{\theta} > \frac{p^e}{p} = \frac{p^e}{p^e\mathcal{P}(M/p^e, 1)} \quad \Longleftrightarrow \quad \overline{\left(\frac{M}{p^e}\right)} > \frac{M}{p^e}.$$

Therefore, the expectations effect is zero and the price contour must be horizontal in that region. The price law is linear in money balances M. Below expectations and money effects are positive and employment and output are below capacity. In other words, the equilibrium set takes the form as shown in Figure 3.12 given by the unit contour of the price law $\mathcal{P}$ with a horizontal section in the cone above the critical

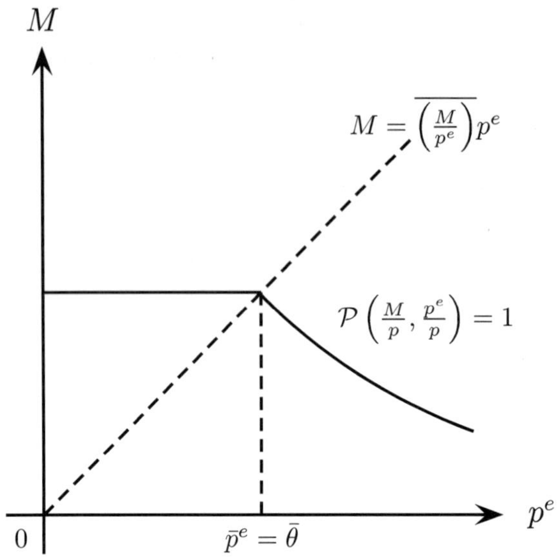

Fig. 3.12 Equilibrium set with linear disutility of labor

line. On the unit contour one must have $\bar{p}^e = \bar{\theta}\mathcal{P}(\frac{\bar{p}^e}{p}\overline{\left(\frac{M}{p^e}\right)}, \frac{\bar{p}^e}{p}) = \bar{\theta}$. All other contours are radial projections of the unit contour.

If $v' = 0$, labor supply is constant and equal to $n_w \ell_{\max}$. In this case, the equilibrium real wage function and the aggregate supply function are constant functions while the $LE(p)$ curve is linear. The equilibrium set is a horizontal line (and all contours of the price law are horizontal lines) which is equivalent to saying that the equilibrium price level is a linear function of money balances. Thus, the special case with constant labor supply implies constant output and employment at constant real wages and a complete absence of expectations effects, unless they reappear in more general situations through the demand side on the commodity market.

Linear Technologies

It is apparent that existence and uniqueness of competitive temporary equilibria will be obtained for such economies if the production function is linear (violating the Inada conditions) or if it is bounded as long as all other properties of Assumption 3.2.1 are maintained. Nevertheless, it is useful and informative to analyze the implications of alternative conditions on the production side for the aggregate supply function and for existence and uniqueness of temporary equilibrium in order to understand the role of each specific modification for the characteristics of the price law and the equilibrium set.

In the first case the equilibrium real wage must be constant while aggregate supply will be determined by labor supply conditions alone at the fixed real wage and respective supporting expected inflation rates. In this case the properties of the aggregate supply function remain unchanged so that competitive temporary equilibria exist and are unique under the remaining restrictions of Assumption 3.2.1.

Bounded Technologies: Capacity Constraints

Assume first that the set of production possibilities of each producer has a finite upper bound $y_{\max} > 0$ defined by a modified production function

$$\tilde{F}(z) := \min\{F(z), y_{\max}.\} \tag{3.2.64}$$

$y_{\max}$ is assumed to satisfy the feasibility condition (v) of 3.2.1 while F satisfies the Inada conditions. As a consequence one obtains a bounded labor demand function for each firm as a function of the real wage $\alpha := w/p$ given by

$$\tilde{h}(\alpha) := \min\{h(\alpha), z_{\max}\} = \begin{cases} z_{\max} & \alpha \leq \bar{\alpha} \\ h(\alpha) & \alpha > \bar{\alpha} \end{cases} \tag{3.2.65}$$

with $z_{\max} := F^{-1}(y_{\max})$ and $\bar{\alpha} := F'(z_{\max})$. This implies an aggregate labor demand function $\tilde{H}(\alpha) := n_f \min\{h(\alpha), z_{\max}\}$. Labor market learing occurs at $\tilde{H}(\alpha) - N(\alpha V(\theta^e)) = 0$ which is equivalent to

$$\frac{N^{-1}(n_f \min\{h(\alpha), z_{\max}\})}{\alpha} = V(\theta^e). \tag{3.2.66}$$

Both sides are strictly decreasing surjective functions of their respective arguments.

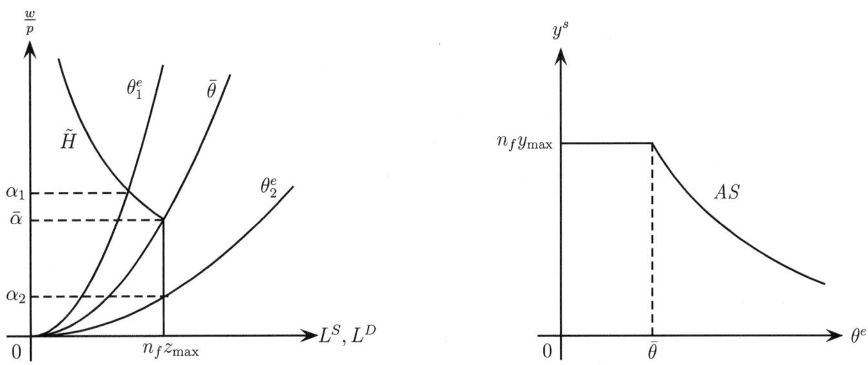

a Equilibrium in labor market: $\theta_1^e > \bar{\theta} > \theta_2^e$ **b** Aggregate Supply

Fig. 3.13 Capacity constraints: labor market equilibrium and Aggregate Supply

Therefore, for every $\theta^e > 0$ there exists a unique positive real wage clearing the labor market $\alpha = \tilde{W}(\theta^e) > 0$ which is a continuous increasing function of expected inflation with a unique non-differentiability at $\bar{\theta}$ satisfying $\bar{\alpha} = \tilde{W}(\bar{\theta})$ (see Figure 3.13). Therefore, the aggregate supply function becomes

$$y^S = AS(\theta^e) := \begin{cases} n_f y_{\max} & \theta^e \leq \bar{\theta} \\ n_f F(h(\tilde{W}(\theta^e))) & \theta^e > \bar{\theta} \end{cases} \tag{3.2.67}$$

Since AS is nonincreasing and aggregate demand is strictly decreasing in p, for

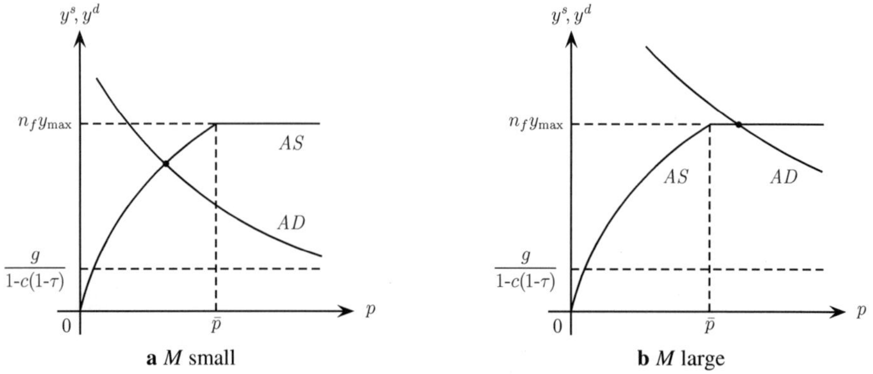

Fig. 3.14 Temporary equilibrium with capacity constraints; p^e given

every $(M, p^e) \gg 0$, there exists a unique positive equilibrium price. In other words, the price law $\mathcal{P}$ is well defined inheriting the homogeneity property from zero homogeneity of aggregate demand.

One obtains two distinct equilibrium configurations (similar to the case with linear disutility) displayed in Figure 3.14. Due to the nondifferentiability of the aggregate supply function at $\bar{\theta}$ where maximum output is reached the price law becomes nondifferentiable as well at all points (M, p^e) where $\bar{p}(p^e) := p^e(\bar{\theta})$ and

$$\frac{1}{1-c(1-\tau)}\left(\frac{M}{\bar{p}(p^e)} + g\right) = n_f y_{\max} \iff M = p^e \frac{n_f y_{\max}(1-c(1-\tau)) - g}{\bar{\theta}}$$

Therefore, for

$$\frac{M}{p^e} > \frac{n_f y_{\max}(1-c(1-\tau)) - g}{\bar{\theta}} \tag{3.2.68}$$

the price law is linear in M and independent of the expected price p^e and defined by

$$\mathcal{P}(M, p^e) = \frac{M}{n_f y_{\max}(1-c(1-\tau)) - g}.$$

The unit contour of the price law takes the form shown in Figure 3.15. Due to the homogeneity of $\mathcal{P}$ all other contours are radial projections. This shows geometri-

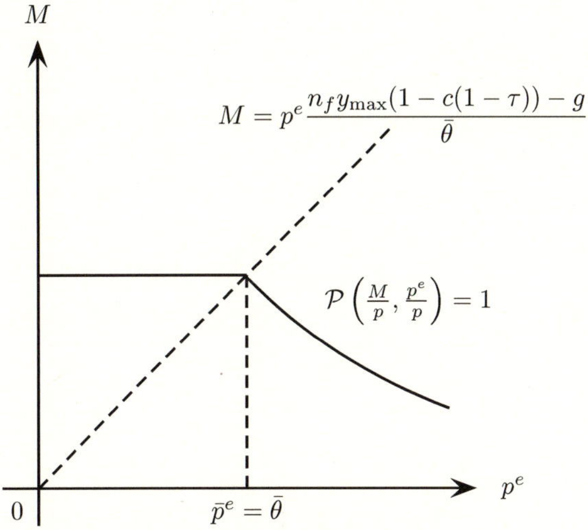

Fig. 3.15 Equilibrium set with capacity constraints

cally that changes in expectations have no influence on equilibrium prices for states above the dashed line. Increases in money balances, however, induce immediate one-to-one proportional increases in equilibrium prices for any given level of expectations p^e.

Since the unit contour of $\mathcal{P}$ is equivalent to the set of pairs of equilibrium values (θ^e, m) for any state (M, p^e), *all dynamics* of money balances and expectations of the economy induce real dynamics on the unit contour, i.e. on the equilibrium set. In contrast, changes of other parameters like government demand g imply displacements of the equilibrium set and of the discontinuity.

Linear Technologies with Capacity Constraints

Additional structural insight into the existence of specific properties of the temporary equilibrium mappings can be found when there are constant returns to scale. Assume that the production function of each producer is of the form

$$y = F(z) = \min\{az, y_{\max}\} \tag{3.2.69}$$

for some positive propductivity $a > 0$. This induces the labor demand correspondence of the producer

$$\tilde{h}(\alpha) := \begin{cases} z_{\max} & \alpha < a \\ [0, z_{\max}] & \alpha = a \\ 0 & \alpha > a \end{cases} \tag{3.2.70}$$

where $z_{\max} = y_{\max}/a$. Let $\tilde{H}(\alpha, \theta^e) := n_f \tilde{h}(\alpha) - N(\alpha V(\theta^e))$ denote the excess demand correspondence on the labor market. Evidently, any zero $0 \in \tilde{H}(\alpha, \theta^e)$ with positive

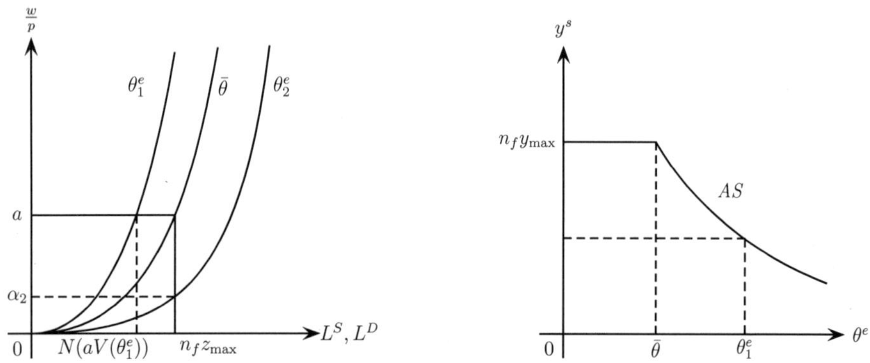

a Equilibrium in labor market: $\theta_1^e > \bar{\theta} > \theta_2^e$ **b** Aggregate Supply

Fig. 3.16 Labor market: Linear technology with capacity constraints

employment level requires that $\alpha \leq a$. In other words, the introduction of constant returns to scale implies an upward rigidity of the real wage under labor market clearing. Let

$$\bar{\theta} := V^{-1}\left(\frac{N^{-1}(n_f z_{\max})}{a}\right) \tag{3.2.71}$$

denote the unique positive solution of $N(aV(\bar{\theta})) = n_f z_{\max}$ which is monotonically decxreasing in $y_{\max}$. Due to the monotonicity of the labor supply function N and of V one obtains for the real wage function clearing the labor market

$$\alpha = W(\theta^e) := \begin{cases} \dfrac{N^{-1}(n_f z_{\max})}{V(\theta^e)} & \theta^e \leq \bar{\theta} \\ a & \theta^e > \bar{\theta} \end{cases} \tag{3.2.72}$$

with associated equilibrium employment level

$$L = N(W(\theta^e)V(\theta^e)), \tag{3.2.73}$$

see Figure 3.16. The real wage function is strictly monotonically increasing for all $\theta^e < \bar{\theta}$ and constant for all larger values. Therefore, one obtains as the aggregate supply function

$$AS(\theta^e) := aN(W(\theta^e)V(\theta^e)) \tag{3.2.74}$$

which has the same properties as in the two cases with linear disutility fro workers and with capacity constraints under Inada conditions. As a consequence, there exist

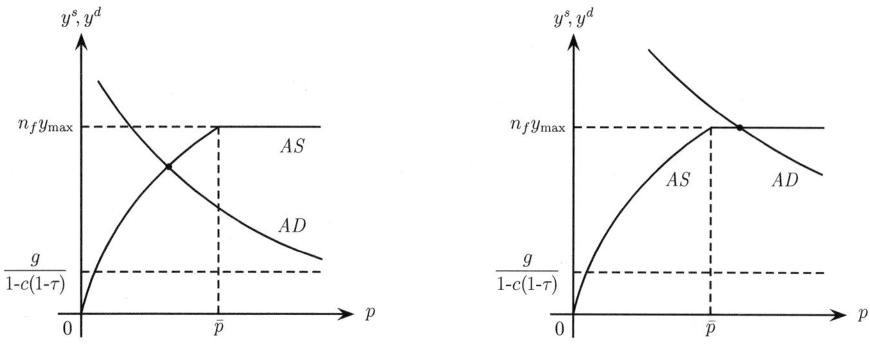

Fig. 3.17 Temporary equilibrium with capacity constraints and linear technology; p^e given

the same two distinct equilibrium configurations as in the two previous cases, see Figure 3.17, implying a price law which is insensitive to expectations for states with large money balances relative to $\bar{\theta}$. Therefore, the characteristics of the equilibrium set with constant returns and constraints is identical to the one displayed in Figure 3.15.

Finally, the critical level $\bar{\theta}$ is a decreasing function converging to zero as the capacity constraint $y_{\max}$ becomes large. Therefore, the results of the constraint case imply that the equilibrium real wage becomes constant for large enough constraints under bounded labor supply as $\bar{\theta}$ goes to zero. The nondifferentiability of the aggregate supply function disappears which is of the form $AS(\theta^e) = aN(aV(\theta^e))$. The price law as the unique solution $p = \mathcal{P}(M, p^e)$ of

$$\frac{M/p + g}{1 - c(1 - \tau)} = aN\left(aV\left(\frac{p^e}{p}\right)\right)$$

shows the ususal properties with respect to money balances and expectations and the typical equilibrium set as in the standard cases (see Figure 3.5) while the equilibrium real wage is constant for all states (M, p^e), i.e. equilibria exhibit full real wage rigidity. In other words, the wage law is linear given by $\mathcal{W}(M, p^e) = a\mathcal{P}(M, p^e)$. Profits are zero and all income goes to wage earners. Output and employment levels are monotonic functions of real expected money balances M/p^e.

Heterogeneous Technologies

Since uniform constant returns technologies and capacity constraints induce endogenous real wage rigidity as well well as price stickiness with respect to expectations it is an interesting exercise to examine their influence on equilibrium characteristics if a number of "smooth" producers, whose technology satisfies the Inada conditions of Assumption 3.2.1, coexist with a subgroup of producers for which the special conditions prevail.

Consider the standard economy given under Assumption 3.2.1 and assume that there exists a finite number of producers n_a with a linear technology and constraint $y_{\max}$ as in (3.2.69). This implies that, for every θ^e, the excess demand correspondence for labor has a closed graph and consists of an interval at the real wage level a. Nevertheless, this yields a unique market clearing real wage $\alpha = W(\theta^e)$ for every θ^e. W must be nondecreasing and of the form portrayed in Figure 3.18 **b** with a nondegenerate compact interval $[\theta_1, \theta_2]$

$$\theta_1 := V^{-1}\left(\frac{N^{-1}(n_f h(a) + z_{\max})}{a}\right) < V^{-1}\left(\frac{N^{-1}(n_f h(a))}{a}\right) =: \theta_2 \qquad (3.2.75)$$

on which the equilibrium real wage equals a, i.e. the real wage is rigid for a strictly positive interval of inflation rates. Thus, there are three distinct intervals: $0 < \theta^e < \theta_1 < \theta^e < \theta_2 < \theta^e$ where the constant returns firms produce at capacity, below capacity, or leave the market respectively when the equilibrium real wage is higher than a, see Figure 3.18 **b**. The labor market equilibrium induces an aggregate supply

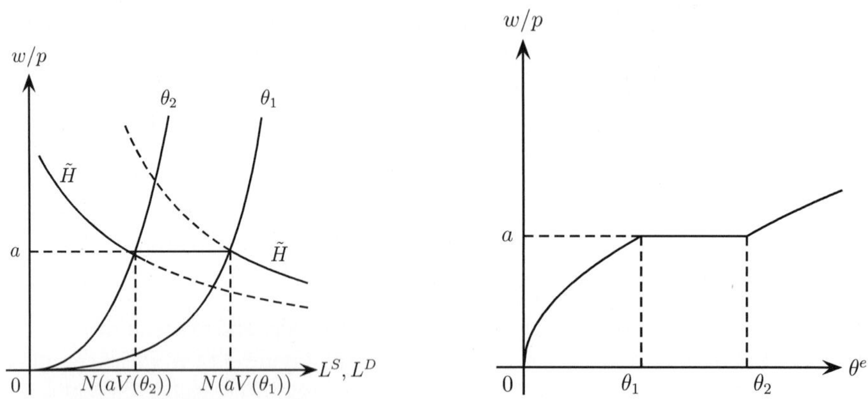

a Equilibrium in labor market: $\theta_1 < \theta_2$

b Equilibrium real wage function

Fig. 3.18 Labor market equilibrium with heterogeneous producers

function of the form

$$y^s = AS(\theta^e) := \begin{cases} n_f F(h(W(\theta^e))) + n_a y_{\max} & \theta^e < \theta_1 \\ n_f F(h(a)) + a(N(aV(\theta^e)) - n_f h(a)) & \theta_1 \leq \theta^e \leq \theta_2 \\ n_f F(h(W(\theta^e))) & \theta_2 < \theta^e \end{cases} \tag{3.2.76}$$

which is strictly decreasing as portrayed in Figure 3.19 **a**. This indicates two non-

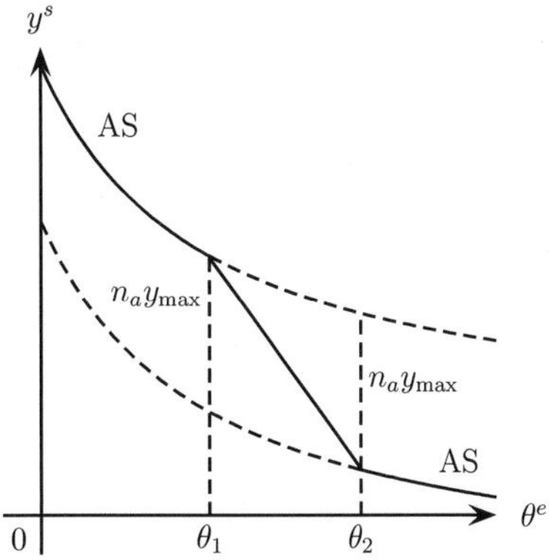

a Aggregate Supply

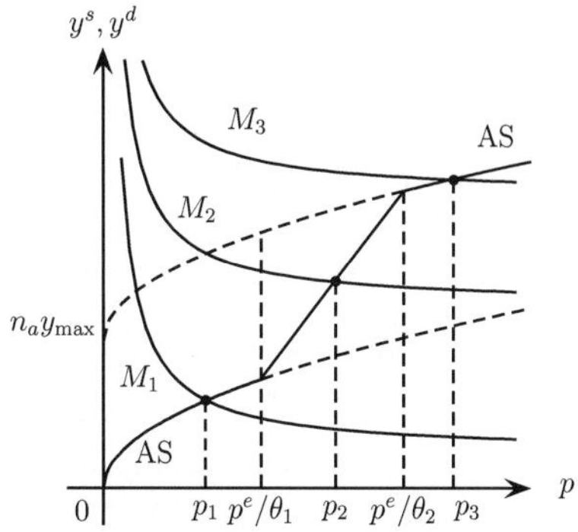

b Three equilibria for $M_1 < M_2 < M_3$; p^e given

Fig. 3.19 Heterogeneous producers: Aggregate Supply and equilibria

differentiabilities and a drop (of size $n_a y_{\max}$ between the two scenarios) in aggregate output as the expected inflation rate increases from θ_1 to θ_2.

Given the remaining assumptions of the standard model a unique market clearing commodity price exists for all pairs $(M, p^e) \gg 0$ inducing a well-defined price law $\mathcal{P}$. Due to the nondifferentiability and the usal properties of aggregate demand, this implies three distinct temporary equilibrium situations which occur depending on the state (M, p^e). The aggregate demand function has been drawn for three levels of money balances $M_1 < M_2 < M_3$, see subfigure **b**. The price law must be nondifferentiable along two rays in state space (marked as $(M/p^e)_1$ and $(M/p^e)_2$ in Figure 3.20) with associated kinks for the unit contour and equilibrium set respectively. The equilibrium real wage is constant for all states between the two rays and employment and output are constant along any ray M/p^e.

The contours of $\mathcal{P}$ are downward sloping in all three sections, i.e. there are no complete equilibrium price rigidities in spite of the endogenous partial real wage rigidity. The elasticities with respect to money balances as well as with respect to price expectations are strictly between zero and one. However, the slope of the unit contour between the two rays - where the firms with constant returns are active - is steeper than at both end points. Thus, the coexistence does not make the unit contour flat or even horizontal. This indicates that the presence of the additional firms increases the effect of a change in money balances and decreases the expectations

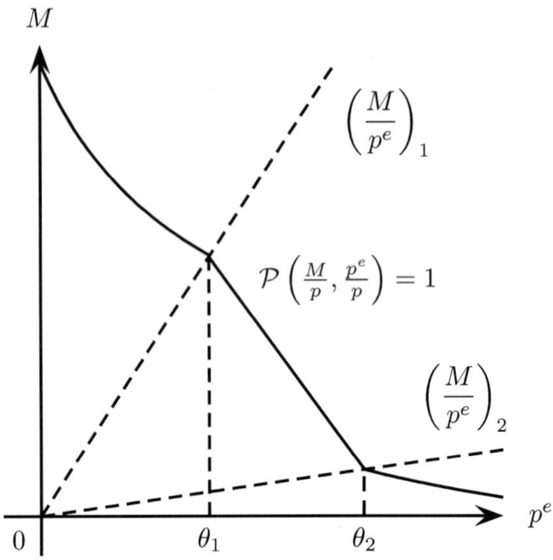

Fig. 3.20 The equilibrium set with heterogeneous producers

effect relative to the smooth case and the converse which one might have expected from the previous result.

Technologies with Fixed Costs

With fixed costs in production a non-convexity of the technology arises causing a discontinuity in competitive supply by a producer. Let the technology of each producer exhibit real fixed costs $c > 0$, such that the production function is of the form

$$\tilde{F}(z) = \max\{0, F(z) - c\} \tag{3.2.77}$$

where the function F remains the production function introduced in Assumption 3.2.1, which is strictly increasing, strictly concave without fixed costs, and which continues to satisfy the Inada conditions. Then, there exists a unique positive input level $\overline{z}$ and a level of real wages $\overline{\alpha} := F'(\overline{z})$ at which maximal profits become nonnegative for $\alpha \geq \overline{\alpha}$. Thus, labor demand of each producer is defined by

$$\tilde{h}(\alpha) = \arg\max_{z} p\tilde{F}(z) - wz = \begin{cases} (F')^{-1}\left(\frac{w}{p}\right) = h(\alpha) & \frac{w}{p} < \overline{\alpha} \\ \{0, h(\overline{\alpha})\} & \frac{w}{p} = \overline{\alpha} \\ 0 & \frac{w}{p} > \overline{\alpha} = F'(\overline{z}) \end{cases} \tag{3.2.78}$$

which is no longer a function but a *two-valued* correspondence at $\overline{\alpha}$. The level $\overline{\alpha}$ is the reservation real wage for the producer above which he does not hire labor to produce. As a consequence, aggregate labor demand is a correspondence of the form

$$\tilde{H}(\alpha) := \begin{cases} 0 & \alpha > \overline{\alpha} \\ \{0, 1, \ldots, n_f\}\, h(\overline{\alpha}) & \alpha = \overline{\alpha} \,. \\ n_f h(\alpha) & \alpha < \overline{\alpha} \end{cases} \tag{3.2.79}$$

To obtain labor market clearing, pairs of (θ^e, α) must satisfy $N(\alpha V(\theta)) \in \tilde{H}(\alpha)$. For $0 < \alpha \leq \overline{\alpha}$, the equality $n_f h(\alpha) = N(\alpha V(\theta^e))$ holds if and only if

$$\theta^e = V^{-1}\left(\frac{N^{-1}(n_f h(\alpha))}{\alpha}\right),$$

which defines a maximal inflation rate

$$0 < \theta^e_{n_f} := V^{-1}\left(N^{-1}\left(\frac{n_f h(\overline{\alpha})}{\overline{\alpha}}\right)\right) \tag{3.2.80}$$

for which labor market clearing is possible with all n_f firms producing and hiring $h(\alpha)$. Thus, there exists a real wage function $\overline{W} : (0, \theta^e] \to [0, \overline{\alpha}]$ valid for all n_f firms which is strictly increasing. For $\overline{\alpha}$, define a list of inflation rates $\theta^e_1 > \theta^e_2 > \ldots > \theta^e_i > \ldots > \theta^e_{n_f}$,

$$\theta^e_i := V^{-1}\left(\frac{N^{-1}(ih(\overline{\alpha}))}{\overline{\alpha}}\right), \quad i = 1, \ldots, n_f - 1, \tag{3.2.81}$$

which imply unique labor market clearing solutions $N(\overline{\alpha} V(\theta^e_i)) = ih(\overline{\alpha})$ with $1 \leq i < n_f$ firms hiring $h(\overline{\alpha})$ at $\overline{\alpha}$. Therefore, the equilibrium set of real wages and inflation rates consists of the union of the graph of $\overline{W}$ and the set of $n_f - 1$ pairs $(\theta^e_i, \overline{\alpha})$, $i = 1, \ldots, n_f - 1$, as shown in Figure 3.21 **a**. As a consequence, the aggregate supply under fixed costs consist of the union of the graph of $\overline{AS} : (0, \theta^e] \to \mathbb{R}_+$ defined by $y^s = \overline{AS}(\theta^e) := n_f F(h(\overline{W}(\theta^e)))$ with the set of $n_f - 1$ pairs $(\theta^e_i, y^s_i) := (\theta^e_i, iF(h(\overline{\alpha})))$, $i = 1, \ldots, n_f - 1$, see Figure 3.21 **b**.

Finally, define $p_i := p^e / \theta^e_i$, $i = 1, \ldots, n_f - 1$ the set of prices supporting supply for fixed expectations p^e. This implies a graph of the aggregate supply function containing also a finite set of discrete points (p_i, y^s_i). Therefore, existence of a temporary equilibrium price depends on the position of aggregate demand relative to the discrete part of aggregate supply as shown in Figure 3.22. For low money balances M, a temporary equilibrium may exist with all firms producing and the equilibrium price $p = \mathcal{P}(M, p^e) < p_{n_f}$. If M is small, the discreteness of the set of equilibrium pairs consistent with the labor market may cause nonexistence, since $(AD(M, p), p) \neq (y^s_i, p_i)$ for all $i = 1, \ldots, n_f$. This is equivalent to the fact that for given (M, p^e) the equation

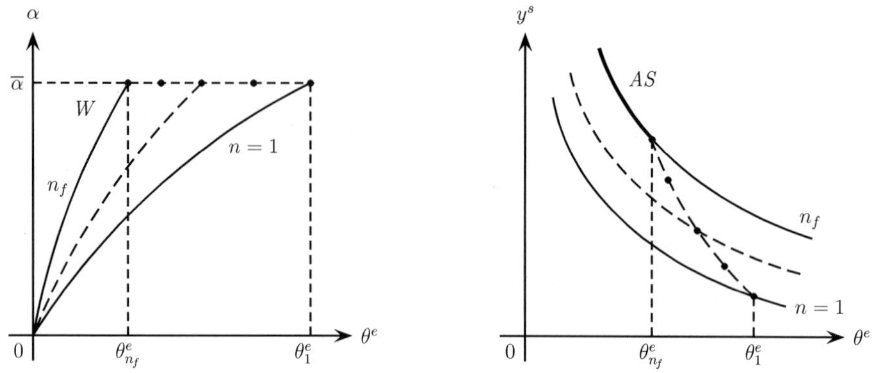

a Labor market equilibrium **b** Aggregate supply

Fig. 3.21 Labor market equilibrium and aggregate supply with fixed costs

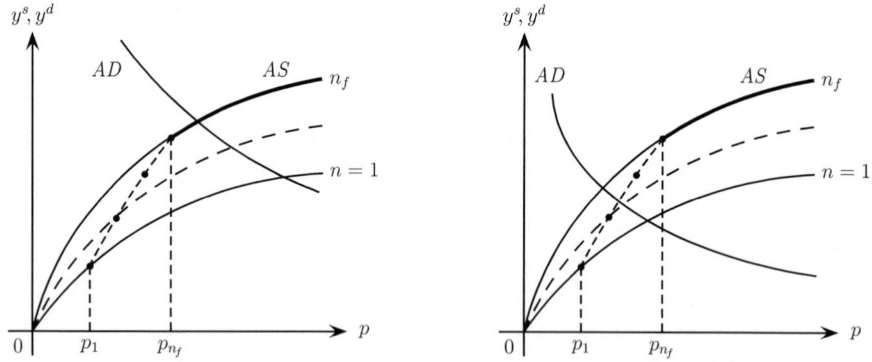

a Existence with $p > p_{n_f}$ **b** Failure of existence

Fig. 3.22 Equilibrium with fixed costs

$$D(p, M) - iF(h(W_i(p^e/p))) \tag{3.2.82}$$

has no zero $p > 0$ for all $i = 1, \ldots, n_f$, where $W_i(\theta^e)$ is the equilibrium real wage function solving $ih(\alpha) = N(\alpha V(\theta^e))$, for $i = 1, \ldots, n_f$ firms, (see Figure 3.22). In other words, for a range of expectations and money balances, no temporary competitive equilibrium exists while existence may occur in some cases if exit or entry of firms is allowed.

3.2.9 Concluding the Competitive Case

The first two sections of this chapter showed that competitive temporary equilibrium configurations are described by a pair of mappings called the *price law* and the *wage law* which incorporate all effects of the simultaneous interaction between the two markets for labor and output under perfect competition in any period. The homogeneity of commodity demand and of labor supply in money balances, prices, wages, and expectations implies that the implications of the *Quantity Theory of Money* hold for all values of the state variables and of the parameters, results which are a mirror image of Walras' Law. These are consequences of intertemporal optimizations under nonsatiation of preferences and income consistency for private agents in the closed economy model.

The specific properties of the homogeneity of intertemporal preferences for consumption and the separability of preferences for workers with respect to leisure-consumption substitution built into Assumption 3.2.1 imply a separation of expectations and income distribution effects in aggregate demand. This allows a conceptual separation of the determination of equilibrium in the output market from that in the labor market. It makes the AS-AD approach a useful analytical device to determine the equilibrium price level in the output market seemingly separate from the equilibrium in the labor market. While the aggregate supply function is a partial equilibrium relation for the labor market, in the end, however, the determination of equilibrium prices and wages has to be achieved simultaneously. The price law always incorporates most but not all features of competitive temporary equilibria with some effects remaining from wage determination, as is evident from the equivalent LE-CE approach. A complete separation of price and wage determination is possible when the strict convexity of Assumption 3.2.1 is relaxed. In more general situations than under Assumption 3.2.1 the joint approach is the only successful one for an equilibrium analysis to determine the price law *and* the wage law as the zero of a two-dimensional mapping.

The two mappings determine the equilibrium values of nominal prices and wages in the monetary competitive economy as a function of the two state variables of the economy, i.e. of aggregate holdings of money balances and of price expectations for *given values* of government parameters and of all other structural parameters which describe the stationary elements of the economy (i.e. the technological parameters, the number of producers or consumers, their preferences, etc.). Their parametric treatment implies in turn that the two equilibrium laws remain the *same mappings* in any period of the evolution of money balances and expectations regardless of changes of the parameters as well. Thus, conceptually it is essential to recognize that the two laws are time invariant mappings which determine the functional restrictions for *any* sequence of prices, wages, money balances, and expectations.

The insight of the structural invariance of the two laws has consequences for an analysis of policy questions for such economies. The variables determining each temporary equilibrium are, on the one hand, money balances and price expectations held by the consumers at the beginning of the respective period, and, on the other hand, the fiscal parameters (g, τ_w, τ_π). A detailed analysis of the comparative statics effects of changes in government demand or tax rates on the temporary equilibrium is straightforward and can be derived in the usual way. The primary effect of such changes occurs through the aggregate demand function which affect the CE curve. Therefore, output and employment effects are accompanied by an associated change of the real wage, when the tax rates remain constant. Thus, higher government demand induces higher output and employment, higher prices and wages, and higher nominal income with lower real wages. Similar implications hold true for changes of specific or of the average tax rate which always induce an effect through the demand multiplier. In addition, changes of the tax rate on wages have an additional supply effect which modify the overall effects.

The comparative statics effects describe the impact of fiscal policy under alternative parametric and static specifications for *given expectations and money balances*. Since these are the primary state variables for the price law, the dynamic analysis of the economy will require the description of the interaction of the government budget and of *changes* of money balances which induce *intertemporal effects*. Such intertemporal effects will also occur through the expectations formation of consumers. The interaction of these two mechanisms induces the dynamics of the two state variables of the economy whose development will be analyzed in Chapter 4.

A parametric form of any given *functional relationships* corresponding to Assumption 3.2.1 assures that the price law and the wage law are a globally defined family of parametrized mappings. This guarantees in general that the mappings are time invariant under changes of all other parameters (production elasticities, time preference, number of consumers or firms, etc.) as well. In other words, any sequence of temporary equilibria, whether experimentally generated or attributed empirically to such an economy even with time varying parameters must be contained in the graph of the two equilibrium laws (or their implied mappings for output, employment, etc.). Thus, for all allowable parametric changes (shocks) of *all parameters* their effect on the temporary equilibrium is described by the same given family of mappings. Therefore, a forward recursive dynamic analysis of sequences of competitive equilibria is obtained whenever the rules of change, i.e. the recursive mappings of the two main state variables money balances and expectations *and* the recursive rules for changes of parametric shocks, whether deterministic or stochastic, are specified. The induced orbits will always be embedded in the graphs of the two families of time invariant mappings.

Finally, it should be emphasized again that the results derived describe the prices, wages, and allocations of a closed competitive monetary economy with *full employment*, allowing no conclusions concerning the occurrence of unemployment when equilibria exist. An analysis of unemployment situations as feasible macroeconomic states in such a closed monetary model requires an *extension* of the equilibrium paradigm to allow disequilibrium allocations as income consistent states when markets do not clear. Some issues related to nonexistence are treated in Section 3.7 at the end of this chapter while the disequilibrium approach is presented in Chapter 6 and beyond.

3.3 Noncompetitive Commodity Pricing

Market power of producers in commodity markets (possibly together with other market imperfections) is often considered as a main reason why underemployment or unemployment in labor markets might occur. Noncompetitive price setting by firms at levels above marginal costs is said to cause a reduction in output and labor demand implying a lower responsiveness of wages and prices to market data, in particular to monetary aggregates (see for example Blanchard & Kiyotaki, 1987; Blanchard & Quah, 1989). This effect is interpreted as a form of rigidity of nominal prices (the price level), implying that monetary aggregates have real effects on the economy via the fiscal multiplier. On the one hand, this is said to create 'Keynesian' effects in the economy contradicting the classical hypothesis of the neutrality of money. On the other hand, such rigidities are then said *not* to induce the proper adjustment effects and prevent full market clearing everywhere – as would be the case when prices were sufficiently flexible to clear all markets. As a consequence, such rigidities are then said to prevent full employment equilibrium to be achieved in the economy. The so-called *New Keynesian Approach* (see Gordon, 1990; Clarida, Galí & Gertler, 1999; Dixon, 2008, and others) pursues this modeling strategy which introduces other imperfections in addition to strategic behavior (for example Akerlof & Yellen, 1985; Calvo, 1983; Taylor, 1993).

In this literature monopolistic competition is identified as a source for the appearance of Keynesian features on the demand side of the economy which are then said to be responsible for rigidities of prices or wages inducing disequilibrium properties in the labor market. It seems that the occurrence and existence of such intertemporal rigidity does not necessarily arise from the properties of imperfect competition within the standard neoclassical model alone under the usual assumptions. The occurrence often requires additional structural or behavioral assumptions on the micro level which lie outside of the descriptive domain of behavioral assumptions within the neoclassical model. This section analyzes the effects of price setting behavior of firms and its implications for the level of prices and wages and on the level of employment and output for the macroeconomy, i.e. for the complete and income consistent allocation of the closed monetary economy. It investigates, in particular, to what extent strategic or monopolistic behavior is responsible for allocations with unemployment in the macroeconomy. From a structural point of view the question is which way noncompetitive elements distort the two equilibrium mappings.

3.3.1 Monopolistic Price Setting: The New Keynesian Approach

Consider first the simplest (and rather unrealistic) case of an economy of the standard type from Section 3.2 with a single producer ($n_f = 1$) who has complete market power on the goods market[11]. In other words, he is a monopolist in the usual sense of partial equilibrium theory while the labor market in the economy remains competitive. He believes to know the aggregate demand behavior of consumers and of the government on the goods market and he will set the commodity price and quantity accordingly, while he takes the wage rate from the labor market as given. Let $D(p, M)$ denote the aggregate demand function which he perceives as being the correct one[12]. Then, the firm maximizes profits subject to aggregate commodity demand, i. e.

$$\max_{p \geq 0; L \geq 0} \left\{ p\, F(L) - w\, L \,\middle|\, F(L) \leq D(p, M) \right\}$$

for any given wage rate $w > 0$. Assuming the Inada conditions for the production function F and any general downward sloping demand function $D(p, M)$ in p, a profit maximizing interior decision of the monopolist $(p, L) \gg 0$ must satisfy the equality $D(p, M) = F(L)$, i.e. a profit maximizing decision is always "on the demand curve". Therefore, one can write the decision problem of the monopolist using the profit function

$$\Pi(p, w) = \underbrace{p\, D(p, M)}_{\text{revenue}} - \underbrace{w\, F^{-1}\left(D(p, M)\right)}_{\text{costs}}$$

known from partial equilibrium monopoly theory as

$$\max_{p \geq 0} \left[p\, D\,(p, M) - w\, F^{-1}\,(D(p, M)) \right]$$

which implies a monopoly price

$$\tilde{p} = P_{\text{monopoly}}(w, M) := \arg\max_{p \geq 0} \left[p\, D\,(p, M) - w\, F^{-1}\,(D(p, M)) \right].$$

The monopoly price P_{monopoly} is homogeneous of degree one in (w, M) if the demand function is homogeneous of degree zero in (p, M). The necessary first order conditions for an interior solution $\tilde{p} > 0$ requires

$$\tilde{p} \left(1 + \frac{1}{E_D(\tilde{p})} \right) F' \left(F^{-1}\,(D(\tilde{p}, M)) \right) \stackrel{!}{=} w < \tilde{p} F' \left(F^{-1}\,(D(\tilde{p}, M)) \right) \tag{3.3.1}$$

where

$$E_D\,(p) := \frac{\partial D}{\partial p} \frac{p}{D(p, M)} < 0$$

is the price elasticity of demand. Existence and interiority of the solution requires $E_D\,(\tilde{p}) < -1$.

---

[11] As an alternative one might assume that there is a larger number $n_f > 1$ of identical producers, but only one of them behaves strategically while all others take prices and wages as given. Such a model creates additional analytical complexities to discuss the monopolistic case, but would not change the main conclusions and implications of the analysis for the investigations.

[12] There exists an extensive literature on which notion of the demand function is the appropriate one in a general equilibrium context, using or defining different concepts of objective, subjective, perceived, conjectural, estimated, ..., etc.

If the elasticity condition holds at $\tilde{p}$ for $(w, M) \gg 0$, then w is less than the marginal value product $\tilde{p}F'\left(F^{-1}(D(M, \tilde{p}))\right)$ which implies that $\tilde{p} = P_{\text{monopoly}}(w, M)$ is larger than marginal costs at $\tilde{y} = D(\tilde{p}, M)$. This implies that the monopolistic labor demand is lower than the respective competitive one at all $w > 0$. In other words,

$$h_{\text{monopoly}}(w, M) := F^{-1}\left(D(P_{\text{monopoly}}(w, M), M)\right) < h\left(\frac{w}{P_{\text{monopoly}}(w, M)}\right). \qquad (3.3.2)$$

Equation (3.3.1) defines an explicit labor market cross effect

$$w = CE_{\text{monopoly}}(p) := p\left(1 + \frac{1}{E_D(p)}\right)F'\left(F^{-1}(D(p, M))\right), \qquad (3.3.3)$$

which is the marginal willingness of the monopolist to pay for his demand for labor (his inverse demand function for labor) if he sets the monopoly price $p > 0$ and chooses the associated monopolistic output level. Under Assumption 3.2.1 and if $E_D(p)$ is nonincreasing in p, (3.3.3) is an increasing and convex function, satisfying

$$CE_{\text{monopoly}}(p) < CE(p). \qquad (3.3.4)$$

Its inverse is the optimal price setting rule as a function of the wage rate.

Caution has to be taken invoking the elasticity condition of the aggregate demand function $D(p, M)$ to guarantee the existence of a finite monopolistic solution. The income-consistent aggregate demand function of Assumption 3.2.1 of the standard multiplier form

$$D(p, M) = \frac{M/p + g}{1 - c(\theta^e)(1 - \tau)}.$$

Since the price setting behavior of the producer implies a change in the income distribution between wage earners and profits (even under an isolelastic production function!), the overall price effect on aggregate commodity demand with monopolistic behavior consists of the direct price effect plus an induced income effect. The standard form above excludes distribution effects from monopolistic pricing on demand (profit income versus labor income), but it clearly includes the aggregate income effect. If the monopolist takes that second effect into account as well – which he should if he has full information on the profit incomes of his customers – he uses the above functional form.

If c is constant, the elasticity of the demand function is greater than minus one everywhere, which violates the necessary condition for existence. Thus, a solution to the monopoly problem does not exist and the attempt to implement a profit maximizing rule leads to prices and profits going to infinity while output goes to zero. However, the propensity to consume $c(\theta^e)$ may depend on the expected inflation rate and therefore on prices. Therefore, without additional specific expectations effects the demand function violates this condition, so that a general equilibrium where a single monopolist chooses a profit-maximizing price does not exist. Conversely, if the propensity to consume is a decreasing function in θ^e, the elasticity condition may hold for given expectations, so that the global monopoly problem has in fact a solution.

The consequences of the failure of the elasticity condition are a well-known phenomenon in models where cross market effects play a role and where the monopolist uses the so-called objective demand function of the competitive part of the rest of the economy. The implications of such cross-market effects have been discussed widely in the literature on general equilibrium models (for example Marschak & Selten, 1974; Arrow & Hahn, 1971; Benassy, 1988; Negishi, 1961, 1987; Bonanno, 1990) and different proposals for the kind of demand function to be chosen by monopolists have beeen made in a microeconomic but also in macroeconomic models (see Silvestre, 1977; Benassy, 1987). In case of a profit maximizing monopolist within a group of other competitive firms the elasticity condition required will be much weaker than under monopoly and existence may prevail.

There are other economic scenarios besides profit maximization against a perceived demand curve describing monopolistic behavior within a macroeconomic setting which involve positive finite solutions to the price setting problem. Some of them do not involve the monopolist's notion of demand and are justified purely from the cost or technology side. For example, pricing rules with fixed markups over marginal or average cost etc. have associated finite solutions. In fact, equation 3.3.1 corresponds precisely to a markup rule of prices over marginal costs equal to $\gamma = 1/E_D(\tilde{p})$. When technological or other non-demand reasons are used to define the markup the left hand side of equation 3.3.1 corresponds to the inverse labor demand function of the noncompetitive firm.

In order to determine the temporary equilibrium of the economy the labor market clearing condition has to be added. With one producer only ($n_f = 1$), a temporary monopolistic equilibrium is a pair $(p, w) \gg 0$ such that

$$w = CE_{\text{monopoly}}(p) \qquad \text{and} \qquad h_{\text{monopoly}}(w, M) = N\left(\frac{w}{p} V\left(\frac{p^e}{p}\right)\right) \tag{3.3.5}$$

which can be written equivalently as

$$w = CE_{\text{monopoly}}(p) \qquad \text{and} \qquad D(p, M) = F\left(N\left(\frac{w}{p} V\left(\frac{p^e}{p}\right)\right)\right). \tag{3.3.6}$$

Both conditions assume equality of supply and demand on the output market, so that the second equation does not correspond fully to the monopolistic analogue of the LE curve from the two-step procedure of Section 3.2.6 in the competitive case. If the monopolists price setting rule is independent of the demand function, the analogous curve LE_{monopoly} is well defined and can be used to determine the temporary equilibrium, see (3.3.16) and Figure 3.25 below.

The second equation of (3.3.6) determines all equilibrium pairs (p, w) which are consistent with competitive demand behavior on the commodity market and competitive supply behavior of the labor market, which is also referred to as the *competitive price-wage frontier* PWF of the economy[13]. Given the invertibility of F and the strict monotonicity of N, it has an explicit representation for each $(M, p^e) \gg 0$ given by

$$w = PWF(p) := \begin{cases} \dfrac{p}{V\left(\frac{p^e}{p}\right)} N^{-1}\left(F^{-1}(D(p,M))\right) & \text{if } F^{-1}(D(p,M)) \leq n_w L_{\max} \\ \in \left[\dfrac{pv'(L_{\max})}{V\left(\frac{p^e}{p}\right)}, +\infty\right) & \text{else} \end{cases} . \tag{3.3.7}$$

As a consequence, a solution $(\tilde{p}, \tilde{w}) \gg 0$ of $\tilde{w} = PWF(\tilde{p}) = CE_{\text{monopoly}}(\tilde{p})$ defines a temporary monopolistic equilibrium. Surely, the equilibrium may occur at the minimal price which solves $D(p_{\min}, M) = F(n_w L_{\max})$ and $w_{\min} = p_{\min} F'(n_w L_{\max})$.

If $p_{\min}$ is not a solution, then $PWF(p_{\min}) - CE_{\text{monopoly}}(p_{\min}) > 0$. Given the elasticity condition, one finds that

$$H(p) := PWF(p) - CE_{\text{monopoly}}(p)$$

is strictly decreasing for $p \geq p_{\min}$ since $0 < E_F \cdot E_N \cdot (1 + E_V(p)) < 1$, using Lemma B.1.1, and

$$\frac{\partial}{\partial p} PWF(p) = \frac{D}{pF'N'} \left(E_D(p) + E_F \cdot E_N \cdot (1 + E_V(p))\right). \tag{3.3.8}$$

Moreover,

$$\begin{aligned} &\lim_{p \to \infty} \frac{H(p)}{p} \leq k < 0 \qquad \text{and} \\ &\lim_{p \to p_{\min}} \frac{H(p)}{p} = \frac{1}{p_{\min}} \left(\frac{p_{\min} v'(L_{\max})}{V\left(\frac{p^e}{p_{\min}}\right)} - CE_{\text{monopoly}}(p_{\min}) \right) > 0 \end{aligned} \tag{3.3.9}$$

Therefore, $H(p)$ has a unique positive zero implying that there exists a unique monopolistic equilibrium. Thus, as in the competitive case, one obtains as associated temporary equilibrium mappings the price and wage laws with monopolistic price setting given by

$$\begin{aligned} p &= \mathcal{P}_{\text{monopoly}}(M, p^e) := P_{\text{monopoly}}(M, \mathcal{W}_{\text{monopoly}}(M, p^e)) \\ w &= \mathcal{W}_{\text{monopoly}}(M, p^e) \equiv E_{PWF}(\mathcal{P}_{\text{monopoly}}(M, p^e)) \equiv CE_{\text{monopoly}}(\mathcal{P}_{\text{monopoly}}(M, p^e)), \end{aligned} \tag{3.3.10}$$

[13] Consumers behave competitively on both markets as price and wage takers. If maximal labor supply is finite, N is not globally invertible and equilibrium prices have a positive lower bound.

inducing associated employment and output laws $\mathcal{L}_{\text{monopoly}}$ and $\mathcal{Y}_{\text{monopoly}}$. Thus, under some mild additional qualifications of the neoclassical conditions assumed in 3.2.1, a temporary equilibrium with a competitive labor market and a monopolist in the commodity market is associated with a *full employment* level and not with unemployment. In other words, structurally monopoly pricing in the macroeconomy does not imply unemployment. With weak additional assumptions which guarantee sufficient concavity of $\Pi(p, w)$ in p, one finds that the monopolist's labor demand $h_{monopoly}$ is decreasing in w and increasing in M, while the monopoly price $P_{monopoly}(w, M)$ is increasing in w and increasing in M.

With these additional assumptions one obtains the comparative statics effects with respect to money balances and price expectations (M, p^e) which can be obtained in the same fashion using the Implicit Funtion Theorem (as for example in Section 3.2.6). They have the same signs as under competition.

$$\begin{aligned} &\frac{\partial}{\partial M}\mathcal{P}_{\text{monopoly}}(M, p^e) > 0 && \frac{\partial}{\partial p^e}\mathcal{P}_{\text{monopoly}}(M, p^e) > 0 \\ &\frac{\partial}{\partial M}\mathcal{W}_{\text{monopoly}}(M, p^e) > 0 && \frac{\partial}{\partial p^e}\mathcal{W}_{\text{monopoly}}(M, p^e) > 0 \\ &\frac{\partial}{\partial M}\mathcal{L}_{\text{monopoly}}(M, p^e) > 0 && \frac{\partial}{\partial p^e}\mathcal{L}_{\text{monopoly}}(M, p^e) < 0 \end{aligned} \tag{3.3.11}$$

Figure 3.23 portrays the equilibrium price and wage under monopoly and competition where the price-wage-frontier has been drawn for the invertible region. Comparing monopoly with perfect competition, one finds that for each $(M, p^e) \gg 0$, monopolistic price setting leads to higher equilibrium prices and lower equilibrium wages, lower real wages, lower employment/lower output than under competition. In summary, one obtains a result for monopolistic equilibria analogous to

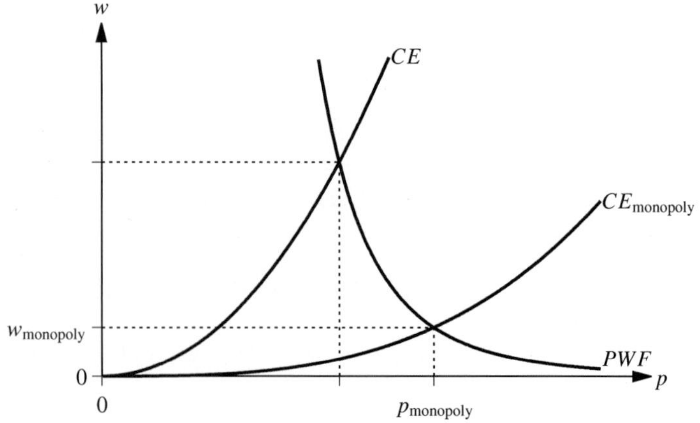

Fig. 3.23 Competition vs. monopoly in the commodity market; (M, p^e) given

Lemma 3.2.1 for the competitive case.

Lemma 3.3.1.
Let the basic Assumption 3.2.1 be given. Then,

(a) The price and the wage law $\mathcal{P}_{monopoly} : \mathbb{R}^2_+ \to \mathbb{R}_+$, $\mathcal{W}_{monopoly} : \mathbb{R}^2_+ \to \mathbb{R}$ *are both monotonically increasing functions in* (M, p^e).

(b) Their partial elasticities with respect to expectations and to money balances are less than one and satisfy

$$0 < E_{\mathcal{P}_{monopoly}}(p^e) < E_{\mathcal{W}_{monopoly}}(p^e) < 1 \tag{3.3.12}$$

$$0 < E_{\mathcal{W}_{monopoly}}(M) < E_{\mathcal{P}_{monopoly}}(M) < 1. \tag{3.3.13}$$

(c) An increase of expected prices decreases output and employment while an increase of money balances increases output and employment.

(d) If aggregate demand is homogeneous of degree zero in (p, M), *then the quantity theory of money holds, i.e. the functions* $\mathcal{P}_{monopoly}$ *and* $\mathcal{W}_{monopoly}$ *are homogeneous of degree one in* (M, p^e) *with*

$$p = M\mathcal{P}_{monopoly}\left(1, \frac{p^e}{M}\right), \quad w = M\mathcal{W}_{monopoly}\left(1, \frac{p^e}{M}\right) \tag{3.3.14}$$

and $\mathcal{Y}_{monopoly}$ *and* $\mathcal{L}_{monopoly}$ *are homogeneous of degree zero in* (M, p^e) *with*

$$L = \mathcal{L}_{monopoly}\left(1, \frac{p^e}{M}\right), \quad y = \mathcal{Y}_{monopoly}\left(1, \frac{p^e}{M}\right). \tag{3.3.15}$$

The lemma confirms that temporary equilibria with a monopoly are described by a time invariant mapping which is homogeneous of degree one in money balances and expectations with the same structural features as under perfect competition in both markets. Therefore, the comparative statics properties are qualitatively the same as under comeptition, but they will surely differ quantitatively in general.

In order to understand the differences and similarities, consider first *the equilibrium set*, i.e. the set of all pairs (θ^e, m) of equilibrium values of expected inflation and of real money balances. These are described by the unit contour of the price law which is a downward sloping geometric object in (θ^e, m)-space implying an identical elasticity condition as (3.2.23) in the competitive case. Therefore, changes of any parameters of the economy (government demand, tax rates, elasticities of production or consumption, production shocks) induce displacements of the unit contour while all changes of intertemporal effects from money balances or expectations induce changes *along* the unit contour.

If labor supply is endogenous and non-constant one has for all $(M, p^e) \gg 0$ that $\mathcal{P}_{\text{monopoly}}(M, p^e) > \mathcal{P}(M, p^e)$. Therefore, because of the homogeneity of both functions the unit contour of the price law for monopoly retracts towards the origin relative to the competitive case, i.e. geometrically speaking the monopolistic unit contour lies below/to the South-West of the corresponding competitive unit contour

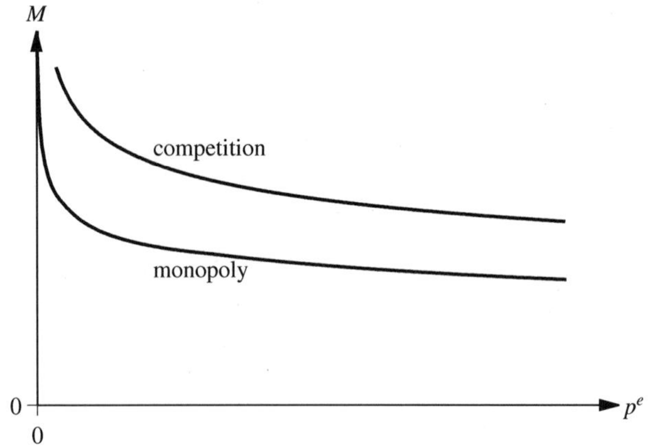

Fig. 3.24 Equilibrium sets under competition and monopoly

(see Figure 3.24). The two contours do not intersect. To verify this property consider any fixed $(M, p^e) \gg 0$ with $1 = \mathcal{P}_{\text{monopoly}}(M, p^e) > \mathcal{P}(M, p^e)$ which implies $1 = M\mathcal{P}_{\text{monopoly}}(1, p^e/M) > M\mathcal{P}(1, p^e/M)$. Therefore, for $\lambda := \frac{\mathcal{P}_{\text{monopoly}}(1,p^e/M)}{\mathcal{P}(1,p^e/M)} > 1$ one has

$$\mathcal{P}(M, p^e) < \mathcal{P}(\lambda M, \lambda p^e) = \lambda M\mathcal{P}(1, p^e/M) = \frac{\mathcal{P}_{\text{monopoly}}(1, p^e/M)}{\mathcal{P}(1, p^e/M)} M\mathcal{P}(1, p^e/M) = 1.$$

The comparative statics properties with respect to fiscal parameters are straightforward. For the two fiscal multipliers one has $\partial\mathcal{P}_{\text{monopoly}}(M, p^e)/\partial g > 0$ and $\partial\mathcal{P}(M, p^e)/\partial g > 0$. In other words, the fiscal policy is effective in both cases and qualitatively the multipliers are the same, which implies that both equilibrium sets shift downwards as government demand g increases (see also Figure 3.26).

Finally, some special properties arise when labor supply in the economy is fixed equal to $L_{\max} > 0$ and independent of expectations. The general equilibrium properties become the most striking when at the same time the monopolist sets prices using a rule depending on cost or technology arguments only and not on properties of the demand function from the output market. Therefore, for explanatory purposes, assume that the monopolist sets the price using a fixed mark-up $\gamma > 0$ over marginal costs. In this case, the monopolists's demand for labor $L = h_{\text{monopoly}}(w, M)$ is independent of money balances M and defined by the condition $w(1+\gamma) = pF'(L)$. This implies the equilibrium conditions under monopoly as

$$w = CE_{\text{monopoly}}(p) = \frac{p}{1+\gamma}F'(F^{-1}(D(M/p))) \qquad \text{and} \qquad w = \frac{p}{1+\gamma}F'(L_{\max}) \tag{3.3.16}$$

defining two monopolistic curves $CE_{\text{monopoly}}(p)$ and $LE_{\text{monopoly}}(p)$ in price-wage-space whose intersection provides the monopolistic equilibrium, see Figure 3.25. A positive solution $(\tilde{p}, \tilde{w})$ of 3.3.16 exists if and only if $\tilde{p}$ solves

$$F(L_{\max}) = D(M/p) = \frac{M/p + g}{1 - c(p^e/p)(1-\tau)}.$$

It is surprising to find that, for any given $(M, p^e) \gg 0$, the solution $\tilde{p}$ must coincide with the solution of the competitive equilibrium, which implies equality of the two price laws $\mathcal{P}_{\text{monopoly}}(M, p^e) = \mathcal{P}(M, p^e)$ for all (M, p^e). To prove this,

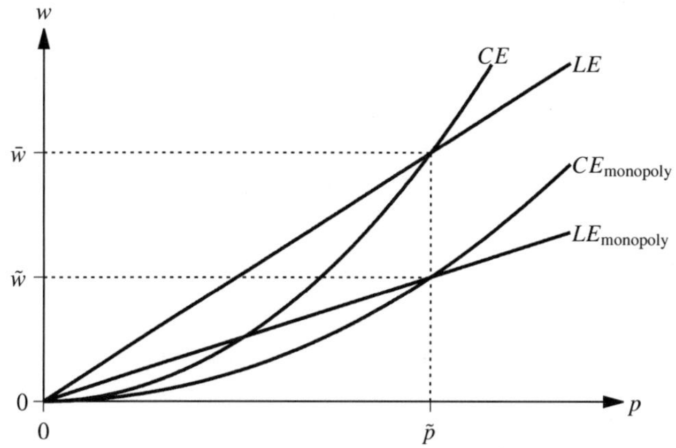

Fig. 3.25 Competition vs. monopoly: constant labor supply and markup pricing

consider the competitive CE curve and $(p, w) \in CE$ which holds if and only if $D(M/p) = F(h(w/p))$. Therefore,

$$w = CE(p) = ph^{-1}(F^{-1}(D(M/p))) = pF'(F^{-1}(D(M/p))).$$

The competitive LE curve is given by $w = LE(p) = pF'(L_{\max})$ since $L_{\max} = h(w/p)$ if and only if $w/p = F'(L_{\max})$. Therefore, the competitive equilibrium price solves $LE(p) = CE(p)$ if and only if $F(L_{\max}) = D(M/p)$ which is equal to $\tilde{p}$. Therefore, the equilibrium price level under monopoly coincides with the one under competition while the equilibrium wage

$$\tilde{w} = \mathcal{W}_{\text{monopoly}}(M, p^e) = \frac{F'(L_{\max})}{1+\gamma}\mathcal{P}_{\text{monopoly}}(M, p^e) = \frac{F'(L_{\max})}{1+\gamma}\tilde{p} \qquad (3.3.17)$$
$$< \tilde{p}F'(L_{\max}) = LE(\tilde{p}) = \mathcal{W}(M, p^e)$$

is lower than the competitive wage by the markup factor $1/(1+\gamma)$ (see Figure 3.25). This somewhat surprising result shows that the features of the price laws and the equilibrium sets for the two cases may depend in ways on structural parameters of the economy which appear only if the full temporary equilibrium is considered and

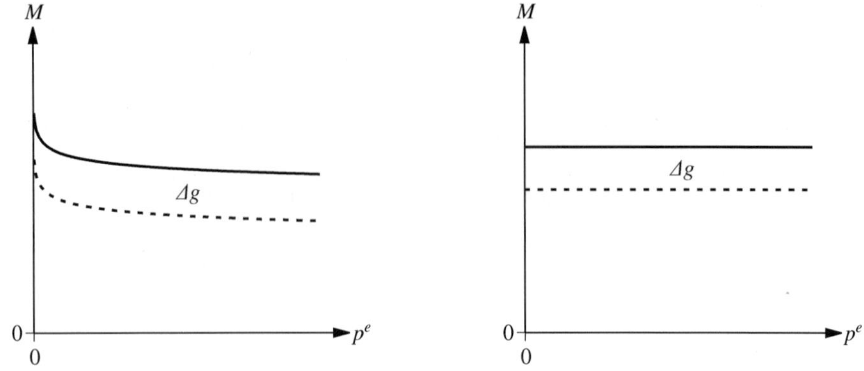

a A positive expectations effect on demand

b No expectations effect on demand

Fig. 3.26 Role of government demand on the equilibrium set: constant labor supply

which would not be expected under standard partial equilibrium reasoning. As a consequence the equilibrium sets under perfect competition and monopoly as unit contours of the price law coincide once the expectations effect of labor supply is eliminated *and* the monopolistic pricing does not use features of aggregate demand. Figure 3.26 displays two cases with and without an expectations effect in aggregate demand. In both cases, the fiscal multiplier is well defined with $\partial \mathcal{P}_{\text{monopoly}}/\partial g > 0$. Since output and employment are constant, additional government demand implies a pure price effect and crowding out/reduction of private consumption by old consumers. Any increase $\Delta g > 0$ induces a downward shift of the equilibrium set.

3.3.2 Oligopolistic Price Setting

Next consider the general situation with $i = 1, \ldots, n_f$ firms producing the same homogeneous output and assume that they chose their output strategically assuming knowledge of the aggregate demand function *and* realizing that other firms are doing the same in the market. In other words, all firms behave as in the standard Cournot oligopoly, while the labor market remains competitive, so that all firms take the wage as given. In this case, the typical firm i choses $y_i = F(L_i)$ for the commodity market to maximize profit

$$py_i - wL_i \quad \text{subject to} \quad \sum_j y_j = D(p, M),$$

taking the decisions $(y_j)_{j \neq i}$ of all other firms as given. The associated Lagrangean is

$$\mathcal{L} = py_i - wF^{-1}(y_i) + \lambda[D(p, M) - \sum_j y_j],$$

which induces the first order conditions of an interior solution for every firm $i = 1, \ldots, n_f$

$$F'(L_i)\left(p + \frac{y_i}{\partial D/\partial p}\right) = w, \qquad L_i = F^{-1}(y_i).$$

Since all firms produce with the same production function F, an interior *Cournot equilibrium* must be symmetric with each firm producing $1/n_f$ of total output and demanding $1/n_f$ of total labor. Therefore, a price p set by symmetric oligopolists under demand consistency (commodity market clearing) implies a unique wage rate w (or an inverse labor demand function) equal to

$$w = pF'(F^{-1}(y_i))\left(1 + y_i \frac{1}{p\partial D/\partial p}\right), \qquad \sum_{j=1}^{n} y_j = D(p, M) \tag{3.3.18}$$

at which each oligopolist is willing to hire the amount of labor $L_i = F^{-1}(y_i)$ required for his production y_i. Thus, commodity market clearing under symmetric best response behavior defines a cross market wage function

$$w = CE_{\text{Cournot}}(p) := pF'\left(F^{-1}\left(\frac{D(p, M)}{n_f}\right)\right)\left[1 + \frac{1}{n_f} \cdot \frac{1}{E_D(p)}\right], \tag{3.3.19}$$

which associates with each oligopolistic price chosen a unique competitive equilibrium wage rate (i.e. the marginal willingness to pay of each oligopolist) for each unit of labor required under Cournot competition. In other words, (3.3.19) defines the oligopolistic equilibrium wage curve CE_{Cournot} in price-wage space which is analogous to the equilibrium wage curve CE under perfect competition in the commodity market.

Existence and positivity of the solution requires that

$$\frac{1}{n_f} \cdot \frac{1}{E_D(p)} > -1,$$

which is the usual elasticity condition for Cournot equilibria. Moreover, for each $p > 0$, because $E_D(p) < 0$, the willingness to hire labor $CE_{\text{Cournot}}(p)$ is less than the corresponding value $CE_{\text{com}}(p) \equiv CE(p)$ under a competitive commodity market (see Section 3.2.6). Thus, the competitive CE curve lies above the Cournot curve CE_{Cournot}.

Finally, a temporary equilibrium with Cournot competition in the commodity market is a pair (p, w) which in addition to (3.3.19) clears the labor market, i.e. it must solve the two equations

$$w = CE_{\text{Cournot}}(p) \qquad \text{and} \qquad F^{-1}\left(\frac{D(p, M)}{n_f}\right) = \frac{1}{n_f} N\left(\frac{w}{p} V\left(\frac{p^e}{p}\right)\right).$$

As under monopoly, the second equation implies the existence of a so-called price-wage frontier under symmetry/equal treatment (compare (3.3.7) in the monoply case), which guarantees aggregate feasibility under competitive labor supply behavior. As a consequence, a Cournot equilibrium is defined by a price-wage pair (p, w) which solves

$$w = PWF_{\text{Cournot}}(p) = CE_{\text{Cournot}}(p). \tag{3.3.20}$$

Given the properties of Assumption 3.2.1, the frontier under symmetry has the analogous explicit form

$$PWF_{\text{Cournot}}(p) := \begin{cases} \dfrac{p}{V\left(\frac{p^e}{p}\right)} N^{-1}\left(n_f F^{-1}\left(\dfrac{D(p,M)}{n_f}\right)\right) & \text{if } n_f F^{-1}\left(\frac{D(p,M)}{n_f}\right) \le n_w L_{\max} \\ \ge \dfrac{p}{V\left(\frac{p^e}{p}\right)} v'(L_{\max}) & \text{else} \end{cases} \tag{3.3.21}$$

which is downward sloping where N is invertible and with identical limiting properties as in (3.3.7) and (3.3.9). Thus, applying the same technique as in the monopoly case guarantees existence and uniqueness of the Cournot equilibrium. There exists a unique positive solution $p > 0$ which defines the price law and the wage law under Cournot competition as

$$p = \mathcal{P}_{\text{Cournot}}(M, p^e) \tag{3.3.22}$$

$$w = \mathcal{W}(M, p^e) := PWF(\mathcal{P}_{\text{Cournot}}(M, p^e)) = CE_{\text{Cournot}}(\mathcal{P}_{\text{Cournot}}(M, p^e)). \tag{3.3.23}$$

Figure 3.27 displays the qualitative characteristics of the Cournot equilibrium as the intersection of the graphs of $CE_{\text{Cournot}} \cap PWF_{\text{Cournot}}$. Thus, the Cournot price-wage pair $(p_{\text{Cournot}}, w_{\text{Cournot}})$ must lie below/to the right of the graph of CE_{com}. Moreover, the intersection of the graphs $PWF \cap CE$ must occur at the competitive price $p_{\text{com}} = \mathcal{P}(M, p^e)$ which also solves

$$CE(p_{\text{com}}) = LE(p_{\text{com}}) = PWF(p_{\text{com}}).$$

Therefore, since the price-wage frontier is downward sloping, one has

$$p_{\text{com}} = \mathcal{P}(M, p^e) < \mathcal{P}_{\text{Cournot}}(M, p^e) = p_{\text{Cournot}}$$

and

$$w_{\text{com}} = \mathcal{W}(M, p^e) > \mathcal{W}_{\text{Cournot}}(M, p^e) = w_{\text{Cournot}}$$

Thus, under Cournot competition, equilibrium prices are higher and wages are lower compared to the situation when both markets are competitive, implying a lower real wage, a lower level of employment, and a lower level of output when producers be-

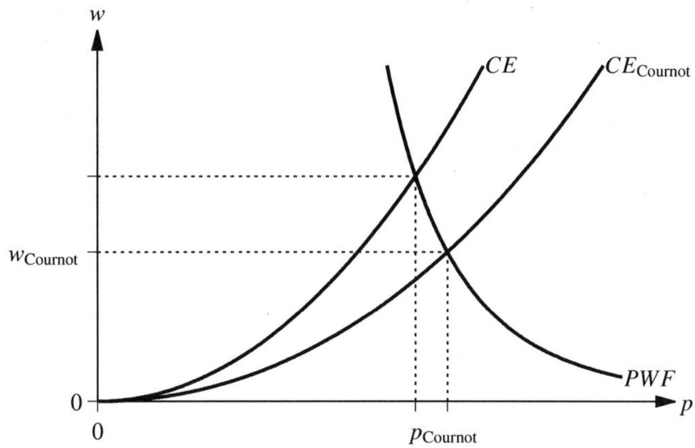

Fig. 3.27 Cournot vs. perfect competition in the commodity market

have as Cournot competitors. Most importantly, however, at the Cournot equilibrium under wage taking behavior of producers full employment prevails at positive nominal wages and prices. Thus, under the standard conditions for which competitive equilibria exist, the noncompetitive behavior of firms in the output market interacting with a competitive labor market does not induce endogenous wage rigidities which might cause unemployment. In other words, strategic price setting does not lead to endogenous wage rigidities which may cause unemployment.

Finally, one also finds that, if aggregate demand is homogeneous of degree zero, that the implications of the quantity theory follow also for the Cournot price and wage law, i. e. the price and wage law are homogeneous of degree one in (M, p^e), while the output law and the employment law are homogeneous of degree zero. In addition, as in the monopoly case, the equilibrium set in (θ^e, m)–space under Cournot competition is defined by a downward sloping unit-contour of the price law to the left/to the Southwest of the corresponding unit-contour of the competitive price law. Thus, all dynamic developments of such economies occur on the equilibrium set while changes of the parameters imply displacements of the equilibrium sets.

3.3.3 Heterogeneity and Product Differentiation

Most models within the framework of the so-called New Keynesian approach discuss the implications of strategic price setting for wage and price rigidities within a model with heterogeneous products or firms (as, for example, in Blanchard & Kiyotaki, 1987; Blanchard & Fischer, 1989). The above analysis shows that such a channel from price setting to wage and price rigidities does not exist when producers use an identical technology producing the same homogeneous output, assuming that firms are wage takers and that the labor market is competitive and satisfies the usual global smoothness conditions. Before discussing the possible implications of non-competitive behavior in an extended model with product differentiation, it seems worth while to investigate whether unemployment with homogeneous production may be possible in other circumstances.

If firms produce the same commodity using different strictly concave production functions all satisfying the Inada conditions, the diversity in technologies under Cournot competition lead essentially to an additional aggregate production inefficiency (see Böhm & Eichberger, 2006) while the labor demand behavior of heterogeneous producers is structurally identical to the competitive case. When the Inada condition does not hold for all producers but production functions are still concave (no nonconvexities), one would still expect inverse factor demands to be sufficiently continuous. As a consequence, technological diversity alone among firms cannot be a source of unemployment in an economy with a competitive labor market under the usual convexity conditions.

Next, let us consider whether price competition à la Bertrand could lead to unemployment. In this case it may be crucial whether price competition with heterogeneous producers in a homogeneous market leads to price diversity or a single equilibrium price. In the latter case, the aggregate income consistent demand function needs to be modified. Under strictly concave production functions, i.e. strictly increasing marginal costs, which are primarily determined by the competitive wage rate, it is evident that production feasibility given a Bertrand price equilibrium would induce a continuous inverse factor demand function (a wage rate equivalent to the marginal willingness to hire labor). In such cases, continuity and the global smoothness of labor supply would imply existence of a temporary Bertrand equilibrium with full employment at a positive wage rate. If, however, the Bertrand equilibrium causes discontinuities in the labor demand, situations of endogenous wage rigidities may arise which depend on the number of active producers, entry or exit of firm, and which are similar to situations with menu costs or fixed costs. However, under regular labor supply conditions, these do not induce unemployment equilibria, but rather exit situations for firms being unable to acquire the necessary labor input or excess supply in the commodity market. Such equilibria are of the type with endogenous wage or price rigidities (as those in Section 3.7) which do not induce *involuntary* unemployment.

Finally, consider the situation with heterogeneous technologies where each firm uses a different markup rule over average costs or marginal costs. In this case again, each producer will follow a smooth inverse factor demand function which, under commodity market clearing, implies a smooth aggregate willingness to pay for labor. As a consequence, the equilibrium cross market wage function together with smoothness of labor supply will imply existence of a full employment equilibrium.

It seems to follow from the above arguments that under the general equilibrium point of view, – which is the undisputed condition for any investigation concerned with the properties of a closed-flow macroeconomic model –, alternative models for commodity pricing in the monetary economy instead of competition or Cournot pricing will generate structurally similar results to the Cournot model, provided that in each case the same closed-flow approach is used. Noncooperative best response behavior of agents in the commodity domain of the economy for a given wage rate lead to mutually consistent actions, trades, and prices which respect macroeconomic income consistency and temporary general equilibrium conditions. In other words, allocations and trades have a Nash equilibrium property for the commodity market parametrized in a competitive wage rate. Under the global smoothness condition of labor supply this implies the existence of a temporary equilibrium with labor market clearing and no unemployment. Therefore, the temporary equilibrium in the monetary economy is described by two mappings, a price law and a wage law, of the same type as under a competitive commodity market or under Cournot competition.

A final question is whether further generalizations of the model to product diversity with several producers may make a difference to these results. Equilibria with monopolistic competition and heterogeneous producers, as in Blanchard & Kiyotaki (1987) in an environment where the labor market is competitive, are a conceptual extension of a temporary Nash equilibrium for each given positive wage rate. Under wage flexibility and standard neoclassical assumptions concerning labor supply a monetary temporary equilibrium with labor market clearing will exist and the multidimensional price and wage maps are homogeneous of degree one in (M, p^e). Their *comparative statics effects* will have the same signs as in the monopoly case or under perfect competition. Whether they will show weaker effects of each firm's price setting rule on the price level is clearly a possibility and could be examined. Their equilibrium set (unit contour of the price law) will be disjoint from those of the monopoly case and of perfect competition. Whether this implies smaller or larger effects (less or more curvature of the equilibrium set) from M for more firms/more product heterogeneity than under competition requires a detailed qualitative analysis. To determine their role on the size of the fiscal multiplier is a main challenge and could be examined in a detailed numerical analysis. However, it seems unlikely that the equilibrium analysis will provide clear answers to the issues on which their analysis concentrates (Blanchard & Kiyotaki, 1987, p. 647):

> "*... how important is monopolistic competition to an understanding of the effects of aggregate demand on economic activity?*", in particular,
> "*... whether monopolistic competition together with some other imperfections,* (can) *generate effects of aggregate demand which perfect competition cannot?*"

As far as these features refer to the fiscal multiplier (or to changes of the tax rates) the model of this chapter indicates that qualitatively the price laws in the two cases respond in the *same* way to changes in the fiscal parameters implying displacements of the two disjoint equilibrium sets. It seems that specific structural assumptions different from those of Assumption 3.2.1 are necessary to derive a generic result of the kind showing that the fiscal multipliers in the two cases differ, one being zero while the other one is different from zero on an open set of parameters.

Therefore, it seems that if the model with monopolistic competition is fully integrated into the temporary equilibrium model of a monetary macroeconomy where the aggregate demand function is appropriately extended to the multi-commodity situation with the associated income consistency, the global smoothness assumption of labor supply must imply existence of a full employment equilibrium at a positive wage rate for the Blanchard–Kiyotaki model. No additional structural elements of the standard model were detected which may cause rigidities or discontinuities of labor demand, which could induce boundary equilibria or non existence of labor market equilibria, and which are present *only* under product differentiation and disappear for homogeneous commodities and under perfect competition. In other words, one concludes that unemployment does not occur in temporary equilibrium as a cause of noncompetitive behavior in commodity markets when the labor market is competitive.

As a consequence the causes for macroeconomic allocations with unemployment in monetary economies will have to be sought through modifications of the behavioral assumption for the labor market scenarios *within* the standard neoclassical model or using extensions *beyond* the standard Assumption 3.2.1. Section 3.4 examines first extensions of the labor market scenarios under noncompetitive wage setting by producers, by a union, and under bargaining between a union and producers, Section 3.5. The so-called efficiency-wage model is integrated into the monetary economy in Section 3.7.3 showing that under this modification strategic behavior may induce *involuntary* unemployment while competitive behavior does not.

Alternatively, one possible extension consists of situations where the respective equilibrium concept *fails to exist* for specific structural reasons like fixed costs or restricted entry (see Section 3.2.8). Another one could arise because of *endogenous* rigidities due to *discontinuities* in labor supply or labor demand, as in Section 3.7. Finally, there may exist structural reasons why price and wage adjustments are *sluggish* implying that they do not lead to a temporary equilibrium of the macroeconomy within the unit time period considered in the model. In other words, the rigidity of prices or wages is temporary, i.e. exogenous and given! In this case, however, a theory of income consistent trading at nonequilibrium prices in the macroeconomy is needed, as is done in Chapter 6. Simultaneously, the adjustments of prices and wages over time need to be described to relax/overcome the temporary exogenous rigidity in order to analyze dynamic development of the macroeconomy (see Chapter 7).

3.4 Noncompetitive Wage Determination

Traditional neoclassical models of noncompetitive wage setting treat market power in the labor market in a perfectly symmetric way to the commodity market. In other words, there exist the two standard cases of *noncooperative* strategic wage setting: 1) when producers are able to influence/set the wage rate in the labor market implying a demand monopoly for labor, or 2) when workers are able to set the wage rate implying a supply side monopoly. In addition, there is a vast literature treating different forms of *cooperative* wage setting models, presenting different forms of wage bargaining among producers and unions of which the efficient bargaining model serves as a benchmark model from a general equilibrium point of view (see Booth, 1995, as a general reference). This section analyzes the three versions of the traditional neoclassical models of labor supply and labor demand behavior and integrates them into the temporary equilibrium model in order to exhibit the consequences for the macroeconomy. Section 3.7 treats some of the extensions of the standard neoclassical framework often analyzed in labor economics, which use alternative microeconomic features, and discusses their consequences for unemployment in the monetary macroeconomic model.

3.4.1 Monopsonistic Wage Setting by Producers

Consider a producer with market power on the labor market, i.e. a monopolist on the demand side of the labor market, often called a *monopsonist* (compare Blanchard, 2003). For simplicity, it is assumed that he is the only producer demanding labor ($n_f = 1$) and that he knows the supply behavior of workers given by the aggregate labor supply function N. However, he is a price taker on the commodity market. Therefore, the producer maximizes profits subject to the labor supply function $N((w/p)V(p^e/p))$, i.e.

$$\max_{w \geq 0; L \geq 0} \left\{ pF(L) - wL \,|\, L \leq N\left(\frac{w}{p}V\left(\frac{p^e}{p}\right)\right)\right\}.$$

In principle, this allows choices (w, L) with $L < N((w/p)V(p^e/p))$, i.e. implying unemployment. However, optimal solutions are always on the labor supply curve, since the profit function $pF(L) - wL$ is strictly decreasing in w and labor supply is assumed to be continuously increasing. Therefore, given $(p, p^e) \gg 0$ and the production function F one has the profit function of the producer

$$\Pi(p, w, p^e) = \underbrace{pF\left(N\left(\frac{w}{p}V\left(\frac{p^e}{p}\right)\right)\right)}_{\text{revenue}} - \underbrace{wN\left(\frac{w}{p}V\left(\frac{p^e}{p}\right)\right)}_{\text{costs}}. \tag{3.4.1}$$

The optimal wage $\tilde{w}$ is defined as

$$\tilde{w} = LE_{\text{mon}}(p, p^e) := \arg\max_{w \geq 0} \left[pF\left(N\left(\frac{w}{p}V\left(\frac{p^e}{p}\right)\right)\right) - wN\left(\frac{w}{p}V\left(\frac{p^e}{p}\right)\right)\right] \tag{3.4.2}$$

which is the producer's marginal willingness to pay for labor (the marginal costs which he is willing to accept), or his wage strategy.

As the first order conditions for an interior optimal solution one obtains

$$F'\left(N\left(\frac{\tilde{w}}{p}V\left(\frac{p^e}{p}\right)\right)\right) \stackrel{!}{=} \frac{\tilde{w}}{p}\left(1 + \frac{1}{E_N\left(\frac{\tilde{w}}{p}V\left(\frac{p^e}{p}\right)\right)}\right) > \frac{\tilde{w}}{p} \tag{3.4.3}$$

$$\text{where} \quad 0 < E_N\left(\frac{\tilde{w}}{p}V\left(\frac{p^e}{p}\right)\right) := \frac{\tilde{w}}{p}V\left(\frac{p^e}{p}\right)\frac{N'\left(\frac{\tilde{w}}{p}V\left(\frac{p^e}{p}\right)\right)}{N\left(\frac{\tilde{w}}{p}V\left(\frac{p^e}{p}\right)\right)}$$

is the elasticity of labor supply. Thus, one finds that at the optimal monopsonistic

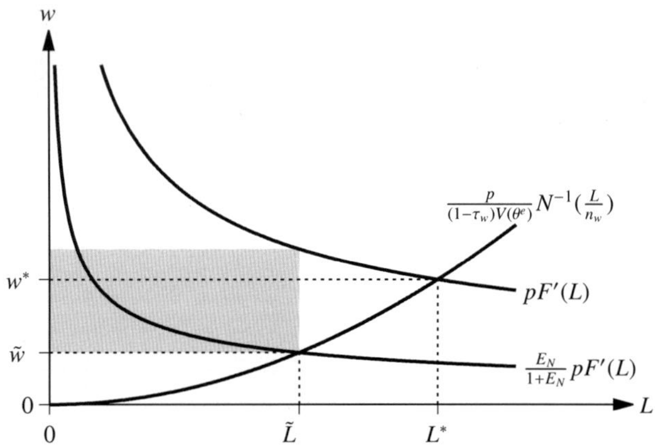

Fig. 3.28 Wages, employment, and surplus in monopsonistic labor market; (p^e, p) given

choice $(\tilde{L}, \tilde{w})$, the marginal product of labor is higher than the real wage paid in the market, implying an extra rent for the producer. He produces at a level where his own sales price is larger than marginal costs, producing less and hiring less labor than under competitive conditions, i.e. $\tilde{L} < L^*$. Figure 3.28 displays the optimal decision $(\tilde{L}, \tilde{w})$ for a given pair (p, p^e). The pair (L^*, w^*) marks the employment-wage pair which would occur at the same price and price expectations under competitive behavior of the producer.

In addition one finds, that the wage setting function LE_{mon} is homogeneous of degree one in (p, p^e) while his associated labor demand function is homogeneous of degree zero. Thus, for all (p, p^e), one has

$$\tilde{L} = h_{\text{mon}}(\theta^e) := N\left(\frac{pLE_{\text{mon}}(1, \theta^e)}{p}V(\theta^e)\right) < h(W(\theta^e)) = L^* \tag{3.4.4}$$

i.e. the level of employment is lower than under competitive labor market clearing, but there is no unemployment. Moreover, $h_{\text{mon}}(\theta^e)$ is a strictly decreasing function.

Finally, for given (M, p^e), a price-wage pair (p, w) is a temporary equilibrium with a monopsonistic labor market if it solves

$$w = LE_{\text{mon}}(p, p^e) \qquad \text{and} \qquad D(p, M) = F\left(h_{\text{mon}}\left(\frac{p^e}{p}\right)\right) =: AS_{\text{mon}}(\theta^e) \tag{3.4.5}$$

Equivalently, one obtains the temporary equilibrium $(p, w) \gg 0$ as a solution of

$$w = LE_{\text{mon}}(p, p^e) \qquad \text{and} \qquad D(p, M) = F\left(N\left(\frac{w}{p}V(\theta^e)\right)\right). \tag{3.4.6}$$

The second equation defines the competitive price-wage frontier in (p, w)–space. There exists a geometrical representation of the temporary equilibrium as the intersection of the graphs of $LE_{\text{mon}} \cap PWF$. Thus, the two equilibrium conditions (3.4.5)

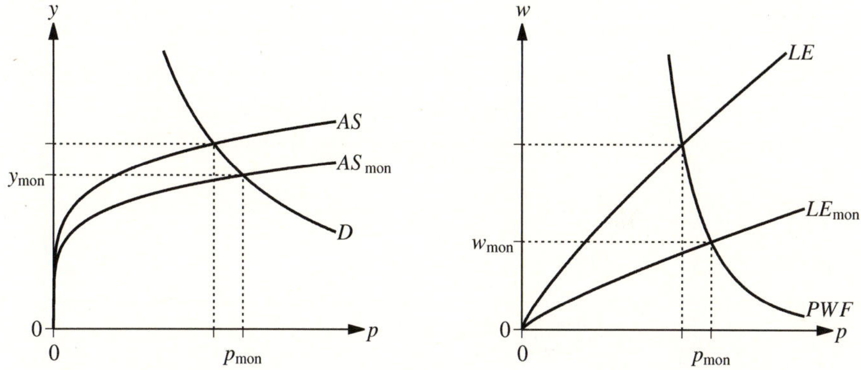

a Prices and output

b Prices and wages

Fig. 3.29 Prices, wages, and output under labor market monopsony

and (3.4.6) provide two alternative geometric characterizations of the monopsonistic equilibrium configuration shown in Figure 3.29.

The second equation of (3.4.5) is independent of wages and has a unique positive solution defining the equilibrium price, which in turn defines the price law and the wage law

$$p = \mathcal{P}_{\text{mon}}(M, p^e) \qquad \text{and} \qquad w = \mathcal{W}_{\text{mon}}(M, p^e) := LE_{\text{mon}}(\mathcal{P}_{\text{mon}}(M, p^e), p^e). \tag{3.4.7}$$

Summarizing, for all (M, p^e), one finds that monopsonistic wage setting leads to a unique temporary equilibrium with full employment which has lower equilibrium wages and higher equilibrium prices, lower real wages, and a lower level of employment and output than under competition, see Figure 3.29. Thus, monopsonistic wage setting behavior is a cause for lower employment and inefficiency compared to equilibria with labor market competition. It does not induce unemployment under the usual global smoothness/regularity assumptions of labor supply.

If aggregate demand is homogeneous of degree zero in (M, p), the quantity theory holds, i.e. $\mathcal{P}_{\text{mon}}$ and $\mathcal{W}_{\text{mon}}$ are homogeneous of degree one in (M, p^e) and $\mathcal{L}_{\text{mon}}$ and $\mathcal{Y}_{\text{mon}}$ are homogeneous of degree zero in (M, p^e). The comparative statics effects have the following properties with respect to the state variables (M, p^e):

$$\begin{aligned}
&\frac{\partial}{\partial M}\mathcal{L}_{\mathrm{mon}}(M,p^e) > 0 && \frac{\partial}{\partial p^e}\mathcal{L}_{\mathrm{mon}}(M,p^e) < 0 \\
&\frac{\partial}{\partial M}\mathcal{W}_{\mathrm{mon}}(M,p^e) > 0 && \frac{\partial}{\partial p^e}\mathcal{W}_{\mathrm{mon}}(M,p^e) > 0\,. \\
&\frac{\partial}{\partial M}\mathcal{P}_{\mathrm{mon}}(M,p^e) > 0 && \frac{\partial}{\partial p^e}\mathcal{P}_{\mathrm{mon}}(M,p^e) > 0
\end{aligned} \tag{3.4.8}$$

In addition, the fiscal multipliers are

$$\frac{\partial}{\partial g}\mathcal{P}_{\mathrm{mon}}(M,p^e) > 0 \qquad \frac{\partial}{\partial g}\mathcal{L}_{\mathrm{mon}}(M,p^e) > 0 \tag{3.4.9}$$

which are both positive. In other words, all demand side effects are qualitatively identical to the competitive case and temporary equilibria are structurally indistinguishable from the competitive ones. Therefore, the equilibrium set under monopsony, i.e. the unit contour of the monopsonistic price law is below/to the Southwest of the corresponding contour of the competitive price law (see Figure 3.30) such that the two sets do not intersect. All intertemporal effects induce movements along the equilibrium set while an increase in government demand induces a downward shift of both equilibrium sets.

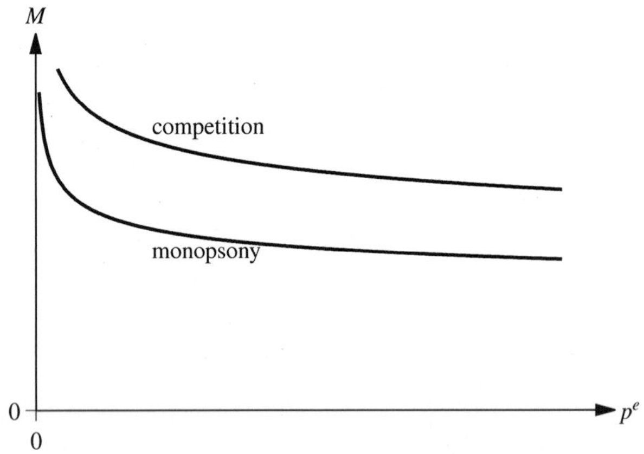

Fig. 3.30 Equilibrium sets under competition and labor market monopsony

3.4.2 Monopolistic Wage Setting by a Labor Union

The analysis of the situation where a powerful union sets wages is the symmetric opposite case to the monopsonistic firm and can be treated in a similar fashion. Therefore, consider a worker's union with market power on the labor market representing n_w homogeneous consumer/workers. Assume that the union knows the aggregate labor demand function $h(w/p)$ of the production sector[14]. Therefore, a unionized monopolistic decision will determine a wage and an aggregate employment level which must be consistent with the labor demand behavior of the producer. Since all consumer-workers are identical, let us assume that they are treated equally for any union decision, i.e. that all receive the same wage and they work the same amount.

In order to define the unions objective function, it is useful to consider the reservation wage of an individual worker which is defined by the lowest wage rate at which he is willing to supply labor, i.e. his participation constraint. For the intertemporal preferences this is given by

$$w_{\text{res}}(\ell, \theta^e) := \frac{p}{(1-\tau_w)V(\theta^e)} \frac{v(\ell)}{\ell} \tag{3.4.10}$$

whose graph defines those pairs (ℓ, w), which provide the level of indirect utility as if not working and receiving no wage income. Observe that the reservation wage is not constant, but endogenously determined by a function which is increasing in the individual working level and in expected inflation.

Assuming equal treatment of workers by the union, one obtains the corresponding function for the reservation wage of the union

$$w_{\text{res}}(L, \theta^e) := \frac{p}{(1-\tau_w)V(\theta^e)} \frac{v(L/n_w)}{L/n_w} \tag{3.4.11}$$

as a function of any aggregate employment level $L > 0$.

Assume that the objective of the union is to maximize the aggregate excess wage bill[15] defined as

$$\Omega(L, w, p, p^e) := wL - w_{\text{res}}(L, p, p^e)L \tag{3.4.12}$$

subject to the constraint $L \leq h(w/p)$. Since Ω is strictly increasing in the wage rate for each employment level L, any optimal solution $(\hat{w}, \hat{L})$ must satisfy the labor demand function with equality. This implies that the optimal wage must be a maximizer of the excess wage bill with $L = h(w/p)$, i.e.

$$W_{\text{union}}(p, p^e) = \arg\max_{w \geq 0} \frac{w(1-\tau_w)V(\theta^e)h(w/p)}{pn_w} - v\left(\frac{h(w/p)}{n_w}\right). \tag{3.4.13}$$

The associated first order condition of an interior solution is given by

$$\frac{w}{p}\left(\frac{ph(w/p)}{wh'(w/p)} + 1\right) = \frac{1}{(1-\tau_w)V(\theta^e)} v'\left(\frac{h(w/p)}{n_w}\right). \tag{3.4.14}$$

---

[14] For simplicity, it is assumed that there is a single producer, i.e. $n_f = 1$.

[15] In principle there may be many different objective functions of the union, which could be independent of labor demand or depend on the given specific context. Here the maximization of the excess is used as an appropriate aggregate indicator of utility which is the one most often applied in the literature (for example Booth, 1995; Landmann & Jerger, 1999; McDonald & Solow, 1981; Böhm & Claas, 2017).

This is equivalent to

$$w = \left(\frac{1+E_h}{E_h}\right)\frac{p}{(1-\tau_w)V(\theta^e)}v'\left(\frac{h(w/p)}{n_w}\right) = \left(\frac{1+E_h}{E_h}\right)\frac{p}{(1-\tau_w)V(\theta^e)}N^{-1}\left(\frac{L}{n_w}\right)$$

where $w = \frac{p}{(1-\tau_w)V(\theta^e)}N^{-1}(\frac{L}{n_w})$ is the marginal willingness to work of each consumer/worker, which coincides with the inverse of the aggregate labor supply function under symmetry. For a positive union wage the typical condition

$$E_h(w/p) := \frac{wh'(w/p)}{ph(w/p)} > -1$$

for the elasticity E_h of the labor demand function has to hold, which is assumed to be satisfied. In case it is violated other markups for the willingness to work can be applied (as in the monopoly case for the commodity market, Section 3.3.1) which do not use information of the labor demand function.

Figure 3.31 portrays the qualitative features of the wage setting behavior of the union. For every (p^e, p), this induces a wage equal to the marginal value product which is, however, larger than the competitive wage and larger than the marginal willingness to work of every worker at the associated level of employment. Thus, the workers obtain an aggregate monopolistic surplus equal to $pF'(\tilde{L}) - \frac{p}{((1-\tau_w)V(\theta^e)}N^{-1}(\frac{\tilde{L}}{n_w})$, see Figure 3.31. Notice further, that at the union wage $\tilde{w}$, the employment level $\tilde{L}$ is lower than the corresponding aggregate labor supply, i.e. $N(\tilde{w}V(\theta^e)/p) > \tilde{L}$, the difference being a measure of voluntary unemployment. Finally, it is worth noticing that the maximization of the aggregate surplus by the union results in a solution which maximizes per capita utility at given prices and price expectations, since workers are identical with separable utility between consumption and leisure and employment being distributed uniformly among them. To see this,

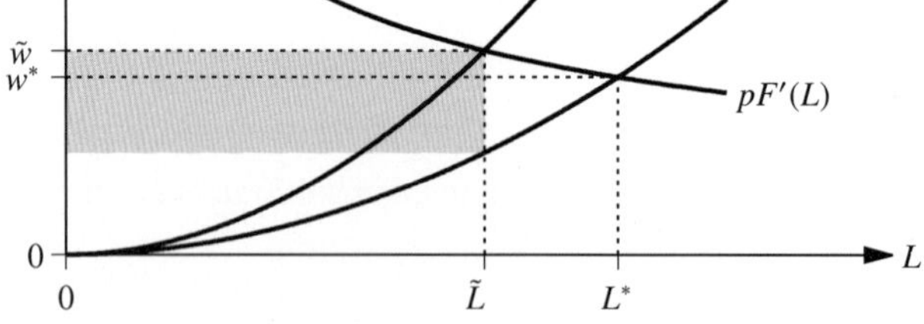

Fig. 3.31 Wages, employment, and the surplus under the monopolistic union; (p^e, p) given

consider the maximization of indirect utility

$$\begin{aligned} W_{\text{union}}(p,p^e) &= \arg\max_{w\geq 0, L\geq 0} \{U(wL/p^e n_w, L/n_w) \mid L \leq h(w/p)\} \\ &= \arg\max_{w\geq 0, L\geq 0} \left\{ \frac{w(1-\tau_w)L}{p^e n_w} - v\left(\frac{L}{n_w}\right) \| L \leq h(w/p) \right\} \\ &= \arg\max_{w\geq 0} \frac{w(1-\tau_w)h(w/p)}{p^e n_w} - v\left(\frac{h(w/p)}{n_w}\right). \end{aligned} \tag{3.4.15}$$

In other words, the optimal solution of equation (3.4.13) maximizes per capita utility of each worker.

Further properties of the union monopoly solution are easily obtained from (3.4.13) and (3.4.14). First one finds that the union wage function must be homogeneous of degree one in (p, p^e), i.e. one can write $\tilde{w} = W_{\text{union}}(p, p^e) = pW_{\text{union}}(1, p^e/p)$. Therefore, for every $\theta^e := (p^e/p)$, the wage decision induces a level of employment and aggregate supply given by

$$\tilde{L} = h_{\text{union}}\left(\frac{p^e}{p}\right) := h\left(W_{\text{union}}\left(1, \frac{p^e}{p}\right)\right) < h\left(W\left(\frac{p^e}{p}\right)\right), \tag{3.4.16}$$

$$\tilde{y} = AS_{\text{union}}\left(\frac{p^e}{p}\right) := F\left(h_{\text{union}}\left(\frac{p^e}{p}\right)\right) < AS\left(\frac{p^e}{p}\right), \tag{3.4.17}$$

since the competitive labor market clearing real wage satisfies $W(\theta^e) < W_{\text{union}}(\theta^e)$. This implies immediately

$$W_{\text{union}}(\theta^e) > W(\theta^e) \tag{3.4.18}$$

the competitive labor market clearing real wage, and $h_{\text{union}}(\theta^e) < h(W(\theta^e))$ inducing

$$AS_{\text{union}}\left(\frac{p^e}{p}\right) < AS\left(\frac{p^e}{p}\right).$$

Thus, for every θ^e, the unionized wage setting implies lower output and lower employment compared to the competitive scenario under identical prices and price expectations.

In order to determine the temporary equilibrium under unionized wage setting the aggregate income consistent demand function has to be analyzed in more detail since cross market effects may appear due to changes in the income distribution. Since labor supply is not competitive income consistency does no longer correspond to the properties of the competitive price-wage frontier as in the previous monopolistic cases. Since the producer, as price and wage taker on the two markets, pays a wage equal to the marginal value product, the wage share of total income

$$\frac{wL}{py} = h^{-1}(L)\frac{L}{F(L)} = \frac{F'(L)L}{F(L)} = E_F(L)$$

still coincides with elasticity of the production function determining the functional income distribution. Thus, there is no separate income distribution effect on aggregate demand from unionized wage setting as compared to competitive factor pricing except through the elasticity. Therefore, one obtains for the income consistent aggregate demand function under unionized wage setting as compared to the competitive one

$$\begin{aligned} D_{\text{union}}\left(\frac{M}{p},\frac{p^e}{p}\right) \\ &= \frac{M/p+g}{1-c_s(\frac{p^e}{p})(1-\tau_\pi)(1-E_F(h_{\text{union}}(\frac{p^e}{p})))-c_w(\frac{p^e}{p})(1-\tau_w)(E_F(h_{\text{union}}(\frac{p^e}{p})))} \\ &\approx \frac{M/p+g}{1-c_s(\frac{p^e}{p})(1-\tau_\pi)(1-E_F(h(W(\frac{p^e}{p}))))-c_w(\frac{p^e}{p})(1-\tau_w)(E_F(h(W(\frac{p^e}{p}))))} \\ &= D\left(\frac{M}{p},\frac{p^e}{p}\right) \end{aligned} \tag{3.4.19}$$

which are close to each other if the elasticities are approximately the same, i.e. if $E_F(h_{\text{union}}(p^e/p)) \approx E_F(h(W(p^e/p)))$. This makes the distribution effects of a second order magnitude under the separability assumption of preferences for workers. When the production function is isoelastic or when net consumer characteristics are identical the functional income distribution does not matter, so that the union wage setting has no income distribution effect on aggregate demand and the two functions coincide. As a consequence, a temporary equilibrium with unionized wage setting will be determined by a price-wage pair $(p, w) \gg 0$ which solves

$$D_{\text{union}}(M/p, p^e/p) = AS_{\text{union}}(p^e/p) \qquad \text{and} \qquad w = W_{\text{union}}(p, p^e) \tag{3.4.20}$$

or equivalently, using the two-step procedure of Section 3.2.6,

$$\begin{aligned} w &= LE_{\text{union}}(p, p^e) := pW_{\text{union}}(p^e/p) \quad \text{and} \\ w &= CE_{\text{union}}(p) := ph^{-1}\left(F^{-1}\left(D_{\text{union}}(M/p, p^e/p)\right)\right). \end{aligned} \tag{3.4.21}$$

The solution of the first equation of (3.4.20) defines the equilibrium price law $p = \mathcal{P}_{\text{union}}(M, p^e)$ while the second one induces the wage law, i.e.

$$p = \mathcal{P}_{\text{union}}(M, p^e) \quad \text{and} \quad \mathcal{W}_{\text{union}}(M, p^e) := W_{\text{union}}(\mathcal{P}_{\text{union}}(M, p^e), p^e). \tag{3.4.22}$$

These satisfy, for all (M, p^e),

$$\begin{aligned} &\mathcal{P}_{\text{union}}(M, p^e) > \mathcal{P}(M, p^e) \quad \text{and} \quad \mathcal{W}_{\text{union}}(M, p^e) > \mathcal{W}(M, p^e) \\ &\frac{\mathcal{W}_{\text{union}}(M, p^e)}{\mathcal{P}_{\text{union}}(M, p^e)} > \frac{\mathcal{W}(M, p^e)}{\mathcal{P}(M, p^e)}. \end{aligned} \tag{3.4.23}$$

The first equation of (3.4.20) is independent of the wage rate, providing an independent determination of the equilibrium price. This allows a geometric representation in price-output space (see Figure 3.32 **a**), while the two equations in (3.4.21) provide the alternative representation in the price-wage space, panel **b**. Both display the competitive and the unionized equilibrium, showing the position of the unionized equilibrium relative to the competitive equilibrium. In other words, one finds

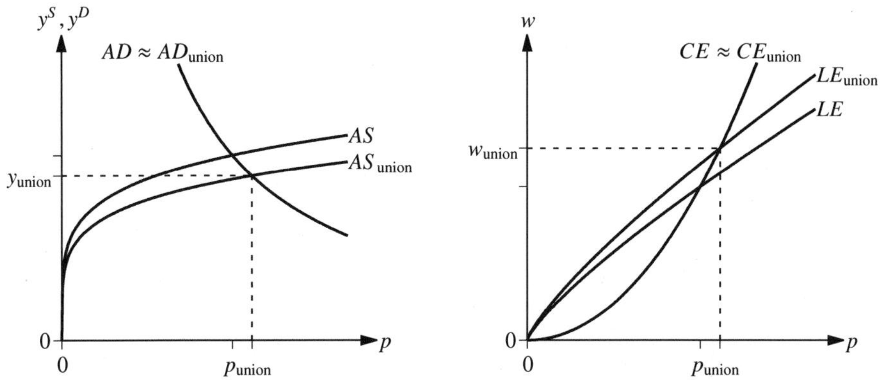

a Prices and output

b Prices and wages

Fig. 3.32 Prices, wages, and output under unionized wage setting; (M, p^e) given

that unionized wage setting leads to higher wages, higher prices, and higher real wages compared to the competitive equilibrium

The price law $\mathcal{P}_{union}$ induces the associated employment law

$$L = \mathcal{L}_{union}(M, p^e) := h_{union}\left(\frac{p^e}{\mathcal{P}_{union}(M, p^e)}\right) \tag{3.4.24}$$

and output law

$$\begin{aligned} y = \mathcal{Y}_{union}(M, p^e) &:= F\left(h_{union}\left(\frac{p^e}{\mathcal{P}_{union}(M, p^e)}\right)\right) \\ &\equiv AS_{union}(\mathcal{P}_{union}(M, p^e), p^e) \equiv D_{union}\left(\frac{M}{\mathcal{P}_{union}(M, p^e)}, \frac{p^e}{\mathcal{P}_{union}(M, p^e)}\right). \end{aligned} \tag{3.4.25}$$

These yield a lower level of employment and lower level of output compared to the competitive case. Thus, the unionized wage rate coincides with the marginal value product of labor payed the producer. Workers work less than their respective notional labor supply at the unionized wage. In this sense they suffer from *voluntary* unemployment. Thus, complete monopolistic union power in the labor market induces a temporary equilibrium with a strategically chosen level of unemployment which maximizes each worker's excess over his reservation wage implying a

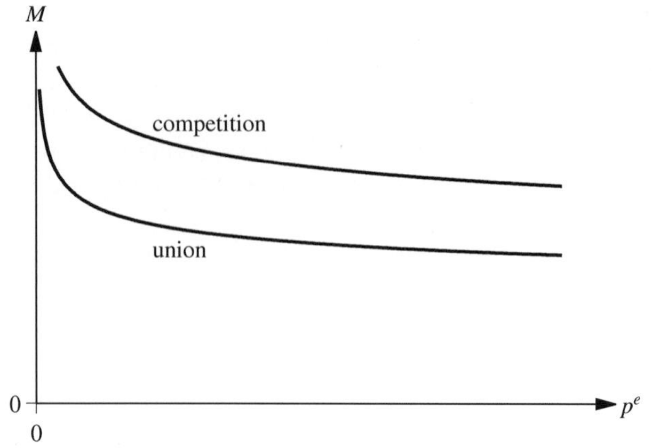

Fig. 3.33 Equilibrium sets under competition and union wage setting

monopoly rent under voluntary unemployment. Finally, the equilibrium set under union wage setting is displaced downward relative to the corresponding set under competition, see Figure 3.33.

3.4.3 Reappraisal of Noncompetitive Wage Models

The analysis of models where neoclassical firms operate in the labor market and set wages in a non-competitive optimizing way has revealed the known inefficiencies from imperfect competition (relative to perfect competition). However, within the class of models considered, such behavior does not lead to macroeconomic allocations with unemployment. The standard strategic optimal behavior of producers induces market clearing for the non-strategic agents (the workers) with an associated adjustment of their notional labor supply possibly to low levels of employment, but these are *on the supply curve*. There is no evidence of a structural explanatory element for unemployment or disequilibrium. Thus, wage and price flexibility as a market clearing device induce equilibria and not disequilibria.

The macroeconomic implications of wage setting behavior are characterized by specific features of the price law and of the wage law. These deviate from those of the competitive case with different elasticities but without rigidities under the basic Assumption 3.2.1. They determine the time-invariant properties of the equilibrium set of the economy. As a consequence, the dynamics of prices, wages, and of macroeconomic allocations of such models, i.e. their observable properties in time series, are structurally equivalent to the competitive case whether under adaptive or rational expectations. Therefore, it seems safe to say that no specific Phillips curve tradeoffs or endogenous cycles are to be expected within the class of macroeconomic models with non-competitive labor markets which cannot occur in comparable competitive environments, unless they are due to behavioral rules which are imposed excluding the possibility of flexible prices and wages by assumption or structural modifications or deviations from the neoclassical model, see Section 3.7. This argument extends to situations with random perturbations and policy shocks implying that stochastic cycles and/or the transient behavior arising from demand or supply shocks are structurally identical to those of the competitive case. The exogenous changes induce the same type of modifications (displacements) for the associated price laws and their equilibrium sets (as shown in Sections 3.3 to 3.7) or as in Chapters 4 and 5 for the dynamics.

3.5 Efficient Bargaining in the Labor Market

The previous sections analyzed noncompetitive wage setting within the standard *noncooperative* monopolistic models where total market power is assigned to one side of the market while the other one behaves competitively. In bargaining situations[16], however, labor unions and producers typically negotiate the outcomes in a *cooperative* approach where both sides agree on a decision for the wage rate or the level of employment, sometimes jointly for both variables in a binding agreement. Thus market forces for the labor market given the competitive behavior of both sides provide the reference framework for which the ultimate comparison between outcomes and their macroeconomic impact will be analyzed.

The literature on bargaining within a consistent macroeconomic framework is not too sizable with some excellent presentations most notably by McDonald & Solow (1981), Blanchard & Fischer (1989), Booth (1995), and Landmann & Jerger (1999). The presentation here is adapted from Böhm & Claas (2017) which contains a consistent general equilibrium approach analyzing all cross market price and income spillover effects under more general assumptions than in the literature. Attention will be payed to additional allocative implications under efficient bargaining in comparison to alternative labor market scenarios.

3.5.1 Efficient Bargaining of Wages and Employment

Consider the economic configuration as in the previous section with unionized wage setting. There are $n_w \geq 1$ homogeneous consumer-workers represented by a labor

[16] The material of this section is taken from Böhm & Claas (2017). It is part of an extended joint project using a cooperative bargaining approach investigating the role of the distribution of power between a producer syndicate and a workers' union on wages and employment in a macroeconomic setting. Claas (2017) analyzes additional bargaining scenarios as well as results on perfect foresight dynamics.

union and a single producer[17], representing the producers' association or employers' syndicate. However, in contrast to the unilateral asymmetric decision by the union alone, consider now the situation where the wage and the employment level will now be determined as the outcome of a joint bargaining decision. Both sides assume that the commodity market is competitive, for which they take the price and future price expectations as given.

Since the firm can hold no inventory a production decision $L \geq 0$ implies current profits $\Pi(p, w, L) := pF(L) - wL$ for any nominal wage rate $w \geq 0$ for labor and a sales price $p \geq 0$ for the commodity. Individual rationality for the producer under bargaining implies that bargaining agreements (L, w) of employment and wage levels have to guarantee nonnegative profits $\Pi(p, w, L) \geq 0$. Therefore, the zero-profit contour implies the *participation constraint* for the producer, i.e.

$$w \leq p\frac{F(L)}{L} =: W_\Pi(p, L). \tag{3.5.1}$$

This defines the producer's reservation wage as a function of the employment level $L > 0$ and the output price as being equal to the average return.

In order to define the union's objective function, we follow the procedure laid out in the previous section. Let the reservation wage of an individual worker be defined by the lowest wage rate at which he is willing to supply a certain amount of labor as defined in (3.4.10) given by

$$w_{\text{res}}(\ell, p, p^e) := \frac{p}{(1-\tau_w)V(\theta^e)}\frac{v(\ell)}{\ell} = \frac{1}{E_v(\ell)}\frac{p}{(1-\tau_w)V(\theta^e)}v'(\ell). \tag{3.5.2}$$

In other words, each worker's participation constraint is defined by those pairs (ℓ, w) which induce the level of indirect utility as if not working. This defines an endogenous functional relationship which depends on the worker's disutility of work *and* on market data. Specifically, for given (p, p^e), the reservation wage function is homogeneous of degree one in (p, p^e). It is an increasing function in ℓ and it satisfies the well known relationship between the inverse competitive labor supply (giving by the marginal acceptable wage or willingness to work derived from the first order conditions under competitive behavior) and the reservation wage (see Böhm & Claas, 2017, for details) (which corresponds to the minimal acceptable wage or average willingness to work), i.e.

$$w = \frac{p}{(1-\tau_w)V(\theta^e)}v'(\ell) = E_v(\ell)\frac{p}{(1-\tau_w)V(\theta^e)}\frac{v(\ell)}{\ell} = E_v(\ell)w_{\text{res}}(\ell, p, p^e), \tag{3.5.3}$$

with $\theta^e = p^e/p$. The elasticity E_v of the disutility function v determines the associated markup for the competitive willingness to work over the reservation wage.

Under equal treatment of workers by the union, one obtains the reservation wage as a function of aggregate union employment $L \equiv n_w\ell$ in a multiplicative separable

[17] This assumption is made for ease of exposition only, the extension to multiple homogeneous firms organized in a producers association is straightforward.

form as in (3.4.11)

$$\begin{aligned} w_{\text{res}}(L, p, p^e) &= \frac{p}{V(\theta^e)} S(L) := \frac{p}{V(\theta^e)} \frac{v(L/n_w)}{L/n_w(1-\tau_w)} \\ &= \frac{1}{E_v(L/n_w)} \frac{p}{V(\theta^e)} \frac{1}{(1-\tau_w)} v'\left(\frac{L}{n_w}\right) \\ &:= \frac{1}{E_v(L/n_w)} \frac{p}{V(\theta^e)} S_{\text{com}}(L). \end{aligned} \tag{3.5.4}$$

Here $S_{\text{com}}(L)$ denotes the inverse of the aggregate competitive labor supply function

$$N((1-\tau_w)wV(\theta^e)/p) := n_w \left(v'\right)^{-1} ((1-\tau_w)wV(\theta^e)/p),$$

which also has a separable multiplicative form under Assumption 3.2.1. This implies a useful relationship between the reservation wage function S and the competitive inverse labor supply function S_{com} given by

$$S_{\text{com}}(L) = E_v(L/n_w)\, S(L) \qquad \text{for all } L, \tag{3.5.5}$$

which is independent of (p, p^e), and where S satisfies

$$E_S(L) = E_v(L/n_w) - 1. \tag{3.5.6}$$

Finally, in order to complete the description of the bargaining characteristics of the union, let us now assume that the objective of the union is to maximize the aggregate excess wage bill[18] defined as

$$\Omega(L, w, p, p^e) := wL - w_{\text{res}}L = wL - \frac{p}{V(\theta^e)} S(L)L. \tag{3.5.7}$$

Given the characteristics of each individual young worker, the union is perceived of as an aggregate agent representing all consumer-workers. Since workers have identical characteristics, the union's bargaining will be concerned with the determination of the wage level w and the aggregate level of employment L, assuming that all workers are treated equally, i.e. for each aggrement (L, w) each worker is paid the wage w and asked to work at the level L/n_w.

The bargaining framework chosen between the union representing the consumer-workers and the producer serves as a wage and employment determination device for the temporary equilibrium in an arbitrary period. It consists of an application of a

[18] Among labor economists one finds a discussion of what should be the objective function of a union, for example Pencavel (1985) or Blanchard & Fischer (1989), offering, however, no clear cut choice or definition. Booth (1995), Landmann & Jerger (1999), and McDonald & Solow (1981) use the maximization of the excess as an appropriate aggregate indicator of utility. This will be followed here as in Böhm & Claas (2017). Under the separability and homotheticity assumption of preferences here, maximizing the excess wage bill is indeed equivalent to maximizing per capita utility of workers.

cooperative bargaining solution[19] to the *simultaneous* determination of an aggregate level employment L and of a wage rate w under the assumption that the negotiating parties, the union and the producer, face a competitive market where both of them are price takers. In other words, the producer and the union treat the commodity price and price expectations as given, implicitly assuming that their bargaining decision has no influence on the induced equilibrium price. It is evident and known from other applications of noncompetitive equilibrium analysis within subsectors of an economy attached to a competitive output market in the economy, that this is a wrong perception in general equilibrium. For any closed monetary economy of the type presented here, there exist spillover effects between the labor market and the output market channeled through prices. The impact from wage negotiations on the income distribution will have effects on aggregate demand and therefore on output and income. Thus, the bargaining parties ignore such interdependencies – or misinterpret the nature of the competitive commodity price in an economy with interacting markets. On the other hand, it is one of the central macroeconomic challenges to determine to what extent the spillover effects induced by competitive prices in the output market influence the equilibrium allocations and their descriptive and normative characteristics. Thus, one of the goals in this section will be to reveal and characterize the nature and size of the spill-over effects from prices generated by the solution under bargaining, i.e. their allocative implications.

Under efficiency considerations, the Nash bargaining solution is a common and generally accepted choice[20] which is often used in the literature. Its application within an intertemporal monetary macroeconomic model makes it a static bargaining solution, which ignores all aspects of intertemporal or repeated negotiation. The union and the producer behave essentially in a myopic way, without concern for an intertemporal objective function. By using a notion of a bargaining concept in a static (myopic) way, the parties ignore that the current bargaining result may influence future bargaining positions, for example that the size of the future cake to be divided may depend on the division of the cake today. Except for one early contribution by Güth & Selten (1982) there seem to be no other contributions which have addressed sequential strategic issues in fully specified macroeconomic models, although the existence and relevance stemming from a long term relationship between a union and producers have been recognized and discussed informally to some extent. Güth and Selten emphasize the intertemporal aspect of repeated bargaining explicitly in a non-monetary business cycle model balancing off myopic principles against farsighted behavior using subgame consistency.

The main objective here consists in deriving the allocative implications on the macroeconomic level, in particular to unveil the cross effects between the outcomes of bargaining and the commodity market and discuss their implications for the economic laws for prices, wages, output, and employment. A second objective of the

[19] The bargaining decision is taken as a given outcome of an agreement among the two parties, leaving noncooperative considerations aside as to how the outcome can be supported or generated as an equilibrium of a sequential game, as for example by Binmore, Rubinstein & Wolinsky (1986).

[20] From a game-theoretic point of view, the generalized Zeuthen solution for half-space games can be applied which is less specific than Nash, see Böhm & Claas (2017).

analysis is to examine in detail the comparative statics effects of different levels of union power on the macroeconomy. Böhm & Claas (2017) also present the dynamics for the economy under constant union power over time. Additional intertemporal aspects regarding considerations of sequential bargaining or endogenous bargaining power are left for further research. Therefore, a *temporary equilibrium with efficient wage and employment bargaining* is defined by a competitive price level p which equalizes aggregate supply and aggregate demand of the commodity market at which the levels of employment and wages induce the desired efficient bargaining solution between the union and the producer obtained in a sistuation of a given distribution of power between the union and the producer.

Let $\Pi(p, w, L)$ denote the net profit and $\Omega(p, p^e, w, L)$ the excess wage bill. Given the commodity price and price expectations and $(p, p^e) \gg 0$, a bargaining agreement (L, w) is called *individually rational* if Π and Ω are nonnegative. An *efficient bargaining agreement* between the union and the employer is defined in the usual way.

Definition 3.5.1. Given $(p^e, p) \gg 0$, an employment–wage pair $(L, w) \in \mathbb{R}^2_+$ is called *efficient* if there exists no other pair (L', w') such that

$$\Pi(p, w', L') \geq \Pi(p, w, L) \qquad \text{and} \qquad \Omega(p, p^e, w', L') \geq \Omega(p, p^e, w, L)$$

with at least one strict inequality.

To characterize efficient agreements, one may use the maximizer of the associated Lagrangean function

$$\Lambda(w, L, \kappa) = \Omega(p, p^e, w, L) + \kappa\left(\Pi(p, w, L) - \bar{\Pi}\right)$$

and obtains the first-order conditions of an interior solution $(L, w) \gg 0$ as

$$F'(L) = \frac{1}{V(\theta^e)} S(L)(1 + E_S(L)) = \frac{1}{V(\theta^e)} S_{\text{com}}(L), \qquad L > 0, \tag{3.5.8}$$

using (3.5.5) and (3.5.6). Any positive solution $L > 0$ determines the same level of employment for all levels of net profit $\bar{\Pi}$. Thus, the first component of an efficient employment-wage pair is independent of the the determination of the second one. Moreover, a solution of (3.5.8) is identical with that level of employment which would clear the labor market under conditions of perfect competition between the union and the producer for any given pair $(p^e, p) \gg 0$.

This property is well-known from the literature. It occurs in situations of bargaining/cooperative decision making between any two agents who are the only participants trading in an industry which is vertically integrated, as in a cartel or a bilateral monopoly. In such cases, under efficiency, the two traders internalize all potential net gains. They decide on a level of trade and price between them which maximizes the sum of their net gains. If they are both facing competitive markets upstream and downstream, the resulting level of activity between them under efficiency is identical to that level of trade which would result under competitive trading, with some mild assumptions. This level guarantees that there are no further joint gains to share. In other words, the level of trade equalizes marginal cost to marginal revenue between the two players and maximizes the cake to share. For the model here between the union and the producer, this implies that the determination of an efficient bargaining solution can be divided into two steps: the choice of the level of employment which depends on the market data upstream and downstream, and the determination of the wage which then turns out to become the central point in the bargaining procedure of sharing the net gains.

Employment as in the Bilateral Monopoly

For a given pair of price expectations and commodity price $(p^e, p) \gg 0$, define the joint net gain as

$$\begin{aligned}(\Pi + \Omega)(p, p^e, L) &:= \Pi(p, w, L) + \Omega(p, p^e, w, L)\\ &= pF(L) - wL + wL - \frac{p}{V(\theta^e)}S(L)L = pF(L) - \frac{p}{V(\theta^e)}S(L)L\end{aligned}$$

which is a function of the employment level alone. For an an efficient bargaining agreement (L, w) it is necessary that L maximizes the total net gain $(\Pi + \Omega)$. Let

$$(\Pi + \Omega)_{\text{bar}}(p, p^e) := \max_{L \geq 0}(\Pi + \Omega)(p, p^e, L) = \max_{L \geq 0} p\left\{F(L) - \frac{1}{V(\theta^e)}S(L)L\right\} \tag{3.5.9}$$

and

$$h_{\text{bar}}\left(\frac{p^e}{p}\right) := \arg\max_{L \geq 0}(\Pi + \Omega)(p, p^e, L) \tag{3.5.10}$$

Existence and uniqueness of an efficient employment agreement follow from the properties of Assumption 3.2.1. Since $F(L)$ and $-S(L)L$ are strictly concave functions satisfying the Inada conditions, $pF(L) - \frac{p}{V(\theta^e)}S(L)L$ is a strictly concave function as well, and the set of individually rational employment levels L is compact. The maximization (3.5.9) induces the first-order condition

$$\begin{aligned}pF'(L) &= \frac{p}{V(\theta^e)}S(L)(E_S(L) + 1)\\ &\overset{(3.5.6)}{=} \frac{p}{V(\theta^e)}S(L)(E_v(L/n_w) - 1 + 1) \overset{(3.5.5)}{=} \frac{p}{V(\theta^e)}S_{\text{com}}(L),\end{aligned} \tag{3.5.11}$$

which coincides with (3.5.8). Thus, the solution of equation (3.5.8) defines an employment function $h_{\text{bar}} : \mathbb{R}_{++} \to \mathbb{R}_{++}$, $\theta^e \mapsto h_{\text{bar}}(\theta^e)$, which satisfies the equivalence

$$h_{\text{bar}}(\theta^e) = h(W(\theta^e)) \qquad \text{for all} \qquad \theta^e > 0, \tag{3.5.12}$$

where h is the competitive labor demand by the producer and W is the equilibrium real wage function for the competitive labor market defined in (3.2.25). This implies all the structural properties of $h_{\text{bar}} = h \circ W$ derived in Section 3.2.5. It is differentiable and strictly decreasing with elasticity greater than minus one. It has a globally defined explicit inverse induced by (3.5.8) given by

$$\theta^e = (h_{\text{bar}})^{-1}(L) := V^{-1}\left(\frac{S_{\text{com}}(L)}{F'(L)}\right) \tag{3.5.13}$$

Finally, using the two reservation wage functions, one obtains an intuitive and interesting relationship from rewriting the first order condition (3.5.8) as

$$W_\Omega(p, p^e, L) = \frac{p}{V(\theta^e)} S(L) = \frac{E_F(L)}{E_S(L)+1} \frac{pF(L)}{L} = \frac{E_F(L)}{E_S(L)+1} W_\Pi(p, L). \tag{3.5.14}$$

This shows that for $L = h_{\text{bar}}(\theta^e)$ the relative income shares of workers and share holders depend on the elasticities of the reservation wage functions, which also characterizes the bargaining level of employment. This stipulates that the ratio between the two status-quo values should correspond to the ratio of their respective elasticities.

Summarizing, under efficient bargaining, the level of employment $h_{\text{bar}}(\theta^e)$ is a well-defined, strictly monotonically decreasing, and invertible function of the expected inflation rate θ^e. It is homogeneous of degree zero in price expectations and prices, it is decreasing in expected prices and increasing in the current output price. The employment level chosen by the two bargaining parties is the same as the one which would result in equilibrium under a perfectly competitive labor market *and* it coincides with the one in a bilateral monopoly maximizing joint net gain against the rest of the economy. Thus, the employment decision to yield the maximal joint net gain can be separated from the wage decision of how this gain is to be distributed under efficient bargaining. With this perspective, the labor market has been eliminated, the employment decision L corresponds to an internal decision of a union-producer monopoly, while the agreement for a specific wage rate becomes a pure division "profit allocation issue".

The separability of the employment and the wage decision can be portrayed geometrically in an associated employment-wage-space diagram (see Figure 3.34). For $L > 0$, an acceptable wage must be such that $\Pi \geq 0$ *and* $\Omega \geq 0$, i. e.

$$w \leq p\frac{F(L)}{L} = W_\Pi(p, L) \qquad \text{and} \qquad w \geq \frac{p}{V(\theta^e)} S(L) =: W_\Omega(p, p^e, L),$$

inducing the two status-quo wage functions W_Π and W_Ω which correspond to the reservation wage of the producer and of the union respectively. The curved triangular area between the two functions in Figure 3.34 defines the set of individually rational employment–wage pairs.

The set of efficient employment-wage choices under bargaining are those on the contract curve shown as the bold red line. Geometrically speaking, each point on the contract curve must be a tangency point of an iso-utility and of an iso-profit curve

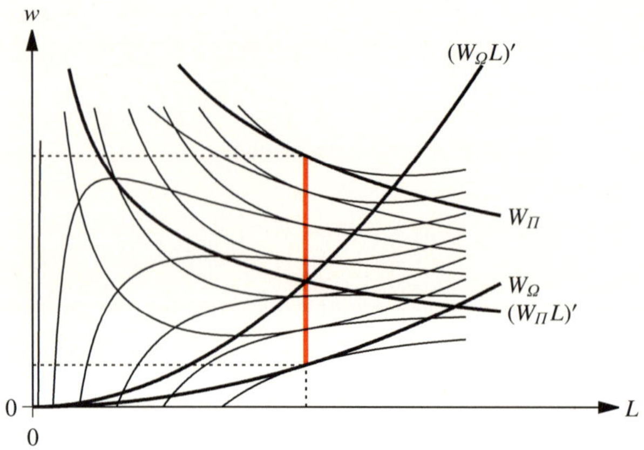

Fig. 3.34 The set of efficient employment-wage choices

(the thin lines). Since all iso-utility/iso-profit curves are of the form

$$W_{\bar{\Pi}}(L) = \frac{pF(L) - \bar{\Pi}}{L} \quad \text{resp.} \quad W_{\bar{\Omega}}(L) = \frac{p}{V(\theta^e)} S(L) + \frac{\bar{\Omega}}{L}$$

for all levels $\bar{\Pi}$ and $\bar{\Omega}$, the tangency condition $\partial W(L)/\partial L$ implies

$$\frac{pF'(L)L - W(L)L}{L^2} \stackrel{!}{=} \frac{p}{V(\theta^e)} S'(L) - \frac{W(L) - \frac{p}{V(\theta^e)} S(L)}{L}.$$

The Wage Rate under Efficient Bargaining

Given $(p^e, p) \gg 0$ and $L = h_{\text{bar}}(p^e/p) > 0$, the decision between the two parties concerning the wage rate now constitutes a *bargaining game with constant transfers* since

$$\begin{aligned}(\Pi + \Omega)_{\text{bar}}(p^e, p) &= p\left(F(h_{\text{bar}}(p^e/p)) - \frac{1}{V(\theta^e)} S(h_{\text{bar}}(p^e/p))h_{\text{bar}}(p^e/p)\right) \\ &= p\left[\frac{1 + E_S(h_{\text{bar}}(\theta^e)) - E_F(h_{\text{bar}}(\theta^e))}{E_F(h_{\text{bar}}(\theta^e))}\right] \frac{S(h_{\text{bar}}(\theta^e))h_{\text{bar}}(\theta^e)}{V(\theta^e)}\end{aligned} \tag{3.5.15}$$

is a constant sum for each (p^e, p). Thus, one obtains a special case of a bargaining problem, to which the generalized Zeuthen solution applies (see Rosenmüller, 2000). For such games the bargaining power between the two parties is usually measured by a number $0 \leq \lambda \leq 1$, which defines the relative share of the total cake to be alloted to the party having "bargaining power" λ. Thus, for a constant total gain $\Pi + \Omega = W_\Pi(p, L)L - W_\Omega(p, p^e, L)L$, the weights $(\lambda, 1-\lambda)$ determine a linear redistribution of the total net gain among the two agents.

Therefore, with $L > 0$ and $0 \le \lambda \le 1$ given, an application of the generalized Zeuthen solution[21] to the total gain implies choosing the bargaining wage as a convex combination of the two reservation wage levels W_Π and W_Ω with the weights λ and $1-\lambda$, i.e.

$$\begin{aligned} W(p^e, p, \lambda, L) &:= \lambda W_\Pi(p, L) + (1-\lambda) W_\Omega(p, p^e, L), \qquad L = h_{\text{bar}}(\theta^e), \\ &= p\left[\lambda \frac{F(L)}{L} + (1-\lambda)\frac{S(L)}{V(\theta^e)}\right], \qquad L = h_{\text{bar}}(\theta^e) \end{aligned} \tag{3.5.16}$$

or more compactly after using the first order condition (3.5.8)

$$V(\theta^e)F'(L) = S(L)(1 + E_s(L))$$

$$\begin{aligned} W(p^e, \lambda, p) &:= p\left[\lambda \frac{F(h_{\text{bar}}(\theta^e))}{h_{\text{bar}}(\theta^e)} + (1-\lambda)\frac{S(h_{\text{bar}}(\theta^e))}{V(\theta^e)}\right] \\ &\equiv p\left[\lambda \frac{F(h_{\text{bar}}(\theta^e))}{h_{\text{bar}}(\theta^e)} + (1-\lambda)\frac{F'(h_{\text{bar}}(\theta^e))}{1 + E_S(h_{\text{bar}}(\theta^e))}\right] \\ &\equiv p\left[\lambda + (1-\lambda)\frac{1 + E_F(h_{\text{bar}}(\theta^e))}{1 + E_S(h_{\text{bar}}(\theta^e))}\right]\frac{F(h_{\text{bar}}(\theta^e))}{h_{\text{bar}}(\theta^e)} \\ &\equiv pW(p^e/p, \lambda, 1). \end{aligned} \tag{3.5.17}$$

Observe that the bargaining wage is homogeneous of degree one in (p, p^e) which allows the convenient definition of the bargain real wage $W(\theta^e, \lambda, 1)$ as a function of the inflation rate and of union power. As a consequence one finds immediately that the real wage is an increasing function of union power λ, since $0 < E_F < E_S$, i.e.

$$\frac{\partial}{\partial \lambda} W(\theta^e, \lambda, 1) > 0. \tag{3.5.18}$$

The homogeneity of W in (p, p^e) implies

$$\frac{\partial W}{\partial p} = W(\theta^e, \lambda, 1) - \theta^e \frac{\partial W}{\partial \theta^e}(\theta^e, \lambda, 1) \qquad \text{and} \qquad E_W(p) = 1 - E_W(\theta^e).$$

Thus, the role of the expected inflation rate θ^e on the two multiplicative terms in (3.5.17) determine the sign and size of the price and of the expectations effects. Due to the concavity of the production function and the Inada conditions, $F(h_{\text{bar}}(\theta^e))/h_{\text{bar}}(\theta^e)$ is increasing in θ^e. Therefore, if the term in square brackets is nondecreasing, the bargaining real wage is increasing in expected inflation, while

[21] Note that the generalized Zeuthen solution which can only be applied to half-space games coincides with the generalized Nash solution, yet requiring fewer properties.

the nominal bargaining wage responds to prices with an elasticity less than one, i.e.

$$\frac{\partial}{\partial \theta^e} W(\theta^e, \lambda) > 0 \qquad \text{and} \qquad E_W(p) = \frac{\partial p W(\theta^e, \lambda)}{\partial p} \frac{p}{p W(\theta^e, \lambda)} < 1. \tag{3.5.19}$$

This property will hold in particular when the two elasticities E_F and E_S are constant.

Substituting (3.5.17) into the utility and into the profit functions yields the payoff vector $(\Pi(p^e, \lambda, p), \Omega(p^e, \lambda, p))$ of the bargaining solution

$$\begin{aligned}
\begin{pmatrix} \Pi(p^e, \lambda, p) \\ \Omega(p^e, \lambda, p) \end{pmatrix} &:= \begin{pmatrix} pF(L) - W(p^e, \lambda, p, L)L \\ W(p^e, \lambda, p, L)L - p^e S(L)L \end{pmatrix}, \qquad L = h_{\text{bar}}(\theta^e) \\
&= (pF(L) - \frac{p}{V(\theta^e)} S(L)L) \begin{pmatrix} 1 - \lambda \\ \lambda \end{pmatrix}, \qquad L = h_{\text{bar}}(\theta^e) \qquad (3.5.20) \\
&= (\Pi + \Omega)_{\text{bar}}(p^e, p) \begin{pmatrix} 1 - \lambda \\ \lambda \end{pmatrix}.
\end{aligned}$$

For given (p^e, p), Figure 3.35 displays the range of the mapping (3.5.20) for different

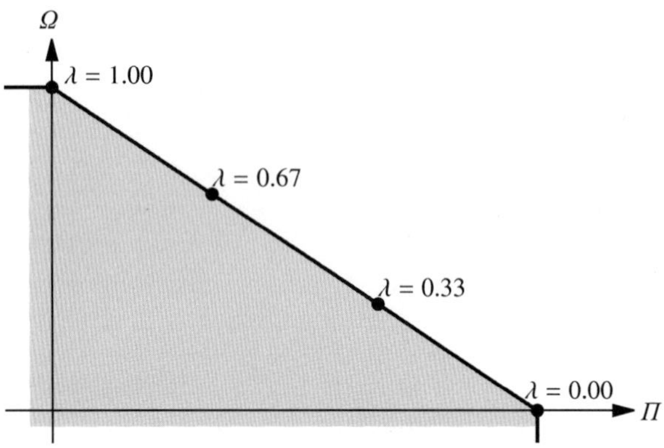

Fig. 3.35 The impact of the bargaining power λ on the equilibrium payoff; $(p^e, p) \gg 0$ given

values of the parameter λ, revealing its linear impact on the payoff distribution. A similar linear relationship holds for the role of λ on the bargaining wage. Finally, substituting (3.5.14) into the bargaining wage function (3.5.17), one finds that the bargaining wage

$$W(p^e, \lambda, p) = \left(\lambda + (1-\lambda)\frac{E_F(L)}{E_S(L)+1}\right)\frac{pF(L)}{L}, \qquad L = h_{\text{bar}}(\theta^e)$$

$$= \left(\frac{E_F(L)}{E_S(L)+1} + \lambda\frac{E_S(L)+1-E_F(L)}{E_S(L)+1}\right)\frac{pF(L)}{L}, \qquad L = h_{\text{bar}}(\theta^e)$$

is a multiple of average productivity, and that the bargaining real wage

$$\frac{W(p^e, \lambda, p, L)}{p} = \frac{1}{E_F(L)}\left(\frac{E_F(L)}{E_S(L)+1} + \lambda\frac{E_S(L)+1-E_F(L)}{E_S(L)+1}\right)F'(L), \quad L = h_{\text{bar}}(\theta^e) \tag{3.5.21}$$

is a positive multiple of the marginal product of labor (with $L = h(p^e/p)$). Both equations show clearly how the bargaining parameter interacts with the elasticities of the two reservation wage functions

Relative Union Power

An efficient bargaining solution $(L, w) = (h_{\text{bar}}(p^e/p), W(p^e, \lambda, p, h(p^e/p)))$ is defined for a given (p^e, p) and parametrically for a given $0 \le \lambda \le 1$ measuring the "bargaining power". Since λ is taken as a parameter, the model does not provide a fully endogenous determination of the bargaining power between the union and the producer. However, the efficient level of employment is independent of λ, implying that union–employer negotiations do guarantee productive efficiency. Therefore, the bargaining parameter λ determines exclusively the redistribution of the joint surplus or net gain between the two parties, i. e. the share of wages and profits in total revenue.

It is intuitively clear (and also evident from the geometry of Figure 3.34) that for each (p^e, p) there must be a unique bargaining level for which the parties agree on the competitive wage. This one equalizes marginal cost resp. marginal revenue ($(W_\Pi L)'$ resp. $(W_\Omega L)'$). Geometrically speaking (see Figure 3.34), this corresponds to the wage where the respective iso-utility and iso-profit curves are horizontal. Let the unique λ for which this condition holds be denoted by λ_{nat}, the "natural" λ. It is the solution of either

$$W(p^e, \lambda, p, L) \overset{!}{=} \frac{\partial(W_\Pi(p, L)L)}{\partial L} \qquad \text{or} \qquad W(p^e, \lambda, p, L) \overset{!}{=} \frac{\partial(W_\Omega(p^e, L)L)}{\partial L},$$

where $L = h_{\text{bar}}(\theta^e)$. Inserting the definition of $W(p^e, \lambda, p, L)$ into the second equation gives

$$\lambda_{\text{nat}} W_\Pi(p, L) + (1-\lambda_{\text{nat}})W_\Omega(p^e, L) = \frac{\partial(W_\Pi(p, L)L)}{\partial L} = pF'(L) = E_F(L)W_\Pi(p, L).$$

Exploiting (3.5.14) then gives

$$
\begin{aligned}
E_F(L)W_\Pi(p,L) &= \lambda_{\text{nat}} W_\Pi(p,L) + (1-\lambda_{\text{nat}})W_\Omega(p^e,L), \qquad L = h_{\text{bar}}(\theta^e)\\
&= \lambda_{\text{nat}} W_\Pi(p,L) + (1-\lambda_{\text{nat}})\frac{E_F(L)}{E_S(L)+1}W_\Pi(p,L)\\
&= \left(\lambda_{\text{nat}} + (1-\lambda_{\text{nat}})\frac{E_F(L)}{E_S(L)+1}\right)W_\Pi(p,L)\\
&= \left(\frac{E_F(L)}{E_S(L)+1} + \lambda_{\text{nat}}\frac{E_S(L)+1-E_F(L)}{E_S(L)+1}\right)W_\Pi(p,L), \quad L = h_{\text{bar}}(\theta^e)
\end{aligned}
$$

which implies

$$
\lambda_{\text{nat}}(L) = \frac{E_F(L)E_S(L)}{E_S(L)+1-E_F(L)}, \qquad L = h_{\text{bar}}(\theta^e). \tag{3.5.22}
$$

In other words, $\lambda_{\text{nat}}(L)$ is determined by the elasticities E_S and E_F of the labor supply function and of the production function respectively. Therefore, with isoelastic functions $\lambda_{\text{nat}}(L)$ is constant and independent of prices and price expectations.

The wage share of total income can be computed in a similar manner.

$$
\begin{aligned}
\frac{wL}{py} &= \frac{W(p^e,\lambda,p,L)}{W_\Pi(p,L)} = \lambda + (1-\lambda)\frac{W_\Omega(p^e,L)}{W_\Pi(p,L)} \overset{(3.5.14)}{=} \lambda + (1-\lambda)\frac{E_F(L)}{E_S(L)+1}\\
&= \frac{E_F(L)}{E_S(L)+1} + \lambda\left(1 - \frac{E_F(L)}{E_S(L)+1}\right) \in \left[\frac{E_F(L)}{E_S(L)+1}, 1\right], \qquad L = h_{\text{bar}}(\theta^e).
\end{aligned} \tag{3.5.23}
$$

Therefore, the profit share of total income is

$$
\frac{\pi}{py} = 1 - \frac{wL}{py} = (1-\lambda)\left(1 - \frac{E_F(L)}{E_S(L)+1}\right), \qquad L = h_{\text{bar}}(\theta^e). \tag{3.5.24}
$$

Note that the wage share resp. the profit share for $\lambda_{\text{nat}}(L)$ is $E_F(L)$ resp. $1 - E_F(L)$, as expected, since at $\lambda_{\text{nat}}(L)$ the factor shares in total output must be equal to the respective elasticities of the production function F.

Underemployment and Overemployment

Since the bargaining solution $(L, w) = (h_{\text{bar}}(\theta^e), W(p^e, \lambda, p, h_{\text{bar}}(\theta^e)))$ is a joint negotiated agreement between the producer and the union, in the current context it is conceptually questionable to define a measure of *involuntary* unemployment or overemployment, since any difference between $L = h_{\text{bar}}(\theta^e)$ and the desired labor supply $N(h_{\text{bar}}(\theta^e)V(\theta^e)/p)$ is part of the agreement. It has to be interpreted as a measure of a *voluntary* deviation from the competitive labor supply of the workers, as a *supply side* indicator of dissatisfaction (or of utility, of rent foregone, or of shadow unemployment) between bargaining and competition.

Similarly, any difference between L and the desired competitive employment of the producer $h(W(p^e, \lambda, p, h_{\text{bar}}(\theta^e))/p)$ by the producer would be a *demand side* measure of *voluntary* (or shadow) deviation relative to the competitive regime. Either measure would be primarily of theoretical interest under bargaining and only of limited additional information from an empirical point of view. Nevertheless, let us define the voluntary (or shadow) *underemployment rate* in the usual way as the relative deviation of aggregate notional labor supply from actual employment. For given (p^e, p) and any bargaining agreement $(L, w) = (h_{\text{bar}}(\theta^e), W(p^e, \lambda, p, h_{\text{bar}}(\theta^e)))$, define

$$U = \mathcal{U}\left(L, \frac{w}{p}V(\theta^e)\right) := \frac{N\left(\frac{w}{p}V(\theta^e)\right) - L}{N\left(\frac{w}{p}V(\theta^e)\right)} = 1 - \frac{L}{N\left(\frac{w}{p}V(\theta^e)\right)}. \tag{3.5.25}$$

which measures the gap between the amount of labor which is actually traded (i.e. worked) and which would be supplied by the workers under competitive conditions at the given wage level. Since the rate of unemployment is defined for all expected real wages and all levels of labor, U as defined in (3.5.25) can also be negative. This occurs for example if $\frac{w}{p}V(\theta^e)$ is relatively low or L is relatively high. As usual, negative rates of underemployment are interpreted as overemployment (or overtime).

3.5.2 Temporary Equilibrium with Efficient Bargaining

Aggregate Supply and Aggregate Demand under Efficient Bargaining

It is now straightforward to close the model in order to determine the properties of a temporary equilibrium under wage bargaining. Since the employment decision h_{bar} under bargaining is independent of the bargaining power λ and coincides with the competitive equilibrium configuration of the labor market, i.e. $h_{\text{bar}}(\theta^e) = h(W(\theta^e))$, aggregate supply under bargaining coincides with the competitive configuration as well. Therefore, one has the same aggregate commodity supply function as in Section 3.2.5 defined by

$$AS : \mathbb{R}_{++} \to \mathbb{R}_{++}, \quad AS(\theta^e) := F(h(W(\theta^e))),$$

which is differentiable and globally invertible with $AS'(\theta^e) < 0$, so that, for any given price expectation $p^e > 0$, aggregate supply is a strictly increasing bijective function of temporary commodity prices.

In contrast, the bargaining wage $W(p^e, \lambda, p, h(p^e/p))$ has a decisive influence on the income distribution, so that there must be also an influence on aggregate demand. Let the bargaining wage $W(p^e, \lambda, p, L)$ and the associated employment level $L = h_{\text{bar}}(p^e/p)$ be given as derived in the previous section. Then, given money balances, price expectations, the bargaining weight, and prices (M, p^e, λ, p), the income consistent aggregate demand y must be the solution of

$$
\begin{aligned}
y &= m + g + c_s(\theta^e)(1-\tau_\pi)\frac{\pi}{p} + c_w(\theta^e)(1-\tau_w)\frac{wL}{p} \\
&\overset{(3.5.24)}{=} m + g + c_w(\theta^e)(1-\tau_w)y \\
&\quad + \left[c_s(\theta^e)(1-\tau_\pi) - c_w(\theta^e)(1-\tau_w)\right](1-\lambda)\left(1 - \frac{E_F(L)}{E_S(L)+1}\right)
\end{aligned}
$$

with $L = h_{\text{bar}}(\theta^e)$ and a profit share

$$
\frac{\pi}{py} = \left(1 - \frac{E_F(L)}{E_S(L)+1}\right)
$$

as defined in (3.5.24). Therefore, one obtains as the income-consistent aggregate demand function (with $L = h_{\text{bar}}(\theta^e)$)

$$
\begin{aligned}
y &= D_{\text{bar}}(m, \theta^e, \lambda) \\
&:= \frac{m+g}{1 - c_w(\theta^e)(1-\tau_w) + [c_s(\theta^e)(1-\tau_\pi) - c_w(\theta^e)(1-\tau_w)](1-\lambda)\left(1 - \frac{E_F(L)}{E_S(L)+1}\right)},
\end{aligned}
\tag{3.5.26}
$$

which is of the usual multiplier form with respect to money balances and government demand. Observe that aggregate demand is homogeneous of degree zero in (M, p^e, p). Therefore, for given λ, it is a function of real money balances and of the expected rate of inflation, as in all previous cases. Obviously, $\partial D/\partial m > 0$, i.e. real balances have a positive effect on demand. In addition,

$$
\frac{\partial D_{\text{bar}}}{\partial \lambda} < 0 \quad \text{if and only if} \quad c_s(\theta^e)(1-\tau_\pi) - c_w(\theta^e)(1-\tau_w) > 0. \tag{3.5.27}
$$

In other words, if the net expenditure effect from share holders is larger than that from workers, an increase of union power decreases aggregate demand. If consumer characteristics are identical, any change of the income distribution due to a change in bargaining power λ neutralizes on the demand side, leaving the aggregate demand function unaffected by bargaining. In all other cases, the bargaining power exerts a positive or negative influence on aggregate demand.

Therefore, given a bargaining weight $0 \leq \lambda \leq 1$ and any pair $(M, p^e) \gg 0$ of money balances and expected prices, the price law is defined by a solution $p > 0$ such that

$$
D_{\text{bar}}\left(\frac{M}{p}, \frac{p^e}{p}, \lambda\right) = AS\left(\frac{p^e}{p}\right). \tag{3.5.28}
$$

In other words, one obtains as in all other previous cases, that the temporary equilibrium under efficient bargaining is characterized by two mappings homogeneous of degree one in (M, p^e), a price law $\mathcal{P}_{\text{bar}}$ i.e.

$$p = \mathcal{P}_{\text{bar}}(M, p^e, \lambda) \equiv p^e \mathcal{P}_{\text{bar}}(m^e, 1, \lambda), \quad m^e := M/p^e \tag{3.5.29}$$

and a wage law $\mathcal{W}_{\text{bar}}$,

$$\begin{aligned} w =: \mathcal{W}_{\text{bar}}(M, p^e, \lambda) &= \{pW(p^e/p, \lambda) | p = \mathcal{P}_{\text{bar}}(M, p^e, \lambda)\} \\ &= p^e \mathcal{P}_{\text{bar}}(m^e, 1, \lambda) W\left(\frac{1}{\mathcal{P}_{\text{bar}}(m^e, 1, \lambda)}, \lambda\right) \equiv p^e \mathcal{W}_{\text{bar}}(m^e, 1, \lambda), \qquad m^e := M/p^e. \end{aligned} \tag{3.5.30}$$

If $\partial D_{\text{bar}}/\partial \theta^e \geq 0$, then the demand is strictly decreasing in the commodity price p, i. e. $\partial D_{\text{bar}}(M/p, \theta^e, \lambda)/\partial p < 0$ is negative. This property holds in particular when the savings proportion by shareholders is nondecreasing and when the reservation wage and the production function are isoelastic. In this case, the equilibrium is unique and the two equilibrium laws are well defined. Figure 3.36 portrays the equilibrium situation in the usual aggregate demand–aggregate supply diagram of the commodity market.

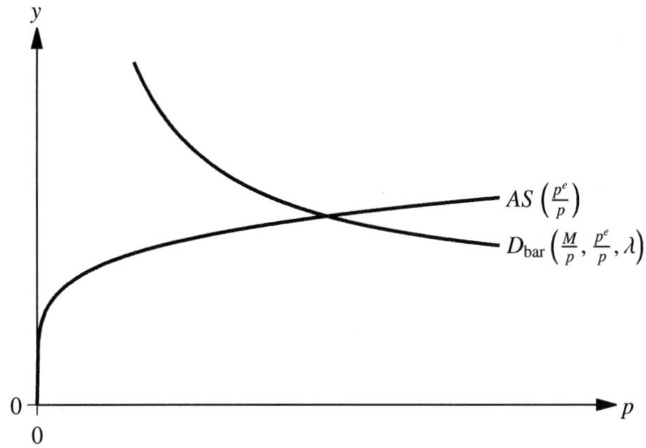

Fig. 3.36 The temporary equilibrium price under efficient bargaining

Properties of the Price Law

Applying the Implicit Function Theorem to (3.5.28) one obtains the effect of an increase of money balances

$$\frac{\partial \mathcal{P}_{\text{bar}}}{\partial M} = \frac{\partial \mathcal{P}_{\text{bar}}}{\partial m^e} = \frac{\mathcal{P}_{\text{bar}} \frac{\partial D_{\text{bar}}}{\partial m}}{-AS' + m^e \frac{\partial D_{\text{bar}}}{\partial m} + \frac{\partial D_{\text{bar}}}{\partial \theta^e}} > 0$$

with an elasticity

$$0 < E_{\mathcal{P}_{\text{bar}}}(M) = E_{\mathcal{P}_{\text{bar}}}(m^e) = \frac{\partial \mathcal{P}_{\text{bar}}}{\partial m^e}\frac{m^e}{\mathcal{P}_{\text{bar}}} = \frac{m^e \frac{\partial D_{\text{bar}}}{\partial m}}{-AS' + m^e \frac{\partial D_{\text{bar}}}{\partial m} + \frac{\partial D_{\text{bar}}}{\partial \theta^e}} < 1. \quad (3.5.31)$$

Thus, the mapping $\mathcal{P}_{\text{bar}}(m^e, 1, \lambda)$ is a strictly increasing and strictly concave function of expected money balances m^e satisfying $m^e \partial \mathcal{P}_{\text{bar}}/\partial m^e < \mathcal{P}_{\text{bar}}$. Together with the homogeneity of $\mathcal{P}_{\text{bar}}(M, p^e, \lambda)$ this implies for the derivative with respect to p^e

$$\begin{aligned} 0 < \frac{\partial \mathcal{P}_{\text{bar}}(M, p^e, \lambda)}{\partial p^e} &= \mathcal{P}_{\text{bar}}(m^e, 1, \lambda) - m^e \frac{\partial \mathcal{P}_{\text{bar}}(m^e, 1, \lambda)}{\partial m^e} \\ &= \frac{p^e \mathcal{P}_{\text{bar}}(m^e, 1, \lambda)}{p^e}\left[1 - \frac{\partial \mathcal{P}_{\text{bar}}(m^e, 1, \lambda)}{\partial M}\frac{M}{p^e \mathcal{P}_{\text{bar}}(m^e, 1, \lambda)}\right] \end{aligned} \quad (3.5.32)$$

which yields as an elasticity with respect to p^e

$$\begin{aligned} E_{\mathcal{P}_{\text{bar}}}(p^e)(M, p^e, \lambda) &= \frac{p^e}{p^e \mathcal{P}_{\text{bar}}(m^e, 1, \lambda)}\frac{\partial \mathcal{P}_{\text{bar}}(M, p^e, \lambda)}{\partial p^e} \\ &= \left[1 - \frac{\partial \mathcal{P}_{\text{bar}}(m^e, 1, \lambda)}{\partial M}\frac{M}{p^e \mathcal{P}_{\text{bar}}(m^e, 1, \lambda)}\right] < 1, \end{aligned} \quad (3.5.33)$$

which is also less than one. Therefore, $\mathcal{P}_{\text{bar}}(M, p^e, \lambda)$ is a strictly increasing and strictly concave function in price expectations as well. Together this implies that the price law $\mathcal{P}_{\text{bar}}$ is homogeneous of degree one and strictly concave in (M, p^e).

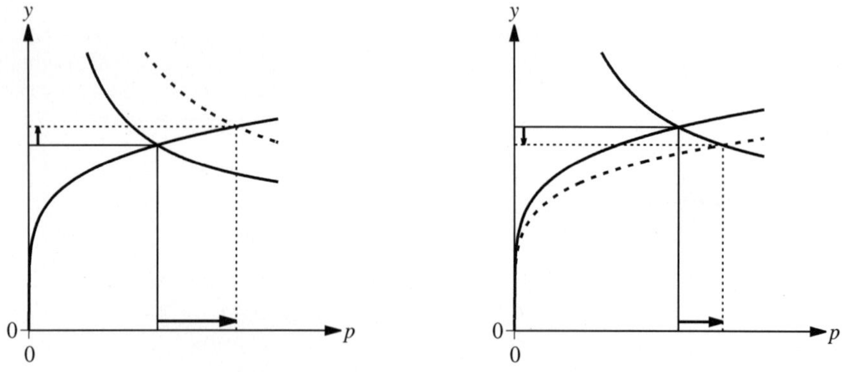

a Increase of money balances **b** Increase of p^e (for $\partial D/\partial \theta^e = 0$)

Fig. 3.37 Changes of money balances and price expectations

Properties of the Wage Law

The comparative statics effects of the wage law with respect to the state variables (M, p^e) cannot be signed for both variables in general since diverse effects interact in a nonlinear way. From (3.5.30) one has

$$w = \mathcal{W}_{\text{bar}}(M, p^e, \lambda) := \mathcal{P}_{\text{bar}}(M, p^e, \lambda) W\left(\frac{1}{\mathcal{P}_{\text{bar}}(M/p^e, 1, \lambda)}, \lambda\right) \tag{3.5.34}$$

as a multiplicative expression of the price law times the bargaining real wage function W. This implies

$$\frac{\partial w}{\partial p^e} = \frac{\partial \mathcal{P}_{\text{bar}}(M, p^e, \lambda)}{\partial p^e} W\left(\frac{1}{\mathcal{P}_{\text{bar}}(M/p^e, 1, \lambda)}\right) + \mathcal{P}_{\text{bar}}(M, p^e, \lambda) \frac{\partial W}{\partial \theta^e} \frac{\partial \mathcal{P}}{\partial m^e} \frac{M}{(p^e)^2 \mathcal{P}^2} > 0, \tag{3.5.35}$$

which shows a positive expectations effect, since $\frac{\partial W}{\partial \theta^e}$ is positive (see (3.5.19)) and $\mathcal{P}_{\text{bar}}$ is monotonically increasing in (M, p^e). Together with homogeneity one has

$$\begin{aligned} 0 < \frac{\partial}{\partial p^e} \mathcal{W}_{\text{bar}}(M, p^e, \lambda) &= \mathcal{W}_{\text{bar}}(m^e, 1, \lambda) - m^e \frac{\partial}{\partial m^e} \mathcal{W}_{\text{bar}}(m^e, 1, \lambda) \\ &= \mathcal{W}_{\text{bar}}(m^e, 1, \lambda) \left(1 - E_{\mathcal{W}_{\text{bar}}}(m^e)(m^e, 1, \lambda)\right) \\ &= \mathcal{W}_{\text{bar}}(m^e, 1, \lambda) \left(1 - E_{\mathcal{W}_{\text{bar}}}(M)(M, p^e, \lambda)\right), \end{aligned}$$

implying that the real wage elasticity with respect to money balances is less than one. However, this does not imply that the equilibrium wage increases with money balances. One sees from

$$\begin{aligned} \frac{\partial w}{\partial M} &= \frac{\partial \mathcal{P}_{\text{bar}}(M, p^e, \lambda)}{\partial M} W\left(\frac{1}{\mathcal{P}_{\text{bar}}(M/p^e, 1, \lambda)}\right) - \frac{\partial W}{\partial \theta^e} \frac{\partial \mathcal{P}_{\text{bar}}(M, p^e, \lambda)}{\partial m^e} \frac{1}{(\mathcal{P}(M, p^e, \lambda))^2} \\ &= p^e \frac{\partial}{\partial m^e} \mathcal{P}_{\text{bar}}(m^e, 1, \lambda) W\left(\frac{1}{\mathcal{P}_{\text{bar}}(m^e, 1, \lambda)}\right) [1 - E_W(\theta^e)], \end{aligned} \tag{3.5.36}$$

that the nominal wage (or real wage) effect is positive if and only if the elasticity of the bargaining wage function W is positive and less than one (see the discussion for (3.5.19)), which holds when the production function and the labor supply function are isoelastic.

Output and Employment

Associated with the price law are the output law and the employment law given by

$$y = \mathcal{Y}_{\text{bar}}(M, p^e, \lambda) := F\left(h_{\text{bar}}\left(\frac{1}{\mathcal{P}_{\text{bar}}(m^e, 1, \lambda)}\right)\right) \tag{3.5.37}$$

$$L = \mathcal{L}_{\text{bar}}(M, p^e, \lambda) := h_{\text{bar}}\left(\frac{1}{\mathcal{P}_{\text{bar}}(m^e, 1, \lambda)}\right), \tag{3.5.38}$$

which are homogeneous of degree zero in (M, p^e). Using (3.5.31) and $0 < E_F(L) < 1$, one obtains the corresponding elasticities of money balances on employment and output as

$$\begin{aligned} E_{\mathcal{L}_{\text{bar}}}(M) &= -E_{h_{\text{bar}}}(\theta^e)E_{\mathcal{P}_{\text{bar}}}(M) > 0 \\ E_{\mathcal{L}_{\text{bar}}}(M) &> E_F(L)E_{\mathcal{L}_{\text{bar}}}(M) = E_{\mathcal{Y}_{\text{bar}}}(M) > 0. \end{aligned} \tag{3.5.39}$$

Thus, higher money balances imply higher equilibrium prices but also higher levels of employment and output. Similarly, applying property (3.5.33), $0 < E_F(L) < 1$, and the relationship

$$E_{\mathcal{L}_{\text{bar}}}(p^e) = \underbrace{E_{h_{\text{bar}}}(\theta^e)}_{<0}\underbrace{(1 - E_{\mathcal{P}_{\text{bar}}}(p^e))}_{\in(0,1)} < 0 \tag{3.5.40}$$

yields

$$E_{\mathcal{L}_{\text{bar}}}(p^e) < E_F(L)E_{\mathcal{L}_{\text{bar}}}(p^e) = E_{\mathcal{Y}_{\text{bar}}}(p^e) < 0.$$

Thus, output and employment decline with higher price expectations. Therefore, combined with the zero-homogeneity of the employment law and output law, this confirms the tradeoff between money balances and expectations for a constant level of output and employment. Figure 3.37 displays the comparative statics results for changes of price expectations and of real money balances. These results correspond qualitatively to those under perfect competition in both markets, i.e. the price and output contours under efficient bargaining display the same features as those shown in Figure 3.2.

It is possible, however, in some special situations to determine the effects under more restricted conditions. Writing the wage as the associated mark-up over the reservation wage of the workers (or equivalently as a mark-down from the reservation wage of the producer)

$$\begin{aligned} w &= \left(1 + \lambda\frac{E_S(\mathcal{L}(M, p^e, \lambda)) + 1 - E_F(\mathcal{L}(M, p^e, \lambda))}{E_F(\mathcal{L}(M, p^e, \lambda))}\right)W_\Omega(p^e, \mathcal{L}(M, p^e, \lambda)) \qquad (3.5.41) \\ &= \left(\lambda + (1 - \lambda)\frac{E_F(\mathcal{L}(M, p^e, \lambda))}{E_S(\mathcal{L}(M, p^e, \lambda)) + 1}\right)W_\Pi(\mathcal{P}(M, p^e, \lambda), \mathcal{L}(M, p^e, \lambda)), \end{aligned}$$

one observes that the state variables exert their influence on wages via a primary effect through the price and employment laws and a secondary effect through the respective elasticities, which determine the mark-up. Therefore, in situations where the effect of the state variable on the mark-up is small and can be neglected, the wage effect has the same sign as the employment effect, i. e.

$$\begin{aligned} \operatorname{sgn} E_{\mathcal{W}}(M) &= \operatorname{sgn} E_S(L) E_{\mathcal{L}}(M) > 0 \\ \operatorname{sgn} E_{\mathcal{W}}(p^e) &= \operatorname{sgn} (E_{\mathcal{P}}(p^e) - (1 - E_F(L)) E_{\mathcal{L}}(p^e)) > 0 \end{aligned} \tag{3.5.42}$$

In this case, wages increase with money balances and with price expectations. This indicates, however, that wages can also fall when employment increases.

The effect of the state variables on the real wage can be determined using the same procedure. Writing the real wage as

$$\begin{aligned} \frac{w}{p} &= \left(\lambda + (1-\lambda)\frac{E_F(\mathcal{L}(M,p^e,\lambda))}{E_S(\mathcal{L}(M,p^e,\lambda)) + 1}\right)\frac{F(\mathcal{L}(M,p^e,\lambda))}{\mathcal{L}(M,p^e,\lambda)} \\ &= \left(\lambda + (1-\lambda)\frac{E_F(\mathcal{L}(M,p^e,\lambda))}{E_S(\mathcal{L}(M,p^e,\lambda)) + 1}\right)\frac{F'(\mathcal{L}(M,p^e,\lambda))}{E_F(\mathcal{L}(M,p^e,\lambda))} \\ &= \left(\frac{\lambda}{E_F(\mathcal{L}(M,p^e,\lambda))} + \frac{1-\lambda}{E_S(\mathcal{L}(M,p^e,\lambda)) + 1}\right) F'(\mathcal{L}(M,p^e,\lambda)), \end{aligned} \tag{3.5.43}$$

one finds that it can be written as a positive multiple of average labor productivity or of the marginal product of labor respectively. Therefore, *for given* λ, due to the concavity of the production function with average productivity declining in L, output and employment always move in the opposite direction as the real wage with respect to the state variables (M, p^e), provided that the elasticities do not change too much. Section 3.6 contains a detailed analysis of the wage law for a specific parametric example.

The Role of Union Power and Efficiency

Since the parameter λ does not influence aggregate supply, assuming $\partial D_{\text{bar}}/\partial \theta^e \geq 0$ implies that

$$\operatorname{sgn} \frac{\partial \mathcal{P}_{\text{bar}}}{\partial \lambda} = \operatorname{sgn} \frac{\partial D_{\text{bar}}}{\partial \lambda} = \operatorname{sgn} \frac{\partial \mathcal{Y}_{\text{bar}}}{\partial \lambda}.$$

Therefore, an increase of union power has a negative effect on the temporary equilibrium price and on output, if and only if the net expenditure effect from share holders is larger than that from workers (see (3.5.27)). Thus, while an increase in union power increases the share of workers in output, i.e. improving their *relative income position*, its impact on output is favorable only if the income effect also induces a demand effect which points in the same direction requiring that workers must spend more than share holders after the relative income gain. In other words, the sign of the spillover effect of an increase in union power through the commodity market which controls the *size* of the cake to be shared, depends on demand conditions which are typically independent of the supply conditions of the labor market.

This price feedback through the commodity market, for which the bargaining parties assume that the price is given *and* unaffected by their decisions, defines a cross market impact in temporary equilibrium, which is typically neglected in most discussions of the role of bargaining in macroeconomic models. From a game theoretic point of view it stipulates that the bargaining parameter not only defines *how* the cake is shared by the bargaining parties, but that it also has an influence on the *size* of the cake. In such cases a complete analysis of the effects of efficient bargaining requires a sequential solution to obtain intertemporally optimal decisions as attempted in Güth & Selten (1982), which still seems to be the only exception in the literature. Any such solution requires a correct strategic balance between the intratemporal tradeoffs presented here and the intertemporal options induced by each temporary decision. Conversely, if consumption characteristics of workers and share holders are identical, there exists no union induced price feed back, so that the myopic static solution induces the optimal intertemporal solution.

From now on the analysis here derives and discusses the allocative features of the myopic behavior of the union and the producer. The two cases of a positive or a negative price/output feed back effect from union power are analytically symmetric. In order to bring out the main features as strikingly as possible we will analyze the case of a negative effect in detail for the isoelastic example already introduced in Section 3.2.7. These allow, at the same time, a consistent comparison of the allocative features of all other temporary equilibrium labor market scenarios discussed so far in Sections 3.4 and 3.5 with the competitive temporary equilibrium.

From now on assume that the net propensity to consume of shareholders is larger than the one of workers. Then, an increase in union power induces a reduction of prices, output, and employment. Using the properties of the employment law (3.5.38) one has

$$E_{\mathcal{L}_{\text{bar}}}(\lambda) = -E_{h_{\text{bar}}}(\theta^e)E_{\mathcal{P}_{\text{bar}}}(\lambda) < 0 \qquad E_{\mathcal{L}_{\text{bar}}}(\lambda) < E_F(L)E_{\mathcal{L}_{\text{bar}}}(\lambda) = E_{\mathcal{Y}_{\text{bar}}}(\lambda) < 0. \tag{3.5.44}$$

Figure 3.38 portrays the effects of changes of union power on equilibrium prices, showing that there exists a strong nonlinear feedback from the bargaining power on the equilibrium prices, which translates directly into associated output, and employment changes. Thus, while the wage bargaining procedure assumes price-taking behavior on behalf of both parties inducing a perceived wage increase under increased union power, the level λ of union power has a negative indirect or spillover effect on the equilibrium price which operates through a negative income/expenditure effect on aggregate demand.

The bargaining power λ enters in multiple but opposite ways into the wage equation (3.5.30), similar to money balances and price expectations (M, p^e). This implies that, in general, the overall effect of union power on the equilibrium wage cannot be signed. However, the effect of λ on the real wage can be determined using the same technique as above. Rewriting the real wage equation as in (3.5.21) as

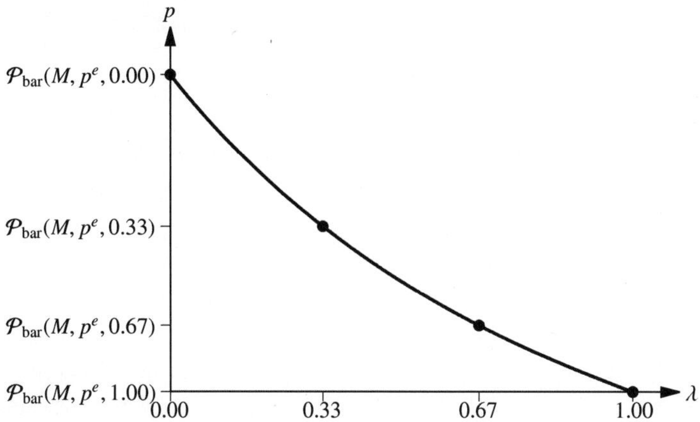

Fig. 3.38 Range of equilibrium prices $\mathcal{P}_{\text{bar}}(M, p^e, \lambda)$ for λ from 0 to 1

$$\frac{w}{p} = \left(\frac{E_F(\mathcal{L}_{\text{bar}}(M, p^e, \lambda))}{E_S(\mathcal{L}_{\text{bar}}(M, p^e, \lambda)) + 1} + \lambda \left(1 - \frac{E_F(\mathcal{L}_{\text{bar}}(M, p^e, \lambda))}{E_S(\mathcal{L}_{\text{bar}}(M, p^e, \lambda)) + 1} \right) \right) \frac{F(\mathcal{L}_{\text{bar}}(M, p^e, \lambda))}{\mathcal{L}_{\text{bar}}(M, p^e, \lambda)},$$

one finds that it must increase with union power whenever the wage is nonincreasing or when the effect of λ on the elasticities can be neglected. Section 3.6 also contains a detailed study of the role of union power for a parametrized version of the model.

3.5.3 Comparing Bargaining and Competition

The results in the previous section indicate that the level of prices, output, and employment vary positively or negatively with union power λ depending on features of aggregate demand. It is somewhat surprising that such fairly strong comparative statics properties hold in general. With such clear influence on output and employment from powerful but efficient wage bargaining, it is particularly desirable to investigate the role of bargaining in its general relationship to competitive allocations.

To carry out a systematic comparison between temporary equilibria under competition and under efficient wage bargaining, the impact of bargaining on aggregate demand and aggregate supply relative to the competitive case has to be examined. Given the labor demand function of the competitive producer (see equation (3.1.1)) $h_{\text{com}}(w/p) = (F')^{-1}(w/p)$, the labor market clearing condition

$$N_{\text{com}}\left(\frac{w}{p} \Big/ \frac{p^e}{p}\right) = h_{\text{com}}\left(\frac{w}{p}\right)$$

implies the usual equilibrium relationship between expected inflation and the real wage

$$\frac{p^e}{p} = \theta^e = \frac{w/p}{N_{\text{com}}^{-1}(h_{\text{com}}(w/p))} = \frac{w/p}{S_{\text{com}}(h_{\text{com}}(w/p))} =: W_{\text{com}}^{-1}\left(\frac{w}{p}\right).$$

Using equation (3.5.11) with $L = h(\theta^e)$, this induces

$$W_{\text{com}}^{-1}\left(h_{\text{com}}^{-1}(L)\right) = \frac{h_{\text{com}}^{-1}(L)}{S_{\text{com}}\left(h_{\text{com}}\left(h_{\text{com}}^{-1}(L)\right)\right)} = \frac{F'(L)}{S_{\text{com}}(L)} \overset{(3.5.45)}{=} \theta^e.$$

Therefore, for all $p^e/p = \theta^e$,

$$h_{\text{com}}\left(W_{\text{com}}(\theta^e)\right) = h(\theta^e), \tag{3.5.45}$$

the equilibrium employment decisions in the labor market under bargaining and under competition are identical. This in turn implies that the two aggregate supply functions are the same, i. e. for all θ^e,

$$AS_{\text{com}}(\theta^e) = F\left(h_{\text{com}}\left(W_{\text{com}}(\theta^e)\right)\right) = F(h(\theta^e)) = AS(\theta^e).$$

To define income-consistent aggregate demand under competition, let prices and wages (p, w) be given. The competitive firm chooses its labor input according to the marginal product rule $w = pF'(L)$, implying that the profit share of total revenue is

$$\frac{py - wL}{py} = 1 - \frac{F'(L)L}{F(L)} = 1 - E_F(L).$$

Thus, income-consistent aggregate demand in the competitive case must satisfy

$$y = m + g + c(\theta^e)(1 - \tau_\pi)(1 - E_F(L))\, y,$$

leading to the aggregate demand function under perfect competition in the labor market

$$y = D_{\text{com}}(m, \theta^e) = \frac{m + g}{1 - c(\theta^e)(1 - \tau_\pi)(1 - E_F(L))}, \quad L = h_{\text{com}}(W_{\text{com}}(\theta^e)) \overset{(3.5.45)}{=} h(\theta^e),$$

as compared to the aggregate demand function under bargaining derived from

$$y = m + g + c(\theta^e)(1 - \tau_\pi)(1 - \lambda)\left(1 - \frac{E_F(L)}{E_S(L) + 1}\right) y$$

in (3.5.46) as

$$y = D(m, \theta^e, \lambda) = \frac{m + g}{1 - c(\theta^e)(1 - \tau_\pi)(1 - \lambda)(1 - \frac{E_F(L)}{E_S(L)+1})}, \qquad L = h(\theta^e). \tag{3.5.46}$$

Thus, the two aggregate demand functions differ essentially only by the size of the multiplier, which depends on λ and on the values of the respective elasticities. Therefore, one finds that, for all (M, p, p^e), aggregate demand under bargaining is strictly decreasing in λ with

$$D(m, \theta^e, 1) < D_{\text{com}}(m, \theta^e) < D(m, \theta^e, 0)$$

and, since aggregate supply is independent of λ and identical in the two cases, that

$$\mathcal{P}(M, p^e, 1) < \mathcal{P}_{\text{com}}(M, p^e) < \mathcal{P}(M, p^e, 0).$$

As a consequence, for given (M, p^e), by the continuity and monotonicity of the price law under bargaining as a function of λ, there must exist a unique value $0 < \lambda_{\text{com}} < 1$, where the temporary equilibrium price at the bargaining equilibrium coincides with that of the competitive equilibrium, i.e. one has

$$\mathcal{P}_{\text{com}}(M, p^e) = \mathcal{P}(M, p^e, \lambda_{\text{com}}).$$

Thus, given the equivalence $\mathcal{P}_{\text{com}}(M, p^e) = p = \mathcal{P}_{\text{bar}}(M, p^e, \lambda_{\text{com}})$ of the equilibrium price under competition and under bargaining for λ_{com}, aggregate supply and aggregate demand at equilibrium must be the same

$$D_{\text{com}}(M/p, p^e/p) = AS_{\text{com}}(p^e/p) = AS(p^e/p) = D(M/p, p^e/p, \lambda_{\text{com}})$$

so that the level of output, employment, and of wages

$$\mathcal{Y}_{\text{com}}(M, p^e) = D_{\text{com}}\left(\frac{M}{p}, \frac{p^e}{p}\right) = D\left(\frac{M}{p}, \frac{p^e}{p}, \lambda_{\text{com}}\right) = \mathcal{Y}(M, p^e, \lambda_{\text{com}}),$$

$$\mathcal{L}_{\text{com}}(M, p^e) = F^{-1}\left(D_{\text{com}}\left(\frac{M}{p}, \frac{p^e}{p}\right)\right) = F^{-1}\left(D\left(\frac{M}{p}, \frac{p^e}{p}\right), \lambda_{\text{com}}\right) = \mathcal{L}(M, p^e, \lambda_{\text{com}}),$$

$$\mathcal{W}_{\text{com}}(M, p^e) = \mathcal{W}(M, p^e, \lambda_{\text{com}})$$

are equalized as well. Therefore, the competitive temporary equilibrium is a special case of the possible equilibria under efficient bargaining for a specific value λ_{com} of union power.

While the coincidence of the two equilibria does not seem surprising at first sight, one should note that this results depends crucially on the fact that the reservation wages for workers and for the firm are defined by the zero-activity level of workers and producers and by the fact that they are common knowledge in the bargaining procedure. These assumptions imply a symmetric no-participation constraint (or threat point) for both sides which induces the specific equilibrium characteristics with no loss in production-effort efficiency, equalizing the real marginal product to the competitive marginal willingness to work. Thus the employment choice corresponds to the competitive one, making the aggregate supply function under bargaining equivalent to the competitive one. Thus, the bargaining equilibrium not only provides an *efficient redistribution* of value added, but it also eliminates interparty inefficiencies leading to an optimal tradeoff between marginal disutility of effort and marginal productivity of labor. In this sense, the temporary equilibrium with bargaining satisfies conditional Pareto optimality at any level $\lambda > 0$ of bargaining power. Yet, the total value added could always be improved by setting $\lambda = 0$. Combined with a lump-sum redistribution of the surplus, an improvement in payoffs could be obtained[22].

If, however, the reservation wages of either side had been chosen to be the levels of the corresponding competitive inverse demand or supply functions, i.e. their *marginal willingness* to work or hire at given prices and price expectations, conditional Pareto optimality could not be obtained under bargaining since total net value would not have been maximized in equilibrium. In such cases, the bargaining equilibrium would generate allocations with prices and wages, levels of employment and output which are continuous deformations between the two cases of one-sided full market power for the union, i.e. the union monopoly, and the producer monopsony, which were discussed in Section 3.4. As was shown there, these would suffer from additional inefficiencies and the competitive temporary equilibrium could not be achieved as an equilibrium under efficient bargaining.

3.5.4 Inefficient Redistribution under Efficient Bargaining

The negative feedback of union power on prices, output, and employment derived in (3.5.44) indicates that, from a macroeconomic point of view, a strong union under efficient bargaining may not guarantee an overall efficient allocation in temporary equilibrium. In other words, given the data of the economy (M, p^e, λ), output is maximal when $\lambda = 0$ and minimal when $\lambda = 1$. This suggests that the bargaining procedure will never attain the global maximal surplus in the economy unless $\lambda = 0$.

To investigate the role of the bargaining power more closely, consider the payoff vector (Π, Ω) in temporary equilibrium, which is obtained by substituting the price law $\mathcal{P}(M, p^e, \lambda)$ from (3.5.29) and the wage law from (3.5.30) into the payoff vector (3.5.20). This yields

$$\begin{pmatrix} \Pi(M, p^e, \lambda) \\ \Omega(M, p^e, \lambda) \end{pmatrix} = \Big(\mathcal{P}(M, p^e, \lambda) F(\mathcal{L}(M, p^e, \lambda)) - p^e S(\mathcal{L}(M, p^e, \lambda))\mathcal{L}(M, p^e, \lambda)\Big) \begin{pmatrix} 1-\lambda \\ \lambda \end{pmatrix}.$$

Thus, the efficient bargaining solution at the temporary equilibrium is a linear one-to-one redistribution of the total net surplus

[22] Note that this discussion argues only about efficiency in terms of the payoff between the firm and the union and not in welfare terms with respect to the two groups of consumers and their indirect utility. A welfare comparison should use their utility functions. In this case, the effects stemming from underemployment/overemployment would have to be accounted for as well. Moreover, the intertemporal structure of overlapping generations requires additional criteria between old and young consumers and their position in the temporary equilibrium, for which a Pareto criterion is not universally defined.

$$\begin{aligned}&\Pi(M,p^e,\lambda)+\Omega(M,p^e,\lambda)\\&\quad=\mathcal{P}(M,p^e,\lambda)F(\mathcal{L}(M,p^e,\lambda))-p^eS(\mathcal{L}(M,p^e,\lambda))\mathcal{L}(M,p^e,\lambda),\end{aligned}\tag{3.5.47}$$

implying a marginal rate of substitution between $\Pi(M,p^e,\lambda)$ and $\Omega(M,p^e,\lambda)$ equal to minus one. Taking the derivative of (3.5.47) with respect to λ, one finds that

$$\begin{aligned}&\frac{\mathrm{d}}{\mathrm{d}\lambda}(\Pi(M,p^e,\lambda)+\Omega(M,p^e,\lambda))\\&=\frac{\mathrm{d}}{\mathrm{d}\lambda}(\mathcal{P}(M,p^e,\lambda)F(\mathcal{L}(M,p^e,\lambda))-p^eS(\mathcal{L}(M,p^e,\lambda))\mathcal{L}(M,p^e,\lambda))\\&=F(\mathcal{L}(M,p^e,\lambda))\frac{\partial\mathcal{P}(M,p^e,\lambda)}{\partial\lambda}+\underbrace{\frac{\mathrm{d}}{\mathrm{d}L}(pF(L)-p^eS(L)L)}_{\overset{(3.5.45)}{=}0}\frac{\partial\mathcal{L}(M,p^e,\lambda)}{\partial\lambda}\\&=F(\mathcal{L}(M,p^e,\lambda))\frac{\partial\mathcal{P}(M,p^e,\lambda)}{\partial\lambda}<0\end{aligned}\tag{3.5.48}$$

has a negative sign. Therefore, higher union power λ also induces a lower aggregate equilibrium surplus. Thus, the aggregate surplus is a strictly decreasing function with a global maximum at $\lambda = 0$. Geometrically speaking, this implies that the bargaining possibility frontier for all $0 < \lambda \leq 1$ in temporary equilibrium is strictly below the minus one tradeoff line at $\Pi(M,p^e,0) - \Omega(M,p^e,0)$.

It is obvious that the profit term of the payoff $\Pi(M,p^e,\lambda)-\Omega(M,p^e,\lambda)$ is decreasing in λ while the influence on the wage bill cannot be signed in all cases. In fact,

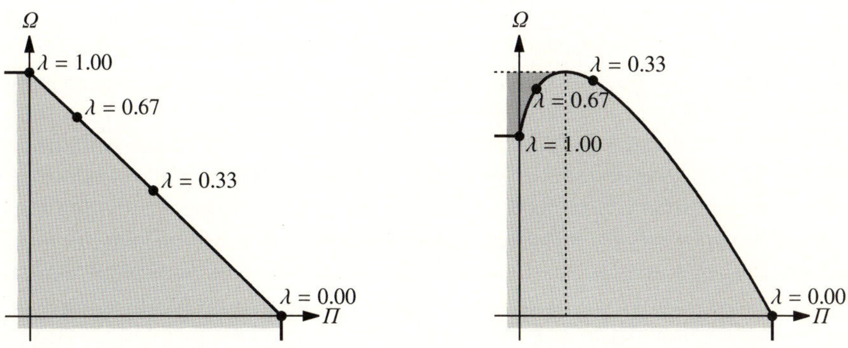

a High government consumption **b** Low government consumption

Fig. 3.39 Net wage bill and profit with price feedback

it may be increasing or decreasing depending on the data. Figure 3.39 displays the payoff frontier in equilibrium for two different levels of government consumption, taking the feedback into account. Both panels show that the distribution of wealth is

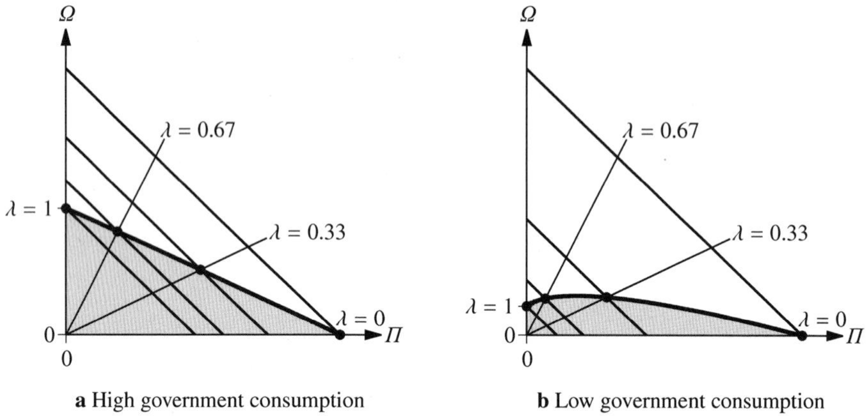

Fig. 3.40 The role of government demand for $\lambda = 0.00$, $\lambda = 0.33$, $\lambda = 0.67$, and $\lambda = 1.00$

not linear in λ. While the equilibrium profit always decreases with union power, the right panel clearly shows that even the wage bill may be declining with union power in some circumstances. Figure 3.40 combines Figure 3.35 and Figure 3.39 displaying the equilibrium payoffs for four levels of union power ($\lambda = 0.00$, $\lambda = 0.33$, $\lambda = 0.67$, and $\lambda = 1.00$) as intersections of the sharing ratios $\lambda/(1-\lambda)$ and the corresponding associated linear tradeoff frontier (thin downward-sloping lines with prices assumed to be fixed at the respective levels). Finally, the two properties of declining aggregate surplus (3.5.48) and the linearity of the payoffs *for given* λ imply that the bargaining solution is inefficient at the equilibrium price for all $\lambda > 0$. This follows directly from the fact that the slope of the bargaining frontier must be smaller than one in absolute value at any λ. The argument is given geometrically in Figures 3.40 and 3.41. The bargaining frontier is given by the bold downward-sloping curve. To provide the intuition for this result, it is useful to reconsider the bargaining problem. Since both groups are price takers in the commodity market, they assume that its price is given and unaffected by their wage setting for given λ. Thus, the negotiating parties have a perceived payoff frontier with slope minus one while the slope of the bargaining frontier is less in absolute value. In addition to the frontier shown in Figure 3.40, Figure 3.41 contains the level curve of the Nash bargaining solution, which must have slope minus one at the equilibrium payoff. Since the slope of the bargaining frontier is flatter, there exists a lower λ and a redistribution at the equilibrium price $p = \mathcal{P}(M, p^e, \lambda)$ which improves the Nash product. The possibility of such improvements is indicated geometrically by the red regions, the feasible upper contour set.

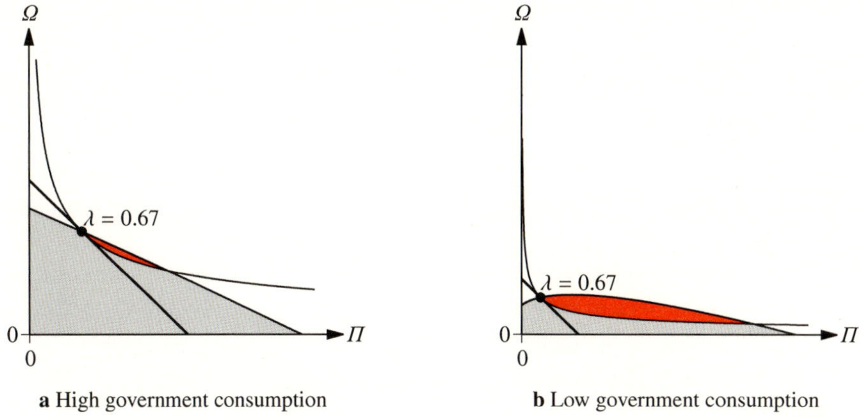

Fig. 3.41 No efficient Nash bargaining solution under price feedback: the better set (red)

3.5.5 The Second-Best Nature of Efficient Bargaining

For a general discussion of the role of bargaining as a wage determination device, note first that it was shown that temporary equilibria with efficient bargaining exist and that they are unique under the same set of assumptions as in other cases of wage setting with price flexibility and market clearing. Thus, efficient bargaining by itself cannot be the cause for involuntary unemployment.

From a macroeconomic point of view, however, the most striking result is that higher union power directed toward a desired and successful redistribution from profits to wages in temporary equilibrium may cause uniformly lower employment and lower output. This structural negative impact of union power on employment and total output has additional allocative consequences. With constant exogenous demand (government demand plus money balances), an increase of union power implies lower profits and lower effective demand by young shareholders. Production becomes less attractive to producers even if the income distribution (i.e. the profit share in output) stays constant, but the multiplier decreases. In other words, aggregate output to be distributed for private and public consumption declines with higher union power.

Therefore, if total output or aggregate private consumption in temporary equilibrium is considered as a welfare proxy, it would not be desirable to have a strong union imposing a high level of λ. However, the redistribution due to a higher wage bill implies higher savings and demand for money by workers inducing higher expected consumption in the second period. Thus, higher union power also induces an increase of real wealth for workers and higher expected indirect utility. Thus, young shareholders partly pay the bill of high union power through reduced consumption in both periods. Nevertheless, this increase always incurs a macroeconomic cost of lower total.

Finally, it was shown that an efficient bargaining procedure between the participants in the labor market alone does not lead to an efficient outcome with respect to the objective of the bargaining when the remaining market is competitive. Generally speaking, this reconfirms the typical features of results known from Second-best Theory, which say that noncompetitive or deviant behavior in one market alone while all others are competitive does not guarantee Second-best allocations if there are spillovers between markets. Notice that this result equally applies to the competitive temporary equilibrium. In other words, even the fully competitive temporary equilibrium is not efficient with respect to the bargaining criterion, due to the price feedback. Thus, the exogenous parametric setting of the negotiating power of one side of the market induces only an efficient allocation with respect to the *perceived* feasible bargaining set, and which is inefficient with respect to general equilibrium feasibility. Thus, an efficient level of bargaining power would have to be determined endogenously.

From a general welfare perspective, however, it is not clear whether this inefficiency implies also suboptimality and failure to satisfy a Second-Best property since both criteria are applied to a comparative-statics analysis of allocations in temporary equilibrium at given money balances and expectations. Therefore, for the *dynamic* macroeconomic perspective taken here with overlapping generations of consumers, the Second-Best failure may not seem to be of such primary importance. Moreover, the welfare issue becomes even more complex for sequences of temporary equilibria and requires further criteria and investigations, also with respect to stationary states.

With constant union power it is surprisingly straightforward to show that the dynamics under perfect foresight is essentially identical to the one under competition. In other words, there are no systematic effects arising from efficient wage bargaining on the qualitative dynamic properties under perfect foresight: all existence issues of attracting balanced paths are structurally identical to the competitive case (as derived in Sections 4.1.3 to 4.2). Different but constant levels of union power invoke distributional effects only, but no changes in the nature of the long-run behavior. However, when union power becomes an endogenous variable depending on measures of performance or success for union members additional effects appear modifying the dynamic characteristics (see Böhm & Claas, 2017; Claas, 2017).

3.6 Prices, Wages, and Payoffs: The Isoelastic Case Revisited

Specializing now from the general qualitative characteristics to the parametric isoelastic case introduced in section 3.2.7, let us restate in brief the main features of the isoelastic model. The shareholder's propensity to consume is constant and equal to $0 < c < 1$. Workers are asssumed to derive no utility from consumption when young (i.e. u_w is linear), while their disutility of effort of the young worker is given by

$$v(\ell) = \frac{C}{C+1}\ell^{1+\frac{1}{C}}, \qquad 0 < C < 1.$$

Moreover, let the production function be isoelastic and of the form

$$F(L) = \frac{A}{B} L^B, \qquad A > 0, \quad 0 < B < 1.$$

Solving the young worker's first-order condition of optimality$(1 - \tau_w)\frac{w}{p^e} = \ell^{1/C}$ yields the individual utility-maximizing labor supply as

$$\ell = \left((1 - \tau_w) \frac{w}{p^e} \right)^C,$$

implying an isoelastic competitive aggregate labor supply function

$$N_{\text{com}} \left(\frac{w}{p^e} \right) = n_w \left((1 - \tau_w) \frac{w}{p^e} \right)^C.$$

Its inverse is given by

$$S_{\text{com}}(L) = \frac{1}{1 - \tau_w} \left(\frac{1}{n_w} L \right)^{1/C}.$$

This is a strictly convex isoelastic function measuring the *aggregate marginal willingness* to work at the aggregate level L when n_w homogeneous workers are employed equally. This is the inverse of the competitive aggregate labor supply function.

The *individual reservation wage* of each worker is the solution of

$$\frac{w}{p^e} = \frac{1}{1 - \tau_w} \frac{v(\ell)}{\ell} = \frac{1}{1 - \tau_w} \frac{C}{C + 1} \ell^{1/C}.$$

Thus, the maximal amount of labor each worker is willing to supply at a given wage w is given by

$$\ell = \left((1 - \tau_w) \frac{C + 1}{C} \frac{w}{p^e} \right)^C. \qquad (3.6.1)$$

Therefore, the *aggregate reservation wage* function of the union is given by

$$S(L) = \frac{C}{C + 1} \frac{1}{1 - \tau_w} \left(\frac{1}{n_w} L \right)^{1/C},$$

which has the same constant elasticity as the aggregate marginal willingness to work of the union. Therefore, one finds that

$$S(L) = \frac{C}{C + 1} S_{\text{com}}(L) \qquad \text{and} \qquad N\left(\frac{w}{p^e} \right) = \left(\frac{C + 1}{C} \right)^C N_{\text{com}} \left(\frac{w}{p^e} \right).$$

The functions S and S_{com} have the same elasticity $1/C$, which coincides with the elasticity of the individual marginal willingness to work, while N and N_{com} have the same elasticity C.

The inverse of the demand for labor (3.5.14) can be computed explicitly in elasticity form as

$$\theta^e = h^{-1}(L) = \frac{E_F(L)}{E_S(L)+1}\frac{F(L)}{S(L)L} = \frac{BC}{C+1}\frac{F(L)}{S(L)L}$$

$$= A(1-\tau_w)n_w^{1/C}L^{\frac{BC-(C+1)}{C}}.$$

This yields the labor demand function under bargaining as

$$L = h(\theta^e) = \left(\frac{\theta^e}{A(1-\tau_w)\,n_w^{1/C}}\right)^{\frac{C}{BC-(C+1)}} \tag{3.6.2}$$

$$= A^{\frac{C}{C+1-BC}}(1-\tau_w)^{\frac{C}{C+1-BC}}\,n_w^{\frac{1}{C+1-BC}}\,(\theta^e)^{\frac{C}{BC-(C+1)}}, \tag{3.6.3}$$

which has a constant elasticity satisfying

$$-C < E_h(\theta^e) = \frac{C}{BC-(C+1)} = -\frac{C}{C(1-B)+1} < 0. \tag{3.6.4}$$

Therefore, aggregate labor demand under bargaining is an isoelastic, strictly monotonically decreasing function in expected inflation. For a given $p^e > 0$, it is also isoelastic, strictly monotonically increasing, and concave in the price. Substituting labor demand (3.6.2) into the production function implies a strictly decreasing isoelastic aggregate supply function in expected inflation given by

$$AS(\theta^e) = \frac{1}{B}A^{\frac{C+1}{C+1-BC}}(1-\tau_w)^{\frac{BC}{C+1-BC}}\,n_w^{\frac{B}{C+1-BC}}\,(\theta^e)^{\frac{BC}{BC-(C+1)}}, \tag{3.6.5}$$

making it an isoelastic, strictly increasing, and strictly concave function of the commodity price p for any given price expectation p^e.

Regarding the income distribution, equation (3.5.24) implies that, for any given union power $0 \le \lambda \le 1$, the profit share in output is a given constant

$$\frac{\pi}{py} = (1-\lambda)\left(1-\frac{BC}{C+1}\right). \tag{3.6.6}$$

Thus, with isoelastic production and preferences, the *profit share* under efficient bargaining becomes a linear, decreasing function in λ, independent of the expected inflation rate.

The two properties, an isoelastic utility of shareholders together with an inflation-independent profit distribution (3.6.6), imply that there is no inflation feedback into aggregate commodity demand under bargaining. Thus, one obtains from (3.5.46) as the income-consistent aggregate demand function

$$D(m,\lambda) = \frac{m+g}{1 - c(1-\tau_\pi)(1-\lambda)(1-\frac{BC}{C+1})}, \tag{3.6.7}$$

which is strictly decreasing in λ and independent of expected prices. Equating aggregate demand (3.5.26) and aggregate supply (3.6.5), one obtains a unique positive equilibrium price $p = \mathcal{P}(M, p^e, \lambda)$ where the price map $\mathcal{P}$ has the usual properties, i. e. it is increasing and linear homogeneous in (M, p^e). Due to the isoelasticity of aggregate supply given in (3.6.5), its inverse with respect to price expectations $\mathcal{P}^e$ is given explicitly by

$$\begin{aligned} p^e = \mathcal{P}^e(p, M, \lambda) &:= pAS^{-1}(D(M/p, \lambda)) \\ &= pAS^{-1}(1)\left(\frac{M/p+g}{1 - c(1-\tau_\pi)(1-\lambda)(1-\frac{BC}{C+1})}\right)^{\frac{BC-(C+1)}{BC}}, \end{aligned} \tag{3.6.8}$$

which is one-to-one, strictly increasing, and strictly convex in p. Notice that the inverse of the price law is an isoelastic function in $(M/p + g)$, which becomes an isoelastic function in p only when exogenous government demand g is equal to zero. Thus, the price law itself is an isoelastic function in M/p^e only when $g = 0$.

In addition to the bounds derived in the general setting of Section 3.5.2, one obtains upper and lower bounds for the respective elasticities of the employment function using the isoelasticity of the labor supply function (3.6.4).

$$\begin{aligned} 0 < E_{\mathcal{L}}(M) &\overset{(3.5.39)}{=} -E_h(\theta^e)E_{\mathcal{P}}(M) = \frac{C}{C(1-B)+1}E_{\mathcal{P}}(M) < E_{\mathcal{P}}(M), \\ -C < E_{\mathcal{L}}(p^e) &\overset{(3.5.40)}{=} E_h(\theta^e)(1 - E_{\mathcal{P}}(p^e)) = -\frac{C}{C(1-B)+1}(1 - E_{\mathcal{P}}(p^e)) < 0, \\ 0 > E_{\mathcal{L}}(\lambda) &\overset{(3.5.44)}{=} -E_h(\theta^e)E_{\mathcal{P}}(\lambda) = \frac{C}{C(1-B)+1}E_{\mathcal{P}}(\lambda) > E_{\mathcal{P}}(\lambda). \end{aligned} \tag{3.6.9}$$

Since the output function $\mathcal{Y}(M, p^e, \lambda) = F(\mathcal{L}(M, p^e, \lambda))$ is simply the composition of the production function with the employment function, its elasticities are the same expressions as in (3.6.9) each multiplied by B, the elasticity of the production function F. Observe again that all equilibrium maps will be isoelastic functions only if government demand g is equal to zero.

Lower bounds for $E_{\mathcal{W}}(M)$ and $E_{\mathcal{W}}(p^e)$ have been found in (3.5.42). In order to establish upper bounds, note that the wage law can be written as a multiple of the workers' reservation wage, neither of which depends on M nor p^e, using the constant elasticities of production and labor supply. From (3.5.41) one has

$$\mathcal{W}(M, p^e, \lambda) = \left(1 + \lambda\frac{C(1-B)+1}{BC}\right)W_{\Omega}(p^e, \mathcal{L}(M, p^e, \lambda)) \tag{3.6.10}$$

which, using (3.6.9) and again (3.6.4), implies both

$$0 < E_{\mathcal{W}}(M) = E_S(L)E_{\mathcal{L}}(M) = \frac{E_{\mathcal{P}}(M)}{C(1-B)+1} < E_{\mathcal{P}}(M) < 1$$

and

$$0 < E_{\mathcal{W}}(p^e) = 1 - E_S(L)E_{\mathcal{L}}(p^e) = 1 - \frac{1 - E_{\mathcal{P}}(p^e)}{C(1-B)+1} < 1.$$

Therefore, we conclude that the wage elasticity with respect to money balances and price expectations are positive and less than unit-elastic.

3.6.1 The Role of Union Power

While union power determines uniquely the relative share $\lambda/(1-\lambda)$ of labor income to profits as a monotonically increasing function in λ, its impact on the other employment–wage related equilibrium values is not necessarily monotonic due to the price feedback. For the wage law

$$\mathcal{W}_{\text{bar}}(M, p^e, \lambda) = W\left(p^e, \lambda, \mathcal{P}_{\text{bar}}(M, p^e, \lambda), \mathcal{L}_{\text{bar}}(M, p^e, \lambda)\right),$$

one finds from (3.5.43) that the nominal wage is proportional to the firm's average nominal labor productivity,

$$\mathcal{W}_{\text{bar}}(M, p^e, \lambda) = \left(\frac{BC}{C+1} + \lambda\frac{C+1-BC}{C+1}\right)\frac{pF(L)}{L}. \tag{3.6.11}$$

While the term in parenthesis is monotonically increasing in λ and independent of the state variables (M, p^e), the nominal labor productivity itself with $p = \mathcal{P}_{\text{bar}}(M, p^e, \lambda)$ and $L = \mathcal{L}_{\text{bar}}(M, p^e, \lambda)$ is not necessarily increasing in λ. Therefore, due to the price feedback, the nominal wage is not necessarily an increasing function in union power λ. However, from the above equation it follows that the equilibrium real wage

$$\alpha = \frac{w}{p} = \frac{\mathcal{W}_{\text{bar}}(M, p^e, \lambda)}{\mathcal{P}_{\text{bar}}(M, p^e, \lambda)} = \left(\frac{BC}{C+1} + \lambda\frac{C+1-BC}{C+1}\right)\frac{1}{B}F'(\mathcal{L}_{\text{bar}}(M, p^e, \lambda))$$

is a constant multiple of the marginal product of labor, where the constant is an increasing linear function of λ and independent of demand parameters. Thus, in the isoelastic case, the parameter λ determines the *mark-up of the real wage over the marginal product of labor*, which is independent of the state variables M and p^e and of all fiscal and demand parameters. Nevertheless, the latter do affect the temporary equilibrium prices and wages as well as the allocation.

Concerning the nominal payoff, an increase in union power always increases the payoff of the union while decreasing the firm's profit, as shown in Figure 3.42. There the ranges of the firm's profits, the union's utilities, and the total wage bill (both in nominal and in real terms) are depicted as functions of union power. Notice that the

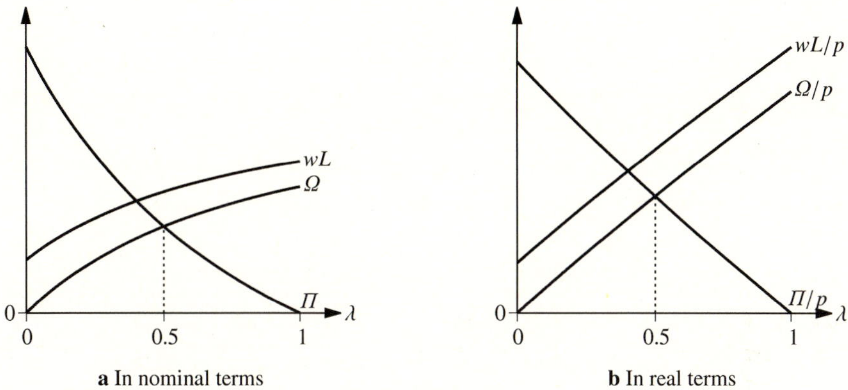

Fig. 3.42 Range of profits, utilities, and wage bill for λ from 0 to 1

share in total output Π/py is linear in λ while the real profit Π/p is not (panel **b**).

Finally, the rate of underemployment can be calculated explicitly using the wage law and the price law. Because of

$$\frac{w}{p^e} \overset{(3.5.41)}{=} \left(1 + \lambda\frac{C(1-B)+1}{BC}\right) S(L) = \left(1 + \lambda\frac{C(1-B)+1}{BC}\right) \frac{C}{C+1} S_{\text{com}}(L), \tag{3.6.12}$$

the rate of underemployment can be simplified since N and S are isoelastic. This implies

$$\begin{aligned} \mathcal{U}(M, p^e, \lambda) = \mathcal{U}\left(L, \frac{w}{p^e}\right) &= 1 - \frac{L}{N_{\text{com}}(w/p^e)} \\ &= 1 - \left(\left(1 + \lambda\frac{C(1-B)+1}{BC}\right) \frac{C}{C+1}\right)^{-C} \frac{L}{N_{\text{com}}(S_{\text{com}}(L))} \\ &= 1 - \left(\left(1 + \lambda\frac{C(1-B)+1}{BC}\right) \frac{C}{C+1}\right)^{-C}. \end{aligned} \tag{3.6.13}$$

Thus, with isoelastic production and utility functions, the equilibrium rate of underemployment is a constant determined by union power and by labor market parameters, i.e. by supply side factors only. It is totally independent of the state of the economy (M, p^e) and of fiscal and demand parameters. It is an increasing function of union power. Therefore, high λ imply positive voluntary underemployment and low imply negative voluntary underemployment. Its range is given by the interval

$$\left[1-\left(\frac{C+1}{C}\right)^C, 1-B^C\right].$$

In addition, one obtains that for the bargaining weight

$$\lambda_{\text{nat}} \equiv \frac{B}{C(1-B)+1},$$

for which the competitive equilibrium is obtained, as the zero of (3.6.13), i. e.

$$\mathcal{U}(M, p^e, \lambda_{\text{nat}}) = 1 - \left(\left(1 + \frac{B}{C(1-B)+1}\frac{C(1-B)+1}{BC}\right)\frac{C}{C+1}\right)^{-C} = 0.$$

Thus, $\lambda_{\text{com}} \equiv \lambda_{\text{nat}}$ is independent of the state (M, p^e) and of all demand parameters.

The next diagrams portray the influence of union power on output, prices, and wages for the isoelastic case using the values of the parameters given in Table 3.1 which were chosen as a benchmark. All diagrams of this section are drawn to scale.

Table 3.1 Standard parameterization

A	B	C	$\tau_\pi = \tau_w$	λ	M	g	p^e	c	n_w	n_f
1	0.5	0.5	0.25	0.5	1	1	1	0.5	1	1

Figure 3.43 **a** depicts the equilibrium situation as the intersection of aggregate demand and aggregate supply, exploiting the fact that the union power has no effect on the aggregate supply curve. Thus, provided that there is no additional expectations feedback in aggregate demand, the influence of higher λ on the temporary equilibrium operates exclusively through the income distribution which causes a negative (downward) shift of the aggregate demand function (see equation (3.5.26)). This induces lower prices which then lead to lower employment and lower output.

3.6.2 Union Power and Wages

It is quite a remarkable fact that the isoelastic situation allows to determine a number of the equilibrium outcomes as explicit formulas. In particular, they reveal that the isoelastic relationships of the labor market (i.e. the supply side of the economy) under efficient bargaining determine constant relative results (i.e. markups, unemployment rates, etc.) which are independent of the parameters of the demand side of the economy and most importantly of the state variables (M, p^e). Unfortunately, such a property does not carry through to the determination of the nominal bargaining wage, since the impact of union power on the nominal wage is more involved than all of the previous comparisons.

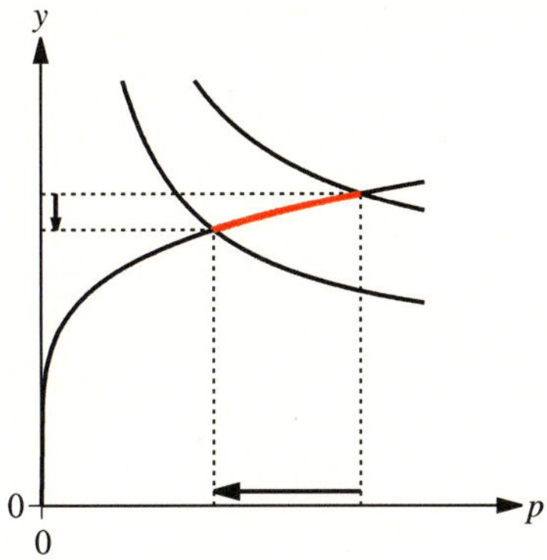

a Prices and output

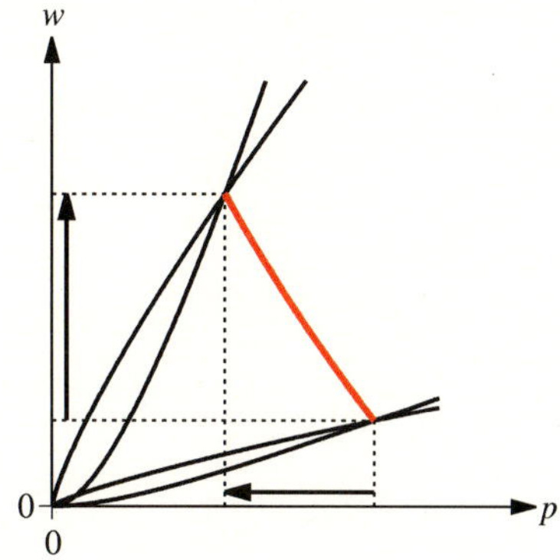

b Prices and wages

Fig. 3.43 Range of output, prices, and wages for λ from zero to one

Figure 3.43 **b** shows the range of the equilibrium price and of the bargaining wage (the red curve) in temporary equilibrium for λ between zero and one, for values of the parameters where wages are monotonically increasing. The diagram has been augmented by the graphs of two functions (the black curves) which represent the market clearing conditions under bargaining for the labor market and the commodity market separately, each parametrized by the commodity price p. To derive their properties, consider first the wage equation (3.6.11) in the isoelastic case with *employment consistency* (labor market equilibrium) only, i.e. with $L = h(\theta^e)$. For given p^e, this implies the bargaining wage

$$LE(p,\lambda) := \left(\frac{BC}{C+1} + \lambda\frac{C+1-BC}{C+1}\right)\frac{pF(h(\theta^e))}{h(\theta^e)}, \tag{3.6.14}$$

for each commodity price, which is taken as given by workers as well as by the producer. The properties of F and h imply that the function LE is strictly increasing and strictly concave in p. In addition, since h is independent of λ, the employment-consistent bargaining wage LE is strictly increasing in λ as well.

Similarly, for *commodity-market* consistency, $F(L) = D(M/p,\lambda)$ must hold. Therefore, inserting the aggregate demand function for the isoelastic case from (3.5.26), one obtains an induced price–wage relation under commodity market equilibrium

$$CE(p,\lambda) := \left(\frac{BC}{C+1} + \lambda\frac{C+1-BC}{C+1}\right)\frac{pD(M/p,\lambda)}{F^{-1}(D(M/p,\lambda))}. \tag{3.6.15}$$

With isoelastic functions of consumers and the producer, one finds that the function CE is increasing and convex in p and it is also increasing in λ. Clearly, the intersection of the graphs of the two functions LE and CE defines the temporary equilibrium pair (p, w), which follows also from the equality of aggregate supply and aggregate

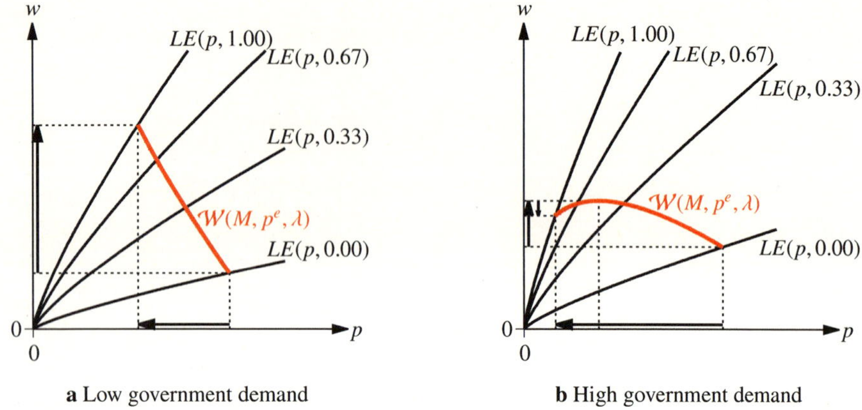

Fig. 3.44 The role of government demand on prices and wages for λ from zero to one

demand

$$AS(\theta^e) = F(h(\theta^e)) = D(M/p, \lambda),$$

which is equivalent to equating (3.6.14) and (3.6.15). As shown above, λ shifts both wage functions upward always decreasing the equilibrium price. However, the impact of union power on the equilibrium bargaining wage may still be ambiguous, depending on whether the demand effect dominates the supply effect. Nevertheless, the associated real wage must always be increasing in λ.

Figure 3.43 **b** portrays a situation of a negatively-sloped price–wage curve, implying a monotonic *increase* in nominal wages as λ changes from zero to one. However, there are situations where the equilibrium bargaining wage is not always monotonically increasing in union power λ. Figure 3.44 displays the effect of union power for two different levels of money balances with isoelastic functions and given elasticities. For low levels of money balances or government demand (left panel), the wage is globally increasing whereas for high levels, the wage is increasing initially reaching a maximum for some critical level $0 < \lambda < 1$ and then declines with further increases of union power (right panel). The reason for the reverse effect, arises from the fact that the elasticity of the price law cannot be constant as long as government demand is positive *and* that it is a function of money balances. Thus, the level of money balances and of government demand could be potential reasons for the decline in wages.

In order to understand this effect, one may consider the elasticity of the price law and its impact on the wage law. If one computes the elasticity of the wage law (3.6.10) with respect to union power

$$E_{\mathcal{W}}(\lambda) = \underbrace{\frac{(C+1-BC)\lambda}{(C+1-BC)\lambda + BC}}_{\text{from the mark-up}} + \underbrace{\frac{E_{\mathcal{P}}(\lambda)}{C(1-B)+1}}_{\text{from } W_{\Omega}},$$

one obtains two distinct effects. The parameter λ affects the workers' reservation wage negatively, but it affects the scaling factor positively. For wages to decrease in union power, the latter needs to be outbalanced by the reservation wage effect. Let us first show that this cannot occur when government demand is equal to zero. Using the explicit form of the inverse of the price law (3.6.8), one also obtains an explicit form of the inverse with respect to λ given by

$$\lambda = \Lambda(M, p^e, p) := \frac{1}{\tilde{c}}\left(\left(\frac{p^e}{\tilde{A}}\right)^{\tilde{B}} \frac{M/p + g}{p^{\tilde{B}}} - (1-\tilde{c})\right) \tag{3.6.16}$$

with

$$\tilde{A} = AS^{-1}(1), \qquad \tilde{B} := \frac{BC}{C+1-BC}, \qquad \text{and} \qquad \tilde{c} := c(1-\tau_\pi)\left(1 - \frac{BC}{C+1}\right).$$

The function Λ is strictly decreasing in p with elasticity greater than minus one. Therefore, $|E_{\mathcal{P}}(M, p^e, \lambda,)| = |1/E_{\Lambda}(M, \mathcal{P}(M, p^e, \lambda))| > 1$ in general.

For $g = 0$, one obtains from (3.6.16)

$$E_{\Lambda}(\lambda) = -(1+\tilde{B})\frac{Mp^{-(1+\tilde{B})}}{Mp^{-(1+\tilde{B})} - (1-\tilde{c})(\tilde{A}/p^e)^{\tilde{B}}}.$$

Solving for $Mp^{-(1+\tilde{B})}$ from (3.6.16) and substituting implies

$$E_{\Lambda}(\lambda) = -(1+\tilde{B})\frac{1-\tilde{c}+\lambda\tilde{c}}{\lambda\tilde{c}}$$

and

$$E_{\mathcal{P}}(\lambda) = -\frac{\lambda\tilde{c}}{(1+\tilde{B})(1-\tilde{c}+\lambda\tilde{c})}.$$

Thus, $E_{\mathcal{P}}(\lambda)$ is monotonically decreasing in λ with $E_{\mathcal{P}}(0) = 0$ and

$$-1 < E_{\mathcal{P}}(1) = -\frac{\tilde{c}}{(1+\tilde{B})} < E_{\mathcal{P}}(0) = 0. \tag{3.6.17}$$

Therefore, the wage elasticity is positive for all (M, p^e, λ). Moreover,

$$\begin{aligned} E_{\mathcal{W}}(\lambda) &= \frac{\lambda(C(1-B)+1)}{BC + \lambda(C(1-B)+1)} + \frac{E_{\mathcal{P}}(\lambda)}{C(1-B)+1} \\ &= \frac{\lambda(C(1-B)+1)}{BC + \lambda(C(1-B)+1)} - \frac{1}{C+1-BC}\frac{\lambda\tilde{c}}{(1+\tilde{B})(1-\tilde{c}+\lambda\tilde{c})}. \end{aligned}$$

is the difference of two concave and increasing functions in λ with $E_{\mathcal{W}}(0) = 0$ and

$$E_{\mathcal{W}}(1) = \frac{C(1-B)+1}{C+1} - c(1-\tau_\pi)\frac{C+1-BC}{(C+1)^2} > 0.$$

Thus, by continuity, the wage elasticity is also positive for large λ and for all $g > 0$ small. With this information, one can now identify situations numerically were a higher government demand g may lead to a negative elasticity of wages with respect to union power. The properties shown are qualitatively identical in a large neighborhood of the benchmark values. However, for large government demand, one obtains a negative wage effect as displayed in Figure 3.44**b**. The reason for this effect lies primarily in the impact of g on the elasticity of the aggregate demand function. For $g > 0$, one finds that it is an increasing function which becomes less elastic for higher prices such that

$$-1 < E_D(p) := -E_D(M/p) = -\frac{\partial D(M/p,\lambda)}{\partial(M/p)}\frac{M/p}{D(M/p,\lambda)} = -\frac{M/p}{M/p+g} < 0.$$

It seems that this increase of the price elasticity together with the change of the income distribution as λ increases eventually induces the reversal effect for the wage law.

3.6.3 Comparing Labor Market Scenarios

It may be of interest to further analyze the allocative differences between the cooperative bargaining model with the noncooperative wage determination models discussed so far and compare the consequences of the spillover effects between the commodity market and the scenarios in the labor market for the isoelastic situation. In order to understand the influence of the implicit price feed back which operates in all four cases (including the competitive equilibrium), it is useful to construct the set of feasible (individually rational) bargaining agreements between the union and the producer *including* the price feed back.

Let $(L, w) \geq 0$ denote an arbitrary labor market agreement. Given the restrictions of non negativity of payoffs, (L, w) is called individually rational for a given price and price expectations (p, p^e) if

$$\Omega(L,w,p) = wL - p^e S(L)L \geq 0 \qquad \text{and} \qquad \Pi(L,w,p) = pF(L) - wL \geq 0. \tag{3.6.18}$$

An agreement (L, w) is called income-demand consistent at p if

$$pF(L) = M + pg + c(1-\tau_\pi)(pF(L) - wL), \tag{3.6.19}$$

which imposes further restrictions on feasibility and on the equilibrium price p. Non-negativity of profit implies that feasible employment levels have to satisfy $F(L) - g \geq 0$.

Given the form of the aggregate demand function (3.6.19), one can solve for the associated price explicitly to obtain

$$p(L, w) := \frac{M - c(1 - \tau_\pi)wL}{F(L)(1 - c(1 - \tau_\pi)) - g}, \qquad F(L)(1 - c(1 - \tau_\pi)) - g \neq 0, \quad (3.6.20)$$

which must be positive for any $(L, w) \gg 0$. This implies

$$\begin{aligned} \Pi(L, w) :=& p(L, w)F(L) - wL = \frac{M - c(1 - \tau_\pi)wL}{F(L)(1 - c(1 - \tau_\pi)) - g} F(L) - wL \\ =& \frac{1}{F(L)(1 - c(1 - \tau_\pi)) - g}(MF(L) - wL[F(L) + g]). \end{aligned} \quad (3.6.21)$$

The profit function (3.6.21) has a discontinuity at $L_{\text{crit}} = F^{-1}(g/(1 - c(1 - \tau_\pi)))$,

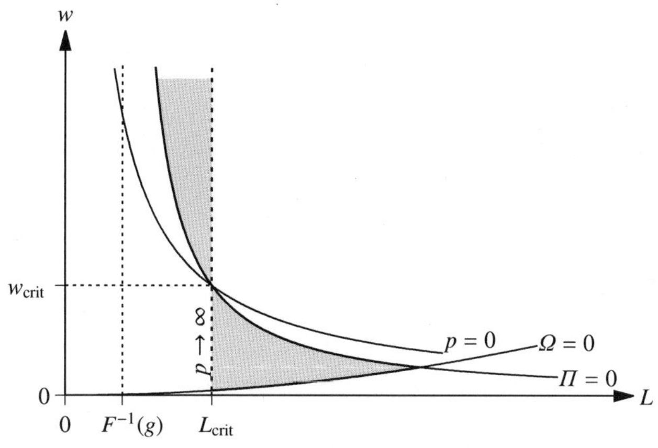

Fig. 3.45 Employment–wage pairs under individual rationality and feasibility

which is the critical level where the denominator of the price function is zero and changes sign, and where the price and profit become infinite. Thus, the set of bargaining pairs (L, w) with *positive* profit consists of the union of two disjoint open regions allowing unbounded wages for $L < F^{-1}(g/(1 - c(1 - \tau_\pi)))$ and unbounded employment levels[23]. As a consequence one finds that the set of individually rational and income-demand consistent employment-wage pairs takes the form of a union of two adjoining sets as depicted in Figure 3.45.

[23] Strictly speaking, the set also contains the boundary point $(F^{-1}(g/(1 - c(1 - \tau_\pi))), w_{crit})$, since there exists an unbounded interval of positive prices which induce positive profits, see Figure 3.45.

The two critical employment levels, which are the same for each state of the economy (M, p^e), are determined by demand features and the production function. They are independent of money balances. However, high price expectations may make the lower compact curvilinear triangle empty, implying that all equilibrium allocations must be in the upper region of feasibility. Since unbounded wages with unbounded prices are feasible income-demand consistent equilibrium allocations for employment levels near the upper critical level, the associated set of pay offs must be unbounded and is equal to all of $\mathbb{R}^2_+$.

Bargaining, Competition, and Monopolies

The set of bargaining equilibria as a projection to any two dimensional space (whether for (L, w), or (p, w), or (Ω, Π)) must consist of a line for $0 \leq \lambda \leq 1$, with the competitive equilibrium being an interior point for $o < \lambda_{\text{com}} < 1$. In contrast, the two one sided strategic monopolistic situations (union and monopsonist) cannot coincide generically with either of the boundary cases of a strong union (i.e. $\lambda = 1$) or of the weak union (i.e. $\lambda = 0$) under bargaining, since each of the strategic monopolistic equilibria induces one sided rents (see Figures 3.28 and 3.31). Both of them induce inefficient employment levels below the efficiency frontier which implies a displacement away from the 'efficiency line'.

This distinction becomes most apparent in Figure 3.46 **a** and **b** which portrays the price-wage configurations for the two boundary cases of a weak union ($\lambda = 0$) compared with the monopsonistic firm and of the strong union ($\lambda = 1$) under bargaining compared with the outcome under the monopolistic union. The intersection of the three black curves marks the competitive equilibrium as the intersection of the LE, CE, and PWF curves respectively, while the intersection of the two red curves is the equilibrium price-wage pair under monopsonistic wage setting. Thus, equilibrium prices will be higher and wages will be lower under monopsonistic wage setting than under perfect competition. The diagrams reveal that the price and wage effects under bargaining show the highest real wage under the powerful union and bargaining, higher than the monopolistic wage, while the lowest real wage is generated for the weakest union under bargaining, which is lower than under the monopsonistic firm.

Since the payoffs are continuous functions of the employment-wage-price allocations, the equilibrium payoffs under the pair-wise comparison between these labor market scenarios can also be drawn as solutions/intersections of the respective 'reaction curves' derived in the respective Section 3.3. Figure 3.47 **a** and **b** show the impact of monopolistic/monopsonistic behavior in comparison to the competitive case. Both diagrams reveal that the one sided strategic behavior of each group increases its own pay off also in the general equilibrium outcome with price feed back. This effect is the most striking for the monopsonistic firm/employer (subfigure **a**), where profits are substantially higher while the wage bill is lower as compared to competition. However, subfigure **b** shows for the union monopoly, that both the wage bill as well as profits are higher compared to the competitive equilibrium.

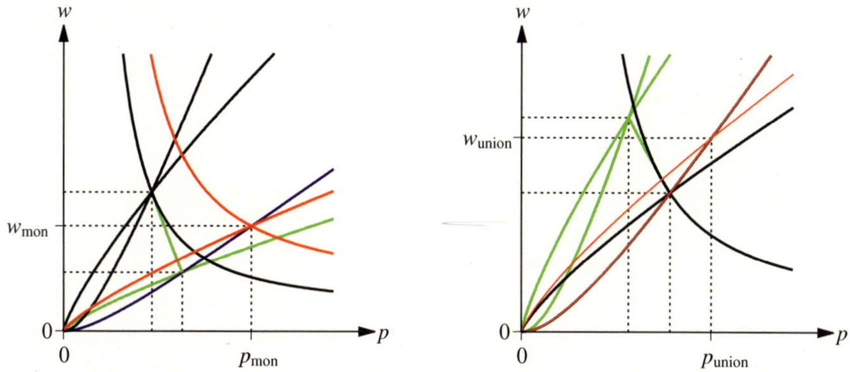

a the monopsonistic firm vs. bargaining: $\lambda = 0$

b the monopsonistic union vs. bargaining: $\lambda = 1$

Fig. 3.46 Comparing prices and wages: Competition (black), bargaining (green)

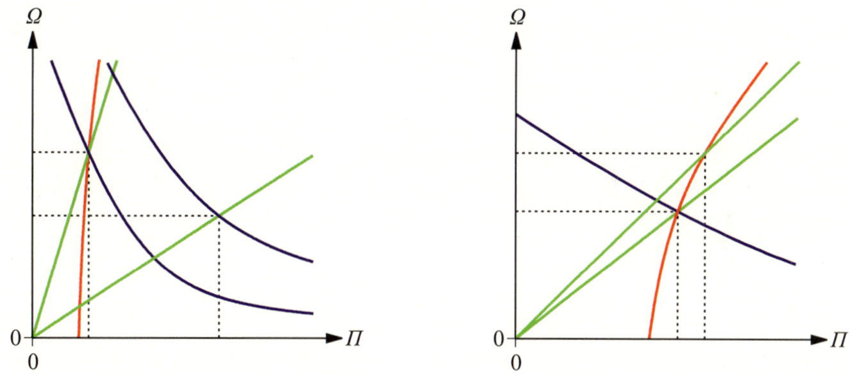

a Monopsonistic employer vs. competition

b Union monopoly vs. competition

Fig. 3.47 Comparing payoffs; *CE* (red), *LE* (green), *PWF* (blue). Three intersecting lines: competition, two intersecting lines: Monopsonistic firm resp. monopolistic union

Thus, the employer gains from union wage setting through a substantial price effect. This occurs in spite of the reduction of employment. However, both monopolistic scenarios induce inefficient employment situations, as shown in Figures 3.28 and 3.31.

By adding the equilibrium points and the λ-efficiency line to the (L, w) consistency diagram Figure 3.45, one obtains in Figure 3.48 a comparison of all scenarios in allocation space and in payoff space. The diagrams are again drawn to scale for the values of the parameters in Table 3.1. For these values, all equilibria are in the compact lower 'triangular' region of the employment-wage space for the chosen

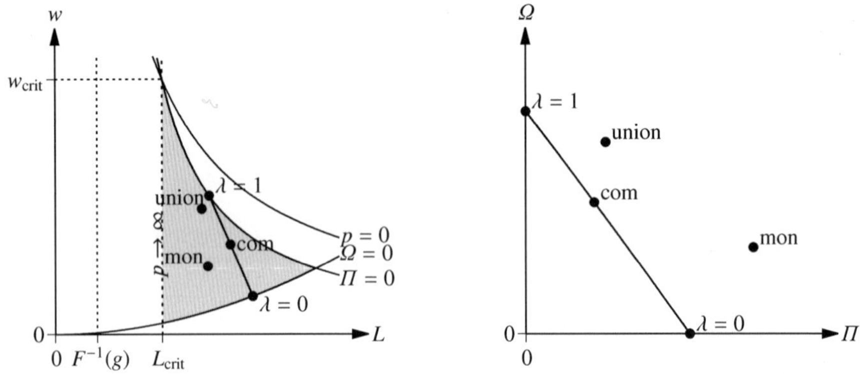

a Employment–wage pairs **b** Payoffs

Fig. 3.48 Union–Monopsony versus efficiency under low price expectations

level of price expectations. The diagram confirms the location of the two one sided monopolistic equilibria *above* the λ-bargaining frontier. In other words, both monopolistic equilibria induce better payoffs which cannot be reached or supported by the cooperative decisions under efficient bargaining. Nevertheless, all equilibrium payoffs can be improved by some employment-wage decision with associated competitive market clearing on the commodity market. When expected prices are high, the lower consistency triangle (see Figure 3.45) is empty, so that all equilibria must belong to the upper region, a situation which is portrayed in Figure 3.49.

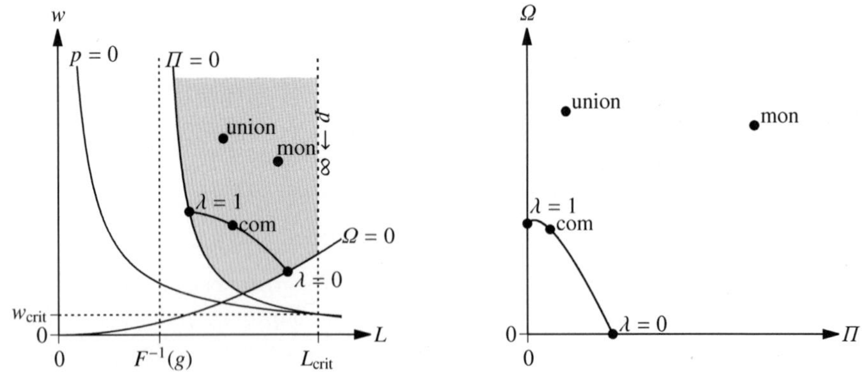

a Employment–wage pairs **b** Payoffs

Fig. 3.49 Union–Monopsony versus efficiency under high price expectations

The results in both cases are drawn for given values of the government parameters and for given values of the state variables money balances and expectations. Because of continuity, these features are locally robust properties and they will be observed for this isoelastic class of models in different magnitudes and possibly also in different relative orders under different parameters and values of the state variables. However, as some numerical experiments have shown, the basic features are preserved for a wide range of values of the parameters and of state variables. The overall homogeneity of the price law and the wage law does not preclude reversals or opposite effects.

While some of these result might seem to be counter intuitive at first sight, it is straightforward to discern the two principal reasons why these effects occur. First of all, the maximization of nominal objectives (profit resp. excess wages) creates spillovers between markets even for static general equilibrium systems, which are primarily due to income effects. Because of these income effects, it is unlikely that the same strong comparative statics results (as often derived in partial equilibrium models with strategic behavior) will persist in general equilibrium models. It is known from general equilibrium theory that such effects are due to price normalization, implying different real allocations, relative prices, and nominal values of incomes (profits and wages) under different choices of a numéraire or of price indexes. These results are well documented and have been recognized in many different contexts in particular in welfare economics, international trade, or oligopoly theory whenever income feed backs are taken into account appropriately with a non constant marginal utility of income for consumers (for example Roberts & Sonnenschein, 1976; Dierker & Grodal, 1986; Böhm, 1994; Gaube, 1997; Kempf, 1999). In temporary equilibrium of a monetary economy these effects clearly do not disappear.

Second, the price feed back, which was shown to be responsible for the inefficiency of the bargaining solution under competitive price taking in temporary monetary equilibrium, operates in each of the three cases endogenously in a different way. There is no structural feature of the model which relates the nominal payoffs, chosen for the bargaining problem neither to the nominal objectives by the monopolist/monopsonist with wage setting and price taking nor to the results induced by the maximization under competitive price and wage taking. Thus, in all three cases, the price feed back and the income feed back have a decisive influence on the nominal values chosen for the payoffs in the monetary economy. For these reasons, the four labor market scenarios whose equilibrium characteristics are compared in the price-wage space and in payoff space, are in general not comparable with respect to real allocations or real payoffs, even under the weak concept of efficiency. Since, in addition, equilibrium prices and allocations depend on the other state variables, an extensive welfare analysis may not lead to conclusive results.

Some features of the results are specific to the isolelastic model since the bargaining parameter λ plays a specific dual role in temporary equilibrium. On the one hand, since there is no impact of union power on aggregate supply, the interaction of the isoelastic structure between production and labor supply shows that the measure of union power λ exerts a direct influence on the real wage markup and on the level

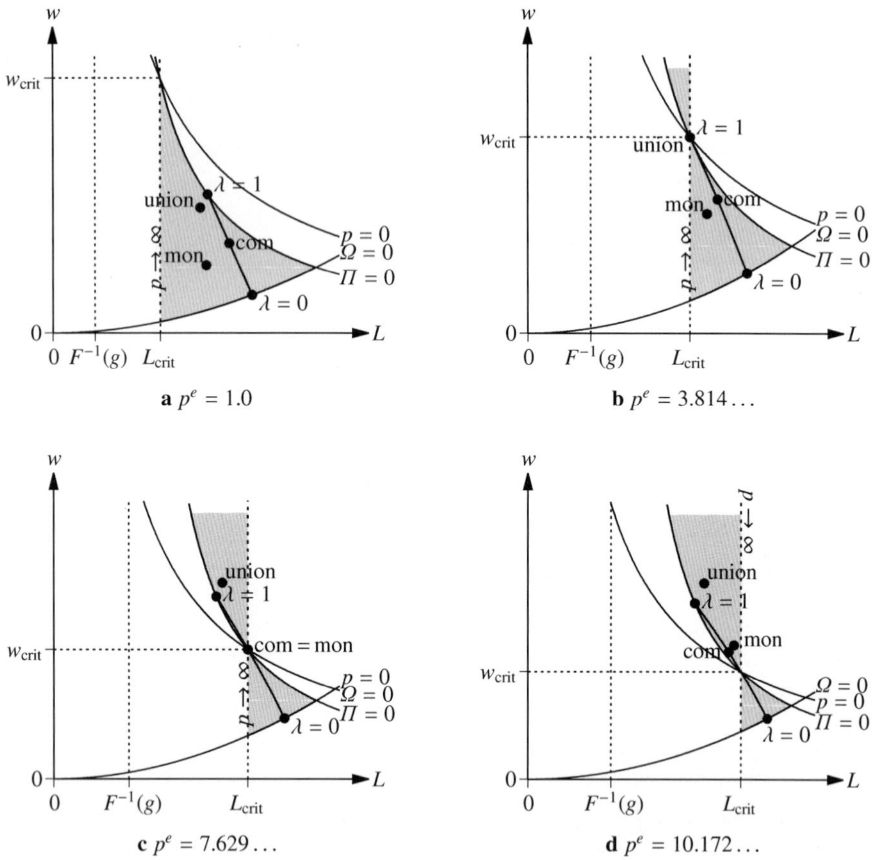

Fig. 3.50 Union–Monopsony versus efficiency: low price expectations increasing from **a** to **d**

of underemployment making both of which are constant in temporary equilibrium. These constants depend on the parameters of the labor markets participants only and they are controllable by the union. Thus, in a dynamic economy, both of them are constant over time (independent of the state (M, p^e)), and they are independent of fiscal and demand parameters in the economy. On the other hand, a powerful union which can chose the parameter λ does not exert absolute control over its seemingly most important endogenous variable the wage rate. Even for the isoelastic case, it seems unclear whether the wage outcome under bargaining dominates the competitive outcome, in some other sense than the efficiency criterion used above. It remains an open question to what extent the inefficiencies will change or disappear if the bargaining agents chose 'real' rather than nominal payoffs as objectives.

One main property of the isoelastic model concerning the outcomes under different noncompetitive behavior consisted in the fact that characteristics of the demand side had little or no influence on the outcomes, i.e. money balances and government

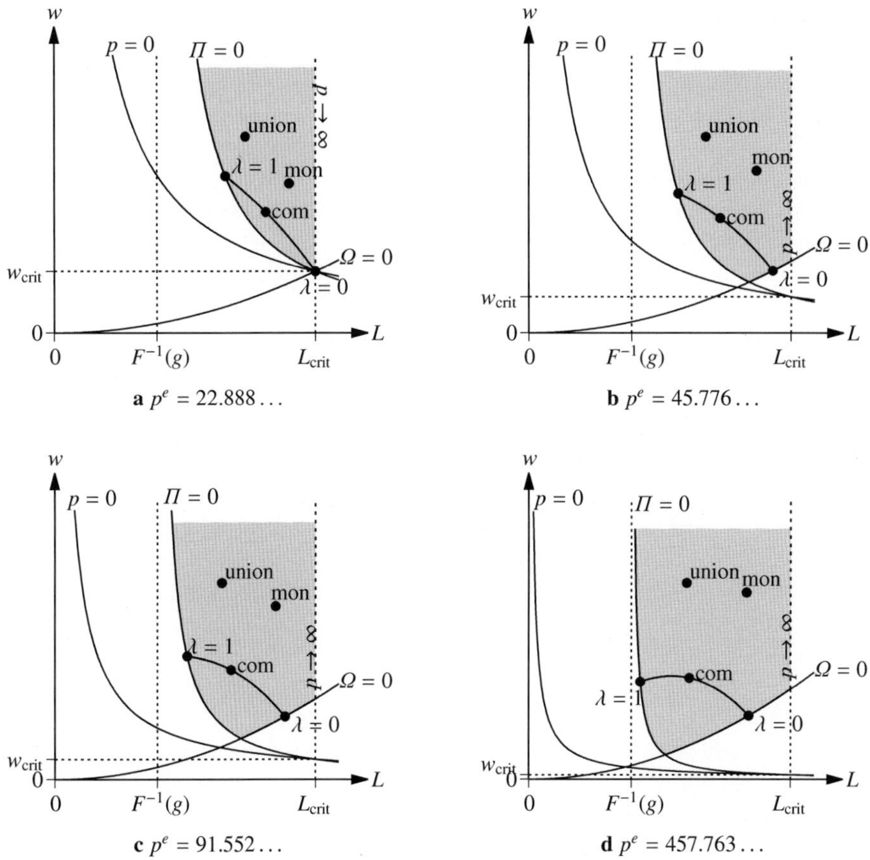

Fig. 3.51 Union–Monopsony versus efficiency: high price expectations increasing from **a** to **d**

parameters play a minor role in the determination of the relative positions of equilibria in different scenarios. However, as shown at the beginning of this section (see Figure 3.45), price expectations are crucial for the size and the position of the two sets of feasible/individually rational employment-wage pairs.

Figures 3.50 and 3.51 show for eight increasing successive levels of price expectations, how the absolute and relative positions of equilibria change and move from the lower triangle to the upper one as expected prices increase. Their values are chosen such that the efficiency line in all eight cases are given by a complete parametric curve in the feasible set for the value of union power $\lambda \in [0, 1]$. The diagrams are drawn to scale for the values of the parameters of Table 3.1 inducing a constant critical employment level L_{crit} as defined by equation (3.6.20). The efficiency line has a negative slope for all lower levels of price expectations, changing its slope only for very high price expectations. The employment-wage configurations of the two noncooperative equilibria (union and monopsony) are below the efficiency line for low price expectations while they are above for high price expectations. In addition, as p^e increases, the efficiency line passes through the critical employment level for the strong union case $\lambda = 1$ imposing zero profit on the producer, which coincides as a boundary solution with the allocation of the monopolistic union (Figure 3.50 **b**). Similarly, Figure 3.51 **a** shows that the allocation for the weak union case ($\lambda = 0$) passes through the critical level of employment when $\Omega = \Pi = 0$, inducing no positive payoff for either group. positive payoffs require a positive λ. Further increases of price expectations may cause the efficiency line to change its curvature/slope (Figure 3.51 **d**) which may be declining for λ near one.

3.7 Wage Rigidities and Unemployment

The analysis of the consequences of one sided strategic behavior of market participants in Sections 3.3 and 3.4 revealed that the deviating behavior alone does not induce temporary allocations with involuntary unemployment. Under the standard convexity and smoothness properties typically assumed in a standard temporary equilibrium model underlying the macroeconomic model of the AS–AD type competitive or monopolistic equilibria exhibit only qualitatively different functional relationships of their respective equilibrium sets and their price laws. However, price and wage flexibility in the temporary equilibrium model without additional specifications or restrictions does preclude the possibility of disequilibria in either the commodity or the labor market. To observe disequilibrium allocations in a temporary situation in spite of the flexibility of prices and wages is equivalent to the fact that no price-wage pair $(p, w) \gg 0$ is a candidate for a competitive equilibrium. This requires weaker assumptions than those postulated. Therefore, except for boundary cases (with zero wages or infinite commodity prices) non-existence of temporary equilibria originates from different structural features not captured so far within the temporary equilibrium setting under the basic Assumption 3.2.1 or those discussed under weaker conditions of Section 3.2.8.

3.7.1 Sources of Rigidities

If prices or wages are restricted by assumption from taking specific ranges of values in a particular period stipulated, for example, by minimum wage laws, price caps or floors, it is easy to conceive of economies and states for which temporary equilibria would imply values outside the restricted ranges only. In such cases, nonexistence of market clearing situations at all allowable prices prevails and allocations under rationing as discussed in Chapter 6 could be described. When such rigidities prevail as temporary phenomena the question arises whether the imposed rigidities persist in every period, whether after the change of the state variables the rigidities are still binding, or whether there are further intertemporal adjustment mechanisms employed to overcome or abolish the rigidities altogether.

When discussing wage rigidities as a cause for unemployment in a particular period (or short run), it is always necessary to simultaneously address the dynamic question of how and why wages or prices adjust intertemporally. Then, in the dynamic context, it is possible to establish whether unemployment is a temporary or a transient phenomenon or whether it persists in the long run. Thus, the occurrence of rigidities in the long run, or more generally, – the non-convergence of states and prices to levels which do not induce trading at market clearing levels – identifies the persistence of disequilibrium as a failure of a *dynamic* adjustment mechanism and *not* as a property of a structural equilibrium configuration. Therefore, in the end, the persistence of rigidities signaling disequilibrium configurations requires a *dynamic* explanation. Chapter 7 analyzes the implications of a class of price and wage adjustments within a general dynamic analysis of disequilibria of the AS–AD model.

It is useful to identify the characteristics of the two causes of disequilibria – nonexistence and sluggish dynamic adjustments – in more detail and isolate them conceptually from a third cause so-called *endogenous rigidities* arising from structural modifications or extensions of AS–AD models while maintaining all features of the neoclassical Assumption 3.2.1.

Lack of existence of a positive equilibrium wage rate excludes full employment under flexibility of wages and prices altogether. In other words, there is no positive price-wage configuration under the market conditions with associated behavioral rules of agents which guarantees a positive level of employment with $L^D(\bar{w}) - L^S(\bar{w}) = 0$. This occurs in the competitive case when weaker conditions than those imposed by Assumption 3.2.1 hold. Then, typically boundary allocations may exist with prices or wages going to zero or infinity. A second reason for nonexistence appears when there is insufficient convexity to guarantee continuity of supply and demand schedules of agents.

Insufficient flexibility of nominal wages is often caused by *institutional* rules or social prescriptions which are exogenously predetermined and imposed on the market for labor.

- There may exist positive lower bounds stipulated by minimum wage laws.
- Real wages may not be sufficiently flexible downwards because of purchasing power guarantees, which under given commodity prices imply endogenous nominal lower bounds for wages.
- Wage or salary indexation according to a consumer price index may exclude the existence of a market clearing wage rate.

In such cases, the nominal wage rate may not exhibit sufficient downward flexibility within the labor market to match supply and demand. Such bounds are essentially institutional rigidities which restrict the allowable range for w to equalize supply and demand on the labor market, excluding the range for which a $\bar{w} > 0$ exists with $L^D(\bar{w}) - L^S(\bar{w}) = 0$. When such restrictions are removed or relaxed full employment allocations are obtained. Therefore, under the standard assumptions, equilibrium configurations coexist with the disequilibrium configurations induced by the insufficient flexibility for most or all states of the economy.

Endogenous rigidities arise from structural restrictions or modifications in an economy which are often unrelated to flexibility or rigidity issues. These are economically the more interesting ones since their ensuing implications are generated indirectly to the labor market through channels or causal effects which seem unrelated to the modification. If Assumption 3.2.1 is maintained the change of the supply or demand behavior of workers or firms induced by the modification implies that prices, wages, and allocations with unemployment appears excluding the competitive outcome. Such structural situations may induce nonconvexities for the optimizing behavior in the relevant ranges implying discontinuities or non-uniqueness of decisions by workers or producers. The discontinuities may also prevent equilibria to exist in spite of full flexibility of wages and prices. Typical situations when such effects may occur are described by the following conditions:

- Labor supply may be zero for some range of low positive wages because of the existence of a positive reservation wage arising

  - from outside options for workers which induce an unwillingness to work at low wages,
  - from the existence of unemployment insurance benefits or private wealth,
  - from expectations of future prices or wages when positive private wealth or assets are held by workers

  causing market wages to have an *endogenously* determined strictly positive lower or upper bound, below which (or above which) positive employment does not occur.
- Informational or behavioral aspects may restrict the individual willingness to exert work effort at low wages, implying a minimal wage to guarantee positive productivity by workers.

  - Setting lower bounds on wages by monopsonistic firms to prevent shirking reducing the productivity of an employed worker,
  - offering wage contracts to guarantee maximal effort or minimal shirking to avoid efficiency losses by firms .

- Conversely, firms may chose to pay higher wages to induce higher productivity of employed workers, as suggested within the class of efficiency-wage models, see Section 3.7.3.

Two modifications of the neoclassical model with endogenous rigidities are analyzed below. The first one shows that in the neoclassical model the existence of a system of unemployment compensation by itself creates a downward wage rigidity and allocations with unemployment while a full employment equilibrium with positive real wages exists when the level of unemployment compensation is reduced to zero. The second one examines the well-known efficiency-wage model for the wage within the AS–AD model. A real-wage rigidity appears endogenously causing involuntary unemployment when monopsonistic wage setting is used in addition to the efficieny-wage rule. Both of these results show that the endogenous rigidity manifests itself as a vanishing partial derivative of the price law or of the wage law in a subset of the state space.

3.7.2 Unemployment due to Unemployment Compensation

Nonexistence of equilibria under competitive behavior occurs most often (or almost exclusively) when competitive behavior induces discontinuities in aggregate demand or supply. These appear typically when positive so-called outside options for consumer-workers exist which carry a positive welfare/utility gain, i.e. when choices can be made against participation in the labor market. Consider the basic model of Chapter 3 and assume now that the government offers a fixed positive income transfer/compensation equal to $\Delta > 0$, if a worker is not employed. In this situation, it is clear that a worker with utility function $c_1 u(c_2/c_1) - v(\ell)$ will offer to work only if under market conditions $(p, w, p^e) \gg 0$ he will receive a utility at least as high as the utility he obtains by not working and receiving the unemployment compensation Δ which is equal to

$$\max\left\{c_1 u\left(\frac{c_2}{c_1}\right) - v(0) \mid pc_1 + p^e c_2 = \Delta\right\} = \frac{\Delta}{p} V(\theta^e) - v(0).$$

Then, for each $(p, w, p^e) \gg 0$, $\alpha := w/p$ and $\theta^e := p^e/p$, his attainable utility levels are given by two functions

$$\tilde{U}\left(\alpha, \theta^e, \ell, \frac{\Delta}{p}\right) := \begin{cases} \dfrac{\Delta}{p} V(\theta^e) - v(0) & \ell = 0 \\ \alpha V(\theta^e)\ell - v(\ell) & \ell \geq 0 \end{cases} \tag{3.7.1}$$

whose maximizers are discontinuous for each positive compensation level $\Delta > 0$ at $\ell = 0$ and some positive level $\alpha V(\theta^e)$. Let

$$\ell^\star(\alpha V(\theta^e), 0) = \arg\max_{\ell \geq 0}\{\alpha V(\theta^e)\ell - v(\ell)\} = (v')^{-1}(\alpha V(\theta^e)) \tag{3.7.2}$$

denote the individual competitive labor supply (for $\Delta = 0$) which is strictly increasing making the associated indirect utility

$$U^\star(\alpha V(\theta^e)) := \alpha V(\theta^e)\,\ell^\star(\alpha V(\theta^e)) - v(\ell^\star(\alpha V(\theta^e), 0)) \tag{3.7.3}$$

a continuous, strictly increasing, strictly convex function in $\alpha V(\theta^e)$, satisfying $U(0) = 0$ and the convex Inada conditions[24]. For $\Delta > 0$, a consumer-worker's utility-maximizing labor-consumption decision maximizes (3.7.1) implying a labor supply given by the correspondence

$$\ell^\star\left(\alpha V(\theta^e), \frac{\Delta}{p}\right) := \arg\max_{\ell \geq 0} \tilde{U}\left(\alpha, \theta^e, \ell, \frac{\Delta}{p}\right)$$

$$= \begin{cases} 0 & \text{if } \frac{\Delta}{p} V(\theta^e) > U(\alpha V(\theta^e)) \\ \{0, \ell^\star(\alpha V(\theta^e))\} & \text{if } \frac{\Delta}{p} V(\theta^e) = U(\alpha V(\theta^e)) \\ (v')^{-1}(\alpha V(\theta^e)) & \text{otherwise} \end{cases} \tag{3.7.4}$$

with continuous indirect utility

$$\max_{\ell \geq 0} \tilde{U}\left(\alpha, \theta^e, \ell, \frac{\Delta}{p}\right) = \max\left(\frac{\Delta}{p} V(\theta^e), U^\star(\alpha V(\theta^e))\right). \tag{3.7.5}$$

Since $U^\star$ is continuous, strictly increasing, and strictly convex there exists for every $(\theta^e, \Delta/p)$ a unique real wage $\alpha_{\min}$ for which $\frac{\Delta}{p} V(\theta^e) = U(\alpha_{\min} V(\theta^e))$ holds given by

$$\alpha_{\min}\left(V(\theta^e), \frac{\Delta}{p}\right) := \frac{(U^\star)^{-1}\left(\frac{\Delta}{p} V(\theta^e)\right)}{V(\theta^e)} \tag{3.7.6}$$

below which the consumer-worker is not willing to supply labor. The reservation real wage (3.7.6) is homogeneous of degree zero and non-decreasing in all three variables (p, p^e, Δ). However, it is not homogeneous of degree zero in (p, p^e) for fixed $\Delta > 0$. In other words, the price level *and* expected inflation θ^e play a separate role for each fixed nominal unemployment compensation $\Delta > 0$. Since individual labor supply (3.7.4) is discontinuous at $\alpha = \alpha_{\min}(V(\theta^e), \Delta/p)$ aggregate labor supply

$$N\left(\alpha V(\theta^e), \frac{\Delta}{p}\right) := n_w \ell^\star\left(\alpha V(\theta^e), \frac{\Delta}{p}\right) \tag{3.7.7}$$

is a discontinuous discrete valued correspondence at $\alpha = \alpha_{\min}(V(\theta^e), \Delta/p)$ implying that for a large range of prices and expectations (p, p^e) the labor market with a fixed number of workers $n_w > 0$ cannot clear in the usual way.

The existence of the reservation real wage $\alpha_{\min}$ from (3.7.6) implies that there exists a minimal nominal wage level

$$w_{\min}\left(\frac{V(\theta^e)}{p}, \Delta\right) := p\alpha_{\min}\left(V(\theta^e), \frac{\Delta}{p}\right) = \frac{(U^\star)^{-1}\left(\frac{\Delta}{p} V(\theta^e)\right)}{\frac{V(\theta^e)}{p}}. \tag{3.7.8}$$

[24] Lemma B.1.1 and the Envelop Theorem imply $(U^\star)'(\alpha V(\theta^e)) = \ell^\star(\alpha V(\theta^e), 0)$, so that $(U^\star)'' = (\ell^\star)'(\alpha V(\theta^e), 0) > 0$. The Inada conditions on the disutility function v induce $(U^\star)'(0) = 0$.

below which aggregate labor supply is equal to zero. Since $U^\star$ satisfies the convex Inada conditions its inverse is concave and satisfies the concave Inada conditions, so that $w_{\min}$ is increasing in p when V is not unit elastic and

$$\lim_{p\to 0} w_{\min}\left(V(\theta^e), \frac{\Delta}{p}\right) > 0, \qquad \Delta > 0.$$

In order to define a temporary equilibrium $(p, w) \in \mathbb{R}^2_+$ it is useful to adjust the labor market clearing condition appropriately and use the approach of Section 3.2.6 ultimately. For any (p, p^e), let $w = LE(p, p^e)$ denote the unique wage clearing the labor market for $\Delta = 0$, i.e. for all (p, p^e), $LE(p, p^e)$ solves

$$n_f h\left(\frac{LE(p,p^e)}{p}\right) = n_w \ell^\star\left(\frac{LE(p,p^e)}{p} V\left(\frac{p^e}{p}\right), 0\right)$$

and define a modified labor market clearing curve

$$LE^\Delta(p, p^e, \Delta) := \max\left\{ w_{\min}\left(\frac{V(\theta^e)}{p}, \Delta\right), LE(p, p^e)\right\}. \tag{3.7.9}$$

$LE^\Delta(p, p^e, \Delta)$ is non-decreasing in p. Because of the properties of $U^\star$, for given p^e, $\Delta > 0$, and small p,

$$LE^\Delta(p, p^e, \Delta) = w_{\min}\left(\frac{V(\theta^e)}{p}, \Delta\right) > LE(p, p^e) \equiv LE^\Delta(p, p^e, 0).$$

A temporary equilibrium is a pair (p, w) such that[25]

$$w = LE^\Delta(p, p^e, \Delta) = CE(M, p) \tag{3.7.10}$$

It is one with unemployment if

$$w = w_{\min}\left(\frac{V(\theta^e)}{p}, \Delta\right) > LE(p, p^e). \tag{3.7.11}$$

Then the number of employed worker-consumers is defined by

$$n := \frac{n_f h\left(\frac{w}{p}\right)}{\ell^\star\left(\frac{w}{p} V\left(\frac{p^e}{p}\right), 0\right)}, \qquad 0 < n < n_w. \tag{3.7.12}$$

In other words, a temporary equilibrium with unemployment is an equilibrium with positive employment for $0 < n < n_w$ workers. Thus, except for the fact that n may not be integer valued (a feature typically ignored in such models) $w_{\min}$ constitutes

[25] It is assumed here that the commodity market clearing relation $CE(M, p)$ defined in (3.2.44) is independent of Δ, increasing and homogeneous of degree one in (M, p) and strictly convex in p with the properties as derived from Assumption 3.2.1.

an equilibrium wage with a positive proportion $(n_w - n)/n_w$ of workers unemployed. On the other hand, the real wage is equal to the marginal product of labor for each producer making the aggregate level of employment equal to $L = n_f h(w/p)$.

The discontinuity of the aggregate labor supply function implies that a temporary equilibrium with *voluntary* unemployment caused by the unemployment compensation occurs at a level of wages endogenously determined as the reservation wage induced by the positive level of unemployment compensation. Assumption 3.2.1 assures that for every (M, p^e, Δ) the equilibrium pair (p, w) is unique. Therefore, the temporary equilibrium is characterized by associated mappings determining prices and wages as well as output and employment, i.e.

$$\begin{aligned} p &= \mathcal{P}^{\Delta}(M, p^e, \Delta), & w &= \mathcal{W}^{\Delta}(M, p^e, \Delta) \\ y &= \mathcal{Y}^{\Delta}(M, p^e, \Delta), & L &= \mathcal{L}^{\Delta}(M, p^e, \Delta). \end{aligned} \tag{3.7.13}$$

For $\Delta = 0$ these coincide with the competitive full employment case. They inherit

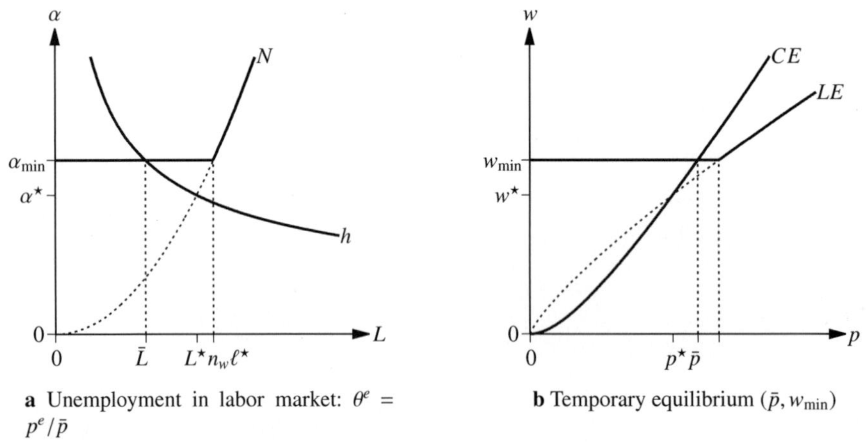

a Unemployment in labor market: $\theta^e = p^e/\bar{p}$

b Temporary equilibrium $(\bar{p}, w_{\min})$

Fig. 3.52 Role of high unemployment compensation $\Delta > 0$: (M, p^e) given

the usual homogeneity properties, i.e. prices and wages are homogeneous of degree one in (M, p^e, Δ) while output and employment are homogeneous of degree zero. Since the homogeneity properties pertain with respect to the vector of all three state variables, one finds that the nominal level of unemployment compensation has real effects which cause the occurrence of unemployment depending on the range of the two other state variables (M, p^e).

An equilibrium (p, w) with unemployment as in (3.7.11) has two main implications. First, the minimal wage rate $w_{\min}$ determines an *endogenous* downward nominal wage rigidity, below which positive employment is impossible given (Δ, p), in spite of the fact that there exists a hypothetical market clearing wage $0 < w^\star < w_{\min}$, which would be the equilibrium wage rate if there were no or a lower income guarantee Δ. This amounts to a failure of existence of a *uniform* labor market equilibrium in spite of the fact of full wage flexibility. Second, the feasible employment level at the given prices and expectations $(\bar{p}, p^e)$ corresponds to the level $\bar{L} = n_f h(w_{\min}/\bar{p}) < L^\star < n_w \ell^\star$. Figure 3.52 portrays a typical situation of an equilibrium $(\bar{p}, \bar{w})$ with voluntary unemployment. At the critical wage $\bar{w} = w_{\min}$, aggregate labor supply at the induced real wage is discontinuous, subfigure **a**. The employment level is $\bar{L} = n\ell^\star < L^\star < n_w \ell^\star$ implying an unemployment rate $(n_w - n)/n_w$.

The determination of the temporary equilibrium depends on a number of interdependent nonlinear effects between the two markets and the behavior of the agents which cannot be signed in all general cases under Assumption 3.2.1. To understand them in more detail and derive comparative statics properties, in particular to determine the regions in the state space where unemployment occurs, it is useful to analyze for the remainder of this section the isoelastic case with a worker-consumer whose propensity to consume when young is zero. Both subfigures of Figure 3.52 have been drawn under that assumption for which $V(\theta^e) = 1/\theta^e$, see Lemma B.1.1. Then, the critical wage (3.7.8) must satisfy

$$U^\star\left(\frac{w_{\min}}{p^e}\right) = \Delta \frac{w_{\min}}{p^e}.$$

This defines a linear relationship between Δ and the fixed point of the indirect utility at the expected real wage $w_{\min}/p^e$ taking the form

$$w_{\min}(p^e, \Delta) := p\,\alpha_{\min}\left(V\left(\frac{p^e}{p}\right), \frac{\Delta}{p}\right) = \frac{p}{V(\theta^e)}\left((U^\star)^{-1}\left(\frac{\Delta}{p}V(\theta^e)\right)\right) = p^e (U^\star)^{-1}\left(\frac{\Delta}{p^e}\right) \tag{3.7.14}$$

which is independent of the commodity price p. It is strictly increasing in Δ and has an elasticity

$$0 < E_{w_{\min}}(p^e) = \frac{E^\star_U(w_{\min}/p^e) - 1}{E^\star_U(w_{\min}/p^e)} < 1. \tag{3.7.15}$$

Since the isoelastic worker does not consume in the first period[26] aggregate commodity demand is independent of Δ, so that the function $CE(M, p)$ is increasing and homogeneous of degree one in (M, p), and strictly convex in p. It is obvious that the size of the unemployment compensation plays a role whether the associated temporary equilibrium is one with or without unemployment, since the minimum wage and the size of the discontinuity are decreasing functions of Δ. In other words, one finds that lowering the compensation sufficiently would reduce the level of unemployment or even eliminate the occurrence for an economy. Figure 3.53 shows an associated situation for the same isoelastic economy as in Figure 3.52. The two sets

---

[26] In general, aggregate demand would have to be augmented by the consumption expenditure $c_w(n_w - n)\Delta$ by unemployed workers.

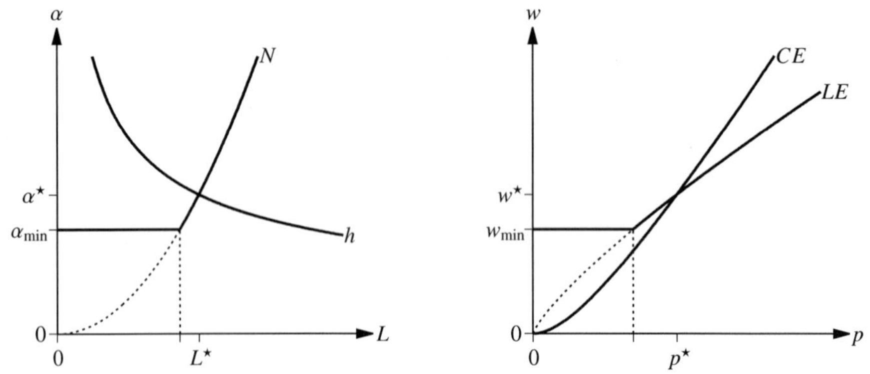

a No unemployment: $\theta^e = p^e/\bar{p}$

b No unemployment: $(\bar{p}, \bar{w}) = (p^\star, w^\star)$

Fig. 3.53 Role of low unemployment compensation $\Delta > 0$: (M, p^e) given

of diagrams are drawn to scale for the same parameter values as before, except for the level of the compensation Δ. They show that the existence of an endogenous downward wage rigidity does not necessarily induce equilibria with unemployment. In such cases the market clearing wage is above the minimum wage rate at the equilibrium price level so that full employment, i.e. positive employment of all workers prevails and no unemployment compensation is paid.

To identify the region of the state space with unemployment, let (p^e, Δ) be given. With $c_w = 0$ for the isoelastic consumer-worker this implies a constant nominal reservation wage $w_{\min}(p^e, \Delta)$ and a unique lowest price level at which full employment is possible (the kink in the LE^Δ-curve). The monotonicity and convexity of the commodity market clearing function $CE(M, p)$ and its independence from Δ guarantee the existence of a unique level of money balances $M = M_\star(p^e, \Delta)$ such that for each pair $(p^e, M_\star(p^e, \Delta))$ full employment prevails. In addition, for each fixed boundary point $(p^e_\star, M_\star(p^e_\star, \Delta))$, increasing $p^e_\star$ or decreasing $M_\star(p^e_\star, \Delta)$ by a small $\varepsilon > 0$ induces unemployment while decreasing $p^e_\star$ or increasing $M_\star(p^e_\star, \Delta)$ by $\varepsilon >$ maintains full unemployment. Formally, this means that

$$\begin{aligned}
\mathcal{P}^\Delta(p^e, M_\star(p^e,\Delta) + \varepsilon, \Delta) &= \mathcal{P}(p^e, M_\star(p^e,\Delta) + \varepsilon) &\Rightarrow&\quad \text{full employment}\\
\mathcal{P}^\Delta(p^e - \varepsilon, M_\star(p^e,\Delta), \Delta) &= \mathcal{P}(p^e - \varepsilon, M_\star(p^e,\Delta)) &\Rightarrow&\quad \text{full employment}\\
\mathcal{P}^\Delta(p^e, M_\star(p^e,\Delta) - \varepsilon, \Delta) &> \mathcal{P}(p^e, M_\star(p^e,\Delta) - \varepsilon) &\Rightarrow&\quad \text{unemployment}\\
\mathcal{P}^\Delta(p^e + \varepsilon, M_\star(p^e,\Delta), \Delta) &> \mathcal{P}(p^e + \varepsilon, M_\star(p^e,\Delta)) &\Rightarrow&\quad \text{unemployment.}
\end{aligned} \tag{3.7.16}$$

Furthermore, the homogeneity of the price law $\mathcal{P}^\Delta$ implies that for all $(p^e_\star, M_\star) = (p^e_\star, M_\star(p^e_\star, \Delta))$ and some small $\delta > 0$ one has

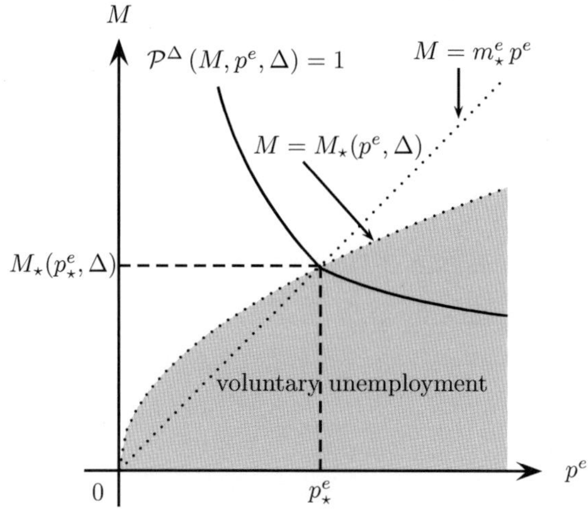

Fig. 3.54 Equilibrium set with unemployment compensation $\varDelta > 0$

$$\begin{aligned} \mathcal{P}^{\varDelta}((1+\delta)p^e_\star, (1+\delta)M_\star(p^e_\star, \varDelta)), \varDelta) \quad &\Rightarrow \quad \text{full employment} \\ \mathcal{P}^{\varDelta}((1-\delta)p^e_\star, (1-\delta)M_\star(p^e_\star, \varDelta)), \varDelta) \quad &\Rightarrow \quad \text{unemployment.} \end{aligned} \tag{3.7.17}$$

In other words, the function $M_\star(p^e, \varDelta)$ is concave (sublinear) in p^e. Its graph describes the boundary of the region between full employment (above) and unemployment (below), see Figure 3.54 where $m^e_\star := M_\star(p^e_\star, \varDelta)/p^e_\star$ denotes an arbitrary ratio of money balances and expected prices. In the shaded region the minimum real

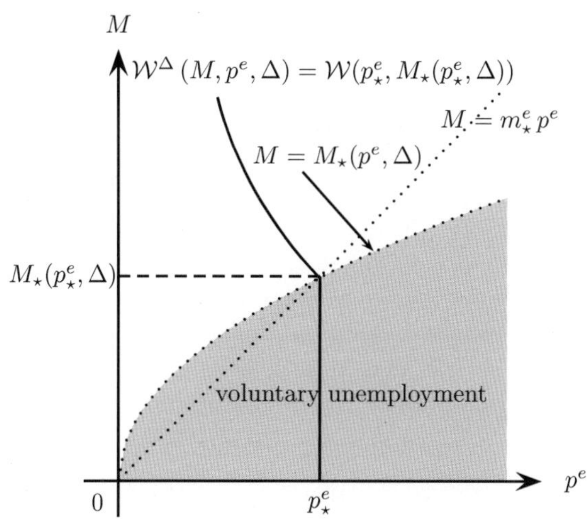

Fig. 3.55 Wage rigidity and unemployment: the contour of the wage law for $\varDelta > 0$

wage is a binding constraint for labor supply, whose value is endogenously determined and in a one-to-one monotonic relationship to the ratio M/p^e associated with full employment. For the isoelastic consumer-worker the real wage rigidity translates into a nominal wage rigidity which is best exhibited by plotting the contours of the wage law rather than the price law, see Figure 3.55. Since the minimum wage is independent of prices the contours of the wage law under unemployment are vertical in the space (p^e, M), i.e. the market wage is linear in money balances and constant for constant price expectations. Since the reservation wage is a decreasing function of the level of compensation the shaded part of the state space shrinks with a decrease in $\varDelta$ and is empty for $\varDelta = 0$.

3.7.3 The Efficiency-Wage Model

One basic characteristic of the standard neoclassical model of the labor market, built into the Assumption 3.2.1 and used throughout this chapter, assumes that a worker's productivity per hour – sometimes referred to as his *efficiency* – is constant, independent of the number of hours worked, and also independent of the wage paid per hour. The model simplifies and abstracts from productivity differentials arising from length of work days, from any worker-producer differences, and from other individual features causing changes of productivity per hour. The underlying microeconomic characteristics of productivity given the technological conditions of a producer as well as of the disutility of a worker are measured linearly within the framework of production functions and preferences (utility functions) respectively implying that the wage paid in the market is a wage per hour for the labor service of any individual worker being used in the given technology. While there are many models of the labor market assuming different varieties (skilled versus unskilled) and therefore different (heterogeneous) productivities of workers in order to analyze and explain the occurrence of differential wages these are modeled primarily as modifications of the technological conditions (production functions, usage of inputs, etc.) or differentials across worker types (skilled versus unskilled labor) maintaining the assumption of a constant productivity within each group.

The central idea of the so-called efficiency-wage model (as developed by Solow, 1979; Shapiro & Stiglitz, 1984; Akerlof & Yellen, 1985, 1986) consists of a specific modification that a worker's productivity per hour increases with an increase of the real wage paid due to a higher motivation or willingness to exert more effort when paid better. Formally, this is described by postulating the existence of an *efficiency function* $e : \mathbb{R}_+ \to \mathbb{R}_+$ which is assumed to be differentiable, strictly increasing, and bounded satisfying

$$e = e(\alpha), \qquad e(0) = 0, \qquad e'(\alpha) > 0 \tag{3.7.18}$$

where $\alpha := w/p$. Often a standard sigmoid function (convex-concave with a unique inflection point) is used which implies that average efficiency has a unique interior maximum to the right of the inflection point with the additional properties $1 > \lim_{\alpha\to 0} e(\alpha)/\alpha = e'(0) > 0$ and $\lim_{\alpha\to\infty} = 0$, see Figure 3.56 (a). The existence of such a function is taken as given while the other microeconomic features of the neoclassical model are maintained. No attempt is made here to derive the efficiency function from an extended model from underlying microeconomic principles.

Given a typical concave neoclassical production function F satisfying the usual conditions, an employment level $z \geq 0$ of the producer induces an output level

$$y = F(e(\alpha)z), \qquad (z, \alpha) \in \mathbb{R}^2_+$$

which is now a function of the two variables (z, α). Therefore, the profit for the producer for given prices and wages $(p, w) \gg 0$, becomes

$$\tilde{\Pi}(z, w, p) := pF\left(e\left(\frac{w}{p}\right)z\right) - wz = p\left[F(e(\alpha)z) - \alpha z\right] =: p\Pi(z, \alpha)$$

where Π is profit in terms of output.

Assume that the commodity market is competitive and each producer takes the commodity price $p > 0$ as given. Consider first the situation where the labor market is also competitive, so that $w > 0$ and therefore $\alpha > 0$ is given as well entering as parameters into the decision of each producer. For any given real wage $\alpha > 0$, the labor demand

$$h(\alpha) := \arg\max_{z\geq 0} p\left[F(e(\alpha)z) - \alpha z\right] \tag{3.7.19}$$

and the maximal profit

$$p\Pi(\alpha) := p\left[F(e(\alpha)h(\alpha)) - \alpha h(\alpha)\right] \tag{3.7.20}$$

are well defined positive functions, since F is strictly concave and satisfies the Inada conditions. Applying the envelope theorem, one finds that

$$\Pi'(\alpha) = \left(\frac{\alpha e'(\alpha)}{e(\alpha)} - 1\right)h(\alpha). \tag{3.7.21}$$

This shows that maximal profit must be first increasing and then decreasing in the real wage α, i.e. it must be a unimodal function since the elasticity of e is monotonically decreasing from values larger than one to zero making average efficiency unimodal as well. In other words, Π does not satisfy the usual duality properties, i.e. Π is not convex and $\Pi'(\alpha) \neq -h(\alpha)$.

The elasticity of the labor demand function

$$E_h(\alpha) = -\frac{1}{E_{F'}}\left(\frac{\alpha e'(\alpha)}{e(\alpha)}(1 + E_{F'}) - 1\right) \tag{3.7.22}$$

is obtained by applying the Implicit Function Theorem to the first order condition $e(\alpha)F'(e(\alpha)z) = \alpha$ for the maximization of equation (3.7.19). Thus, labor demand is unimodal as well but with a maximum at a real wage $\bar{\alpha} < \alpha^*$ before maximal real profit is reached. Thus, real profits as well as labor demand are decreasing functions only for real wages beyond a critical positive level determined by the efficiency function, see Figure 3.56. In order to obtain a more detailed understanding of the

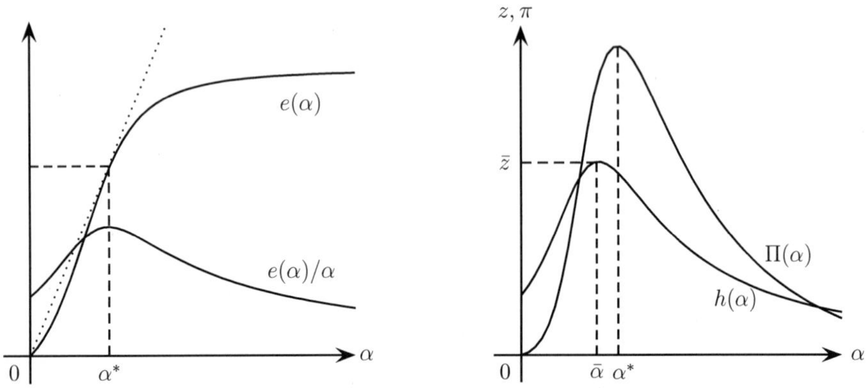

Fig. 3.56 The competitive producer with efficiency-wages

reasons for the different behavior of labor demand chose the wage bill (labor cost) $x := \alpha L$ as the maximizing variable and rewrite the profit function as

$$\Pi(\alpha) = \max_x F\left(\frac{e(\alpha)}{\alpha}x\right) - x. \tag{3.7.23}$$

This shows that profit maximizing *labor costs* depend uniquely and monotonically on the average efficiency $e(\alpha)/\alpha$. F being concave satisfying the Inada conditions, induces the first order condition and explicit global solution

$$\frac{e(\alpha)}{\alpha}F'\left(\frac{e(\alpha)}{\alpha}x\right) = 1 \qquad \text{if and only if} \qquad x = (F')^{-1}\left(\frac{1}{\frac{e(\alpha)}{\alpha}}\right)\frac{1}{\frac{e(\alpha)}{\alpha}}. \tag{3.7.24}$$

In other words, optimal labor costs are a monotonic function of average efficiency and exclusively determined by second order properties of the production function. Let $E_{F'}$ denote the elasticity of the marginal product F', then one obtains for the elasticity of the cost function

$$E_x\left(\frac{e(\alpha)}{\alpha}\right) = -\frac{1}{E_{F'}} - 1 = \frac{1 + E_{F'}}{-E_{F'}} > 0, \tag{3.7.25}$$

showing that the cost elasticity is in one-to-one correspondence with the elasticity of the production function. In particular, the cost elasticity is constant whenever the production function has a constant elasticity.

For example, let the production function be isoelastic with elasticity $0 < B < 1$. Then, (3.7.24) yields the optimal cost function

$$x(\alpha) = \left(\frac{e(\alpha)}{\alpha}\right)^{\frac{1}{1-B}} \quad \text{with} \quad \lim_{\alpha\to 0} x(\alpha) > 0 \tag{3.7.26}$$

which implies the labor demand function $h(\alpha) = x(\alpha)/\alpha$ given by

$$h(\alpha) = \frac{1}{\alpha}\left(\frac{e(\alpha)}{\alpha}\right)^{\frac{1}{1-B}} \quad \text{with} \quad \lim_{\alpha\to 0} h(\alpha) = +\infty \quad \text{and} \quad \lim_{\alpha\to\infty} h(\alpha) = 0. \tag{3.7.27}$$

Therefore, the cost function, the profit function and the labor demand function are all unimodal functions in the real wage since average efficiency is unimodal. Figure 3.56 displays the graphs of the associated functions where the form $s(x) = x/\sqrt{1+x^2}$ has been chosen for the efficiency function $e(\alpha)$ with appropriate shift to obtain $e(0) = 0$ and inflection point at $\alpha = 1$. Observe that maximal average efficiency occurs at $\alpha^* > 1$ at which point maximal profit occurs while maximal labor demand is obtained at a real wage level $\bar{\alpha} < \alpha^*$.

A competitive labor market equilibrium is obtained at a real wage such that

$$n_f h(\alpha) - N(\alpha V(\theta^e)) = 0. \tag{3.7.28}$$

Assumption 3.2.1 assures that there exists at least one equilibrium real wage for each positive inflation rate θ^e since the excess demand function is continuous and takes on positive values for small α and negative values for large α, since the Inada conditions hold for F. In general, however, uniqueness of labor market equilibrium

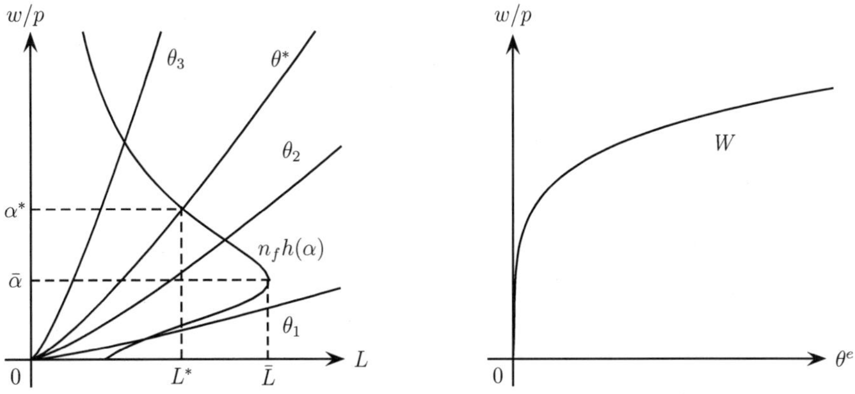

a $\theta_1 < \theta_2 < \theta^* < \theta_3$ **b** Equilibrium real wage

Fig. 3.57 Labor market equilibrium with isoelastic consumers cannot be guaranteed due to the unimodality of labor demand. Multiple equilibria may occur for low levels of expected inflation. Since $N' > 0$ and $V' < 0$, there always exists a consistent expected inflation rate associated with an equilibrium for each positive real wage α, i.e. there is a unique inverse function $\theta^e = g(\alpha)$ implying labor market equilibrium

$$n_f h(\alpha) - N(\alpha V(g(\alpha))) = 0.$$

However, when producer and consumer characteristics are isoelastic with $E_N > 1$, then uniqueness of labor market equilibrium occurs[27] and one finds a regular well-defined real wage function which is monotonically increasing in expected inflation, Figure 3.57 **b**. Given the equilibrium real wage function $\alpha = W(\theta^e)$ one obtains as

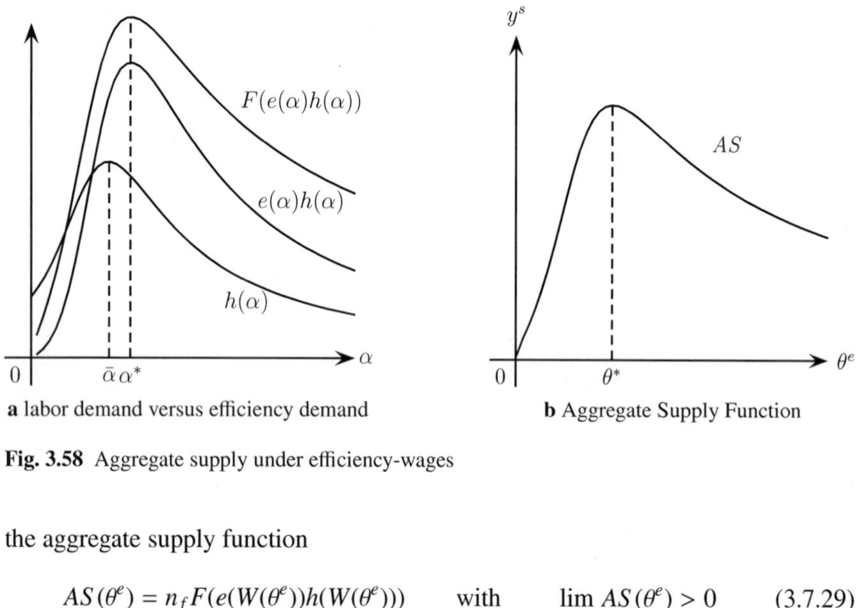

a labor demand versus efficiency demand **b** Aggregate Supply Function

Fig. 3.58 Aggregate supply under efficiency-wages

the aggregate supply function

$$AS(\theta^e) = n_f F(e(W(\theta^e))h(W(\theta^e))) \quad \text{with} \quad \lim_{\theta^e \to 0} AS(\theta^e) > 0 \tag{3.7.29}$$

which turns out to be unimodal even in the isoelastic situation since the function of efficiency hours $e(\alpha)h(\alpha)$ demanded is unimodal with a maximum at $\bar{\alpha} > \alpha^*$, at which total profit is also maximal, Figure 3.58. Therefore, there exists a second source for multiple temporary equilibria arising in the commodity market. Figure 3.59 portrays the situation of unique temporary equilibria for alternative levels of money balances. As a consequence, there exists a unique positive equilibrium price level for each pair of state variables implying a well-defined price law $p = \mathcal{P}(M, p^e)$. The price law is homogeneous of degree one in (M, p^e). However, it is not uniformly

[27] See Lemma B.1.1 and its proof. For example, when consumer characteristics are $C = 1$ and constant $0 < \delta < 1$, labor supply is linear in α. Then, $h(\alpha)/\alpha$ becomes a decreasing function for isoelastic production implying a unique equilibrium real wage.

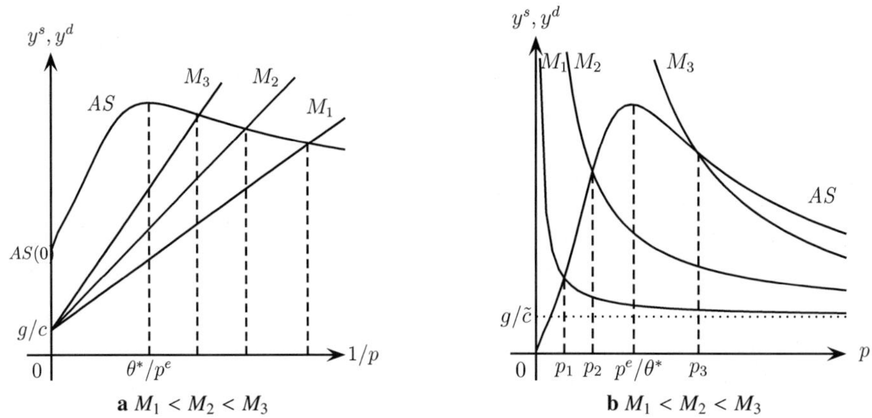

Fig. 3.59 Temporary equilibrium under efficiency-wages; p^e given

monotonically increasing in both state variables. To see this consider the comparative statics effects derived from the equilibrium condition $D(M/p) - AS(p^e/p) = 0$ which yields

$$\begin{aligned} \frac{\partial \mathcal{P}(M, p^e)}{\partial M} &= \frac{pD'}{MD' - p^e AS'} \\ \frac{\partial \mathcal{P}(M, p^e)}{\partial p^e} &= \frac{-pAS'}{MD' - p^e AS'}. \end{aligned} \tag{3.7.30}$$

From the uniqueness of equilibrium one must have $(MD' - p^e AS') > 0$, so that the effect of an increase of money balance must be positive. However, the expectations effect changes sign depending on whether aggregate supply is locally increasing or decreasing. Given the homogeneity, one can write

$$\mathcal{P}(M, p^e) \equiv p^e \mathcal{P}(M/p^e, 1) \tag{3.7.31}$$

which implies that the inflation rate $p^e/p = p^e/\mathcal{P}(M, p^e) = 1/\mathcal{P}(M/p^e, 1)$ in equilibrium depends on the ratio M/p^e only. By the same token, the partial equilibrium effects also depend on the ratio only, since

$$\begin{aligned} \frac{\partial \mathcal{P}(M, p^e)}{\partial M} &= \frac{\partial \mathcal{P}((M/p^e), 1)}{\partial (M/p^e)} \\ \frac{\partial \mathcal{P}(M, p^e)}{\partial p^e} &= \mathcal{P}(M/p^e, 1) - p^e \frac{\partial \mathcal{P}((M/p^e), 1)}{\partial (M/p^e)}. \end{aligned}$$

There exists a unique critical level of real expected money balances $(M/p^e)^* > 0$ such that

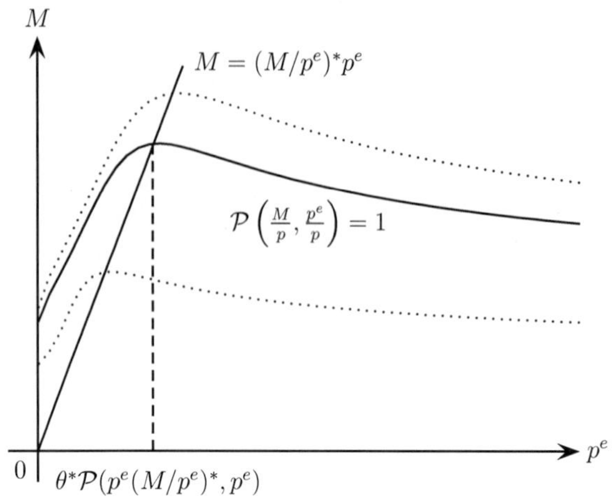

Fig. 3.60 The equilibrium set under competition and efficiency-wages

$$\theta^* := \frac{p^e}{\mathcal{P}(M, p^e)} = \frac{1}{\mathcal{P}((M/p^e)^*, 1)}. \tag{3.7.32}$$

The monotonicity of $\mathcal{P}(M/p^e), 1)$ in M/p^e implies that

$$(AS)'\left(\frac{1}{\mathcal{P}((M/p^e)^*, 1)}\right) < 0 \qquad \text{if and only if} \qquad \frac{M}{p^e} > \left(\frac{M}{p^*}\right)^*. \tag{3.7.33}$$

Therefore, the state space is subdivided into two regions (see Figure 3.60) by the ray $M = (M/p^e)^* p^e$, where the expectations effect is negative (positive) for $M/p^e >$ larger (smaller) than the critical level while the elasticity with respect to money balances is larger (smaller) than one. As a consequence the level sets of the price law take the form as shown in the diagram each of which attaining its maximum on the critical ray. The equilibrium set (the unit contour) has been drawn as a solid curve. All equilibrium pairs of expected inflation rates and real money balances (θ^e, m) are restricted to this unimodal curve. Thus, in contrast to all other equilibrium models discussed so far, competitive equilibrium configurations under efficiency-wages generate a nonmonotonic correlation between real money balances and expected inflation for high real expected money balances. As in all other models changes of the state-space variables (p^e, M) induce equilibrium movements along the unit contour as depicted in the diagram, while parametric changes of all other parameters imply displacements of the unit contour as the equilibrium set.

Monopsonistic Wage Setting and Efficiency-Wages

When there is a single producer active in the labor market with efficiency-wages profit maximizing wage-setting leads to distinctly different wage and employment configurations with potential for involuntary unemployment in a macroeconomic context (as recognized by Solow, 1979; Shapiro & Stiglitz, 1984; Akerlof & Yellen, 1985, 1986, in their contributions). This section completes their analysis by extending the analysis to the income consistent macroeconomic situation and by closing the model from the demand side of the commodity market.

Suppose that the single producer[28] operating in the efficiency-wage situation is a price taker on the commodity market, but sets the nominal wage and his employment (supply on the commodity market) decision in such a way as to maximize profits using the properties implied by the efficiency function e and the aggregate labor supply function N as postulated in Assumption 3.2.1. Keeping with the notation of the previous section, for any given commodity price $p > 0$, let $w := p\,\alpha$ denote the nominal wage set by the producer when chosing a real wage $\alpha \geq 0$.

Since the efficiency-wage model induces globally bounded real profits $\tilde{\Pi}(L, \alpha)$ in the space of choices by the producer function independent of the price level p it is evident that there must be labor supply configurations which have *involuntary* unemployment by assuming that labor supply is larger than the profit maximal employment level chosen at the corresponding real wage. To obtain the full temporary equilibrium with income consistency and to provide a general macroeconomic decription deriving the market clearing price level for all states $(p^e, M) \gg 0$ for the economy requires some detailed analysis.

Consider the decision problem of the monopsonistic firm (as in Section 3.4.1) given by

$$\Pi_{\text{eff}}(\theta^e) := \max_{\alpha, L} \{F(e(\alpha L)) - \alpha L \,|\, L \leq N(\alpha V(\theta^e))\} \tag{3.7.34}$$

which imposes feasibility in the labor market.

Given the fact that real profit $F(e(\alpha)L) - \alpha L$ has a unique global maximum at (α^*, L^*) (see 3.7.20 and Figure 3.57), continuity and monotonicity of the labor supply function implies that, for $\theta^e \leq \theta^*$, labor supply is not binding for the maximization (3.7.34) and both maximizers are constant functions on $(0, \theta^*]$.

As shown in the previous section the first order conditions at the global maximum are

$$e(\alpha^*)F'(e(\alpha^*)L^*) \qquad \text{and} \qquad F'(e(\alpha^*)L^*)e'(\alpha^*) = 1, \tag{3.7.35}$$

i.e. α^* induces maximal average efficiency $e(\alpha^*)/\alpha^*$attained where the efficiency-wage function has unit elasticity

$$\frac{\alpha^* e'(\alpha^*)}{e(\alpha^*)} = 1 \qquad \text{and} \qquad L^* = \frac{1}{e(\alpha^*)}(F')^{-1}\left(\frac{\alpha^*}{e(\alpha^*)}\right). \tag{3.7.36}$$

---

[28] This assumption is made for ease of presentation which still allows displaying the essential consequences of the efficiency-wage model with the help of two-dimensional diagrams. One could, off course, analyze the situation with several producers with different scenarios in the commodity market. As long as they consider the commodity price as given and decide jointly on a uniform nominal wage the analysis would have the same qualitative properties as presented here.

The unit elasticity requirement is often referred to as the Solow rule. Note that the maximizing real wage is exclusively determined by the features of the efficiency-wage function e and that labor demand is codetermined by the production function. Both conditions, however, are independent of labor supply conditions.

Since the global maximizer (α^*, L^*) is unique it follows from the continuity and monotonicity of the labor supply function, for $\theta^e > \theta^*$, that $L^* - N(\alpha^* V(\theta^e)) > 0$. Therefore, any profit maximizing solution $(\alpha_{\text{eff}}, L_{\text{eff}})$ must imply a binding constraint $L_{\text{eff}} = N(\alpha_{\text{eff}} V(\theta^e))$. With a binding constraint the maximization becomes

$$\Pi_{\text{eff}}(\theta^e) = \max_{\alpha} F(e(\alpha)N(\alpha V(\theta^e))) - \alpha N(\alpha V(\theta^e)). \tag{3.7.37}$$

The first order condition for an interior solution under a binding constraint is

$$(e'N + eN'V(\theta^e))\, F'(eN) - (N + \alpha_{\text{eff}} V(\theta^e) N') = 0 \tag{3.7.38}$$

which can be written as

$$\frac{F'e}{\alpha_{\text{eff}}}\left(\frac{\alpha_{\text{eff}} e'}{e} + \frac{\alpha_{\text{eff}} V(\theta^e)N'}{N}\right) = \left(1 + \frac{\alpha_{\text{eff}} V(\theta^e)N'}{N}\right) \iff \frac{F'e}{\alpha_{\text{eff}}} = \frac{1 + E_N}{E_e + E_N}. \tag{3.7.39}$$

Using the envelop theorem on (3.7.37) one finds

$$\Pi'_{\text{eff}}(\theta^e) = (e(\alpha_{\text{eff}})F' - \alpha_{\text{eff}})\alpha_{\text{eff}} N'V' < 0. \tag{3.7.40}$$

Since $\Pi_{\text{eff}}(\theta^e)$ must be decreasing this yields $(e(\alpha_{\text{eff}})F' - \alpha_{\text{eff}}) > 0$ showing that the marginal product of labor must exceed the monopsonistic wage. Therefore, the elasticity of effort must be $E_e(\alpha_{\text{eff}}) < 1$, which implies in turn that $\alpha_{\text{eff}} > \alpha^*$. In other words, the binding demand constraint for labor implies that the monopsonist must pay an efficiency-wage which is higher than he would prefer without the constraint. Nevertheless, he adjusts the employment downward.

The ratio $(1 + E_N)/(E_e + E_N) > 1$ becomes the proportional mark-up factor by which the marginal product of labor exceeds the real efficiency-wage

$$e(\alpha_{\text{eff}})F'(e(\alpha_{\text{eff}})N(\alpha_{\text{eff}} V(\theta^e))) = \alpha_{\text{eff}} \frac{1 + E_N}{E_e + E_N} > \alpha_{\text{eff}}. \tag{3.7.41}$$

The formula reveals a close relationship to the standard model of monopsonistic wage setting (see Section 3.4.1, equations (3.4.2), (3.4.3); also Figure 3.28). Putting the elements of the optimal solution $(\alpha_{\text{eff}}, L_{\text{eff}})$ together one obtains a real wage function of the monopsonistic firm given by

$$\alpha_{\text{eff}} = W_{\text{eff}}(\theta^e) := \max\left\{\alpha^*, \widetilde{W}(\theta^e)\right\} = \begin{cases} \alpha^* & \theta^e \le \theta^* \\ \widetilde{W}(\theta^e) & \theta^* < \theta^e \end{cases} \tag{3.7.42}$$

where $\widetilde{W}(\theta^e)$ is the unique solution of (3.7.38). Clearly, $\widetilde{W}(\theta^e) < W(\theta^e)$ due to the mark-up feature of (3.7.41). The wage function induces the employment function

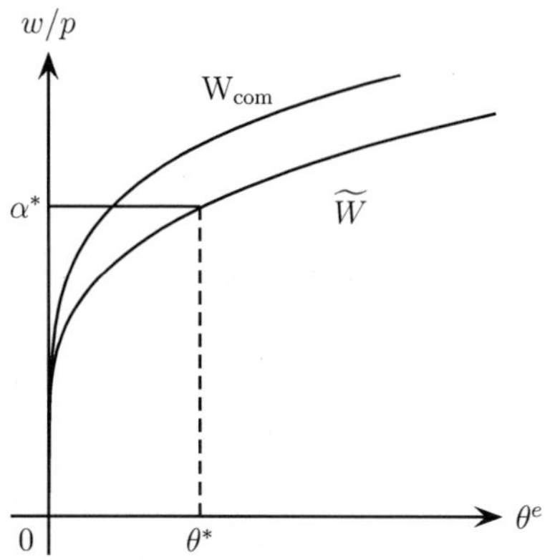

Fig. 3.61 Monopsonistic efficiency-wage function

$$L_{\text{eff}} = L_{\text{eff}}(\theta^e) := \begin{cases} L^* & \theta^e \leq \theta^* \\ N\left(\widetilde{W}(\theta^e)\, V(\theta^e)\right) & \theta^* < \theta^e \end{cases} \tag{3.7.43}$$

and an aggregate supply function of the form

$$y^s_{\text{eff}} = AS_{\text{eff}}(\theta^e) := \begin{cases} F(e(\alpha^*)\, L^*) & \theta^e \leq \theta^* \\ F\left(\alpha(\widetilde{W}(\theta^e)\, N\left(\widetilde{W}(\theta^e)\, V(\theta^e)\right)\right) & \theta^* < \theta^e. \end{cases} \tag{3.7.44}$$

These functions of the supply side confirm the essential property of the solution under monopsonistic behavior, i.e. up to the critical level θ^* the monopsonist maintains constant levels of all three decision variables, the real wage, the employment level, and the supply of output since labor supply is not binding which guarantees the global profit maximum. For all $\theta^e > \theta*$ the global maximum is no longer attainable due to a labor supply restriction. The firm becomes demand constraint on the labor market and the optimal decision is best against the labor supply function.

The above properties must hold in the general case for a unique interior solution under the basic Assumption 3.2.1. Nevertheless, it seems impossible to identify additional general properties for he interaction between the sigmoid function describing the efficiency-wage mechanism combined with the convexity/Inada properties to prove global concavity or uniqueness of the monopsonistic solution. Thus, to derive further properties for the temporary equilibrium configuration additional assumptions on labor supply and the technology have to be made to obtain sufficient concavity of the maximization problem of the firm under a binding labor demand restriction to guarantee uniqueness. Fortunately, the combination of the assumption for the efficiency-wage function introduced above with the isolelastic version of consumers and producers used so far provide a a computationally valid and efficient set of parameters for which typical global relations are obtained.

For the remainder of this section let the production function and the intertemporal preference be isoelastic with values $0 < B < 1$, $0 < \delta < 1$, and $C > 0$ implying the functional forms, (see Lemma B.1.1)

$$
\begin{aligned}
F(L) &= \frac{1}{B}L^B, && 0 < B < 1, \\
c(\theta^e) &= \delta \quad \text{and} \quad V(\theta^e) = \delta^\delta(1-\delta)^{1-\delta}\left(\frac{1}{\theta^e}\right)^{1-\delta}, && 0 < \delta < 1, \\
\ell^\star(\alpha V(\theta^e)) &= (\alpha V(\theta^e))^C, && 0 < C.
\end{aligned}
\tag{3.7.45}
$$

In this case the efficiency-wage function is a strictly increasing function in expected inflation under the condition (3.7.38). For an isoelastic labor supply with $C = 1$ one obtains explicit numerically tractable results. Labor supply is linear and (3.7.38) takes the form

$$
(\alpha e'(\alpha) + e(\alpha))\, F'(\alpha e(\alpha)V(\theta^e)) - 2\alpha = 0 \iff F'(e(\alpha)\alpha V(\theta^e)) = \frac{2\alpha}{e'(\alpha)\alpha + e(\alpha)}
$$

inducing

$$
V(\theta^e) = \frac{1}{e(\alpha)\alpha}(F')^{-1}\left(\frac{2\alpha}{e'(\alpha)\alpha + e(\alpha)}\right) = \frac{1}{e(\alpha)\alpha}\left(\frac{2\alpha}{e'(\alpha)\alpha + e(\alpha)}\right)^{\frac{1}{B-1}}. \tag{3.7.46}
$$

This leads to

$$
(\delta^\delta(1-\delta)^{1-\delta})(e(\alpha)\alpha) = (\theta^e)^{1-\delta}\left(\frac{2\alpha}{e'(\alpha)\alpha + e(\alpha)}\right)^{\frac{1}{B-1}} \tag{3.7.47}
$$

with an explicit inverse solution for $\widetilde{W}(\theta^e)$ given by

$$
\theta^e = (\delta^\delta(1-\delta)^{1-\delta})^{1/(1-\delta)}\,(e(\alpha)\alpha)^{1/(1-\delta)}\left(\frac{2\alpha}{e'(\alpha)\alpha + e(\alpha)}\right)^{\frac{1}{(1-B)(1-\delta)}} \tag{3.7.48}
$$

displayed in Figure 3.61. Since $\alpha_{\text{eff}} = \max(\alpha^*, \widetilde{W}(\theta^e))$ the real wage is constant in the interval $0 < \theta \leq \theta^*$ in which *involuntary* unemployment occurs. As shown in Figure 3.63 **b** aggregate supply is also constant in that interval and decreasing for larger θ^e implying a non-increasing aggregate supply function.

Putting labor market conditions together with income consistent aggregate demand, for every state of the economy $(M, p^e) \gg 0$ one obtains a unique positive price as a solution of

$$
D(M/p) - AS_{\text{eff}}(p^e/p) = 0
$$

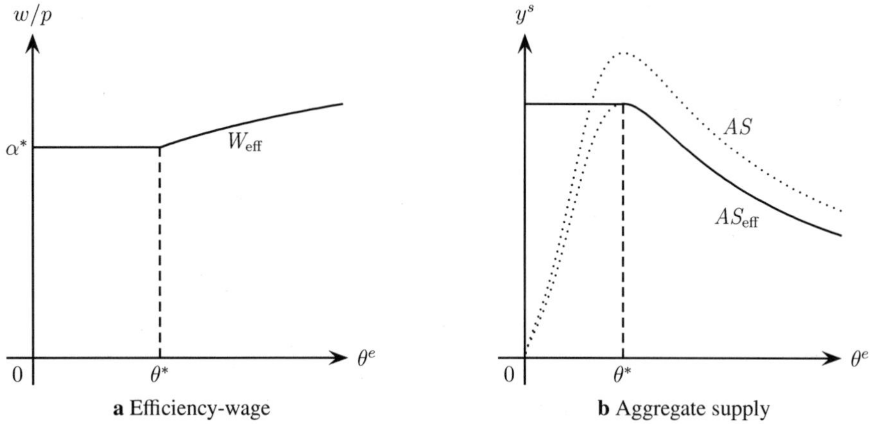

Fig. 3.62 Real wage and aggregate supply with monopsonistic efficiency-wages

defining a continuous price law $p = \mathcal{P}_{\text{eff}}(M, p^e)$. It is straight forward to see (under the remaining features of Assumption 3.2.1) that the price law is homogeneous of degree one and non-decreasing in (M, p^e). Thus, the respective partial elasticities of $\mathcal{P}_{\text{eff}}$ with respect to money balances and expected prices are between zero and one.

Figure 3.63 displays the equilibrium configurations being characterized by two distinct situations, one with full employment at low levels of money balances or one with involuntary unemployment at high levels of money balances. The comparative statics analysis reveals that the price law is strictly monotonically increasing in money balances M. The homogeneity implies that $p = \mathcal{P}_{\text{eff}}(M, p^e) = p^e \mathcal{P}_{\text{eff}}(M/p^e, 1)$. Therefore, unemployment occurs if and only if

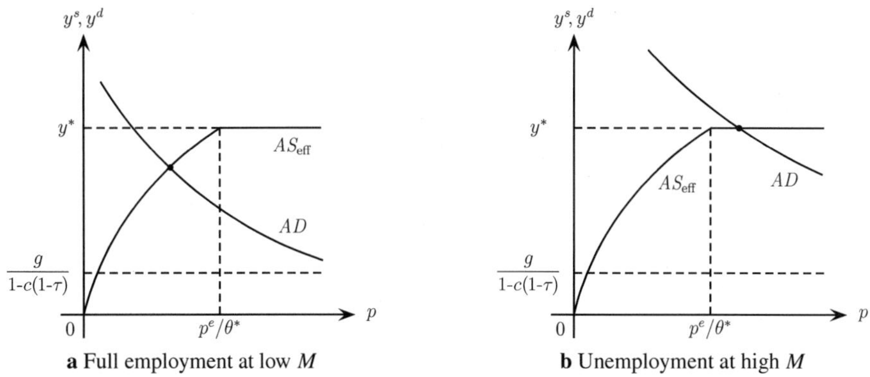

Fig. 3.63 Equilibrium with monopsonistic efficiency-wages; p^e given

$$\frac{p^e}{p} = \frac{1}{\mathcal{P}_{\text{eff}}(M/p^e, 1)} < \theta^*.$$

There exists a unique positive critical level $(M/p^e)^*$ satisfying $\mathcal{P}_{\text{eff}}((M/p^e)^*, 1) = 1/\theta^*$ such that

$$\frac{p^e}{p}\mathcal{P}_{\text{eff}}\left(\left(\frac{M}{p^e}\right)^*, 1\right) < \theta^* \quad \text{if and only if} \quad \frac{M}{p^e} > \left(\frac{M}{p^e}\right)^*. \tag{3.7.49}$$

Thus, unemployment occurs if and only if $M/p^e > (M/p^e)^*$. Therefore, the state space $\mathbb{R}^2_+$ can be divided into two regions by the line $M = (M/p^e)^* p^e$, above which unemployment occurs while below full employment prevails. Moreover, the price law has no expectations effect above the critical line making the price law linear in money balances, i.e. $p = M\mathcal{P}_{\text{eff}}(1, 1/(M/p^e)^*)$. As a consequence one obtains as the equilibrium set under monopsonistic efficiency-wages the graph of the explicit function

$$\frac{M}{p} = (1 - c(1-\tau))AS_{\text{eff}}(p^e/p) - g \tag{3.7.50}$$

which is the unit contour of the price law $\mathcal{P}_{\text{eff}}$ since

$$p = \mathcal{P}_{\text{eff}}(M, p^e) = p\mathcal{P}_{\text{eff}}\left(\frac{M}{p}, \frac{p^e}{p}\right) \iff \mathcal{P}_{\text{eff}}\left(\frac{M}{p}, \frac{p^e}{p}\right) = 1,$$

as shown in Figure 3.64. In other words, income consistent macroeconomic allocations with *involuntary unemployment* occur for values of real expected money

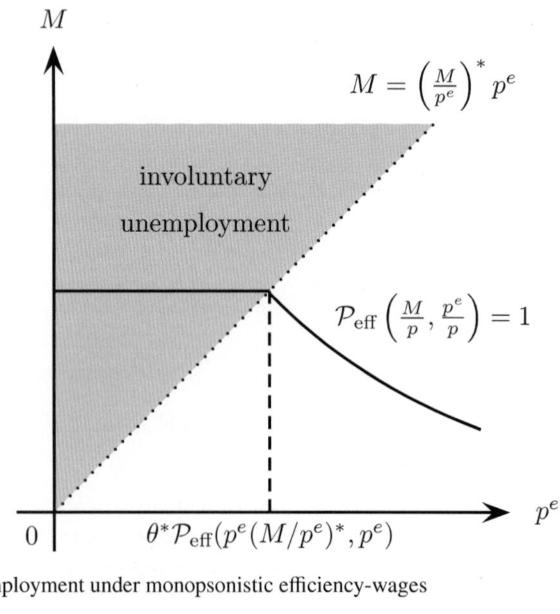

Fig. 3.64 Unemployment under monopsonistic efficiency-wages

balances M/p^e larger than a fixed critical level which is determined by the interaction of the parameters of the efficiency function, the technology, and of labor supply conditions. Thus, the introduction of monopsonistic competition combined with the efficiency-wage structure induces involuntary unemployment in a large open subset of the state space. It appears as a non-concavity of the price law and a linearity of equilibrium prices in money balances in the state space. One concludes that the strategic behavior of the producer is the primary source for the generic occurrence of unemployment under the basic Assumption 3.2.1, a result which does not occur when the labor market operates competitively and producers take both wages and prices as given. Therefore, the efficiency-wage model with strategic wage setting presents one independent situation for which *involuntary unemployment* may occur under otherwise standard neoclassical assumptions for a closed income-consistent macroeconomic model for a large open set of the state variables (M, p^e). The model shows the appearance of an endogenous rigidity of the real wage and of the level of employment which depend on the prevailing state determined by aggregate demand conditions. Here these are taken at their simplest case (originating from specific consumer preferences) when all effects from the income distribution and from expectations in demand on the commodity market are eliminated. At this level of investigation the efficiency-wage mechanism has been introduced as a given rule determining the efficiency of labor input. No microeconomic decision model of consumer behavior or procedures of wage setting by producers has been provided. To develop such microeconomic foundations seems to be an area for further research. There exist other economic variations within partial equilibrium models to extend and enrich the economic structure, for example elements from contract theory with asymmetric information. In such cases, the challenge consists of deriving consistently the cross market effects within the macroeconomic model, the occurrence of quantity restrictions in order to exhibit the causal channels for inefficiencies and unemployment.

3.7.4 On Endogenous Rigidities and Unemployment

The two examples of Section 3.7 assume a competitive commodity market while structural additions or frictions 'disturb' the mechanism of determining a market clearing wage of an otherwise competitive labor market. In both cases *endogenous wage rigidities* appear inducing unemployment, voluntary in one case and involuntary in the other one, for a subset of the state space. These are conceptually different in nature from those related to behavioral or strategically chosen rigidities, as discussed in Sections 3.3 to 3.6.

- With unemployment compensation *voluntary* unemployment occurs in temporary equilibrium as a consequence of an existing positive insurance guarantee which creates a real-wage threshold below which consumers are not willing to offer labor. Its level is determined endogenously in equilibrium depending on numerous spill-over and cross-market effects. Formally, this causes a non-convexity

for the worker's decision problem which induces a discontinuity of his optimal labor market behavior. An *endogenous downward wage rigidity* occurs whose level depends on the structural parameters of the economy. As the compensation becomes small the rigidity disappears and equilibria with full employment are the only ones to exist under the usual neoclassical assumptions.
– The model with the efficiency-wage rule (Section 3.7.3) *and* with a monopsonistic labor market was found to be the only candidate with explanatory power of *involuntary* unemployment causing a real-wage rigidity. This is associated with a non-concavity of the price law.

Other modifications of the neoclassical market paradigm implying endogenous rigidities exist in the literature (e.g. extensions creating different outside options or participation constraints derived in models of search, matching, or shirking) which could be described within this framework as well. Often, their analysis concentrates on the equilibrium features of the labor market only neglecting the consequences of unemployment or overemployment for the spillovers between markets on incomes and prices. Their integration into an income-consistent macroeconomic model implies a variety of additional models determining the nature of endogenous rigidities leading to voluntary or involuntary unemployment. If these induce endogenous rigidities, their implications become part of the time-invariant stationary environment of the economy, i.e. of the functional forms of the price and the wage law. Therefore, the issue of whether unemployment is a transient or a permanent phenomenon depends on the dynamics of the two state variables (M, p^e), i.e. on whether the evolution of money balances and of price expectations leads to attractors in the region with unemployment. Their adjustment mechanisms are typically not governed by criteria of unemployment. The open questions remain which structural characteristics of the underlying assumptions cause convergence to steady states or attractors in the regions with unemployment.

Within the class of neoclassical models analyzed here non-competitive behavior in markets as such has little to say about observable rigidities when logically price and wage flexibility prevails in the temporary situation. In other words, strategic deviations by themselves do not induce endogenous rigidities, i.e. they do not imply time-invariant regions with unemployment in the state space.

Finally, temporarily *exogenous* rigidities of prices or wages do not have an influence on the implication of Assumption 3.2.1 on the price law. When they exist in every period the model of *equilibrium trading* is no longer valid and *feasible disequilibrium trading* has to be defined. This is the subject of Chapter 6. Explanations to the issue of transient versus permanent unemployment within disequilibrium models has to come from the interaction of the neoclassical features of Assumption 3.2.1 and the dynamic rules or mechanisms determining the *changes* of rigid prices or wages which is the subject of Chapters 7 and 8.

Chapter 4
Dynamics of Monetary Equilibrium Models

The analysis of temporary equilibrium allocations in Chapter 3 showed that the structural and behavioral parameters defining the type of equilibrium induce a well defined result in real and monetary terms for the macro economy in any period. In other words, for such a closed economy containing a public sector made up of a government with an integrated central bank as monetary authority supporting the system with fiat money, any *feasible temporary allocation* under income consistency at given money balances and expectations has an impact on the government deficit. Since government expenditure and tax revenue are endogenously determined the size and the sign of the government deficit in every period is price dependent. Therefore, changes of total money balances held in the private sector occur *if and only if* the government budget is not balanced. As a consequence, changes of money balances are determined simultaneously with the temporary equilibrium making them endogenous between periods. This chapter derives the main results for a recursive analysis of the development of prices, price expectations, and of money balances. The emphasis will be on characterizing and investigating the conditions under which recursive forecasting schemes exist which induce equilibrium orbits with perfect foresight and rational expectations. In addition, conditions of uniqueness of balanced paths, their stability and their basins of attraction will be examined. Some discussion will be provided for these issues with general adaptive forecasting rules.

4.1 Dynamics of Money, Prices, and Expectations

Given the basic Assumption 3.2.1, the state of the economy $(M_t, p^e_{t,t+1})$ in period t and government parameters (g, τ)[1], the associated temporary equilibrium is defined by four time invariant mappings given by the price law and the wage law together with the two mappings inducing feasible allocations of employment and output, i.e.

[1] For simplicity, a uniform income tax τ is assumed.

$$\begin{aligned} p_t &= \mathcal{P}(M_t, p^e_{t,t+1}), & w_t &= \mathcal{W}(M_t, p^e_{t,t+1}) \\ L_t &= \mathcal{L}(M_t, p^e_{t,t+1}), & y_t &= \mathcal{Y}(M_t, p^e_{t,t+1}). \end{aligned} \tag{4.1.1}$$

These define a fixed subset in $\mathbb{R}^6_+$ given by the graph of the mapping $(\mathcal{P}, \mathcal{W}, \mathcal{L}, \mathcal{Y}) : \mathbb{R}^2 \to \mathbb{R}^4$. If the structural parameters and the behavioral assumptions do not change between periods, *any* dynamics of the associated temporary equilibria will evolve on this set. Since the price and the wage law are structurally uncoupled from the allocations, it is clear that the dynamics of the economy is determined once the evolution of aggregate money balances and of expectations are specified. When no income distribution effects arise, all necessary information can be derived from the price law alone.

4.1.1 Dynamics of Money Balances

Given the overlapping-generations structure of consumers, Assumption 3.2.1, and the fact that in this simple economy money is the only store of value, total savings by young consumers S_t must be equal to final money holdings by all consumers. Since monetary transfers by the government to be received by young consuers at the beginning of next period are excluded this implies that $M_{t+1} = S_t$ holds in every period t. Feasible aggregate private savings in each period t is defined by

$$S_t := Y^{net}_t - p_t c_t = (1 - \tau_w) W_t - C^w_t + (1 - \tau_\pi) \Pi_t - C^s_t.$$

Together with income/expenditure consistency $Y_t = M_t + G_t + C^w_t + C^s_t$ this implies

$$M_{t+1} = W_t + \Pi_t - (C^w_t + C^s_t) - (\tau_w W_t + \tau_\pi \Pi_t) = M_t + G_t - (\tau_w W_t + \tau_\pi \Pi_t).$$

In other words, the value of total final asset holdings in each period is equal to the level of old holdings plus the government deficit of the period. Therefore, under a proportional income tax rate $0 \le \tau_w = \tau_\pi \equiv \tau \le 1$ and a temporary equilibrium price given by $p_t = \mathcal{P}(p^e_{t,t+1}, M_t)$, the deficit becomes

$$\Delta_t = \Delta(M_t, p^e_{t,t+1}) := \mathcal{P}(M_t, p^e_{t,t+1}) \big(g - \tau \mathcal{Y}(M_t, p^e_{t,t+1}) \big). \tag{4.1.2}$$

This defines the *endogenous change of money balances* as a mapping

$$M_{t+1} = \mathcal{M}(M_t, p^e_{t,t+1}) := M_t + \mathcal{P}(M_t, p^e_{t,t+1}) \big(g - \tau \mathcal{Y}(M_t, p^e_{t,t+1}) \big) \tag{4.1.3}$$

which is homogeneous of degree one in $(M_t, p^e_{t,t+1})$. If, in addition, the propensities to consume for workers and shareholder are the same, i.e. if $c = c_w = c_s$, then the mapping has the equivalent form

$$M_{t+1} = \mathcal{M}(M_t, p^e_{t,t+1}) = \frac{\tilde{c} - \tau}{\tilde{c}} \big(M_t + g \mathcal{P}(M_t, p^e_{t,t+1}) \big) \tag{4.1.4}$$

where $0 < \tilde{c} := 1 - c(1 - \tau) < \tau < 1$. Summarizing, for a given proportional income tax system and given demand behavior of the government, the endogeneity of equilibrium prices and allocations implies that the deficit of the government is determined *simultaneously* and endogenously. Thus, the change in asset holdings by the private sector, i.e. the change in money balances is determined endogenously as well. Under a fiscal policy characterized by (g, τ), government expenditures and tax revenues are endogenous entities since prices and incomes are endogenous, inducing the resulting deficit as a consequence for/at the end of the period. Therefore, under such a policy a government cannot guarantee a balanced budget at any one time. Nonzero deficits and surpluses will be the rule rather than the exception in temporary equilibrium. Moreover, (4.1.2) holds at every time period t, so that the deficit is described by a function of money balances and expectations which is constant over time for any stationary government policy (g, τ), making it the proper object of empirical estimation. From the dynamic perspective of the economy, the endogenous mechanism of money creation/destruction implied by equation (4.1.3) or (4.1.4) is one of the major structural features for the dynamic development of the economy in equilibrium.

4.1.2 Dynamics of Expectations

All four mappings in (4.1.1) which describe the temporary equilibrium configuration (price-wage-allocation) have an expectations feed back with an *expectational lead*, i.e. current equilibrium prices are determined by expectations held currently for prices one period ahead. Therefore, an analysis of the evolution of the economy under the given equilibrium concept requires that the prediction $p^e_{t,t+1}$ for any period *t has to be determined in period t*, while forward recursion (in time) implies that the value of the prediction can only be made on the basis of data available up to time t. For example, let $I_{t-1} := (M_{t-T}, \dots, M_{t-1}, p_{t-T}, p^e_{t-T,t+1-T}, \dots, p^e_{t-1,t}, \dots, p_{t-1}) \in \mathbb{R}^{3T}_+$ denote the finite list of statistics available at time t defined by the vector of time series data of length T for the economy. Based on this information, forecasts will be made by agents (or by a forecasting agency which then are used by the agents in the economy). Thus, the principles or rules of forecasting become an integral and determining factor of the dynamic development of an economy. In more formal language, the principles of forecasting will be described by a forecasting rule or a *predictor* given by a function.

Definition 4.1.1. A function $\psi : \mathbb{R}^k_+ \to \mathbb{R}_+$ such that

$$p^e_{t,t+1} = \psi(I_{t-1})$$

is called a predictor or a forecasting rule.

It is called *stationary* if the rule is independent of time t, implying that the same rule is used in each period on data from different but successive time intervals.

The stationarity stipulates that the same forecasting principles are applied in each period to structurally identical data sets, where I_t is a component-wise update of I_{t-1} of the same length. This clearly covers many cases of so-called adaptive expectations, adaptive learning, error correction models, or statistical or recursive updating[2]. Some examples are given below. Stationarity in this sense stipulates in a natural way that the technology (or knowledge) used for forecasting is constant over time and used by the cohorts of agents (or by a central forecasting agency) in the same way in every period.

Stationary predictors, once they are given for an economy, define a time-invariant interaction between the price law, the money law (4.1.3), and the predictor which induces a uniquely defined evolution of the economy. To be specific, for any given vector $I_{t-1} \in \mathbb{R}^k$ of data and a mapping $\psi : \mathbb{R}^k \to \mathbb{R}_+$, the prediction $p^e_{t,t+1} = \psi(I_{t-1})$ together with the money law (4.1.3) and the price law define the one-step recursion of money, price expectations, and prices $(M_t, p^e_{t,t+1}, p_t) \in \mathbb{R}^3_+$ for the economy given by

$$
\begin{aligned}
M_t &= \tilde{\mathcal{M}}(I_{t-1}) := M_{t-1} + \mathcal{P}(M_{t-1}, p^e_{t-1,t})(g - \tau \mathcal{Y}(M_{t-1}, p^e_{t-1,t})) \\
p^e_{t,t+1} &= \psi(I_{t-1}) = \psi(M_{t-T}, \dots, M_{t-1}, p^e_{t-T,t+1-T}, \dots, p^e_{t-1,t}, p_{t-T}, \dots, p_{t-1}) \qquad (4.1.5) \\
p_t &= \mathcal{P}_\psi(I_{t-1}) := \mathcal{P}(M_t, \psi(I_{t-1})) = \mathcal{P}(\tilde{\mathcal{M}}(I_{t-1}), \psi(I_{t-1})).
\end{aligned}
$$

In other words, these three equations define a time-one map of an associated dynamical system with delay T on $\mathbb{R}^3_+$.

The stationarity of the predictor implies that the system of (4.1.5) is autonomous. Its orbits depend in a structural way on the form of the predictor. In other words, the dynamic evolution of an economy under these circumstances depends in an essential way on the *forecasting rule*. Thus, the dynamic evolution of an economy depends not only on *which* forecasts are made, but also on *how* these are made. Or put differently, the future depends partly on how agents *perceive* the interaction between forecasts and actual observations. Technically speaking the forecasting mechanism becomes a control function of the dynamic economic systems. This interaction defines an *expectations feed back* in economic dynamical systems, a feature which is absent from physical or meteorological systems. There, at least in most experimental work, it is assumed that the prediction does not matter for the future realization.

Minimum State Variable Predictors

Clearly, the type of the mapping as well as the size of the data space taken as input by a predictor will matter for the characteristics of the predictions made. In macroeconomic applications of low dimensional models the so-called mini-

[2] Most recursive models examining statistical or econometric time series as well as many so-called learning mechanisms satisfy the stationarity assumption. Those with infinite memory often have an equivalent finite-dimensional recursive representation.

mal state space predictors are discussed widely. Such predictors are defined to be those which use non expectational variables only as inputs. Applied to the macroeconomic model here this implies that predictions are based exclusively on data on the quantity of money. In other words, one would chose $I_{t-1} = M_{t-1}$ or $I_{t-1} = (M_{t-1}, M_{t-2}, \ldots, M_{t-T})$ for some delay of order T. When $I_{t-1} = M_{t-1}$, this implies

$$M_t = \mathcal{M}_\psi(M_{t-1}) := \mathcal{M}(M_{t-1}, \psi(M_{t-1}))$$

$$p_t = \mathcal{P}_\psi(M_{t-1}) := \mathcal{P}(\mathcal{M}_\psi(M_{t-1}), \psi\left(\mathcal{M}_\psi(M_{t-1})\right).$$

This induces one dimensional dynamics with money balances as the state variable of a dynamical system. Prices are a function of money balances alone, *as if* there is no expectations feed back. In other words, such a predictor eliminates the price feed back and is likely to induce systematic deviations of predictions from future realizations.

Adaptive or Recursive Predictors

Given the fact that the price law has an expectational lead, recursive predictors which take the lead structure into account imply a subtle modification of the general adaptive system 4.1.5. They are especially useful and successful, as will be seen below. Let $I_{t-1} = (M_t, p^e_{t-1,t})$ (or possibly a vector with delay $T > 1$) with $p^e_{t,t+1} = \psi(M_t, p^e_{t-1,t})$. One obtains

$$\begin{aligned} M_t &= \mathcal{M}(M_{t-1}, p^e_{t-1,t}) \\ p^e_{t,t+1} &= \psi(M_t, p^e_{t-1,t}) = \psi(\mathcal{M}(M_{t-1}, p^e_{t-1,t}), p^e_{t-1,t}) =: \phi(M_{t-1}, p^e_{t-1,t}). \end{aligned} \tag{4.1.6}$$

This implies that $(\mathcal{M}, \psi)$ induces a dynamical system $(\mathcal{M}, \phi)$ for the evolution of money balances and price expectations of dimension two. Thus, for any given predictor ψ of this form, the orbits $\left\{(M_t, p^e_{t,t+1})\right\}_{t=0}^{\infty}$ are uniquely defined by the state space dynamics in $\mathbb{R}^2_+$, which imply associated sequences of prices, wages, output and employment

$$\begin{aligned} p_t &= \mathcal{P}(M_t, p^e_{t,t+1}) \quad \text{and} \quad w_t = \mathcal{W}(M_t, p^e_{t,t+1}) \\ L_t &= \mathcal{L}(M_t, p^e_{t,t+1}) \quad \text{and} \quad y_t = \mathcal{Y}(M_t, p^e_{t,t+1}). \end{aligned}$$

Summarizing the role of expectations formation, the dynamic evolution of temporary equilibria in a given economy will depend in a significant structural way on 1) the information sets used (which data, what time length of past data, etc.) and 2) on the forecasting system used, i.e. on the type of the predictor. Recursive equilibrium dynamics without the specification of predictors and their data sets are not well defined and the usage of different forecasting principles induce different dynamics. In other words the dynamics are driven *jointly* by the characteristics of the forecasting rule used and by the parametric structure embodied in the price law and the money law.

As pointed out before, macroeconomic time series data (originating from an equilibrium system with expectations feed back) are influenced by the forecasting principles used, a feature absent in physical systems. This means, conversely, that data from macroeconomic systems with an expectations feed back can never be fully understood without a hypothesis of the expectation formation principles acting in the economy. Orbits of such economic systems do not share the independence property of orbits of physical systems from the perception of a rational outside observer. As a consequence, understanding the role of the underlying forecasting structure becomes an important macroeconomic problem which raises a number of different issues. Among them are the following.

One of them concerns the issue of correctness or exactness of the forecasts, i.e. how reliable or good are the predictions? Can they be 'perfect' or 'rational', i.e. does $p^e_{t-1,t} = p_t$ hold for some t, in the limit for $t \to \infty$, or for all t? Hicks (1939) seems to have been one of the first to introduce explicitly such properties as part of his concept of *Equilibrium over Time* when discussing sequences of temporary equilibria[3]. Are there stationary predictors which guarantee perfect foresight or rational expectations? Can they be found/identified from time series data of the economy? Do they generate converging orbits? It seems clear that a successful forecasting rule for any economy must depend on the specific characteristics of the price law of that economy but also possibly on some general principles of recursive forecasting which can be applied to all economies. The next section investigates the conditions for existence and their characteristics.

Additional criteria and issues arise also when recursive predictors with perfect foresight/rational expectations do not exist or when they cannot be found. Then, the predictors take on a different role as control functions for which additional criteria (e.g. stability, welfare augmenting, optimality in consumption, etc.) have to be defined. The essential dynamic question is always to what extent the forecasting rule is responsible for specific dynamical features of the economic development? What is its impact on the orbits, on stability, on recurrence, on the cycle? The literature of partial equilibrium models of the Cobweb type provide a wide array of different results, from stable environments to chaotic orbits. Such issues become essentially policy question which enter into the design of predictors. For example, should/could the government intervene successfully, if the system is not stable? Is stabilization a desirable governmental policy and which policies can stabilize the economy?

From an empirical point of view it may turn out to be extremely difficult to identify the expectations or the forecasting principles used by agents. In other words, can they be identified from data alone? Finally, if data from a stationary rule are inconsistent with perfect foresight, how can predictions be improved? What are statistical

[3] "This is the condition that the prices realized on the second Monday are the same as those which were previously *expected* to rule at that date", (p. 132, Hicks, 1939). He actually uses the term perfect foresight, (p. 140), but seems to be skeptical as to the possibility of economies actually evolving under such conditions "No economic system ever does exhibit perfect equilibrium over time;" (p. 133).

learning procedures which improve the forecast errors and/or the performance desired?

4.1.3 Perfect Foresight

When there are no random perturbations in the economy, perfect foresight is considered as one of the most desirable macroeconomic properties to hold for the entire evolution of the economy, i.e. for all possible orbits. In other words, the question arises whether recursive forecast rules exist which guarantee that equality between the prediction and the associated realization one period later holds at all times, independent of initial conditions. Imposing this property for orbits implies the following useful definitions.

Definition 4.1.2. An orbit $\left\{(M_t, p^e_{t,t+1}, p_t)\right\}_{t=0}^{\infty}$ of an economy with predictor (4.1.6) is said to have the perfect foresight property if

$$p_t = p^e_{t-1,t} \qquad \text{for all } t. \tag{4.1.7}$$

Therefore, a *stationary forecasting rule* which induces the perfect foresight property along an orbit of a system when used will be called a *perfect predictor.*

Definition 4.1.3. A predictor ψ^* with associated information set I_{t-1} is said to be *perfect* if it induces an orbit $\left\{(M_t, p^e_{t,t+1}, p_t)\right\}_{t=0}^{\infty}$ with the perfect foresight property, i.e. if for all t

$$\begin{aligned} M_t &= \mathcal{M}(M_{t-1}, p^e_{t-1,t}) \\ p^e_{t,t+1} &= \psi^*(I_{t-1}) \\ p_t &= \mathcal{P}(\mathcal{M}(M_{t-1}, p^e_{t-1,t}), \psi^*(I_{t-1})) = p^e_{t-1,t}. \end{aligned} \tag{4.1.8}$$

It is called globally perfect if all orbits have the perfect foresight property.

Since wage expectations do not enter in any of the mappings for the basic economy, a systematic search for conditions guaranteeing existence of a perfect predictor needs an examination of the price law only. Consider the temporary equilibrium in any period t under the associated price law $p_t = \mathcal{P}(M_t, p^e_{t,t+1})$. Given the *expectational lead* of the price law, the current price is determined by the prediction for the next period. Therefore, in any period t the *current* forecast determines whether the *previous* forecast $p^e_{t-1,t}$ was correct. In other words, the current prediction becomes a control variable determining the error of the previous forecast. Define the *forecasting error* e_t in period t as

$$e_t := p_t - p^e_{t-1,t}, \tag{4.1.9}$$

inducing the associated *error function* $e : \mathbb{R}^3_+ \to \mathbb{R}$ defined by

$$e(M, p^e, p^e_{-1}) := \mathcal{P}(M, p^e) - p^e_{-1}. \tag{4.1.10}$$

Then, perfect foresight in period t (with respect to the previous prediction) occurs if and only if the forecast error is zero. For a given price law $\mathcal{P}$ the error function is itself a time invariant function, i.e. it is the same function in every period t and it is determined exclusively by the price law. Hence, existence of a perfect predictor will depend on the properties of the price law. Finding a perfect prediction in any period amounts to finding a zero of the *same* function in every period given an arbitrary previous prediction. Therefore, for any M one may define the following set

$$\mathcal{V}(M) := \left\{(p^e, p^e_{-1}) \,|\, e(M, p^e, p^e_{-1}) = \mathcal{P}(M, p^e) - p^e_{-1} = 0\right\} \tag{4.1.11}$$

which, geometrically, describes the zero contour of the error function for any level of money balances M. In general, such a set $\mathcal{V}$ may have a rather complex form with singularities. However, it includes and describes all the information which is necessary to make a perfect prediction. In particular, it defines the space for the information set to be used for a possible perfect predictor. For the case here, it stipulates that a perfect predictor must use information on current money balances and on the previous prediction *and nothing else*, while any information about previous prices is not useful and potentially misleading, not required, and in fact redundant. As a consequence one has the following lemma (built on the results from Böhm & Wenzelburger, 1999, 2004) which characterizes the structural requirements necessary for perfect predictors to exist. This contrasts with earlier attempts to characterize stationary predictors designed to guarantee perfect foresight orbits in non-monetary economies (for example Grandmont, 1988; Grandmont & Laroque, 1986, 1991) or to characterize competitive business cycles as equilibria with perfect foresight without defining predictors (Woodford, 1984; Grandmont, 1985). Their findings do not apply here.

Lemma 4.1.1.
There exists a unique globally perfect predictor ψ^* if and only if for any $M > 0$:

(a) the error function has a unique zero (is globally invertible) for every M;

(b) $\mathcal{V}(M)$ is the graph of the function ψ^*.

Applying the criteria of the lemma to the specific situation of the price law (3.2.14) with an expectational lead, the condition for a prediction error equal to zero at any t becomes

$$p_t = \mathcal{P}(M_t, p^e_{t,t+1}) \stackrel{!}{=} p^e_{t-1,t}. \tag{4.1.12}$$

From the lemma one obtains the following global characterization of perfect predictors.

Proposition 4.1.1.
Let $\mathcal{P}$ denote the price law of the economy and ψ^* a predictor for $\mathcal{P}$. Then:

(a) ψ^* is globally perfect ***if and only if*** *ψ^* solves*

$$\mathcal{P}(M, \psi^*(M, p^e_{-1})) = p^e_{-1} \quad \textit{for every} \quad (M, p^e_{-1}) \tag{4.1.13}$$

(b) *ψ^* is globally perfect* ***if and only if*** *ψ^* is the inverse of the price law with respect to p, i.e.*

$$p^e = \psi^*(M, p^e_{-1}) := \mathcal{P}^{-1}(M, p^e_{-1}). \tag{4.1.14}$$

(c) *A globally perfect predictor ψ^* exists* ***if and only if*** *the price law is globally invertible for each level of money balances M.*

Therefore, given the parameters of the economy, the only globally perfect predictor for prices is a function of *past expectations* and of *current money balances* only. For fixed M, the price dynamics is one dimensional and determined by the behavior of the perfect predictor ψ^* alone. Observe that the arguments of the perfect predictor are defined from the information sets of two successive dates, requiring the knowledge of past predictions and not of past prices. Since the price law $\mathcal{P}$ is a time invariant function, so is the perfect predictor. Therefore, its properties as well as those of its inverse can be estimated or approximated from time series data under perfect foresight, since any orbit will be contained in the graph of $\mathcal{P}$. In other words, from an empirical point of view, it is straightforward in principle to determine the perfect predictor by standard (non linear) approximation techniques. As a consequence, the time-one map generating the perfect foresight orbits of money balances and expectations can now be defined in the following way. Let $(M_{t-1}, p^e_{t-1}) \equiv (M_{t-1}, p^e_{t-1,t})$ denote the state at the beginning of period $t-1$ implying $p_{t-1} = \mathcal{P}(M_{t-1}, p^e_{t-1})$ and $M_t = \mathcal{M}(M_{t-1}, p^e_{t-1})$. Then, perfect foresight $p^e_{t-1} = p_t = \mathcal{P}(M_t, p^e_t)$ in period t holds if and only if

$$p^e = \mathcal{P}^{-1}(M, p^e_{-1}) = \mathcal{P}^{-1}(\mathcal{M}(M_{-1}, p^e_{-1}), p^e_{-1}) =: \psi^*(M, p^e_{-1}).$$

Therefore, the updating rule for the perfect prediction must be

$$p^e_1 = \phi^*(M, p^e) := \psi^*(\mathcal{M}(M, p^e), p^e). \tag{4.1.15}$$

Hence, (4.1.15) combined with (4.1.3) induces the two dimensional dynamical system

$$\begin{aligned} M_t &= \mathcal{M}(M_{t-1}, p^e_{t-1}) \\ p^e_t &= \phi^*(M_{t-1}, p^e_{t-1}) := \psi^*(\mathcal{M}(M_{t-1}, p^e_{t-1}), p^e_{t-1}) \end{aligned} \tag{4.1.16}$$

which exhibits the perfect foresight property

$$p_t = \mathcal{P}(M_t, p^e_t) = \mathcal{P}(\mathcal{M}(M_{t-1}, p^e_{t-1}), \phi^*(M_{t-1}, p^e_{t-1})) = p^e_{t-1} \quad \text{for all} \quad t. \tag{4.1.17}$$

Figure 4.1 describes the functional structure of the dynamical system (4.1.16) under perfect foresight with an expectational lead. It shows that the predictor ψ^* is perfect if and only if $(\mathcal{M}, \phi^*) \circ \mathcal{P} = \text{proj}_2$, where $\text{proj}_2\,(M_{-1}, p^e_{-1}) = p^e_{-1}$. Observe that the orbits are defined in the space of money balances and price expectations $\mathbb{R}^2_+$, while the perfect foresight property is verified from the projection property on the graph of the price law with a one period lead. Alternatively, as the diagram suggests,

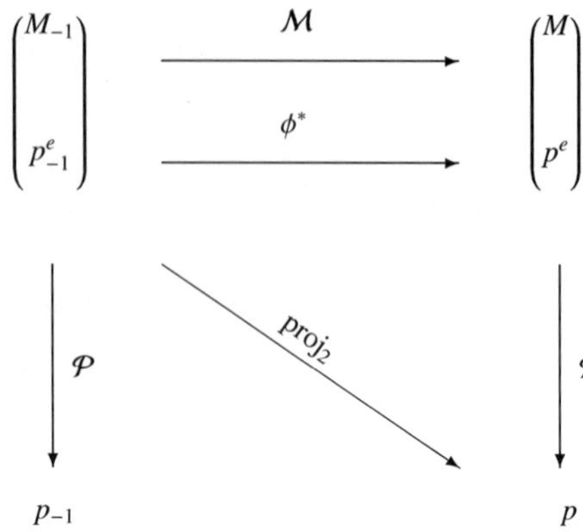

Fig. 4.1 Dynamics with perfect foresight: $(\mathcal{M}, \phi^*) \circ \mathcal{P} = \text{proj}_2$

the system (4.1.16) induces an equivalent mapping on graph $\mathcal{P} \subset \mathbb{R}^3_+$ with unique perfect foresight orbits $\{(M_t, p^e_t, \mathcal{P}(M_t, p^e_t))\}$, $(M_t, p^e_t, \mathcal{P}(M_t, p^e_t)) \in \mathbb{R}^3_+$.

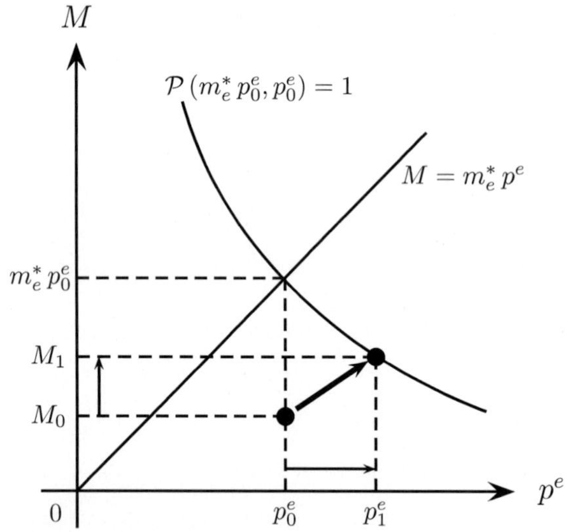

Fig. 4.2 One-step change in state space under perfect prediction

Figure 4.2 illustrates the mechanism of identifying a perfect forecast in any period using a geometric construction which is equivalent to finding the value of the inverse of the price law for a previous state (M_0, p^e_0). Consider the level of expected real money balances $m^*_e := (M/p^e)^*$ solving $\mathcal{P}(m^*_e, 1) = 1$. From homogeneity one has the identity $\mathcal{P}(m^*_e p^e, p^e) = p^e$ for all p^e. Then, for any given state $(M_0, p^e_0) \in \mathbb{R}^2_+$ a prediction p^e_1 is perfect if (M_1, p^e_1) belongs to the level set of $p^e_0 = \mathcal{P}(M, p^e)$, i.e. if

$$\mathcal{P}(\mathcal{M}(M_0, p_0^e), p_1^e) = p_0^e.$$

Therefore, after choosing the one-step change of money balances $M_1 = \mathcal{M}(M_0, p_0^e)$, (which is independent of p_1 !), p_1^e satisfies $\mathcal{P}(M_1, p_1^e) = \mathcal{P}(m_e^*, p_0^e) = p_0^e$ by construction. In other words, $p_1^e = \psi^*(M_1, p_0^e)$, so that the one-step change $(M_0, p_0^e) \mapsto (M_1, p_1^e)$ has the perfect foresight property. This procedure shows that it is sufficient for the forward recursion to know the level set associated with the previous prediction p_0^e and its local inverse at M_1. Global properties of the perfect predictor ψ^* are not required.

4.1.4 Money Balances and Prices under Perfect Foresight

Given the perfect predictor ψ^*, an orbit $\left\{(M_t, p_{t,t+1}^e, p_t)\right\}_{t=0}^{\infty}$ of the extended dynamical system (4.1.16) with the perfect foresight property for all t, can also be rewritten as

$$\begin{aligned} M_{t+1} &= \mathcal{M}(M_t, p_{t,t+1}^e) = M_t + p_t(g - \tau D(M_t, p_t)) \\ p_{t,t+1}^e &= \psi^*(M_t, p_{t-1,t}^e) \\ p_t &= \mathcal{P}(M_t, p_{t,t+1}^e). \end{aligned} \tag{4.1.18}$$

Since ψ^* is perfect, i.e. $\mathcal{P}(M, \psi^*(M, p)) = p$ holds for all (M, p), the orbit satisfies $p_{t-1,t}^e = p_t$ and $p_{t,t+1}^e = p_{t+1}$, for all t, so that

$$p_{t+1} = \psi^*(M_t, p_t). \tag{4.1.19}$$

holds for all t implying the time-one map

$$\begin{aligned} M_{t+1} &= \mathcal{M}(M_t, \psi^*(M_t, p_t)) \\ p_{t+1} &= \psi^*(M_t, p_t) \end{aligned} \tag{4.1.20}$$

which satisfies perfect foresight[4]. Therefore, given the structural assumptions for the model, the dynamics under perfect foresight is also uniquely described by *a two-dimensional dynamical system* in *nominal money balances* and *nominal prices* (M, p), which are the observable data in each period, requiring no information on expectations (predicted values) along equilibrium orbits with perfect foresight. As a consequence any orbit of money balances and prices for (4.1.20) induces sequences of wages, output and employment $\{w_t, y_t, L_t\}$ generated by the associated maps

$$w_t = \mathcal{W}(M_t, p_{t,t+1}^e), \quad y_t = \mathcal{Y}(M_t, p_{t,t+1}^e), \quad L_t = \mathcal{L}(M_t, p_{t,t+1}^e)$$

[4] In fact the orbits of (4.1.20) are the projection onto the first and third component of the orbits of the extension of system (4.1.16) to a time-one map on graph $\mathcal{P}$ which is equivalent to both systems (4.1.16) and (4.1.22).

such that

$$p^e_{t,t+1} = \psi^\star(M_t, p_t) = p_{t+1}.$$

Since there is no expectations feed back in aggregate demand, the equilibrium condition for the commodity market under the restriction of perfect foresight implies that output is uniquely determined by the state (M_t, p_t)

$$\mathcal{Y}(M_t, p_{t+1}) = D(M_t, p_t).$$

Therefore, money balances evolve according to

$$M_{t+1} = \mathcal{M}(M_t, \psi^*(p_t, M_t)) = M_t + p_t(g - \tau D(M_t, p_t)). \tag{4.1.21}$$

Finally, since $p = \mathcal{P}(M, p^e)$ if and only if $AS(p^e/p) = D(M, p)$, it follows that $\psi^*(M_t, p_t) = p_t AS^{-1}(D(M_t, p_t))$. Combined with (4.1.21) one obtains an explicit reduced form of the dynamical system (4.1.20) under perfect foresight as

$$\begin{aligned} M_{t+1} &= \mathcal{M}(M_t, \psi^*(M_t, p_t)) = M_t + p_t(g - \tau D(M_t, p_t)) \\ p_{t+1} &= \psi^*(M_t, p_t) = p_t AS^{-1}\left(D(M_t, p_t)\right). \end{aligned} \tag{4.1.22}$$

This exhibits the dominant influence of the aggregate demand function on money balances while the aggregate supply function plays an essential role for the evolution of prices, namely determining the level of inflation. The functions are defined in terms of the two observable state variables, money balances and prices, no expected prices appear. They are integrated (eliminated by substitution) which is possible because of the lack of an expectation effect in aggregate demand. Notice that both mappings are homogeneous of degree one in (M, p) if the aggregate demand function is homogeneous of degree zero.

4.1.5 Stationary Monetary States

Let us first examine the possibility of steady states under perfect foresight of system (4.1.22) (or equivalently system (4.1.20)). Consider a *stationary monetary steady state* $(\bar{M}, \bar{p}) \gg 0$ defined by a positive fixed point of the system (4.1.22). From the first equation, constant money balances $M_t = M_{t+1} = \bar{M}$ imply a balanced government budget

$$\bar{p}(g - \tau\bar{y}) = 0,$$

which can be obtained if and only if $g/\tau = \bar{y} = D(\bar{M}, \bar{p})$. Since g and τ are given parameters, it follows that the stationary output level $\bar{y}$ is determined exclusively by the demand side and independently of the features of production, the labor market, and the commodity market. To understand the implications of this condition, consider the function $\Delta M(p, M) = M_{t+1} - M_t := pg - \tau D(M, p)$ and define its zero contour

$$\Delta M_0 := \{(M, p) \in \mathbb{R}^2_+ \,|\, \Delta M(M, p) = 0\} = \{(M, p) \,|\, g/\tau = D(M, p)\}$$

which is a contour (a level set) of the demand function in the state space $\mathbb{R}^2_+$. Therefore, ΔM_0 partitions the state space into regions where money balances increase and decrease, according to

$$\frac{\partial}{\partial M}\Delta M(M, p) = -\tau \frac{\partial D}{\partial M} < 0$$

(see Figure 4.3). ΔM_0 is a straight line in $\mathbb{R}^2_+$, whenever $D(M, p)$ is homogeneous of degree zero in (M, p). In other words, given the parameters of the economy constant money balances occur for a specific value of real money balances only.

To investigate conditions of stationary prices, define the function of price change as

$$\Delta p(M, p) := p\left(AS^{-1}(D(M, p)) - 1\right). \tag{4.1.23}$$

Thus, from the second equation of (4.1.22) one finds that $p_t = p_{t+1} = \bar{p}$, i.e. constant prices (=zero inflation) and perfect foresight $\bar{p}^e = \bar{p}$, imply $AS(1) = D(\bar{p}, \bar{M}) = \bar{y}$, inducing steady state output to be $\bar{y} = AS(1)$. In other words, the set

$$\Delta p_0 := \{(p, M) | \Delta p(p, M) = 0\} = \{(p, M) \,|\, AS(1) = D(p, M)\}$$

is also a contour of the demand function, partitioning the space into price increases and decreases according to

$$\frac{\partial}{\partial p}\Delta p(p, M) = \frac{\partial \psi^\star}{\partial p} - 1 > 0.$$

Thus, Δp_0 is also a straight line in $\mathbb{R}^2_+$, whenever $D(p, M)$ is homogeneous of degree zero in (p, M), but determined by the parameters of demand and of supply. Therefore, $(\bar{M}, \bar{p})$ is a positive stationary state, if and only if $(\bar{M}, \bar{p}) \in \Delta p_0 \cap \Delta M_0 \subset \mathbb{R}^2_{++}$ and $\Delta p_0 \cap \Delta M_0$ is non-empty if and only if $AS(1) = g/\tau$, which implies $\Delta p_0 = \Delta M_0$. In this case, there exists a continuum of stationary points, since every $(M, p) \in \Delta p_0 = p\Delta M_0$ is a fixed point of (4.1.22).

Evidently, equality of $AS(1)$ and g/τ cannot hold in general for arbitrary choices of the parameters. Therefore, stationary states $(\bar{M}, \bar{p}) \gg 0$ fail to exist, whenever $AS(1) \neq g/\tau$. They exist if and only if $AS(1) = g/\tau$, which implies a continuum of stationary states of the system (4.1.22).

$$\left\{(\bar{M}, \bar{p}) \mid AS(1) = \frac{g}{\tau} = D(\bar{M}, \bar{p})\right\} \in \mathbb{R}^2_+$$

which is a half line in $\mathbb{R}^2_+$, defined by $M/P = (1 - c(1 - \tau))(AS(1)) - g$.

Although stationary monetary steady states are non-generic, some further insight into the qualitative features of the dynamics of the system (4.1.22) for the excep-

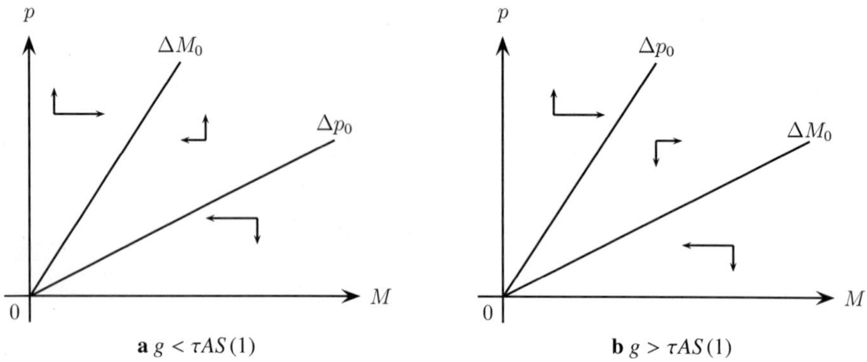

Fig. 4.3 Qualitative dynamics of money and prices under perfect foresight

tional case $AS(1) = g/\tau$ can be obtained from the eigenvalues of the Jacobian of (4.1.18) evaluated at $(\bar{M}, \bar{p})$.

Lemma 4.1.2.
Let $(\bar{p}, \bar{M}) \gg 0$ *denote a fixed point of* (4.1.18). *If the aggregate supply function is globally invertible and if*

$$0 < \bar{p}\,\tau \frac{\partial}{\partial M} D(M, \bar{p}) < 1 \qquad \textit{and} \qquad \frac{\partial}{\partial p} D(M, p) < 0, \tag{4.1.24}$$

the Jacobian has positive real eigenvalues, one of which is always unity.

Proof. When aggregate supply is invertible, the system (4.1.22) takes the form

$$\begin{aligned} M_{t+1} &= M_t + p_t\,(g - \tau\, D(M_t, p_t)) \\ p_{t+1} &= p_t AS^{-1}\,(D(M_t, p_t))\,. \end{aligned} \tag{4.1.25}$$

Its Jacobian evaluated at $(\bar{M}, \bar{p})$ is

$$\frac{\partial(M_{t+1}, p_{t+1})}{\partial(M_t, p_t)} = \begin{pmatrix} 1 - \bar{p}\,\tau \dfrac{\partial D}{\partial M} & -\bar{p}\,\tau \dfrac{\partial D}{\partial p} \\[2ex] \bar{p} \dfrac{1}{AS'(1)} \dfrac{\partial D}{\partial M} & 1 + \bar{p} \dfrac{1}{AS'(1)} \dfrac{\partial D}{\partial p} \end{pmatrix} \tag{4.1.26}$$

with trace

$$\mathrm{tr} = 2 + \bar{p} \left(\frac{\partial D/\partial p - \tau\, AS'(1)(\partial D/\partial M)}{AS'(1)} \right) > 1$$

and determinant

$$\det = 1 + \bar{p} \left(\frac{\partial D/\partial p - \tau\, AS'(1)(\partial D/\partial M)}{AS'(1)} \right) = \mathrm{tr} - 1.$$

Therefore, the system has two positive real roots, given by

$$\lambda_{1,2} = \begin{cases} 1 \\ \mathrm{tr} - 1 \end{cases} = \begin{cases} 1 \\ 1 + \bar{p}\left(\dfrac{\partial D/\partial p - \tau AS'(1)(\partial D/\partial M)}{AS'(1)}\right) \end{cases}. \qquad \square$$

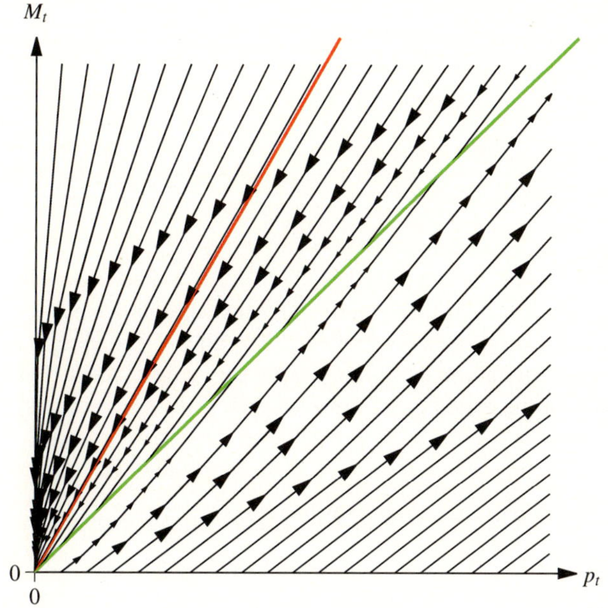

Fig. 4.4 Convergence to continuum of stationary monetary states (= green halfline)

Summarizing one finds that stationary monetary states are rare and non-hyperbolic. They are asymptotically stable when $-1 < \lambda_2 < 1$ if and only if $0 < \mathrm{tr} < 2$. This condition (4.1.24) implies that the trace is larger than one, so that both eigenvalues are positive, which excludes cycles. The second eigenvalue λ_2 is less than 1 if $\mathrm{tr} < 2$. Let E_p and E_{AS} denote the elasticities of the demand and of the supply function. Rewriting the trace as

$$\mathrm{tr} = 2 + \frac{\bar{p}\partial D/\partial p}{AS'(1)} - \tau\bar{p}\frac{\partial D}{\partial M} = 2 + \frac{E_p}{E_{AS}} - \tau\bar{p}\frac{\partial D}{\partial M},$$

shows that the trace will be larger than 2 if the ratio $E_p/E_{AS} > 1$. Since the properties determining the aggregate demand and aggregate supply functions are independent of each other, any size of the ratio $E_p/E_{AS} > 0$ is economically possible. Therefore, divergence occurs if the elasticity of aggregate supply is sufficiently small which, for example, holds for a large range of parameters in the isoelastic case (see Section 3.2.7). Thus, asymptotic stability of the stationary state is not determined universally, sincein general, the critical size of the trace ($tr = 2$) depends on specific features of demand and supply, which cannot be determined in general. Figure 4.4 displays the stable situation with a continuum of monetary steady states.

The stability analysis of the non-generic situation could be misleading to suggest that there might exist stable situations of perfect foresight dynamics of prices alone with constant money balances. This is not the case, since the price dynamics with constant money balances turns out to be unstable in all cases. Notice that the global properties of the price map $\mathcal{P}$ determine the stability of the price dynamics for any value of money balances. Since $\mathcal{P}$ is monotonically increasing in p^e with elasticity less than one (see (3.2.19)), predicted prices are an increasing function of past predictions with an elasticity greater than one, i.e.

$$\frac{\partial \psi^*(p^e_{t-1}, M_t)}{\partial p^e_{t-1}}(\bar{p}) = \frac{1}{\dfrac{\partial \mathcal{P}}{\partial p^e}(\psi^*(\bar{p}, M_t), M_t)} > 1. \tag{4.1.27}$$

Therefore, for constant money balances, there exists a unique positive perfect foresight steady state $\bar{p}$, which is asymptotically unstable under perfect foresight dynamics. All price orbits are monotonically diverging. Depending on the boundary behavior of the price and wage laws, output and employment diverge always to boundary levels of the economy as expected prices diverge to infinity or to zero.

4.1.6 Money: A Constant Veil for Stationary Economies?

The system describing the dynamical evolution of the monetary economy under perfect foresight given by (4.1.18), by (4.1.20), or by (4.1.22), has no steady states for all government parameters $g \neq \tau AS(1)$. Therefore, for almost all policy parameters $(g, \tau) \gg 0$ stationary monetary equilibria do not exist and divergence of prices and money balances occurs with probability one. In other words, one finds that the main ingredients of this basic monetary macroeconomic model, consisting of competition in both markets, market clearing, and perfect foresight are mutually inconsistent with stationarity, except on a set of measure zero of fiscal parameters of such economies. Even in the non-generic case, asymptotic convergence does not prevail always and with constant money balances the price dynamics are unstable. Thus, taken as a descriptive theory, the model of such a competitive economy under perfect foresight dynamics in monetary units could neither be tested nor estimated on data, since stationary time series data from such economies are empirically unobservable with probability one.

It seems evident that these structural properties extend to monetary economies with a richer set of financial assets, with government bonds, private nominal debt or credit, stocks or shares, or insurance. Observe that some of these play no role in a world of certainty and perfect foresight. Thus, the issue of nonexistence has to be investigated again when monetary equilibria are analyzed in a world with random perturbations (see Section 4.3).

These general structural features extend also to economies under non-competitive behavior in markets. The results for the standard non-competitive equilibrium models, as presented in Section 3.3, suggest that the consequences of strategic behavior of firms or consumers generating monopolistic equilibria in the standard monetary model does not modify the general findings of the competitive case substantially. In each case considered, temporary equilibria confirm primarily the known *distortions* of equilibrium allocations relative to the competitive case, while the intertemporal equilibrium configuration is described by a time-invariant price law which has the same structural properties, in particular the homogeneity, as the one for the competitive case. Thus, changes resulting in environments from noncompetitive equilibria are of a qualitative nature of the equilibrium mappings with no essential changes for the dynamics. As a consequence, under perfect foresight these models would imply the same non-stationarity features as in the competitive case. Most likely the dynamics of *imperfectly competitive equilibria*, no matter of which kind, will also induce non-stationary paths under perfect foresight diverging to non monetary equilibria. Thus, for most of the perceived variants of the model, it seems to be impossible for stationary monetary equilibria in such economies to exist in the long run under dynamics with perfect foresight. Other forms of descriptions of markets *and* different forms of the intertemporal linkages for behavior and expectations formation seem to be necessary for such features.

The reason for this generic nonexistence of monetary stationary equilibrium states with perfect foresight lies in the homogeneity property of the time one maps (4.1.18) to (4.1.22), which essentially arises from consequences of Walras' Law, market clearing, and from intertemporal budget consistency. The homogeneity property implies that such economies are either contracting or expanding in monetary entities, which forces the analyst of a monetary macroeconomic system to revise or modify the mathematical methods appropriately in order to understand the dynamics, i.e. to separate out the effects of contraction/expansion, which are essentially scaling properties from those of the structural stability/instability in a reduced or compactified state space determining convergence in 'real terms'. The need to do so has been known by economists for a long time and followed primarily in growth theory in one way or another. This requires a careful choice and evaluation of the space of real economic entities for any economy with more than two real markets or assets. In addition, the methods of dynamic analysis have to be adjusted accordingly, in order to make sure that the structural long-run results derived for the monetary system as a whole do not depend on the specific choice of the compactification chosen. The analysis in the following Sections 4.2 and 4.3 provides answers for the monetary model proposed here (see also Appendix A).

4.2 Balanced Paths of Inflation and Deflation

One reason for the non-existence result shown in the previous section is that the concept of stationary monetary equilibria is too restrictive. The homogeneity of the two mappings is an indication of the fact that temporary equilibria in such economies could be expanding or contracting in nominal terms while real allocations may be constant. In such a case, the investigation should be directed toward examining the structural conditions which may lead to or guarantee the possibility of *balanced orbits* with constant rates of inflation or deflation in monetary economies which induce constant allocations, rather than requiring stationary (constant) monetary states.

If aggregate demand is homogeneous of degree zero in (M, p), additional properties of the dynamics can be deduced when $AS(1) \neq g/\tau$. Since the quantity theory of money holds in this case, there may exist orbits $\{M_t, p_t\}_t$ under perfect foresight where money balances and prices grow or decrease at a constant non-zero rate $\theta = M_{t+1}/M_t = p_{t+1}/p_t$, implying the following definition.

Definition 4.2.1. An orbit $\gamma(M_0, p_0) := \{(M_t, p_t)\}_{t=0}^{\infty}$ of the two dimensional dynamical system (4.1.22) is called *balanced* if there exists a growth factor $\bar{\theta} > 0$, such that

$$M_{t+1} = \bar{\theta} M_t \quad \text{and} \quad p_{t+1} = \bar{\theta} p_t \quad \forall t. \tag{4.2.1}$$

Any balanced growth factor $\bar{\theta} > 0$ implies that real money balances remain constant along a balanced orbit, i.e. for all t, $M_{t+1}/p_{t+1} = M_t/p_t = \bar{m} > 0$. In other words, a balanced path with growth factor $\bar{\theta} > 0$ can always be identified with the associated level of real balances $\bar{m} = M_t/p_t > 0$, $t = 0, 1, \ldots$ defining a half line $L(\bar{m}) := \left\{(M, p) \in \mathbb{R}_+^2 \mid M = \bar{m}p\right\}$ in the state space[5]. Balanced orbits induce constant levels of real balances, real wages, employment, and output. In other words, balanced growth of nominal prices, wages, and money – if it exists – induces constant real money balances. These induce states of the economy with constant levels of output and employment, associated with constant real money balances, constant rates of inflation or deflation, and constant real wages. Therefore, two issue arise, i.e. whether such *balanced paths exist* and whether they are *asymptotically stable* or attracting in the space nominal levels of money balances, prices, and price expectations.

---

[5] Neither the half line nor the vector $\bar{m}$ should be confused with a balanced orbit itself. The half line is the closure of the union of all balanced orbits, while each orbit $\{(M_t, p_t)\}_0^{\infty}$ is the countable set of pairs $(M_t, p_t) \in \left\{(M, p) \in \mathbb{R}_+^2 \mid M = \bar{m}p\right\}$, $t = 0, 1, \ldots$ contained in the half line $L(\bar{m})$. Here the term balanced path will be used both for the half line $L(\bar{m})$ and its associated defining intensity $\bar{m}$. No confusion should arise distinguishing it from a balanced orbit as a subset of the balanced path.

4.2.1 Dynamics of Real Balances under Perfect Foresight

Before discussing the convergence issue of orbits to balanced paths, Consider first the existence issue of balanced paths in the space of nominal values. This is associated with and completely determined by the dynamics of real money balances along an orbit with perfect foresight. The homogeneity of the price law and the perfect predictor in fact guarantee the existence of a one-dimensional dynamical system for real balances alone.

Let aggregate supply be globally invertible (as for example in the isoelastic case, see Section 3.2.7) and assume that aggregate demand is homogeneous of degree zero in (M, p) so that $D(M, p, g, \tau) \equiv D(M/p, g, \tau)$. Define $m_t := M_t/p_t$ and $\theta_t := p_{t+1}/p_t$. Using (4.1.22) one obtains a one dimensional dynamical system

$$m_{t+1} = \mathcal{F}(m_t) := \frac{m_t + (g - \tau D(m_t, g, \tau))}{AS^{-1}(D(m_t, g, \tau))} = \frac{(\tilde{c} - \tau) D(m_t, g, \tau)}{AS^{-1}(D(m_t, g, \tau))} \tag{4.2.2}$$

describing the dynamics of real money balances under perfect foresight. An orbit of the system (4.1.20) induces a perfect foresight orbit since $p^e_{t-1,t} = p_t$ for all t and a positive fixed point $\bar{m} > 0$ induces constant real wages, constant employment, and constant output. Let $\bar{m} > 0$ denote a fixed point of the mapping (4.2.2) which implies an associated inflation factor $\bar{\theta} = AS^{-1}(D(\bar{m}, g, \tau))$. Thus, $(\bar{m}, \bar{\theta})$ together solve

$$\begin{aligned} \bar{m}(\bar{\theta} - 1) &= g - \tau AS(\bar{\theta}) \\ AS(\bar{\theta}) &= D(\bar{m}, g, \tau) \end{aligned} \tag{4.2.3}$$

from which one obtains the usual relationship between inflation/deflation and the government budget at steady states

$$\begin{aligned} \bar{\theta} > 1 \quad &\Leftrightarrow \quad g > \tau AS(\bar{\theta}) \\ \bar{\theta} = 1 \quad &\Leftrightarrow \quad g = \tau AS(1) \\ \bar{\theta} < 1 \quad &\Leftrightarrow \quad g < \tau AS(\bar{\theta}). \end{aligned} \tag{4.2.4}$$

Then, any initial state (p_0, M_0) with $M_0 = \bar{m} p_0$ induces a perfect foresight orbit of prices and nominal money balances with constant rate of inflation (or deflation) equal to $\bar{\theta} - 1 := AS^{-1}(D(\bar{m}, g, \tau)) - 1$. These are necessary one-dimensional conditions of balanced paths which seem independent of demand. However, because of market clearing there exists an equivalent one-dimensional condition with respect to aggregate demand since $AS(\bar{\theta}) = D(\bar{m}, g, \tau)$ if and only if $\mathcal{P}(\bar{m}, \bar{\theta}) = 1$, implies a one-one-relationship between $\bar{m}$ and $\bar{\theta}$, the equilibrium set $\mathcal{E}_{\mathcal{P}}$, see (3.2.41) and Figure 3.5.

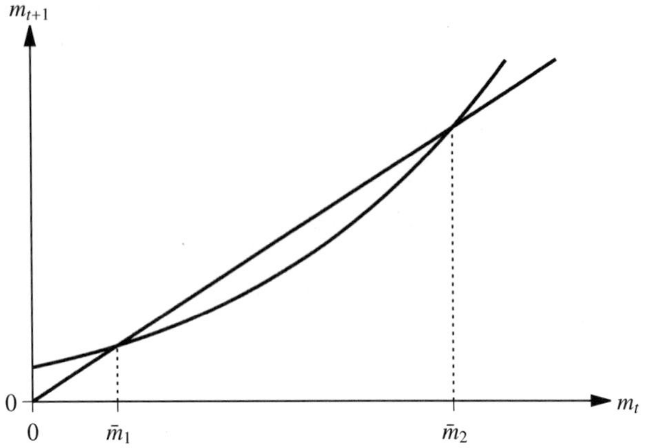

Fig. 4.5 Time one map of real money balances (4.2.2) with perfect foresight

Existence of Balanced Monetary Paths

In general, there may exist no or multiple steady states of (4.2.2). It is apparent that the size of government demand g relative to the tax rate plays a decisive role. Figure 4.5 portrays the graph of the time one map for a typical situation. Let AS be surjective and let $\tilde{\theta}$ denote the solution of $\tau AS(\tilde{\theta}) = g > 0$. Rewriting the equations (4.2.3) for the fixed point as

$$m = \frac{g - \tau AS(\theta)}{\theta - 1},$$

$$\theta = AS^{-1}(D(m, g, \tau))$$

the right hand side of each equation provides global information to determine the steady states $(\bar{m}, \bar{\theta})$.

If $\tilde{\theta} > 1$, then there must exist at least two solutions $(\bar{m}_1, \bar{\theta}_1)$ and $(\bar{m}_2, \bar{\theta}_2)$ with $\bar{m}_1 > \bar{m}_2$ and $\bar{\theta}_2 < 1 < \bar{\theta}_1$. Thus, the path with higher real money balances has deflation while the lower real balances have positive inflation.

If $\tilde{\theta} < 1$, deflationary steady states are unlikely while there may exist multiple or no inflationary paths. In particular, if g is large there is no balanced path. When aggregate demand is a linear function in $m + g$, then some universal properties of the dynamical system in real balances can be obtained, as stated in the following theorem.

Theorem 4.2.1.
Let the aggregate demand function[6] *be given by*

[6] This is the form given in Assumption 3.2.1. The additional parameters have been suppressed as arguments since the discussion of the dynamics under perfect foresight will center on the role of government demand g.

$$D(m, g) := \frac{1}{1 - c(1 - \tau)}(m + g), \qquad 0 < c, \tau < 1,$$

and assume that the aggregate supply function is globally invertible, strictly decreasing, and strictly convex. For all $g \geq 0$ and $0 < c, \tau < 1$:

(a) $\mathcal{F} : \mathbb{R}_+ \to \mathbb{R}_+$ is strictly monotonically increasing and strictly convex.

(b) There exists a critical level $g^\star > 0$, such that there is no steady state for $g > g^\star$.

(c) If $0 \leq g < g^\star$, there are exactly two steady states, $0 \leq \bar{m}_1 < \bar{m}_2$ where the lower one is always asymptotically stable.

(d) If $g = 0$, then $\bar{m} = 0$ is stable while the positive steady state is unstable.

The essential feature which is responsible for this surprisingly general result[7] originates from the structure of the aggregate demand function, which is linear in real balances with an additive shift caused by government demand g. For $g = 0$, one observes directly that $\mathcal{F}$ is strictly convex and strictly monotonically increasing with $\mathcal{F}(0) = 0$. If $g > 0$, the convexity and monotonicity are preserved, so that the dynamics is monotonic. This also implies that there are at most to fixed points inducing the respective stability conditions.

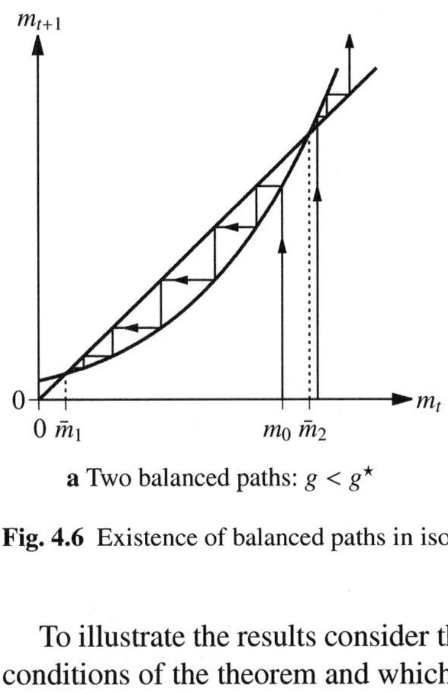

a Two balanced paths: $g < g^\star$

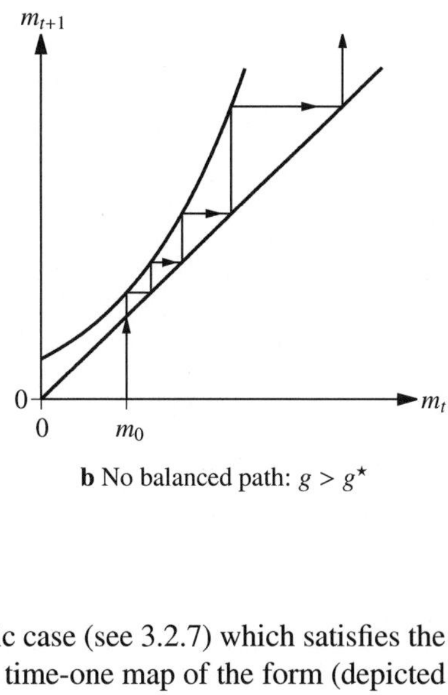

b No balanced path: $g > g^\star$

Fig. 4.6 Existence of balanced paths in isoelastic case

To illustrate the results consider the isoelastic case (see 3.2.7) which satisfies the conditions of the theorem and which induces a time-one map of the form (depicted in Figure 4.5) with explicit analytical form

$$m_{t+1} = \mathcal{F}(m_t) := \tilde{C}^{\frac{1}{b}} (1 - c)(1 - \tau) \left(\frac{m_t + g}{1 - c(1 - \tau)} \right)^{\frac{1+b}{b}}. \qquad (4.2.5)$$

[7] Appendix B contains the proof and more details.

Here

$$b = \frac{BC}{1 + C(1 - B)} > 0 \quad \text{and} \quad \tilde{C} = (B/A_1A_2) > 0$$

are positive constants of the aggregate supply function (derived in Section 3.2.7). For $g > 0$, $\mathcal{F}$ is strictly increasing and strictly convex with $\mathcal{F}(0) > 0$. Hence, there exist at most two positive steady states $0 < \bar{m}_1 < \bar{m}_2$, where

- $\bar{m}_1$ is stable with associated positive inflation $\bar{\theta}_1 > 1$,
- $\bar{m}_2$ is unstable with $\bar{\theta}_2 \lesseqgtr 1$.
- Figure (4.6) displays two alternative situations with two positive fixed points in panel **a**.
- The interval $[0, \bar{m}_2]$ is a forward invariant set with $m_0 \to \bar{m}_1$ for all $m_0 \in [0, \bar{m}_2)$.
- For some critical value $0 < g^\star$ the two steady states coincide $\bar{m}_1 = \bar{m}_2$.
- If $g > g^\star$ no steady state exists and the dynamics of real money balances are globally unstable with all orbits diverging (see panel **b**).

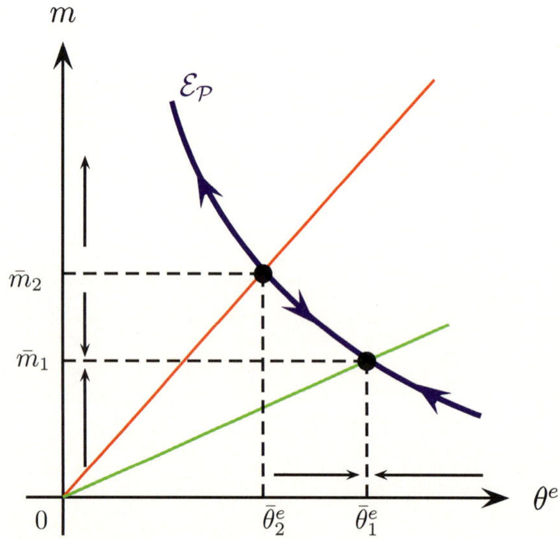

Fig. 4.7 Real dynamics with perfect foresight on equilibrium set with two balanced paths

The dynamics in real space, i.e. on the equilibrium set $\mathcal{E}_\mathcal{P}$ for $0 < g < g^\star$ is shown in Figure 4.7 which indicates that the one-dimensional dynamics of real balances determines a one-to-one dynamics of price inflation inducing the allocative dynamics of output, consumption, and employment. Figure 4.8 shows the role of government demand for the existence of steady states and their stability for the dynamics of real money balances under perfect foresight.

When production is bounded a globally perfect predictor no longer exists. The boundedness of output implies two restrictions for the dynamics under perfect foresight: real balances in temporary equilibrium including steady states must be

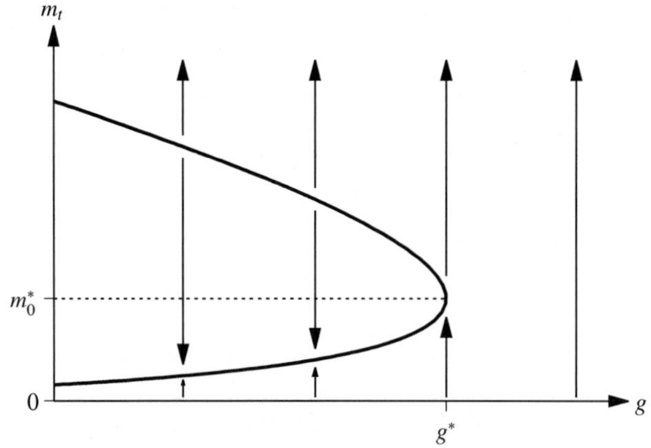

Fig. 4.8 Fixed points $\bar{m}_1(g) < \bar{m}_2(g)$ and stability of real balances: the role of government demand

bounded above and a locally perfect predictor exists only where the aggregate supply function is invertible. Let AS be invertible on its range $[0, AS(0)]$. This implies the maximal level of real money balances consistent with equilibrium as $m_{\max} := (1 - c(1 - \tau))AS(0) - g \equiv \tilde{c}AS(0) - g = 1/\mathcal{P}(1, 0)$. Therefore, the predictor

$$\psi^*(M, p) = p \begin{cases} AS^{-1}\left(\frac{m+g}{\tilde{c}}\right) & m + g < \tilde{c}AS(0) \\ 0 & \text{otherwise} \end{cases} \tag{4.2.6}$$

inducing perfect foresight for $m + g < \tilde{c}AS(0)$, since $\mathcal{P}(M, \psi^*(M, p_{-1})) = p_{-1}$ for $M/p_{-1} + g < \tilde{c}AS(0)$ and $p = M\mathcal{P}(1, 0)$ otherwise. Thus, the system 4.2.2 of the dynamics of real money balances is modified to

$$\mathcal{F}(m) := \min\left(\tilde{c}AS(0) - g, \frac{\tilde{c} - \tau}{\tilde{c}} \frac{m + g}{AS^{-1}\left(\frac{m+g}{\tilde{c}}\right)}\right).$$

Theorem 4.2.2 states the result for existence and stability of the real dynamics under perfect foresight with bounded production which is similar to the one of Theorem 4.2.1 under global invertibility.

Theorem 4.2.2.
Let the aggregate demand function be given by

$$D(m, g) := \frac{1}{1 - c(1 - \tau)}(m + g), \qquad 0 < c, \tau < 1,$$

and assume that the aggregate supply function is strictly decreasing and strictly convex with $AS(0) < +\infty$ *and* $\lim_{\theta^e \to \infty} AS(\theta^e) = 0$. *For all* $g \geq 0$ *and* $0 < c, \tau < 1$*:*

The predictor ψ^* *given by* (4.2.6) *induces dynamics of real balances with perfect foresight given by*

$$\mathcal{F}(m) := \min\left(\tilde{c}AS(0) - g, \frac{\tilde{c} - \tau}{\tilde{c}} \frac{m + g}{AS^{-1}\left(\frac{m+g}{\tilde{c}}\right)}\right). \tag{4.2.7}$$

on a bounded subset of the interval of real money balances $[0 \leq \tilde{c}AS(0) - g]$.

(a) $\mathcal{F} : \mathbb{R}_+ \to \mathbb{R}_+$ *is strictly increasing and strictly convex on* $(0, \tilde{c}AS(0) - g)$.

(b) *There exists a critical level* $g^\star > 0$, *such that* $\bar{m} = \tilde{c}AS(0) - g$ *is the only steady state for* $\tilde{c}AS(0) \geq g > g^\star$.

(c) *If* $\tilde{c}AS(0) > g \geq 0$, *there are three positive steady states*

$$0 < \bar{m}_1(g) < \bar{m}_2(g) < \bar{m}_3(g) = \tilde{c}AS(0).$$

$\bar{m}_1(g)$ *and* $\bar{m}_3(g)$ *are asymptotically stable while* $\bar{m}_2(g)$ *is unstable.*

(d) *All orbits from the basin of attraction* $[0, \bar{m}_2(g))$ *of* $\bar{m}_1(g)$ *have perfect foresight and converge monotonically to* $\bar{m}_1(g)$.

(e) *All orbits starting from* $m > \bar{m}_2(g)$ *converge in finite time to* $\bar{m}_3(g) = \tilde{c}AS(0) - g$ *which does not have the perfect foresight property.*

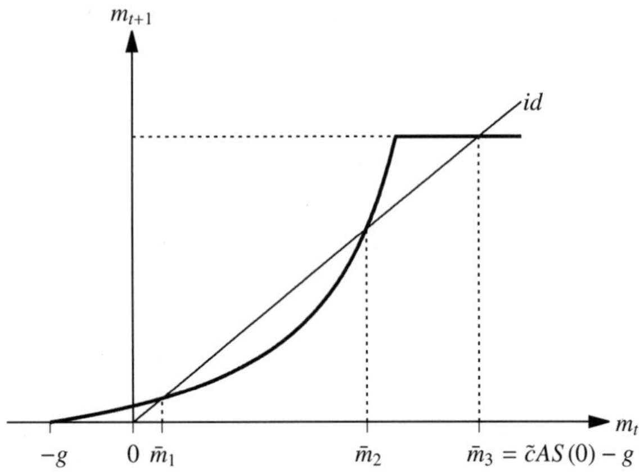

Fig. 4.9 Time one map (4.2.2) with perfect foresight and bounded production: $g < g^\star$

The proof of the properties (a) to (d) follow from Lemma B.2.1 and the proof of Theorem 4.2.1, (e) and (f) are implications of the monotonicity of $\mathcal{F}$. Figure 4.9 portrays the typical situation with three steady states.

Stability of Balanced Paths

It is well known for two dimensional systems defined by maps which are homogeneous of degree one, that convergence/divergence of the system in intensity form is only a necessary condition for convergence in the state space (see Deardorff, 1970; Jensen, 1994; Böhm, 2009; Böhm, Pampel & Wenzelburger, 2005; Pampel, 2009). Convergence of an orbit $\gamma(m_0)$ in real balances for the mapping (4.2.2) to the level $0 < \bar{m} \neq m_0$ does not imply the convergence of an associated orbit $\gamma(p_0, M_0) = \{(p_t, M_t)\}_{t=0}^{\infty}$, $m_0 = M_0/p_0$ of the system (4.1.22) to the half line defined by $\bar{m}$. To examine the convergence define the distance between any state (p, M) from the half line by $\Delta((p, M), \bar{m}) := M - \bar{m}p$. Hence, for an orbit $\gamma(p_0, M_0)$, at any $t \geq 0$, the distance of the state (p_t, M_t) from the half line is given by

$$\Delta_t := M_t - \bar{m}p_t = (m_t - \bar{m})p_t. \tag{4.2.8}$$

Applying the same formula for $t + 1$ one obtains

$$\Delta_{t+1} = (m_{t+1} - \bar{m})p_{t+1} = AS^{-1}(D(m_t))\frac{\mathcal{F}(m_t) - \bar{m}}{m_t - \bar{m}}\Delta_t \tag{4.2.9}$$

which is a function of (m_t, Δ_t). Therefore, along any orbit, the sequence of distances is well defined. Moreover, $\lim_{t\to\infty} \Delta_t = 0$ implies that the orbit converges point wise to the half line when $\lim_{t\to\infty} m_t = \bar{m}$. In this case we will say that the balanced path is asymptotically stable.

Figure 4.10 displays qualitatively the two possible scenarios: convergence (green) vs. divergence (red) in the state space $(p, M) \in \mathbb{R}^2_+$ as defined above analytically. It is assumed that real balances decrease, i.e. $m_t > m_{t+1}$, where $(M_t, p_t) \mapsto (M_{t+1}, p_{t+1})$ portrays the diverging case with an associated increase of $\Delta_t < \Delta_{t+1}$ while $(M_t, p_t) \mapsto (M_{t+1}, p_{t+1})$ shows the converging case with a decrease of $\Delta_t > \Delta_{t+1}$. The reason for

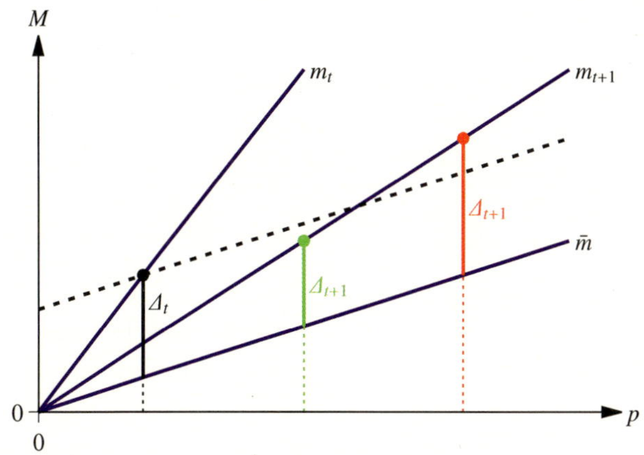

Fig. 4.10 Stability of balanced paths: convergence vs. divergence

the convergence/divergence in the state space $\mathbb{R}^2_+$ for the *same* convergence scenario in real balances arises from the fact that the inflation factor in the convergent case is substatially smaller than the one in the divergence case. Thus, convergence or divergence depends on the rates of expansion, as can also be seen directly from the analytical result presented in Theorem 4.2.3.

Since the mapping $\mathcal{F}$ is monotonically increasing, the sequence $\{\Delta_t\}$ is either positive or negative for all t depending on whether the initial $\Delta_0 := M_0 - \bar{m}p_0$ is positive or negative. Combining (4.2.9) with (4.2.2) one obtains a two-dimensional dynamical system

$$\begin{aligned} m_{t+1} &= \mathcal{F}(m_t) \\ \Delta_{t+1} &= AS^{-1}(D(m_t))\frac{\mathcal{F}(m_t) - \bar{m}}{m_t - \bar{m}}\Delta_t \end{aligned} \tag{4.2.10}$$

whose fixed points are $(\bar{m}, 0)$ with $\bar{m} = \mathcal{F}(\bar{m})$. As eigenvalues one obtains

$$\nu_1 = \mathcal{F}'(\bar{m}) \quad \text{and} \quad \nu_2 = AS^{-1}(D(\bar{m}))\mathcal{F}'(\bar{m}), \tag{4.2.11}$$

which are both positive. Therefore, the fixed point $(\bar{m}_i, 0)$ is asymptotically stable if and only if

$$\max\left(\mathcal{F}'(\bar{m}_i), AS^{-1}(D(\bar{m}_i))\mathcal{F}'(\bar{m}_i)\right) < 1, \qquad i = 1, 2. \tag{4.2.12}$$

Therefore, for the two possible fixed points $0 < \bar{m}_1 < \bar{m}_2$ of $\mathcal{F}$, the higher one $(\bar{m}_2, 0)$ is unstable since $\mathcal{F}'(\bar{m}_2) > 1$. It is either a saddle or a source of (4.2.10). The lower one $(\bar{m}_1, 0)$ is either a saddle or a sink depending on the parameters of the aggregate demand function and the aggregate supply function. Orbits are monotonic, cycles cannot appear.

Using the features of the asymptotic convergence properties of the augmented mapping (4.2.10) one can show that the distance from the balanced path either converges to zero or becomes unbounded with $\lim_{t\to\infty} |\Delta_t| = \infty$, as stated in the following theorem, which is a special case of Theorem A.2.1 in Appendix A.

Theorem 4.2.3.
Let $\mathcal{F}$ be differentiable and let $\bar{m} > 0$ be an asymptotically stable fixed point of (4.2.2) *and $m_0 \in \mathcal{B}(\bar{m})$ the basin of attraction of $\bar{m}$. Let $\gamma(p_0, M_0)$ be an orbit of* (4.1.20) *with $m_0 := M_0/p_0 \neq \bar{m}$, $\Delta_0 := M_0 - \bar{m}p_0 \neq 0$, and $\gamma(m_0, \Delta_0)$ be the associated orbit of* (4.2.10). *Then, the following result holds:*

$$\text{If} \quad \mathcal{F}'(\bar{m})\, AS^{-1}(D(\bar{m})) > 1, \quad \text{then} \quad \lim_{t\to\infty} |\Delta_t| = \infty \tag{4.2.13}$$

$$\text{If} \quad \mathcal{F}'(\bar{m})\, AS^{-1}(D(\bar{m})) < 1, \quad \text{then} \quad \lim_{t\to\infty} |\Delta_t| = 0 \tag{4.2.14}$$

Proof. Let $m_0 > 0$ and $m_0 \neq \bar{m}$. Then

$$\Delta_{t+1} = M_{t+1} - \bar{m}p_{t+1} = AS^{-1}D(m_t)\frac{\mathcal{F}(m_t) - \bar{m}}{m_t - \bar{m}}\Delta_t \qquad \forall t \in \mathbb{N}$$

implies

$$\frac{\Delta_{t+1}}{\Delta_t} = AS^{-1}D(m_t)\,\frac{\mathcal{F}(m_t) - \bar{m}}{m_t - \bar{m}}.$$

Since m_t converges to $\bar{m}$ and $m_0 \in \mathcal{B}(\bar{m})$,

$$\lim_{t\to\infty} \frac{\Delta_{t+1}}{\Delta_t} = AS^{-1}(D(\bar{m}))\,\mathcal{F}'(\bar{m}).$$

This implies $|\frac{\Delta_{t+1}}{\Delta_t} - AS^{-1}(D(\bar{m}))\,\mathcal{F}'(\bar{m})| < \epsilon$ for t larger than some t_0. Therefore,

$$[AS^{-1}(D(\bar{m}))\,\mathcal{F}'(\bar{m}) - \epsilon]|\Delta_t| < |\Delta_{t+1}| < [AS^{-1}(D(\bar{m}))\,\mathcal{F}'(\bar{m}) + \epsilon]|\Delta_t|, \quad t \geq t_0,$$

and by induction

$$[AS^{-1}(D(\bar{m}))\,\mathcal{F}'(\bar{m}) - \epsilon]^{\tau}|\Delta_{\tau+t_0}| < |\Delta_{t+t_0}| < [AS^{-1}(D(\bar{m}))\,\mathcal{F}'(\bar{m}) + \epsilon]^{\tau}|\Delta_{t_0}|, \quad \tau > 0.$$

Therefore, for ϵ sufficiently small,

$$\mathcal{F}'(\bar{m})\,AS^{-1}(D(\bar{m})) < 1 \quad \Rightarrow \quad \mathcal{F}'(\bar{m})\,AS^{-1}(D(\bar{m})) + \epsilon < 1$$

so that $\lim_{t\to\infty} \Delta_t = 0$. Conversely,

$$\mathcal{F}'(\bar{m})\,AS^{-1}(D(\bar{m})) > 1 \quad \Rightarrow \quad \mathcal{F}'(\bar{m})\,AS^{-1}(D(\bar{m})) - \epsilon > 1$$

so that $\lim_{t\to\infty} |\Delta_t| = \infty$. □

Figure 4.11 displays the dynamic behavior for the two possible cases of either two unstable or one stable and one unstable fixed point of (4.2.10). Figure 4.12 shows the implications in either case for convergence and divergence in the state space. In

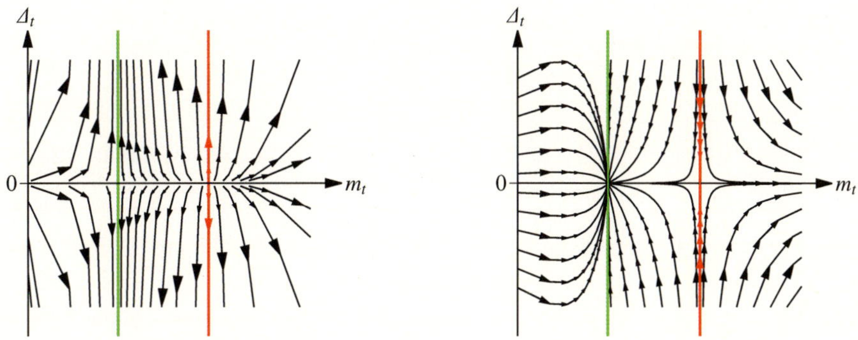

a Two unstable balanced paths **b** One stable – one unstable path

Fig. 4.11 Stability of balanced paths in (Δ, m)-space

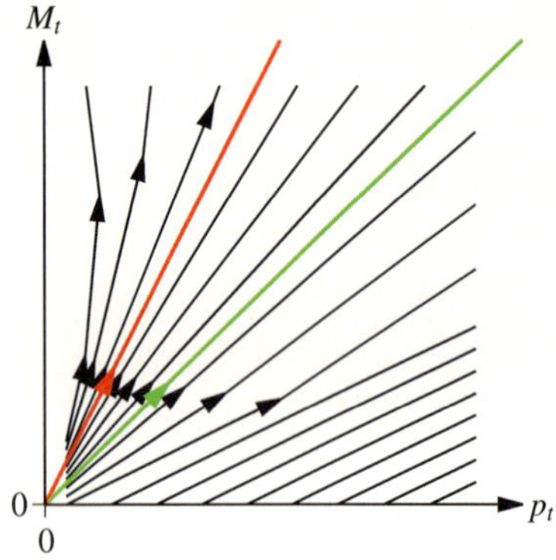

a Two unstable balanced paths

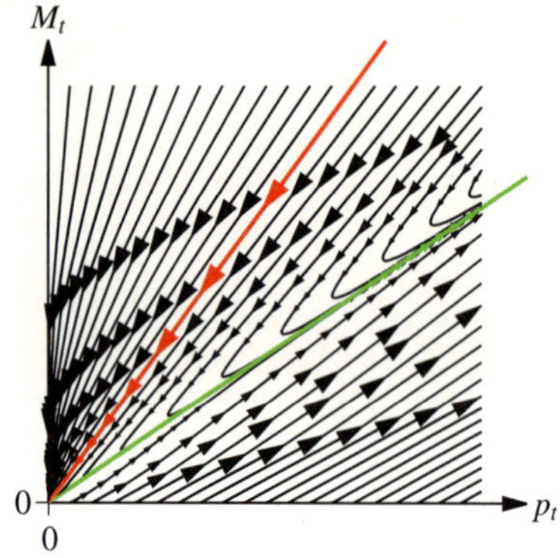

b One stable – one unstable path

Fig. 4.12 Stability of balanced paths in state space

order to understand the stability condition (4.2.12), the value $AS^{-1}(D(\bar{m})) =: \bar{\theta}$ is the value of the inflation factor along the balanced path, which can be larger or smaller than one. Thus, stability prevails if the rate of inflation is not too high relative to the contractivity of the mapping $\mathcal{F}$. The role of the parameters of the aggregate demand and supply functions for stability can be exhibited directly. If aggregate demand is of the form $D(m) = (m + g)/\tilde{c} = D(m)/[1 - c(1 - \tau)]$,$0 < c < 1$ and $0 < \tau < 1$, then

$$\mathcal{F}'(m) = \frac{\tilde{c} - \tau}{\tilde{c}} \cdot \frac{1 - E_{AS^{-1}}(D(m))}{AS^{-1}(D(\bar{m}))}$$

where $E_{AS^{-1}}$ is the elasticity of the inverse of the aggregate supply function. Therefore,

$$AS^{-1}(D(\bar{m}_1))\mathcal{F}'(\bar{m}_1) < 1 \quad \text{if and only if} \quad \frac{\tilde{c} - \tau}{\tilde{c}} \cdot [1 - E_{AS^{-1}}(D(\bar{m}_1))] < 1. \tag{4.2.15}$$

Consider the isolelastic case of Section 3.2.7 with the aggregate supply function given as

$$AS(\theta^e) = AS(1)\,\theta^{1/\bar{A}} \quad \text{with} \quad \bar{A} = -\frac{C + 1 - B}{B} < 0$$

and $0 < B < 1$ and $C > 0$. Then, as a consequence of Theorem 4.2.3, the bifurcation curve/surface in parameter space dividing the stable and the unstable regions is defined by the unit contour of the mapping (4.2.12) given by the set

$$\left\{(c, \tau, B, C) \in \mathbb{R}^4_+ \,\middle|\, \frac{(1 - c)(1 - \tau)}{1 - c(1 - \tau)} \cdot \frac{C + 1}{B} = 1\right\} \tag{4.2.16}$$

for the parameters of the isoelastic example. These conditions neither depend on the level of government demand g nor on the stationary level of real balances $\bar{m}_1$. Moreover, the multiplicative form of the expression (with the first term less than one and the second term larger than one) shows that stable as well as unstable situations occur for each pair of demand parameters or supply parameters, since the demand multiplier term has full range $(0, 1)$ while the supply term has full range $(0, \infty)$ as well. Therefore, continuity implies that pairwise associated open sets of stable and unstable values always exist. Geometrically the stable region of parameters is characterized by quadruples $(c, \tau, B, C) \in \mathbb{R}^4_+$ which satisfy the inequality

$$\frac{(1-c)(1-\tau)}{1-c(1-\tau)}(C+1) < B. \tag{4.2.17}$$

Figure 4.13 displays the stable/unstable regions in two different sections for C against B and for τ against B. These indicate several comparative-dynamics effects

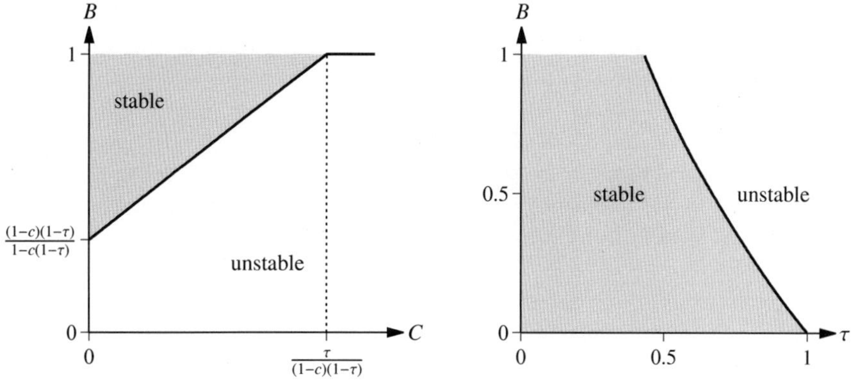

a C versus B: $A = 1$, $c = 0.5$, $\tau = 0.5$ **b** τ versus B: $c = 0.5$, $C = 1.5$

Fig. 4.13 Regions of stability of balanced paths for isoelastic case

of the parameters for stability.

- Crossing the boundary from stable to unstable induces a bifurcation changing the fixed point of the system (4.2.10) from a sink to a saddle. No cyclical movements, i.e. no switching across the balanced path occurs.
- For given values of the supply side (B, C) there always exist open sets of consumption parameters which induce a stable balanced path. Thus, in particular, the government can always find a level of the tax rate $0 < \tau < 1$ high enough which induces stability.
- Conversely, for each given pair of consumption values (c, τ) there exist connected open sets of supply parameters (B, C) for which balanced paths are asymptotically stable.

Summarizing, one finds that the dynamics under perfect foresight exhibits essentially two positive monetary balanced paths depending on the level of government demand. The one with higher real balances and output is always globally unstable in path space $\mathbb{R}^2_+$, while the one with lower real balances and output can be reached asymptotically under positive inflation in some circumstances. For the isoelastic case, the conditions for convergence or divergence depend in a well-structured and systematic way on the parameters of the supply as well as on those from the demand side with clearly defined large open sets of parameters for either case.

4.2.2 Dynamics with Adaptive Expectations

There exists a vast literature discussing the role of adaptive expectations on the dynamics of prices, mostly in partial equilibrium models (see Hommes, 1998; Stiefenhofer, 1999, and references therein). Many of the models are assumed to be of the Cobweb-type (see Böhm & Wenzelburger, 2004, for a definition) which are used to describe prices in markets with delay structure, as originally suggested by Ezekiel (1938). Most applications for financial markets demonstrate that with some degree of heterogeneity of forecasts a variety of bifurcation scenarios of deterministic dynamical systems may occur indicating that diverse cyclical, quasi-periodic as well as chaotic behavior are possible in the long run under *plausible* adaptive rules (see Jensen & Urban, 1984; Brock & Hommes, 1997; Chiarella, Dieci & He, 2009; Böhm, Chiarella, He & Hüls, 2013). However, in particular for the stochastic case there still is no general consensus among researchers whether there exist conditions under which predictors can be found generating converging rational expectations orbits (see Nerlove, 1958; Samuelson, 1976; Pashigian, 1987).

For the stochastic dynamic capital asset pricing model (CAPM) with market clearing, which is a model with an expectational lead, a positive answer was provided by Böhm & Chiarella (2005). The paper provides conditions under which an unbiased prediction rule exists for which rational expectations orbits converge to a stationary solution. The paper also compares standard adaptive rules with unbiased recursive ones. There it is also shown that standard adaptive rules are often unstable under those assumption which guarantee stability of rational expectations equilibria under recursive unbiased forecasts and vice versa. While the methods applied there can also be used here, the results cannot be transferred directly since the CAPM model does not fulfill the closed-flow feature required and desirable for a macroeconomic model.

Many of the stochastic models with adaptive rules in macroeconomic applications examine the convergence of orbits to rational expectations equilibria in the context of learning or bounded rationality (for example Samuelson, 1976; Evans & Honkapohja, 1999, 2001; Marcet & Sargent, 1989; Sargent, 1993; Bray, 1982; Blume, Bray & Easley, 1982; Bray & Savin, 1986). While these discover that convergence to rational expectations equilibria often fails for the mechanisms discussed, few have addressed the issue within the context of statistical learning, as advocated in Zenner (1996); Chen & White (1998); Wenzelburger (1996). There still seems to be disagreement among macroeconomists as to the choices and criteria for adaptive rules to be used in macroeconomic analysis. Existence and stability of the dynamics with recursive rational expectations predictors is examined in Section 4.3.

The following sections analyzing the role of adaptive expectations formation within the AS-AD model is not meant to provide a conclusive final answer to the stability and convergence issue in relation to the literature cited above. The objective is rather to demonstrate that the strict recursive approach, as shown in the perfect foresight case, can be applied successfully for general adaptive recursive predictors, pointing out that in some cases asymptotic convergence or cycles are possibly caused by specific predictors. The discussion is meant to provide additional insights and possible consequences for applications by means of examples and their relationship to the results under perfect prediction within the closed competitive monetary prototype model presented here. At the same time, the issue of path convergence for inflationary economies remains a central problem to be examined.

If price expectations are formed on the basis of past price observations, the dynamics of the price process cannot have the perfect foresight property along orbits of the economy. In addition, if the predictors are not tuned to the possibility of balanced expansion their limiting behavior may not satisfy the perfect foresight property even asymptotically convergence. In particular, the dynamical system will be quite different structurally and the stability may depend heavily on the form of the expectations formation process.

In order to examine the role of the expectations adjustment mechanism, consider first the price dynamics in isolation assuming that money balances are constant. On the one hand, the extreme case of a constant prediction induces constant prices because of an inactive price feed back, implying that the overall dynamics would be money driven alone. On the other hand, one may expect cyclical behavior induced by the expectations process itself, a property which is well known from other models with expectations derived from delay equations.

Two cases will be considered, *a linear error correction predictor* and *an adaptive averaging rule over past data*. Consider a linear error correction predictor defined by the function

$$p^e_{t,t+1} = \psi_\gamma(p_{t-1}, p^e_{t-2,t-1}) := p_{t-1} + \gamma(p_{t-1} - p^e_{t-2,t-1}), \quad 0 \le \gamma \le 1. \qquad (4.2.18)$$

Suppressing the dynamics of money balances for the moment, the price law (3.2.14) together with the error correction predictor (4.2.18) induces a two dimensional dynamical system with state variables in $(p_{t-1}, p^e_{t-2,t-1})$ defined by the two equations

$$\begin{aligned} p_t &= \mathcal{P}(\psi_\gamma(p_{t-1}, p^e_{t-2,t-1}), M_t) \\ p^e_{t,t+1} &= \psi_\gamma(p_{t-1}, p^e_{t-2,t-1}) \end{aligned} \qquad (4.2.19)$$

where the dimension of the dynamical system increases as past states *and* past predictions are included. Substituting the price law into (4.2.18) implies a second order delay equation in expectations

$$p^e_{t,t+1} = (1+\gamma)\mathcal{P}(M_{t-1}, p^e_{t-1,t}) - \gamma p^e_{t-2,t-1}$$

which is equivalent to (4.2.19). The fact that $p_{t-1} = \mathcal{P}(M_{t-1}, p^e_{t-1,t})$ implies that the error correction predictor is a delay equation of order two in expectations, such that the dynamics of (4.2.19) still occurs on the graph of $\mathcal{P} \circ \psi_\gamma$.

Lemma 4.2.1.
For every $M > 0$ and every $0 \leq \gamma \leq 1$, the dynamical system (4.2.19) has a unique steady state $\bar{p}$ with the perfect foresight property which is asymptotically stable.

Proof. The Jacobian matrix of the mapping (4.2.19) has

$$tr = (1+\gamma)\frac{\partial \mathcal{P}}{\partial p^e}(\bar{p}) - \gamma \quad \text{and} \quad det = 0,$$

which implies $\lambda_1 = 0$ and $\lambda_2 = tr$. Therefore, $0 \leq \gamma \leq 1$ and $0 < \frac{\partial \mathcal{P}}{\partial p^e}(\bar{p}) < 1$ yield

$$-1 < (1+\gamma)\frac{\partial \mathcal{P}}{\partial p^e}(\bar{p}) - \gamma = \frac{\partial \mathcal{P}}{\partial p^e} + \gamma\left(\frac{\partial \mathcal{P}}{\partial p^e}(\bar{p}) - 1\right) < 1. \qquad \square$$

Thus, the linear error adjustment forecasts stabilize the price dynamics under constant money balances leading to a perfect foresight steady state. This somewhat surprising stability property of an error adjustment rule contrasts sharply with the global instability of price dynamics under perfect foresight. This feature is part of the common economic folklore, as discussed above, and it appears in many partial equilibrium settings where the conditions for stability under perfect foresight/rational expectations are reversed under adaptive schemes (see for example Böhm & Chiarella, 2005). It is often taken to imply that too much rationality (or perfect knowledge of market data) to adjust prices yields the undesirable property of unstable price adjustments meaning that the final objective of reaching a steady state with perfect foresight cannot be attained under fully rational price adjustments. In other words, the attempt to be perfect becomes self-defeating.

The second adaptive scheme uses past data or statistics from a finite or infinite history. Typically, these operate under the assumption that the price process follows a particular stochastic mechanism with known statistical stationary properties, or that they are drawn as a sample from a statistical distribution (under a maintained hypothesis) without including a 'learning' component. The statistical properties may or may not be confirmed along orbits. But the y will surely violate the prefect foresight property. Such predictors cause an increase of the dimension of the state space as well increasing the potential for expectations generated cycles. Formally, let $\psi_K : \mathbb{R}^K_+ \to \mathbb{R}_+$ denote an adaptive forecasting rule which maps a K vector of past prices $p_{t-K}, p_{t-K-1}, \ldots, p_{t-1}$ to next periods expected price level $p^e_{t,t+1}$, i. e.

$$p^e_{t+1} = \psi_K(p_{t-K}, p_{t-K-1}, \ldots, p_{t-1}) \quad \text{for all } t. \tag{4.2.20}$$

Then, the price process can be written as a K-dimensional delay equation

$$p_t := (\mathcal{P}(\psi_K(p_{t-K}, p_{t-K-1}, \ldots, p_{t-1}), M_t)), \tag{4.2.21}$$

which depends on observed prices only. In both cases, the form of the predictor plays a crucial role for the dynamic behavior since higher order delay equations by themselves combined with non-linear price laws may have complex cycles (see Stiefenhofer, 1999, and references therein).

Finally, consider the full dynamics again with the linear error correction mechanism introduced in Section 4.2.2. Using the relationship $p_{t-1} = \mathcal{P}(M_{t-1}, p^e_{t-1,t})$ and the error correction rule

$$p^e_{t,t+1} = \psi_\gamma(p_{t-1}, p^e_{t-2,t-1}) := p_{t-1} + \gamma(p_{t-1} - p^e_{t-2,t-1}) \qquad 0 < \gamma < 1$$

from (4.2.18), one obtains for the predictor

$$p^e_{t,t+1} = \varphi_\gamma(M_{t-1}, p^e_{t-1,t}, p^e_{t-2,t-1}) := (1+\gamma)\mathcal{P}(M_{t-1}, p^e_{t-1,t}) - \gamma p^e_{t-2,t-1}, \qquad (4.2.22)$$

which is a second order delay equation in expectations. Together with the definition of the deficit

$$\Delta(M_t, p^e_{t,t+1}) = \mathcal{P}(M_t, p^e_{t,t+1})\left[g - \tau D\left(M_t, \mathcal{P}(M_t, p^e_{t,t+1})\right)\right]$$

and the money dynamics equation (4.1.3) this induces a three-dimensional homogeneous dynamical system with state variables $(M_{t-1}, p^e_{t-1,t}, p^e_{t-2,t-1})$ given by the two delay equations

$$\begin{aligned} M_t &= M_{t-1} + \Delta(M_{t-1}, p^e_{t-1,t}) = \frac{\tilde{c} - \tau}{\tilde{c}}\left(M_{t-1} + g\mathcal{P}(M_{t-1}, p^e_{t-1,t})\right) \\ p^e_{t,t+1} &= \varphi_\gamma(M_{t-1}, p^e_{t-1,t}, p^e_{t-2,t-1}). \end{aligned} \qquad (4.2.23)$$

This reveals that structurally all error adjustment predictors induce dynamical systems in money balances and in expectations with delay rather than in prices. If prices are included in the orbits they must obey the price law, so that $p_t = \mathcal{P}(M_t, p^e_{t,t+1})$ inducing orbits on graph $\mathcal{P}$, implying no additional information. There may exist equivalent forms which contain prices, but it is unclear whether these are lower dimensional, more transparent, or easier to work with. Such additional properties sometimes depend on the specific error mechanism or on the so-called 'perceived' law of motion of the underlying statistical model of the forecasting rule[8] used in many models with learning (see for example Evans & Honkapohja, 2001).

By construction the fixed points of adaptive delay mechanisms of the form (4.2.21) or (4.2.23) fulfill perfect foresight. Therefore, if aggregate demand is homogeneous of degree zero in (M, p) existence of steady states fails generically. If they exist, there is a continuum of non-hyperbolic fixed points and their stability is determined essentially by the same restrictions as in the perfect foresight case. Therefore, applications of adaptive schemes to homogeneous systems should be modified to take account of the homogeneity of the macro-system. In other words, the mechanisms should be deflation-inflation adjusted to the fact that, generically, stationary

[8] Such features are known to appear in cobweb models and behavioral financial market models, for example Böhm & Chiarella (2005); Chiarella, Dieci & He (2009).

states with perfect foresight occur along balanced paths with constant rates of expansion or contraction. As a consequence, some care should be taken before the results concerning adaptive principles for non-homogeneous systems (or in partial equilibrium) such as Lemma 4.2.1 can be transposed directly to a macroeconomy. Two cases with adaptive inflationary expectations are discussed below.

Dynamics with Constant Inflationary Expectations

Suppose first that there exists a constant subjectively expected inflation factor with $\bar{\theta} > 0$ which agents believe to be correct in the long run[9] and to which they would like to adapt. Then, an inflation adjusted linear error correction predictor corresponding to (4.2.18) could be written as

$$\begin{aligned} p^e_{t,t+1} &= (1+\gamma)\bar{\theta} p_{t-1} - \gamma(p^e_{t-2,t-1} - \bar{\theta} p_{t-1}) \\ &= \mathcal{P}(M_{t-1}, p^e_{t-1,t})\left((1+\gamma)\bar{\theta} - \gamma\left(\frac{p^e_{t-2,t-1}}{\mathcal{P}(M_{t-1}, p^e_{t-1,t})} - \bar{\theta}\right)\right) \end{aligned} \tag{4.2.24}$$

which is a delay equation of second order in expectations. Such a predictor is called naive if $\gamma = 0$ since in this case the last observed price is taken as the basis for the prediction, regardless of the last observable prediction error. Then, using the homogeneity of $\mathcal{P}$ the prediction process is given by

$$p^e_{t,t+1} = M_{t-1}\bar{\theta}\mathcal{P}\left(1, \frac{p^e_{t-1,t}}{M_{t-1}}\right) \tag{4.2.25}$$

while the dynamics of money balances can be written in the usual way (suppressing the index t) as

$$\begin{aligned} M = M_{-1} + p_{-1}\left[g - \tau\frac{M_{-1}/p_{-1} + g}{1 - c(1-\tau)}\right] &= \left(\frac{\tilde{c} - \tau}{\tilde{c}}\right)(M_{-1} + p_{-1}g) \\ = M_{-1}\left(\frac{\tilde{c} - \tau}{\tilde{c}}\right)\left[1 + g\mathcal{P}\left(1, \frac{p^e_{-1}}{M_{-1}}\right)\right] \end{aligned} \tag{4.2.26}$$

Thus, the dynamical system

$$\begin{aligned} p^e &= M_{-1}\,\bar{\theta}\mathcal{P}\left(1, \frac{p^e_{-1}}{M_{-1}}\right) \\ M &= M_{-1}\left(\frac{\tilde{c} - \tau}{\tilde{c}}\right)\left[1 + g\mathcal{P}\left(1, \frac{p^e_{-1}}{M_{-1}}\right)\right] \end{aligned} \tag{4.2.27}$$

defined by (4.2.25) and (4.2.26) is homogeneous of degree one in (p^e_{-1}, M_{-1}) lacking fixed points generically for the same reason as under perfect predictions.

---

[9] These are so-called unit elastic expectation in the terminology of Chapter 2.

As in the perfect foresight case, balanced paths of the economy are those where all nominal entities change at the same constant factor $\theta > 0$ which, due to the homogeneity of the dynamical system, implies that real allocations, in particular output and employment are constant over time. In other words, a balanced orbit $\left\{(M_t, p^e_{t,t+1}, p_t)\right\}_0^\infty$ under naive expectations moves along a ray in $\mathbb{R}^3_+$ on the graph of $\mathcal{P}$ with $(M_t, p^e_{t,t+1}, p_t) = \theta^t(M_0, p^e_0, p_0)$, $t = 0, 1, \ldots$. In order to examine existence and stability of balanced paths, define $q^e_t := p^e_{t,t+1}/M_t$ and consider the dynamical system in intensive form given by

$$q^e_t = S(q^e_{t-1}) := \frac{\tilde{c}}{\tilde{c} - \tau} \frac{\bar{\theta}\mathcal{P}\left(1, q^e_{t-1}\right)}{\left[1 + g\mathcal{P}\left(1, q^e_{t-1}\right)\right]} \tag{4.2.28}$$

Theorem 4.2.4.
Let the price law $\mathcal{P}$ be differentiable, strictly increasing, concave, and homogeneous of degree one in (M, p^e). Assume that $g \geq 0$ is such that the conditions[10]

$$\lim_{q^e \to 0} \frac{\mathcal{P}(1, q^e)}{q^e} \geq 1 \qquad \textit{and} \qquad \lim_{q^e \to \infty} \frac{\mathcal{P}(1, q^e)}{q^e} \leq \frac{\tilde{c} - \tau}{\tilde{c}} \tag{4.2.29}$$

hold.

(a) There exists a unique positive fixed point $\hat{q}(\bar{\theta})$ of (4.2.28) *which is globally stable on $\mathbb{R}_{++}$.*

(b) All orbits $\left\{(M_{t-1}, M_t, p^e_{t-1,t}, p^e_{t,t+1})\right\}_{t=0}^\infty$ of the associated monetary system (4.2.27) *converge to the balanced path in the sense that the difference $\Delta_t := M_t(q^e_t - \hat{q})$ converges to zero for* $\lim_{t\to\infty} q^e_t = \lim_{t\to\infty} p^e_{t,t+1}/M_t = \hat{q}$.

Proof. The homogeneity of $\mathcal{P}$ implies $\mathcal{P}(1, 0) = 0$, so that $S(0) = 0$. The monotonicity and concavity imply that the elasticity of $\mathcal{P}(1, \cdot)$ and of S are less than one, so that S is concave and strictly increasing. Let

$$H(q^e) := \bar{\theta}\frac{\mathcal{P}(1, q^e)}{q^e} - \frac{\tilde{c} - \tau}{\tilde{c}} \left[1 + g\mathcal{P}(1, q^e)\right].$$

Clearly, H is strictly decreasing, $H(0) > 0$, and $H(q^e) < 0$ for q^e large. Therefore, a unique positive zero $\hat{q}$ exists satisfying

$$\frac{\tilde{c} - \tau}{\tilde{c}} \left[1 + g\mathcal{P}(\hat{q})\right] = \bar{\theta}\frac{\mathcal{P}(\hat{q})}{\hat{q}}. \tag{4.2.30}$$

From (4.2.28) one obtains

$$S'(\hat{q}) = \frac{\bar{\theta}\tilde{c}}{\tilde{c} - \tau} \frac{(\partial\mathcal{P}/\partial p^e)(1, \hat{q})}{\left[1 + g\mathcal{P}(1, \hat{q})\right]^2} = \frac{\hat{q}}{\mathcal{P}(1, \hat{q})} \frac{(\partial\mathcal{P}/\partial p^e)(1, \hat{q})}{\left[1 + g\mathcal{P}(1, \hat{q})\right]} = \frac{E_{\mathcal{P}}(1, \hat{q})}{1 + g\mathcal{P}(1, \hat{q})} < 1.$$

[10] Note that the Inada conditions (4.2.29) hold if $[1 - c(1 - \tau)]\, AS(0) > g$.

Therefore, monotonicity of S implies global stability and monotonic convergence.

To examine the stability of the balanced path associated with $\hat{q}$, define the distance of an orbit from the balanced path (as in Section 4.2) by

$$\varDelta_t = p^e_{t,t+1} - \hat{q}M_t = M_t(q^e_t - \hat{q}) = \frac{M_t}{M_{t-1}} \frac{q^e_t - \hat{q}}{q^e_{t-1} - \hat{q}} \varDelta_{t-1}$$

$$= \frac{M_t}{M_{t-1}} \frac{S(q^e_{t-1}) - \hat{q}}{q^e_{t-1} - \hat{q}} \varDelta_{t-1}.$$

Substituting from (4.2.27) one obtains the two dimensional auxiliary system

$$\begin{aligned} \varDelta_t &= \left(\frac{\tilde{c} - \tau}{\tilde{c}}\right)\left[1 + g\mathcal{P}\left(1, q^e_{t-1}\right)\right] \frac{S(q^e_{t-1}) - \hat{q}}{q^e_{t-1} - \hat{q}} \varDelta_{t-1} \\ q^e_t &= S(q^e_{t-1}) := \frac{\tilde{c}}{\tilde{c} - \tau} \frac{\bar{\theta}\mathcal{P}\left(1, q^e_{t-1}\right)}{\left[1 + g\mathcal{P}\left(1, q^e_{t-1}\right)\right]} \end{aligned} \tag{4.2.31}$$

with unique fixed point $(\bar{\varDelta}, \bar{q}^e) = (0, \hat{q})$ satisfying

$$\hat{q}(\bar{\theta}) = \frac{\tilde{c}\bar{\theta}}{\tilde{c} - \tau} \frac{\mathcal{P}(1, \hat{q}(\bar{\theta}))}{1 + g\mathcal{P}(1, \hat{q}(\bar{\theta}))}. \tag{4.2.32}$$

The two eigenvalues of the characteristic equation of (4.2.31) are

$$\nu_1 = \left(\frac{\tilde{c} - \tau}{\tilde{c}}\right)[1 + g\mathcal{P}(1, \hat{q})]\; S'(\hat{q}) \quad \text{and} \quad \nu_2 = S'(\hat{q}) = \frac{E_{\mathcal{P}}(1, \hat{q})}{1 + g\mathcal{P}(1, \hat{q})} < 1. \tag{4.2.33}$$

Therefore,

$$\nu_1 = \left(\frac{\tilde{c} - \tau}{\tilde{c}}\right)[1 + g\mathcal{P}(1, \hat{q})]\; S'(\hat{q}) = \frac{\tilde{c} - \tau}{\tilde{c}} E_{\mathcal{P}}(1, \hat{q}) < 1$$

which completes the proof. □

For arbitrary subjective inflationary expectations $\bar{\theta}$, however, the stationary rate of growth along the balanced path given by

$$\hat{\mathcal{M}}(\hat{q}(\bar{\theta})) := \frac{\tilde{c} - \tau}{\tilde{c}} [1 + g\mathcal{P}(1, \hat{q})] \neq \bar{\theta}$$

does not coincide necessarily with $\bar{\theta}$ contradicting perfect foresight. Since the fixed point $\hat{q}(\bar{\theta})$ is a strictly increasing function of $\bar{\theta}$, under concavity of $\mathcal{P}$ there exists a unique subjective inflationary rate $\tilde{\theta}$ such that the stationary monetary growth rate coincides with $\tilde{\theta}$, i.e.

$$\frac{\tilde{c} - \tau}{\tilde{c}} \left[1 + g\mathcal{P}\left(1, \hat{q}(\tilde{\theta})\right)\right] = \tilde{\theta}$$

implying perfect foresight along the balanced path. Thus, by choosing the correct expected inflation rate global convergence to the balanced path $\hat{q}(\tilde{\theta})$ can be guaranteed and perfect foresight prevails on the balanced path. Thus, constant naive inflationary expectations have a uniform stabilizing effect for all orbits with naive prediction which converge to balanced paths. If the correct $\tilde{\theta}$ is chosen they exhibit perfect foresight in the limit.

Adaptive Error Correction under Inflationary Expectations

Considering next a more general linear error correction predictor (4.2.18) in inflation rates defined by

$$\frac{p^e_{t,t+1}}{p_{t-1}} =: \theta^e_{t-1,t+1} = \theta_{t-2,t-1} + \gamma\left(\theta_{t-2,t-1} - \theta^e_{t-2,t-1}\right) = \frac{p_{t-1}}{p_{t-2}} + \gamma\left(\frac{p_{t-1}}{p_{t-2}} - \frac{p^e_{t-2,t-1}}{p_{t-2}}\right). \tag{4.2.34}$$

This requires that at time t, when the prediction for the inflation rate $\theta^e_{t,t+1}$ is made, the realization of the price in period t is not known. Therefore, the prediction has to be made on the basis of the last observed inflation rate which is $\frac{p_{t-1}}{p_{t-2}}$ against which the prediction error is calculated. The need for the two period ahead prediction is a consequence of the expectational lead in the price law. The orbits induced by this predictor have endogenously determined expected inflation rates and induce the perfect foresight property along its balanced paths (but not along orbits!). Substituting the price law into (4.2.34) one obtains for the predicted price

$$p^e_{t,t+1} = \left[(1+\gamma)\frac{\mathcal{P}(M_{t-1}, p^e_{t-1,t})}{\mathcal{P}(M_{t-2}, p^e_{t-2,t-1})} - \gamma\frac{p^e_{t-2,t-1}}{\mathcal{P}(M_{t-2}, p^e_{t-2,t-1})}\right]\mathcal{P}(M_{t-1}, p^e_{t-1,t}) \tag{4.2.35}$$

stating that the last observed price p_{t-1} changes at the predicted rate[11] $\theta^e_{t-1,t+1}$.

Consider again the case $\gamma = 0$ when only the last observed inflation rate is used to predict the two-period-ahead price regardless of the observed prediction error. Then, together with the equation for money balances

$$M_t = \frac{\tilde{c} - \tau}{\tilde{c}}\left(M_{t-1} + g\mathcal{P}(M_{t-1}, p^e_{t-1,t})\right)$$

one obtains

[11] One could also take the algebraic expression in square brackets as the one-period-ahead expected inflation rate. In this case, the term in brackets would have to be squared for the two-period-ahead prediction.

$$p^e = \frac{M_{-1}\mathcal{P}\left(1, \frac{p^e_{-1}}{M_{-1}}\right)}{M_{-2}\mathcal{P}\left(1, \frac{p^e_{-2}}{M_{-2}}\right)} M_{-1}\mathcal{P}\left(1, \frac{p^e_{-2}}{M_{-2}}\right),$$
$$M = M_{-1}\left[\frac{\tilde{c}-\tau}{\tilde{c}}\left(1 + g\mathcal{P}\left(1, \frac{p^e_{-1}}{M_{-1}}\right)\right)\right] \tag{4.2.36}$$

which is a homogeneous system in $(M_{-2}, M_{-1}, p^e_{-2}, p^e_{-1})$ with a delay of order two. Defining $q^e := p^e/M$, one obtains a delay equation of order two in intensive form

$$q^e = f(q^e_{-2}, q^e_{-1}) := \frac{1 + g\mathcal{P}(1, q^e_{-2})}{\mathcal{P}(1, q^e_{-2})} \cdot \frac{\left[\mathcal{P}(1, q^e_{-1})\right]^2}{1 + g\mathcal{P}(1, q^e_{-1})} \tag{4.2.37}$$

inducing a two dimensional delay system $S : \mathbb{R}^2_+ \to \mathbb{R}^2_+$ given by

$$S(q^e_{-2}, q^e_{-1}) := (q^e_{-1}, f(q^e_{-2}, q^e_{-1})) \tag{4.2.38}$$

This system has a unique positive fixed point $(\hat{q}, \hat{q})$ with perfect foresight satisfying $\hat{q} = \mathcal{P}(1, \hat{q})$ under identical conditions on the price law as in the previous theorem. However, their stability properties of the balanced monetary expansion are different as stated in the next theorem. Its proof is given in Appendix B.

Theorem 4.2.5.
Let the price law $\mathcal{P}$ be differentiable, strictly increasing, concave, and homogeneous of degree one in (M, p^e). Assume that $g \geq 0$ is such that the conditions

$$\lim_{q^e \to 0} \frac{\mathcal{P}(1, q^e)}{q^e} > 1, \qquad \text{and} \qquad \lim_{q^e \to \infty} \frac{\mathcal{P}(1, q^e)}{q^e} \leq \frac{\tilde{c}-\tau}{\tilde{c}}$$

hold and that the inflation adjusted predictor (4.2.34) *prevails with $\gamma = 0$.*

(a) *There exists a unique positive balanced path of the monetary system* (4.2.36) *with intensity $0 < \hat{q} = \mathcal{P}(1, \hat{q})$ which is a globally stable stationary point with perfect foresight of the system* (4.2.38).

(b) *All orbits $\left\{(M_{t-1}, M_t, p^e_{t-1,t}, p^e_{t,t+1})\right\}_{t=0}^{\infty}$ of the system* (4.2.36) *with $(M_0, p^e_0) \neq (M_0, \hat{q}M_0)$ do not converge to the balanced path in the sense that the difference $|\Delta_t| := |M_t(q^e_t - \hat{q})|$ converges to $+\infty$ for $\lim_{t\to\infty} q^e_t = \lim_{t\to\infty} p^e_{t,t+1}/M_t = \hat{q}$ if the rate of money growth $\hat{M}(\hat{q})$ along the balanced path is nonnegative .*

Statement (a) of the theorem confirms the frequent conjecture in dynamic models of partial equilibrium that error-correction predictors with no or short memory are stabilizing, as in the previous case of Theorem 4.2.4. Due to continuity of the predictor (4.2.35) in γ the stability result extends to situations with small values of the correction parameter γ. However, in their stability properties of monetary expansion (statement (b)) the two cases of Theorems 4.2.4 and 4.2.5 are completely opposite to each other: constant inflationary expectations imply global convergence to the balanced path in all cases whereas with short adaptive memory divergence occurs on

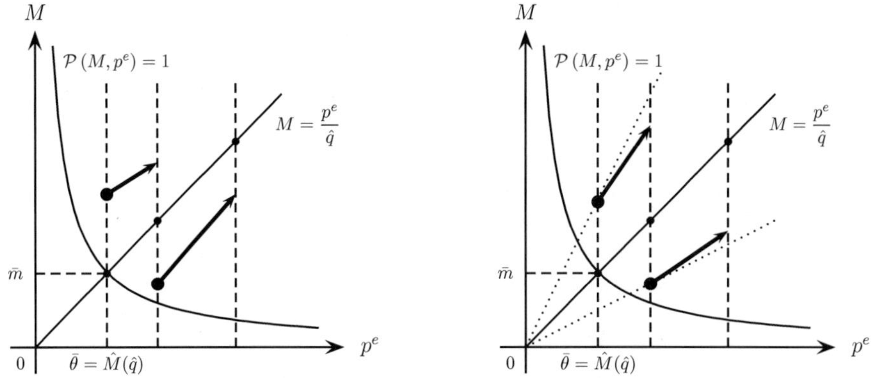

a Constant inflationary expectations **b** Naive inflationary expectations

Fig. 4.14 Convergence-divergence under adaptive expectations

all inflationary balanced paths plus for some situations of deflation. Balanced paths under this form of adaptive expectations formation are only attracting under severe deflation. Since the dynamics of money balances is the same in both cases the difference must be attributed to the form of the predictors. The two results constitute one of the few situations where the causes of stable or unstable monetary growth can be identified clearly, i.e. whether it is due to the predictor or to the nonlinearities in the time-one map. Figure 4.14 displays the qualitative properties of the two cases.

Two additional features with respect to stability are worth noticing. First, the multiplicative form of the delay equation (4.2.38) reveals the possibility of complex eigenvalues for $DF(\hat{q}, \hat{q})$. In this case the dynamics converging to $\hat{q}$ will follow damped oscillations. In this case the convergence to or divergence from the balanced path will also exhibit sequences of Δ_t with alternating signs converging to zero or growing beyond bounds. Delay systems of order greater than or equal to two may also have higher order attracting cycles (see Böhm & Wenzelburger, 1996). This indicates that more general error-correction predictors with $\gamma \neq 0$ could lead to cyclical behavior, periodic, or quasi-periodic orbits. Second, other nonlinearities of the price law may have significant effects on the dynamics. It is known from the theory of economic growth that distribution effects between income groups or their propensities to save have strong effects on the multiplier causing non-concavities or non-monotonicities of the price law (analogous to the results in Böhm & Kaas, 2000) or in noncompetitive environments (see Section 3.3).

A comparison of the characteristics of balanced monetary expansion under adaptive or perfect prediction reveals additional structural information about the long-run behavior of monetary economies and the impact of the structure of the chosen family of predictors. Theorem 4.2.5 states existence of a unique balanced path $\hat{q} > 0$ with perfect foresight while Theorem 4.2.1 states the existence two coexisting balanced paths with perfect foresight for the same range of parameters, as for example for government demand $0 < g < g^\star$ shown in Figure 4.8. This apparent contradiction arises from a fundamental difference between adaptive expectations formation on the basis of past data only as given in (4.1.1) as compared to perfect predictors.

The subtle but important difference consists of the fact that the perfect predictor *requires and uses* explicit information from two successive dates, i.e. *past predictions* and the *next money balances* (as stated in Proposition 4.1.1). The standard adaptive predictor uses past information only, i.e. past prices, expectations, and money balances and *not* the mapping $\mathcal{M}$. It ignores the interdependence created by the expectational lead inducing the specific form of the error function (4.1.10).

This difference can be seen most clearly by comparing the time-one maps of the extended perfect foresight system given of equation (4.1.18)

$$\begin{aligned} M_t =& \mathcal{M}(M_{t-1}, p^e_{t-1}) \\ p^e_t =& \phi^*(M_{t-1}, p^e_{t-1}) := \psi^*(\mathcal{M}(M_{t-1}, p^e_{t-1}), p^e_{t-1}), \end{aligned} \tag{4.2.39}$$

which is equivalent to (4.1.16), with the two dimensional time-one map of an adaptive system with predictor ψ

$$\begin{aligned} M_t =& \mathcal{M}(M_{t-1}, p^e_{t-1}) \\ p^e_t =& \psi(M_{t-1}, p^e_{t-1}) \end{aligned} \tag{4.2.40}$$

which ignores the time-one shift of money. The commutative diagram of Figure 4.1 exhibits the structural difference.

Formally, off course, the perfect predictor ψ^* becomes an adaptive predictor ϕ^* in the sense of Definition (4.1.1) after taking the one-step change of money into account. Both systems are homogeneous of degree one. However, their balanced paths do not coincide since their intensive forms are different time-one maps with different fixed points. Evidence of this deviation is a consequence of the uniqueness result stated in Theorem 4.2.5 as opposed to the result of Theorem 4.2.1 which asserts the existence of two coexisting balanced paths for low government demand.

Since the money feedback under perfect foresight is not part of the adaptive predictor, the system (4.2.40) continues to have a unique positive balanced path $\hat{q}$ for values $g > g^*$ according to Theorem 4.2.5 contradicting one of the results provided in Theorem 4.2.1. The reason is the fact that balanced paths in the state space $(M, p^e) \in \mathbb{R}^2_+$ of homogeneous adaptive predictors and the money adjustment rule simply require the identity of two rates of expansion imposing no condition of perfect foresight or on the time shift. Thus, adaptive rules may have more or fewer balanced paths than those with perfect predictors. For the monotonic rule ψ chosen here this implies a strictly decreasing functional relationship $\bar{m}(g) = 1/\mathcal{P}(1, \hat{q}(g))$ where $\hat{q}(g)$ denotes the balanced path as a function of the parameter g. They satisfy

$$\hat{q} \equiv \hat{q}(g) = \mathcal{P}(1, \hat{q}(g)) \quad \text{and} \quad \bar{m} = \frac{1}{\mathcal{P}(1, \hat{q}(g))}.$$

Therefore,

$$\frac{d\bar{m}}{dg} = -\frac{1}{(\mathcal{P})^2}\left(\frac{\partial \mathcal{P}}{\partial g} + \frac{\partial \mathcal{P}}{\partial q^e}\frac{d\hat{q}(g)}{dg}\right) = -\frac{1}{(\mathcal{P})^2}\frac{\partial \mathcal{P}}{\partial g}\left(\frac{1}{1-\frac{\partial \mathcal{P}}{\partial q^e}}\right) < 0. \tag{4.2.41}$$

Summarizing, homogeneous stationary (time-invariant) predictors induce well defined explicit forward recursive dynamics under perfect foresight as well as under adaptive predictions of the standard (nonrandom) temporary equilibrium model of monetary macroeconomics characterized by a time invariant price law $\mathcal{P}$. Their orbits are induced through forward iteration by a time-one map which is the composition of the price law and a predictor describing the particular forecasting mechanism used in the economy. It is evident that the observable time series of money, prices, and expectations of such economies under any predictor must be sequences of data confined to subsets of the graph of the price law, which is a time invariant global homogeneous cone. Thus, from an empirical point of view, a time series analysis of such models could be directed toward an understanding of the structure of (or parts of) the graph of the price law and the role of specific parameters on the properties of the induced time series.

If the price law is an increasing concave function orbits under perfect foresight are always non-cyclical and often unstable while adaptive expectation mechanisms may induce stabilizing or cyclical effects. It is evident from the definition of the price law that the dimension of the space in which the equilibrium manifold is defined will increase primarily with a higher degree of heterogeneity of agents and/or predictors. However, as long as the predictors are stationary rules, this will only change the dimension of the space in which time series of the economy will be observed, but not the fact that the object to be studied -*the graph of the price law*- is time invariant and accessible to statistical estimation. Therefore, the deterministic temporary market clearing paradigm with perfect foresight does not provide a framework for business cycle theory as such since globally perfect predictors require global monotonicity properties which exclude nonmonotonic time-one maps needed for cycles. However, adaptive predictors and heterogeneity may induce recurrent cyclical properties of time series if the time-one map undergoes associated bifurcations.

The literature on deterministic equilibrium business cycles claims a richness of bifurcation phenomena showing that complex business cycles may be induced under perfect foresight or by specific adaptive predictors. Their results are mostly derived within models of partial equilibrium of specific markets or within real models of growth or exchange for which the associated equilibrium law is a non-concave or non-monotonic mapping. Such features appear in the AS-AD model as well for the same reasons (creating additional non-linearities or non-concavities through non-homothetic preferences, expectations effects, special technological conditions, effects from changing income distributions, or within non-competitive environments, as in Section 3.3). In such cases the existence of globally perfect predictors is lost and cycles are caused by the mix of non-monotonicities from the price law or from the adaptive predictors. The importance of these influences for the occurrence of equilibrium business cycles within the AS-AD-model is not pursued here in more detail. The next section analyzes permanent cyclical (recurrent) properties of macroeconomic time series of output and employment, of real wages, prices, and inflation, when they are generated by *random perturbations* primarily to the technology, i.e. by ongoing recurrent stochastic shocks to productivity. The same methods could be used also in case of perturbations to preferences, policy parameters, or to predictors.

4.3 Dynamics with Random Productivity

Stochastic fluctuations of productivity in an economy are among the most widely discussed sources of random perturbations in macroeconomic models. The study of the observable consequences of mostly unobservable perturbations are important ones to be studied, in order to understand the reasons why empirical time series of macroeconomic entities typically appear to behave in a stochastic way. Such investigations also help to understand why specific statistical regularities or correlations of some variables are typically observed while those of others should not be observed.

Since the early days of the theory of economic growth, technological changes in the macroeconomy were often defined or classified as being neutral or factor augmenting. Such a conceptual separation implies that the source of the unobservable change in the technology would have to be described on the one hand *as if* the change in productivity occurred for the technology *as a whole* with no specific and detectable effect attributable to any of the input factors, or whether, on the other hand, the productivity increase could be attributed to one of the input factors as a so-called *factor augmenting technological effect*, which is tantamount to a situation *as if* more of that particular factor became available.

Consider the first case of a random *factor neutral* technological change which augments the production function of the model of Chapter 3 by a random scale factor. Formally, let $Z \geq 1$ denote a multiplicative scaling factor to the production function used so far implying that total production is now written as $Y = G(Z, L) := ZF(L)$, where the level of Z realized at any one time is drawn and given at the beginning of each period before economic activity starts. It is determined by an outside stochastic process with no connection to any of the elements of the model. Its characteristics will be made more precise at a later stage.

Assume that all of the assumptions of the competitive model of Chapter 3 concerning the production and consumption sectors and the government are maintained. To recap the description of the production sector with n_f producers, profit maximization at given prices and wages (p, w) and technology level Z lead to the first order condition $w/p = ZF'(L)$, implying a labor demand function of each producer

$$h\left(\frac{w/p}{Z}\right) := (F')^{-1}\left(\frac{w}{pZ}\right) \tag{4.3.1}$$

where h is the same labor demand function associated with F as defined in Chapter 3. As a consequence, one obtains the equilibrium real wage as the solution w/p of the labor market clearing condition

$$N\left(\frac{w}{p}V(\theta^e)\right) = n_f h\left(\frac{w}{pZ}\right). \tag{4.3.2}$$

Given the assumptions in Chapter 3, one obtains the following result.

Lemma 4.3.1.
Let $Z \geq 1$ be a multiplicative productivity shock and assume that the production function F satisfies the Inada conditions and that aggregate labor supply N is strictly monotonically increasing. For every $(Z, p, p^e) \gg 0$, the wage solving (4.3.2) is positive and uniquely defined.

(a) The equilibrium wage is homogeneous of degree one in (p, p^e), so one can write

$$w = pW\left(Z, \frac{p^e}{p}\right) \equiv pW(Z, \theta^e). \tag{4.3.3}$$

(b) The real wage function W is strictly increasing in θ^e. For every θ^e, it is strictly increasing in Z with elasticity less than one, implying

$$\frac{W(Z, \theta^e)}{Z} \quad \textit{is strictly decreasing in } Z. \tag{4.3.4}$$

(c) For every Z one has

$$0 < E_W(\theta^e) < -E_V(\theta^e). \tag{4.3.5}$$

(d) The aggregate supply function

$$y^s = AS(Z, p^e/p) := Zn_f F\left(h\left(\frac{W\left(Z, \frac{p^e}{p}\right)}{Z}\right)\right) \tag{4.3.6}$$

is increasing in Z with elasticity $E_{AS}(Z) > 1$, and decreasing in θ^e. In addition,

$$E_{AS}(\theta^e) - E_W(\theta^e) > 0. \tag{4.3.7}$$

Properties (a) – (d) follow from the basic Assumption 3.2.1 and from the multiplicative form of Z in the labor demand in equation (4.3.2); details are given in Section B.2.3 of the Appendix.

When the intertemporal indirect utility function V of workers is unit elastic, i.e. $V(\theta^e) = 1/\theta^e$, the aggregate labor supply function is independent of the current price since

$$N(\alpha V(\theta^e)) = N(\alpha/\theta^e) = N\left(\frac{w}{p^e}\right) =: N_{\text{com}}(w/p^e). \tag{4.3.8}$$

In this special case, the impact of a multiplicative productivity shock factors out in a multiplicative way on aggregate supply operating on the structure of the supply side as if the current price is multiplied by the productivity shock Z. Combined with the remaining assumptions of Chapter 3 the unique positive equilibrium wage is homogeneous of degree one in (p^e, pZ), for every $(p^e, pZ) \gg 0$. Therefore, the equilibrium wage function takes the special form

$$w = pZ\,W(p^e/pZ) \equiv pZW(1, p^e/pZ) \qquad \text{with} \qquad W' > 0,$$

where W inherits all the features from the production function F and the labor supply function N_{com}. The function $pZW(p^e/pZ)$ is increasing and concave in all three variables (p^e, p, Z) separately, implying that the respective partial elasticities are strictly between zero and one, i.e.

$$0 < \frac{p^e}{w}\frac{\partial w}{\partial p^e} = \frac{N'_{\text{com}}}{N'_{\text{com}} - (p^e/pZ)h'} < 1 \quad \text{and} \quad 0 < \frac{p}{w}\frac{\partial w}{\partial p} = \frac{(p^e/pZ)h'}{N'_{\text{com}} - (p^e/pZ)h'} < 1,$$

which also yields

$$0 < \frac{Z}{w}\frac{\partial w}{\partial Z} < 1.$$

As a consequence, the associated aggregate supply function has a factorization with respect to the production shock Z given by

$$y^S = Zn_f F(h(w/pZ)) = Zn_f F\left(h\left(\frac{pZW\,(p^e/pZ)}{pZ}\right)\right) = Z\,AS\,(p^e/pZ). \qquad (4.3.9)$$

Here $AS\,(\cdot) \equiv AS\,(1, \cdot)$ is the same function as determined in Chapter 3 by the properties of F and N for $Z = 1$ which is strictly decreasing and invertible. Equation (4.3.9) indicates that a positive production shock always induces two effects on supply, namely the direct effect on the output itself plus the indirect effect operating through the market for labor stemming from an increase in labor productivity. Therefore, the overall effect must have an elasticity larger than one, which can easily be verified to hold true, since

$$E_{AS}(Z) = 1 - E_F \cdot E_h \cdot E_W(Z) > 1 \qquad (4.3.10)$$

follows from the above results.

To derive the price law and its properties for the general case is now straightforward. Temporary equilibrium is obtained by a price level $p > 0$, such that

$$D(M/p) - AS\,(Z, p^e/p) = 0. \qquad (4.3.11)$$

Given the assumptions there exists a unique positive solution which implies that the price law is given by a function

$$p = \mathcal{P}(M, p^e, Z) \qquad (4.3.12)$$

which is strictly decreasing in the production shock Z. For fixed Z it preserves all the properties of the price law derived before (see Lemma 3.2.1). Summarizing, one obtains the following proposition.

Proposition 4.3.1.
Let $Z \in \Sigma := [Z_{min}, Z_{max}]$ with $Z_{min} \geq 1$. Assume that the aggregate supply function AS is continuously differentiable and strictly decreasing, and that aggregate demand is of the form

$$D\left(\frac{M}{p}\right) = \frac{1}{\tilde{c}}\left(\frac{M}{p} + g\right) \equiv \frac{1}{1 - c(1-\tau)}\left(\frac{M}{p} + g\right).$$

For every $(M, p^e) \gg 0$ and $Z \in \Sigma$, there exists a unique positive temporary equilibrium price $p > 0$ solving equation (4.3.11). There exist differentiable mappings $\mathcal{P} : \mathbb{R}^2_{++} \times \Sigma \to \mathbb{R}_{++}$ and $\mathcal{W} : \mathbb{R}^2_{++} \times \Sigma \to \mathbb{R}_{++}$, such that

(a) the unique positive temporary equilibrium price is given by

$$p = \mathcal{P}(M, p^e, Z), \tag{4.3.13}$$

(b) the unique positive temporary equilibrium wage is given by

$$w = \mathcal{W}(M, p^e, Z) := \mathcal{P}(M, p^e, Z) W\left(Z, \frac{p^e}{\mathcal{P}(M, p^e, Z)}\right). \tag{4.3.14}$$

(c) $\mathcal{P}$ and $\mathcal{W}$ are strictly increasing and homogeneous of degree one in (M, p^e) for $Z \in \Sigma$,
(d) for $M > 0$ and $Z \in \Sigma$ the price law satisfies:

$$\lim_{p^e \to 0} \mathcal{P}(M, p^e, Z) = M\mathcal{P}(1, 0, Z) \quad \text{and} \quad \lim_{p^e \to \infty} \mathcal{P}(M, p^e, Z) = \infty$$

$$\mathcal{P}(M, p^e, Z)/p^e \quad \text{is strictly decreasing in } p^e \text{ with} \tag{4.3.15}$$

$$\lim_{p^e \to 0} \frac{\mathcal{P}(M, p^e, Z)}{p^e} = \infty \quad \text{and} \quad \lim_{p^e \to \infty} \frac{\mathcal{P}(M, p^e, Z)}{p^e} = AS^{-1}\left(Z, \frac{g}{\tilde{c}}\right) \geq AS^{-1}\left(Z_{max}, \frac{g}{\tilde{c}}\right)$$

.

These properties follow from Lemmas 3.2.1-3.2.3 together with the pointwise invertibility of the aggregate supply function for every $Z \in \Sigma$. Figure 4.15 displays the impact of the random productivity on equilibrium output and prices. For given expectations and money balances, the random productivity induces random prices, wages, and output, where high shocks induce higher output, lower prices, and higher real wages (see (4.3.14)). As random variables

$$p = \mathcal{P}(M, p^e, Z) \qquad \text{and} \qquad y = \mathcal{Y}(M, p^e, Z)$$

they are perfectly correlated since they must satisfy the deterministic aggregate demand. The properties of the random price law for a given level of the production shock are the same as those given in Lemma 3.2.1.

It is informative to investigate the role of the production shock in more detail. Since the primary productivity effect enters in a multiplicative way through the term

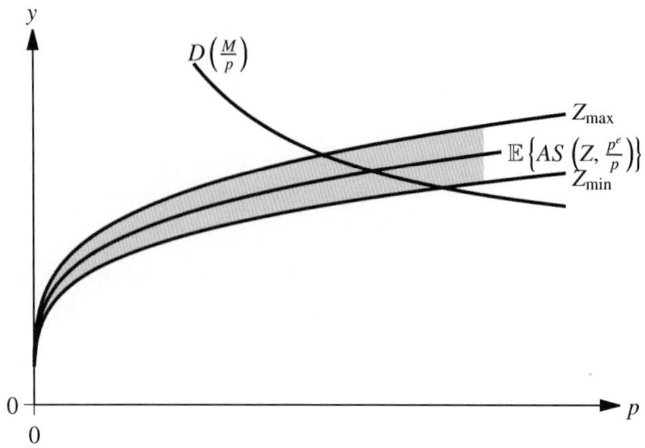

Fig. 4.15 The role of random productivity on prices and output: $Z \in \{Z_{\min}, Z_{\max}\}$

pZ in labor demand, it seems intuitively clear that a positive productivity shock induces lower output prices in equilibrium. Nevertheless, the overall mechanism is much more intricate and it is worthwhile to understand those situations when $Z\mathcal{P}(M, p^e, Z)$ is increasing or decreasing. Using (4.3.11) one obtains

$$1 > 1 + E_{\mathcal{P}}(Z) = 1 - \frac{E_{AS}(Z)}{E_D - E_{AS}(\theta^e)} > 0 \quad \Longleftrightarrow \quad E_D - E_{AS}(\theta^e) > E_{AS}(Z) > 1.$$

In other words, the elasticities of the demand function and of the aggregate supply function must be large enough to dominate the elasticity of aggregate supply with respect to the production shock which is always larger than one (see Lemma (4.3.1)). The first two elasticities are indicators of the commodity market, while the second one is a purely supply side measure which depends on the labor market elasticities. Thus, a universal sign is not to be expected for the joint price-productivity effect in equilibrium.

For the special case when workers do not consume when young (i.e. when the indirect intertemporal utility is unit elastic, see Lemma B.1.1), the multiplicative structure of aggregate supply guarantees a universal positive effect on pZ. From $D(M/p) - AS(p^e/pZ) = 0$ one obtains

$$1 > 1 + E_{\mathcal{P}}(Z) = \frac{E_D}{E_D - E_{AS}} > 0 \tag{4.3.16}$$

which shows a uniformly positive effect on pZ of increases of productivity in equilibrium. At the same time the effect on the nominal wage is also positive, i.e. $E_{\mathcal{W}}(Z) = E_W > 0$, since

$$\mathcal{W}(M, p^e, Z) := Z\mathcal{P}(M, p^e, Z) W\left(\frac{p^e}{Z\mathcal{P}(M, p^e, Z)}\right). \tag{4.3.17}$$

Prices and wages are perfectly negatively correlated. However, as soon as the elasticity is different from minus one, the correlation is no longer perfect. Quite different statistical patterns between the nominal wage and the equilibrium price may occur while the equilibrium real wage will always rise with a positive productivity shock, because of (4.3.14).

4.3.1 Dynamics of Money Balances

The analysis of the impact of random productivity reveals that equilibrium consequences will appear primarily from the supply side of the economy. In order to exhibit the impact of the supply side effects in its purest form and to keep the dynamic analysis as simple as possible the subsequent investigation will employ assumptions on the demand side eliminating all distribution effects assuming that the aggregate demand function is given as in Assumption 3.2.1 by

$$D(M/p) = \frac{M/p + g}{\tilde{c}}$$

with $\tilde{c} := 1 - c(1 - \tau)$ being the constant multiplier. Then, the difference equation for money balances is given by

$$M_{t+1} = M_t + p_t(g - \tau D(M_t/p_t)) \qquad \text{with} \qquad p_t = \mathcal{P}(M_t, p^e_{t,t+1}, Z_t),$$

which yields the random money law

$$\begin{aligned} M_{t+1} &= \mathcal{M}(Z_t, M_t, p^e_{t,t+1}) \\ &:= \mathcal{P}(M_t, p^e_{t,t+1}, Z_t)\left(\frac{M_t}{\mathcal{P}(M_t, p^e_{t,t+1}, Z_t)} + g - \tau D\left(\frac{M_t}{\mathcal{P}(M_t, p^e_{t,t+1}, Z_t)}\right)\right) \qquad (4.3.18) \\ &= \mathcal{P}(M_t, p^e_{t,t+1}, Z_t)\left(\frac{\tilde{c} - \tau}{\tilde{c}}\right)\left(\frac{M_t}{\mathcal{P}(M_t, p^e_{t,t+1}, Z_t)} + g\right) \end{aligned}$$

which is homogeneous of degree one in the two state variables (M, p^e). Notice that (4.3.18) implies random deficits. The random productivity enters only through the price law implying that all demand effects are generated indirectly through prices.

4.3.2 Rational Expectations

In order to describe the dynamics under rational expectations, let $p^e_{t,t+1}$ denote the mean of the subjectively predicted distribution of future commodity prices held by consumers, while the actual distribution of the production shocks in the current period is assumed to be given by some measure μ_t. Then, following the same reasoning and methodology as in the deterministic case, a predictor ψ is a mapping taking past and current data (prior to the realization of the production shock) and making a point prediction for the mean of the future price level conditional on the information available in this period, (see Böhm & Wenzelburger, 2002, for details). Such a predictor is called *unbiased* or rational if the prediction coincides with the conditional mean of the equilibrium price in the next period.

In order to derive the concept of an unbiased predictor, let $(p^e_{t-1,t}, p^e_{t,t+1}) \equiv (p^e_{t-1}, p^e_t)$ denote a pair of predicted mean prices made in two successive dates $(t-1, t)$ and define the prediction error of period t as

$$\mathrm{err}(M_t, p^e_t, Z_t, p^e_{t-1}) := p_t - p^e_{t-1} = \mathcal{P}(M_t, p^e_t, Z_t) - p^e_{t-1}.$$

Then, the prediction $p^e_{t-1,t}$ is called unbiased (or rational) if the conditional mean of the error is equal to zero taken with respect to the true measure μ_t of the production shock. In other words, the condition

$$\mathbb{E}_{\mu_t}\{\mathcal{P}(M_t, p^e_t, Z_t)\} \stackrel{!}{=} p^e_{t-1}$$

of equality of previous mean prediction and the true conditionally expected price induces in a natural way the following definition.

Definition 4.3.1. A mean value predictor $\psi^* : \mathbb{R}^2_+ \times \mathrm{Prob}(\Sigma) \to \mathbb{R}$ is called unbiased if it solves

$$\mathbb{E}_{\mu_t}\{\mathcal{P}(M_t, \psi^*(M, p^e, \mu_t), Z_t)\} = p^e_{-1} \tag{4.3.19}$$

for all $(M, p^e, \mu_t) \in \mathbb{R}^2 \times [0,1] \times \mathrm{Prob}(\Sigma)$, where $\mathrm{Prob}(\Sigma)$ is the set of probability measures on Σ.

This defines an unbiased predictor as a time invariant mapping on the fixed information set $\mathbb{R}^2_+ \times \mathrm{Prob}(\Sigma)$ at the beginning of every period before the occurrence of the production shock. Let

$$(\mathbb{E}\mathcal{P})(M, p^e, \mu) := \mathbb{E}_\mu\{\mathcal{P}(M, p^e, Z)\} := \int \mathcal{P}(M, p^e, Z)\mu(dZ) \tag{4.3.20}$$

denote the conditional expected value of the equilibrium price with respect to the distribution μ. Then, using the definition of the mean price mapping $(\mathbb{E}\mathcal{P})$, ψ^* is unbiased if and only if it solves

$$(\mathbb{E}\mathcal{P})(M, \psi^\star(M, p^e_{-1}, \mu), \mu) = p^e_{-1} \quad \text{for every} \quad (M, p^e_{-1}, \mu). \tag{4.3.21}$$

In other words, an unbiased predictor must be an inverse of the mean price law with respect to the previous expected price. Observe that the current prediction has to be made prior to the realization of the production shock and it becomes a control variable in order for the prediction made in the previous period to become unbiased.

As in the deterministic case, the error function is a time invariant mapping since the random price law is itself time invariant. Hence, the concept of an unbiased predictor as a mapping of money balances, past predictions, and of the measure governing the production shock in each period is also well defined, i.e. it is a t−measurable function. Therefore, unbiased mean predictors can be used as the appropriate forecasting rule in every period, which, by construction, induce rational expectations in the above sense along random orbits. As a consequence one has the following lemma.

Lemma 4.3.2.
Let the random price law $\mathcal{P}$ be continuous and invertible with respect to p^e for every $(M, Z) \gg 0$ and let $\mu \in \text{Prob}(\Sigma)$ denote the distribution of the production shock with compact support Σ. There exists a unique unbiased predictor $\psi^* : \mathbb{R}^2_+ \times [0,1] \times \text{Prob}(\Sigma) \to \mathbb{R}_+$ given by

$$\psi^*(M, p^e_{-1}, \mu) := (\mathbb{E}\mathcal{P})^{-1}(M, p^e_{-1}, \mu) \qquad (4.3.22)$$

which is homogeneous of degree one in (M, p^e_{-1}), and which satisfies

$$\lim_{p^e_{-1} \to 0} \frac{\psi^*(M, p^e_{-1}, \mu)}{p^e_{-1}} = 0, \qquad \lim_{p^e_{-1} \to \infty} \frac{\psi^*(M, p^e_{-1}, \mu)}{p^e_{-1}} \leq AS^{-1}\left(Z_{max}, \frac{g}{\tilde{c}}\right) \qquad (4.3.23)$$

Proof. Applying the methods from Böhm & Wenzelburger (2004) as in Lemma 4.1.1 and Proposition 4.1.1 proves the result. □

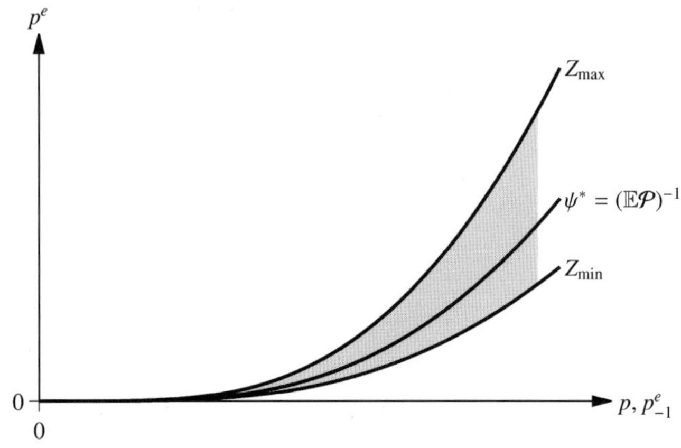

Fig. 4.16 Existence of an unbiased predictor

Figure 4.16 displays the graph of the unbiased predictor ψ^*, i.e. the graph of $(\mathbb{E}\mathcal{P})^{-1}$ for fixed M and the basic geometric features of the range of the random price law with production shocks distributed on a compact interval $\Sigma = [Z_{min}, Z_{max}]$. The mean price law is a concave strictly increasing function with a global inverse which is the unbiased predictor ψ^*, whose existence is guaranteed under the assumptions of Proposition 4.3.1. When the production shocks are identically and independently distributed across time with the same measure μ in every period, then the unbiased predictor is the same *deterministic* function of money balances and expectations alone in each period. Otherwise, for Markovian noise, unbiased predictions depend on the particular distribution of the production shock in each period.

For simplicity, assume from now on that the productivity shock is i.i.d. with a constant measure $\mu_t \equiv \mu$ for all t. Then, combining the unbiased predictor (4.3.22) with the dynamics for money balances (4.3.18) one obtains, analogous to the deterministic situation of (4.1.18), (and after dropping μ_t as an argument of the predictor), a system of two stochastic difference equations in money balances and predictions (M, p^e) given by

$$\begin{pmatrix} M_{t+1} \\ p^e_{t,t+1} \end{pmatrix} = \begin{pmatrix} \mathcal{M}_{\psi^*}(Z_t, M_t, p^e_{t-1,t}) \\ \psi^*(M_t, p^e_{t-1,t}) \end{pmatrix} := \begin{pmatrix} \mathcal{M}(Z_t, M_t, \psi^*(M_t, p^e_{t-1,t})) \\ \psi^*(M_t, p^e_{t-1,t}) \end{pmatrix}. \tag{4.3.24}$$

By construction all orbits generated by $(\mathcal{M}_{\psi^*}, \psi^*)$ exhibit rational expectations. Observe that only the first equation of (4.3.24) is a random difference equation, while the forecasting rule ψ^* is the same function in every period since the measure μ governing the noise process is assumed to be constant through time. Notice, however, that even in more general cases when μ_t is varying through time (for example when the production shock follows a Markov process) the unbiased predictor remains a time invariant deterministic function. Stochastic influences enter through past realizations only which induce random effects on the particular prediction in period t, i.e. on the value of the unbiased predictor at the given state $(M_t, p^e_{t-1,t})$.

In order to derive the main properties of the system (4.3.24), assume that the assumptions of Proposition 4.3.1 hold. Therefore, the homogeneity of the price law in (M, p^e) implies the homogeneity of the unbiased predictor, so that one can write

$$p^e_{t,t+1} = M_t \psi^*(1, p^e_{t-1,t}/M_t) \qquad \text{and} \qquad p_t = M_t \mathcal{P}(1, \psi^*(1, p^e_{t-1,t}/M_t), Z_t). \tag{4.3.25}$$

One can also rewrite the equation for the dynamics of money balances (4.3.18) as

$$\begin{aligned} M_{t+1} &= M_t \mathcal{P}(1, p^e_{t,t+1}/M_t, Z_t) \left(\frac{\tilde{c} - \tau^\star}{\tilde{c}} \right) \left(\frac{1}{\mathcal{P}(1, p^e_{t,t+1}/M_t, Z_t)} + g \right) \\ &= M_t \left(\frac{\tilde{c} - \tau^\star}{\tilde{c}} \right) \Big(1 + g \mathcal{P}(1, \psi^*(1, p^e_{t-1,t}/M_t), Z_t) \Big), \end{aligned} \tag{4.3.26}$$

which together with the unbiased predictor changes the system (4.3.24) equivalently to

$$\begin{pmatrix} M_{t+1} \\ p^e_{t,t+1} \end{pmatrix} = \begin{pmatrix} M_t \left(\frac{\tilde{c} - \tau^\star}{\tilde{c}} \right) \Big(1 + g \mathcal{P}(1, \psi^*(1, p^e_{t-1,t}/M_t), Z_t) \Big) \\ M_t \, \psi^*(1, p^e_{t-1,t}/M_t) \end{pmatrix} \tag{4.3.27}$$

The right hand side is homogeneous of degree one in (M, p^e). Therefore, for fixed Z, deterministic hyperbolic fixed points do not exist generically. As a consequence, stationary solutions typically also do not exist for stationarity production shocks $\{Z_t\}$[12]. Thus, the economically interesting orbits are those inducing positive stationary real allocations while expectations and money balances expand to infinity or contract to zero in a *uniform random way.*

4.3.3 Dynamics of Expected Real Money Balances

As in the deterministic case, the dynamics of the economy 'in real terms' (or in intensive form) is well defined. Obviously, the ratio of the two maps of the system (4.3.27) defines the time-one shift $q_t^e := p_{t-1,t}^e/M_t \mapsto q_{t+1}^e := p_{t,t+1}^e/M_{t+1}$ through the first order stochastic difference equation given by

$$\begin{aligned} q_{t+1}^e &= \left(\frac{\tilde{c}}{\tilde{c} - \tau^\star}\right) \frac{\psi^*(1, p_{t-1,t}^e/M_t)}{1 + g\mathcal{P}(1, \psi^*(1, p_{t-1,t}^e/M_t), Z_t)} \\ &= S(Z_t, q_t^e) := \left(\frac{\tilde{c}}{\tilde{c} - \tau^\star}\right) \frac{\psi^*(1, q_t^e)}{1 + g\mathcal{P}(1, \psi^*(1, q_t^e), Z_t)}, \end{aligned} \tag{4.3.28}$$

inducing a one dimensional stochastic difference equation in the ratio of expectations over money balances $p_{t-1,t}^e/M_t =: q_t^e$ as the state variable[13]. It is the inverse of (ex post) expected real money balances in every period t which is an empirically observable state variable of the random dynamics. Therefore, q_{t+1}^e is the random future value which is measurable with respect to the information at t. Given the assumption of an i.i.d. production shock only the denominator of the mapping is random while the numerator is deterministic. As a consequence, for $g = 0$, the mapping becomes deterministic. Nevertheless, the orbits of prices, output, and employment are stochastic even in the case with zero government consumption. The main features of the time one mapping F which guarantee the properties of the real dynamics are summarized in the following lemma.

Lemma 4.3.3.
Let the conditions of Proposition 4.3.1 hold and assume $Z \in \Sigma = [Z_{min}, Z_{max}]$, $0 < Z_{min} < Z_{max}$, and $g > 0$. Then, the mapping S is monotonically increasing in q^e and satisfies

---

[12] The fact that the system (4.3.27) generically fails to have deterministic fixed points for each level Z implies that for fixed (g, τ) stationary solutions of (M, p^e) will also be non-generic. The possibility of a continuum of stationary rational expectations equilibria in non-generic cases for a related AS-AD-model was observed by Taylor (1977).

[13] For computational reasons the variable $q^e = p_{-1}^e/M$ is defined differently here than in Section 4.2.2. No confusion should arise.

$$\lim_{q^e \to 0} \frac{S(Z,q^e)}{q^e} = \lim_{q^e \to \infty} \frac{S(Z,q^e)}{q^e} = 0 \tag{4.3.29}$$

$$\frac{\partial}{\partial g} S(Z,q^e) < 0 \qquad \textit{and} \qquad \frac{\partial}{\partial Z} S(Z,q^e) > 0. \tag{4.3.30}$$

There exist positive levels of government demand $g^{\star\star} > g^{\star} > 0$ *such that*

$$\begin{aligned} &S(Z,\cdot) \quad \textit{has no positive fixed point for} \quad g > g^{\star\star}, \quad Z \in \Sigma, \\ &S(Z,\cdot) \quad \textit{two positive fixed points for} \quad 0 < g < g^{\star}, \quad Z \in \Sigma \end{aligned} \tag{4.3.31}$$

Proof.

- For all $g \geq 0$ and $Z \in \Sigma$, $\psi^*(q^e) \geq S(Z,q^e)$ and S is continuous. Therefore, the properties (4.3.29) follow from Lemma 4.3.2.
- ψ^* is decreasing in g while $g \cdot \mathcal{P}(\cdot)$ is increasing in g. Hence, S is decreasing in g. $\mathcal{P}$ decreasing in Z implies that S is increasing in Z.
- Since ψ^* is strictly convex with $0 = \psi^*(0) = \lim_{q^e \to 0} \psi^*(q^e)/q^e$, zero is a fixed point of S for every g and $Z \in \Sigma$.
- For $g = 0$, $S = \psi^*$ is non-random and ψ^* has exactly two fixed points, $q_1 = 0$ and $0 < q_2 = \psi^*(q_2)$. Therefore, $Z_{\min} < Z_{\max}$ implies that there exist positive fixed points of S satisfying $0 < q_2(Z_{\max}) < q_2(Z_{\min})$.
- For small $g > 0$, the continuity of S and (4.3.29) imply that there exist two additional fixed points $q_3(Z_{\min}), q_3(Z_{\max})$ such that

$$0 < q_2(Z_{\max}) < q_2(Z_{\min}) < q_3(Z_{\min}) < q_3(Z_{\max}).$$

- The continuity and monotonicity of S in g imply that there exists $g^* > 0$ such that $q_2(Z_{\min}) = q_3(Z_{\min})$ and there exist $g^{**} > g^*$ such that $q_3(Z_{\min}) = q_3(Z_{\max})$.
- For g large enough, $q_1 = 0$ is the only fixed point of S for all $Z \in \Sigma$,

which proves (4.3.31). □

Figures 4.17 and 4.18 portray the implications of Lemma 4.3.3 when S has a sigmoid shape with exactly three fixed points. This occurs in the isoelastic case for appropriate choices of $Z_{\min}$ and $Z_{\max}$, as shown in Figure 4.17 (a). The order of the fixed points $0 < q_2(Z_{\max}) < q_2(Z_{\min}) < q_3(Z_{\min}) < q_3(Z_{\max})$ identifies two coexisting intervals I_2 and I_3, see enlargements **b** an **c**. Figure 4.18 shows the critical role of g when these intervals are degenerate or empty. Since S is monotonically decreasing in g, only the origin will remain as a fixed point when g is sufficiently large. Notice also that in this case the condition (4.3.23) implies that $S(q^e, Z)/q^e$ becomes bounded and less than one for large q^e.

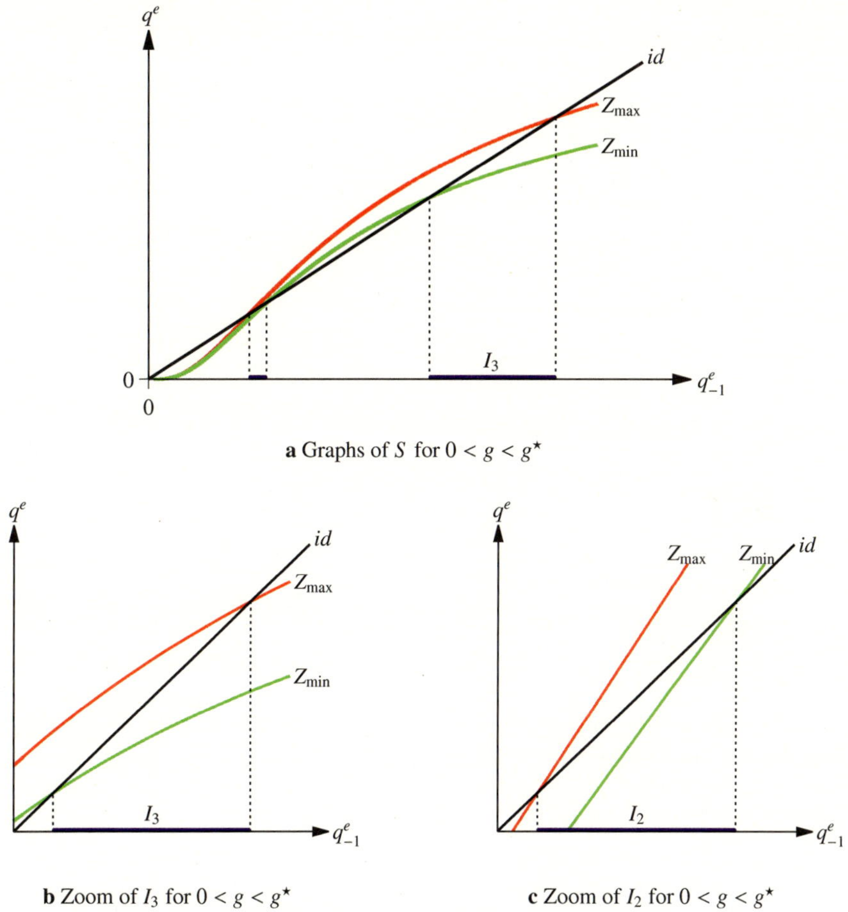

Fig. 4.17 Multiple fixed points and the role of g with $\Sigma = [Z_{min}, Z_{max}]$

4.4 Stochastic Balanced Paths

In deterministic two-dimensional homogeneous dynamical systems (see Section 4.1) the positive fixed points of the one dimensional model in intensity form identify balanced growth paths of the underlying two dimensional system with a constant and identical growth rate of all nominal variables. For the random model an equivalent implication holds, namely random fixed points of the one dimensional random dynamical system (4.3.28) generate or induce *random balanced* orbits of the two dimensional monetary model (4.3.27) where the growth rate of each nominal variable is a stationary random variable. Thus, the notion of what is meant by stochastic balanced expansion of money and expectations is naturally associated with stationary solutions of the reduced system with stationary growth rates along which real

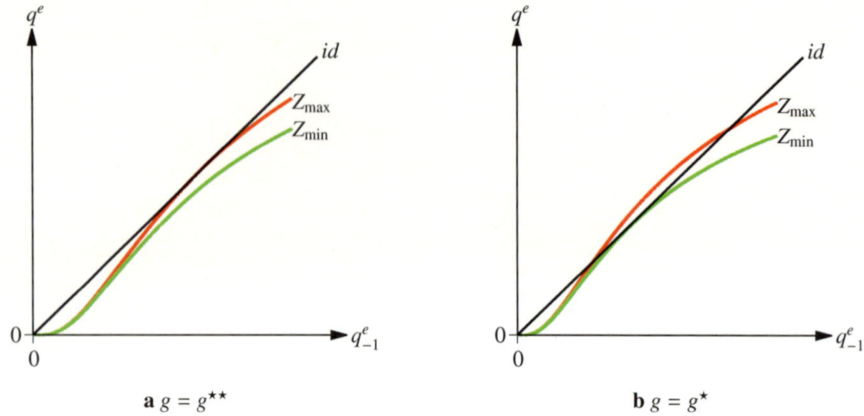

Fig. 4.18 Two critical values of g; $\Sigma = [Z_{\min}, Z_{\max}]$

money balances, output, employment, and inflation are also stationary random variables. A precise definition of a *balanced stochastic* orbit will be given in Definition 4.4.2 below.

In order to establish the mathematical links between the family of parametrized maps S of Lemma 4.3.3 and the theory of stochastic processes on Σ, it is necessary to introduce some additional concepts which allow eventually to discuss existence and stability issues of random balanced orbits or time series of the economy with production shocks[14].

Consider a stationary Markov process on Σ in its canonical representation with associated probability space $(\Omega, \mathcal{F}, \mathbb{P})$ and where $\omega := (\ldots, Z_{-1}, Z_0, Z_1, \ldots) \in \Omega$ denotes a realization (a sample path). Define $\vartheta : \Omega \to \Omega$ to be the so-called left shift on Ω, i.e. $(\vartheta\omega)(s) := \omega(s+1)$ for all $s \in \mathbb{Z}$ and denote by ϑ^t the t–th iterate of ϑ. Then, list $(\Omega, \mathcal{F}, \mathbb{P}, \{\vartheta^t\})$ is an ergodic dynamical system.

Now rewrite the stochastic difference equation as

$$S(\vartheta^t\omega)\, q^e_t := S(Z_t, q^e_t) = q^e_{t+1}. \tag{4.4.1}$$

Then, repeated applications of S under the perturbation ω induce the measurable mapping $\phi : \mathbb{N} \times \Omega \times \mathbb{R} \to \mathbb{R}$ defined by

$$\phi(t, \omega, q^e_0) := \begin{cases} S(\vartheta^{t-1}\omega) \circ \ldots \circ S(\omega)\, q^e_0 & \text{if} \quad t > 0 \\ q^e_0 & \text{if} \quad t = 0 \end{cases} \tag{4.4.2}$$

such that $q^e_t = \phi(t, \omega, q^e_0)$ is the state of the system at time $t \geq 0$ with initial condition q^e_0. The map (4.4.2) (or equivalently (4.4.1)) defines a *random dynamical system* (in the sense of Arnold, 1998, Chapters 1, 2) in forward time, where $S(\omega) := \varphi(1, \omega) : \mathbb{R}_+ \to \mathbb{R}_+$ is referred to as the generator or time one mapping. For

[14] More details and some examples are given in Appendix A.

any initial value q_0^e and perturbation $\omega \in \Omega$, the sequence $\gamma(q_0^e) := \left\{\phi(t,\omega,q_0^e)\right\}_{t=0}^{\infty}$ defines a stochastic orbit of (4.4.1) under the perturbation ω. In order to state the main result on the existence of stochastic balanced orbits the notion of a random fixed point and its asymptotic stability are needed. Both are the appropriate generalizations of the corresponding notions from deterministic systems (see Arnold, 1998, p. 483, also Schmalfuß (1996, 1998) or Appendix A).

Definition 4.4.1. Consider the random dynamical system ϕ induced by the continuous mapping $S(\omega): \mathbb{R}_+^k \to \mathbb{R}_+^k$ over the ergodic dynamical system $(\Omega, \mathcal{F}, \mathbb{P}, (\vartheta^t))$. A *random fixed point* of $S(\omega): \mathbb{R}_+^k \to \mathbb{R}_+^k$ is a random variable $q^*: \Omega \to \mathbb{R}_+^k$ such that

$$q^*(\vartheta\omega) = S(\omega)q^*(\omega) = \varphi(1,\omega,q^*(\omega)) \quad \text{for all} \quad \omega \in \Omega', \tag{4.4.3}$$

where $\Omega' \subset \Omega$ is a ϑ-invariant set of full measure, $\mathbb{P}(\Omega') = 1$.
It is called *asymptotically stable* if there exists a random neighborhood $U(\omega) \subset \mathbb{R}_+^k$ such that $\mathbb{P}$-almost surely

$$\lim_{t\to\infty} \|\varphi(t,\omega,q_0) - q^*(\vartheta^t\omega)\| = 0 \qquad \text{for all} \quad q_0(\omega) \in U(\omega). \tag{4.4.4}$$

With these concepts the main result characterizing the random dynamics of expected real money balances of the system (4.3.28) can be given.

Theorem 4.4.1.
Let the conditions of Lemma 4.3.3 be satisfied for $g \geq 0$. For $(\Omega, \mathcal{F}, \mathbb{P}, \{\vartheta^t\})$ define the associated random dynamical system in expected real balances by

$$q^e(t,\omega,q_0^e) := S(\vartheta^{t-1}\omega) \circ \cdots \circ S(\vartheta\omega) \circ S(\omega) q_0^e. \tag{4.4.5}$$

Then:

(a) $q_1^\star(\omega) \equiv 0$ is an asymptotically stable random fixed point of S for all $g \geq 0$.
(b) There exists $g^{\star\star} > 0$, such that for $g > g^{\star\star}$, $q_1^\star = 0$ is the unique asymptotically stable random fixed point with basin of attraction $\mathbb{R}_+$, i.e. for all $(q_0^e, \omega) \in \mathbb{R}_+ \times [Z_{min}, Z_{max}]^{\mathbb{Z}}$

$$\lim_{t\to\infty} q^e(t,\omega,q_0^e) := \lim_{t\to\infty} S(\vartheta^{t-1}\omega) \circ \cdots \circ S(\vartheta\omega) \circ S(\omega)(q_0^e) = 0, \quad \mathbb{P}\text{-a.s.} \tag{4.4.6}$$

(c) There exists $0 < g^\star < g^{\star\star}$, such that for $0 < g < g^\star$ there exist two nondegenerate disjoint intervals

$$I_2 := [q_2(Z_{max}), q_2(Z_{min})] \subset \mathbb{R}_+ \quad \textit{and} \quad I_3 := [q_3(Z_{min}), q_3(Z_{max})] \subset \mathbb{R}_+ \tag{4.4.7}$$

with $I_2 \cap I_3 = \emptyset$ and associated positive random fixed points $q_2^\star: \Omega \to I_2$ and $q_3^\star: \Omega \to I_3$ satisfying $q_2^\star(\omega) < q_3^\star(\omega)$.
(d) The supports of their associated invariant measures $q_2^\star\mathbb{P}$ and $q_3^\star\mathbb{P}$ are subsets of I_2 and I_3 respectively.

(e) *In addition, $q_3^\star$ is asymptotically stable on I_3 while $q_2^\star$ is stable in backward time, i.e.*

$$\lim_{t\to\infty} \|q^e(t,\omega,q_0^e) - q_3^\star(\vartheta^t\omega)\| = 0, \quad \mathbb{P}\text{-a.s.} \quad q_0^e \in I_3 \tag{4.4.8}$$

$$\lim_{t\to\infty} \|q^e(-t,\omega,q_0^e) - q_2^\star(\vartheta^{-t}\omega)\| = 0, \quad \mathbb{P}\text{-a.s.} \quad q_0^e \in I_2 \tag{4.4.9}$$

Proof. To reduce notation let $q^e \equiv q$.

(a) For all $g \geq 0$ and $Z \in \Sigma$, $\psi^*(q) \geq S(Z,q)$, so that $q^e(1,\omega,q) \leq \psi^*(q)$ for $\mathbb{P}$-a.s. for all $q > 0$. Since ψ^* is strictly convex with $0 = \psi^*(0) = \lim_{q\to 0} \psi^*(q)/q$, ψ^* has two fixed points, $q_1 = 0$ and $\bar{q}_2 = \psi^*(\bar{q}_2) > 0$. Therefore, the interval $[0,\bar{q}_2)$ is a forward invariant set for ψ^* with $\lim_{t\to\infty}(\psi^*)^t(q_0) = 0$ for all $q_0 \in [0,\bar{q}_2)$. As a consequence,

$$\lim_{t\to\infty} q^e(t,\omega,q_0^e) \leq \lim_{t\to\infty}(\psi^*)^t(q_0) = 0 \qquad \mathbb{P}\text{-a.s.} \tag{4.4.10}$$

which implies that $q_1^\star(\omega) \equiv 0$ is a random fixed point of $q^e(1,\omega,q_0)$ for all q_0 in its basin of attraction $U(\omega) \supset [0,\bar{q}_2]$. Thus, for given Σ, $q_1^\star(\omega) \equiv 0$ is an asymptotically stable random fixed point of S for all $g \geq 0$.

(b) According to Lemma 4.3.3, for given Σ there exist $0 < g^* < g^{**}$ such that for $g > g^{**}$ zero is the only fixed point of the mapping $S(Z,q)$ for all $Z \in \Sigma$, see Figure 4.18. Thus, $q^\star \equiv O(\omega) = 0$ is the only random fixed point of $S(\omega) : \mathbb{R}_+ \to \mathbb{R}_+$ with $\lim_{t\to\infty} q^e(t,\omega,q_0) = 0$, $\mathbb{P}$-a.s. for all $q_0 \in \mathbb{R}_+$.

(c) Using Lemma 4.3.3 once more, there exists $0 < g^* < g^{**}$, (see Figure 4.18) such that $S(Z,\cdot)$ has two positive fixed points $q_i(Z)$, $i = 2,3$ for each $Z \in \Sigma$ and $g < g^*$ satisfying

$$q_2(Z_{\max}) < q_2(Z_{\min}) < q_3(Z_{\min}) < q_3(Z_{\max}).$$

Define the associated compact intervals

$$I_2 := [q_2(Z_{\max}), q_2(Z_{\min})] \quad \text{and} \quad I_3 := [q_3(Z_{\min}), q_3(Z_{\max})],$$

see Figure 4.17 displaying the implications of the monotonicity of the mapping S.

(d) Since S is strictly convex near the origin for all Z, $S(\omega)$ is forward invariant and a contraction on I_3. There exists (see Arnold, 1998; Schmalfuß, 1996; Schenk-Hoppé & Schmalfuß, 2001; Schenk-Hoppé, 2005) an asymptotically stable random fixed point $q_3^\star : \Omega \to I_3$ such that

$$\lim_{t\to\infty} |q^e(t,\omega,q_0) - q_3^\star(\vartheta^t(\omega))| = 0 \quad \text{for all} \quad q_0 \in I_3 \quad \mathbb{P} - a.s. \tag{4.4.11}$$

S is convex and strictly monotonically increasing on $I_2 = [q_2(Z_{\max}), q_2(Z_{\min})]$, so that its inverse exists. Since $S(Z,q^e)$ is increasing in Z, its inverse with respect to q^e is decreasing in Z. Moreover, because of

$$S(Z_{\max}, q_2(Z_{\max})) = q_2(Z_{\max}) < S(Z_{\min}, q_2(Z_{\min})) = q_2(Z_{\min}),$$

$(S(\omega))^{-1} : I_2 \to I_2$ restricted to I_2 is a contraction for all ω. Therefore, for all $\omega \in \Omega$, the dynamics on I_2 in backward time is well defined (see p. 52 Arnold, 1998)

$$q^e(-1, \omega) := q^e(1, \vartheta^{-1}\omega)^{-1} = S(\vartheta^{-1}\omega)^{-1} \tag{4.4.12}$$

which implies

$$x_{t-1} = S(\vartheta^{t-1}\omega)^{-1} x_t. \tag{4.4.13}$$

Therefore[15], there exists $q_2^\star : \Omega \to I_2$ such that

$$q_2^\star(\omega) = S(\vartheta\omega)^{-1} q_2^\star(\vartheta\omega) \qquad \text{since} \qquad q_2^\star(\vartheta^{-1}\omega) = S(\omega)^{-1} q_2^\star(\omega). \tag{4.4.14}$$

$q_2^\star$ is a stationary solution and the measure $q_2^\star \mathbb{P} = (q_2^\star \vartheta)\mathbb{P}$ is invariant/stationary since $\mathbb{P}$ is invariant under ϑ. Note that $q_2^\star$ cannot be obtained asymptotically in forward time. □

Out of the three possible scenarios of stochastic balanced expansion under rational expectations, only $q_3^\star$ would be empirically observable. Since $q_2^\star$ is dynamically unstable numerical experiments in forward time are not possible to derive qualitative structural information. Thus, qualitative properties of the unstable random fixed point $q_2^\star$ are difficult to obtain. However, since $I_2 \cap I_3 = \emptyset$ and I_2 is below I_3, its levels of economic performance are uniformly higher than on I_3, i. e. the associated stationary mean output, mean employment, mean money balances are higher than those of the stable solution $q_3^\star$.

The third case is a degenerate stationary solution on the boundary. The fact that zero is an asymptotically stable stationary solution indicates primarily that excessively high initial expectations (low $q_0^e = p_{t-1,t}^e / M_t$) may induce an exploding mechanism with $\lim_t M_t / p_{t-1,t}^e = \infty$ under rational expectations which by itself cannot be halted and which is stationary only in the limit. In such a case money balances grow permanently at a larger rate than price expectations which is essentially an exploding stochastic "monetary bubble". Its implications for the limiting behavior of the real economy and for inflation are an open question at this point and still have to be studied in more detail.

As long as aggregate supply is globally invertible (or larger than $g(1 - c(1 - \tau))$) temporary equilibria exist for each $(M_t, p_{t,t+1}^e)$. The homogeneity of the price law and of the unbiased predictor then indicate that prices and money balances go to infinity, but p_t / M_t must be bounded or tend to zero. This suggests that output and employment increase permanently (or to their upper bounds) without reaching positive stationary levels. However, no stationary solution is reached. Other stochastic scenarios seem possible also.

---

[15] If S is a contraction, the existence of a stationary solution on I_2 follows also from Evstigneev & Pirogov (2007). In many cases weaker assumptions like 'mean contractivity' are sufficient for existence and stability, see Arnold & Crauel (1992); Schmalfuß (1996, 1998).

4.4.1 The Real Economy along Balanced Paths

Under the monotonicity property of the mapping S the balanced invariant behavior of the real stochastic economy is in many ways similar or comparable to the results derivable in the deterministic case. If government demand is not too large the invariant behavior of the economy is characterized by three distinct scenarios, two of which are asymptotically stable phenomena. Convergence to the positive stationary solution $q^\star$ implies the existence of associated stationary solutions of output and employment, of real money balances, and inflation rates on compact sets. These are random variables defined by the associated equilibrium mappings at $q^\star$ inducing associated stationary distributions. Due to ergodicity of the noise process, their properties can be obtained from numerical simulation studies for classes of parametrized models. Such an analysis allows to derive the invariant statistical properties at a macroeconomic level, for example tradeoffs between inflation and employment/output, the role of government demand or fiscal policies.

Real Money Balances, Output and Employment, the Real Wage and Inflation

For the long-run impact of random production on the real economy the effects of $q_3^\star \equiv q^\star : \Omega \to I_3$ on the real variables of he economy are given by the respective temporary equilibrium mapping, for example the price law $\mathcal{P}$, the output law $\mathcal{Y}$, etc.. Let $m_t := M_t/p_t$ denote real money balances. Then, using the fixed point property for the system (4.4.5), its stationary behavior is described by a random variable $m^\star : \Omega \to \mathbb{R}_+$ defined by

$$m^\star(\omega) := \frac{1}{\mathcal{P}(1, \psi^*(1, q^\star(\omega)), Z(\omega))}. \tag{4.4.15}$$

As a consequence the balanced stationary solutions of output $y^\star : \Omega \to \mathbb{R}_+$ and of employment $L^\star : \Omega \to \mathbb{R}_+$ are obtained as[16]

$$y^\star(\omega) := D(m^\star(\omega) + g) = \frac{1}{\tilde{c}}(m^\star(\omega) + g) \quad \text{and} \quad L^\star(\omega) := F^{-1}\left(\frac{D(m^\star(\omega) + g)}{Z(\omega)}\right). \tag{4.4.16}$$

The linearity of the aggregate demand function yields that real balances and output are random variables which differ by a mean shift, implying that real balances and output are perfectly correlated. In addition, the monotonicity of the time one map in government demand g indicates that $q_3^\star$ and its support I_3 undergo a left shift if government demand g increases. Therefore, government demand has a positive effect on stationary output and employment.

Note that stationary employment $L^\star$ is not a one-to-one transformation of stationary output, implying that the two variables will not be perfectly correlated. It

[16] For the remaining analysis of balanced paths the number of firms is set to $n_f = 1$.

is defined as a convex increasing function of the ratio of two stationary random variables. Equivalently,

$$y^\star(\omega) = Z(\omega)F(L^\star(\omega)) \tag{4.4.17}$$

makes random output a product of two stationary random variables. Since the production function F is strictly monotonically increasing and concave the two random variables exhibit a complex nonlinear correlation structure. Surely, the support of their joint distribution is not a product of two intervals. When Z is a discrete random variable with two values $(Z_{\min}, Z_{\max})$, the support of $L^\star\mathbb{P}$ satisfies

$$\operatorname{supp}(L^\star\mathbb{P}) \subseteq F^{-1}\left(\frac{1}{Z_{\min}} \operatorname{supp} y^\star\mathbb{P}\right) \cup F^{-1}\left(\frac{1}{Z_{\max}} \operatorname{supp} y^\star\mathbb{P}\right). \tag{4.4.18}$$

The stationary real wage $\alpha^\star$ is directly obtainable from stationary employment as

$$\alpha^\star(\omega) = Z(\omega)F'(L^\star(\omega)), \tag{4.4.19}$$

which is also the product of two random variables. As a product of two stationary random variables, the stationary real wage and employment are governed by the marginal product rule, which implies that they must be negatively but not perfectly correlated.

For the rate of inflation one obtains from the definition

$$\theta_t := \frac{p_t}{p_{t-1}} = \frac{M_t}{M_{t-1}} \frac{\mathcal{P}(1, \psi^*(1, q_t^e), Z_t)}{\mathcal{P}(1, \psi^*(1, q_{t-1}^e), Z_{t-1})}$$

Using (4.3.27) one obtains for the stationary rate of inflation

$$\begin{aligned}
&\theta^\star(\omega) \\
&:= \frac{\tilde{c} - \tau}{\tilde{c}} \left(1 + g\mathcal{P}\left(1, \psi^*(1, q^\star(\vartheta^{-1}\omega), Z(\vartheta^{-1}\omega))\right)\right) \frac{\mathcal{P}(1, \psi^*(1, q^\star(\omega), Z(\omega)))}{\mathcal{P}(1, \psi^*(1, q^\star(\vartheta^{-1}\omega), Z(\vartheta^{-1}\omega)))} \\
&= \frac{\psi^*(1, q^\star(\vartheta^{-1}\omega))}{S(\vartheta^{-1}\omega)} \frac{\mathcal{P}(1, \psi^*(1, q^\star(\omega), Z(\omega)))}{\mathcal{P}(1, \psi^*(1, q^\star(\vartheta^{-1}\omega), Z(\vartheta^{-1}\omega)))} \\
&= \hat{m}^\star(\omega) \frac{m^\star(\vartheta^{-1}\omega)}{m^\star(\omega)}
\end{aligned} \tag{4.4.20}$$

which equals the stationary rate of money growth $\hat{m}^\star(\omega)$ multiplied by the ratio of stationary real money balances at two successive dates. In other words, it is the product of three stationary random variables. Because of this time shift, an i.i.d process of Z implies that $\theta^\star$ is Markov stationary. Therefore, a state space representation for $(\theta_t^\star, \theta_{t+1}^\star)$ must show significant (nonzero) autocorrelation which could be positive or negative and whose statistical properties can be estimated from time series data.

More importantly, however, substituting (4.4.16) and (4.4.17) into (4.4.20) shows that one obtains an exact expression for the correlation between inflation and employment with a one period delay.

$$\theta^\star(\omega) = \hat{m}^\star(\omega)\,\frac{\tilde{c}Z(\vartheta^{-1}\omega)F(L^\star(\vartheta^{-1}\omega)) - g}{\tilde{c}Z(\omega)F(L^\star(\omega)) - g} \tag{4.4.21}$$

In other words, stationary time series of inflation and employment under an i.i.d. Hicks neutral production shock must satisfy this nonlinear delay restriction making the *empirical Phillips curve* under rational expectations a subset of a two dimensional nonlinear manifold in $\mathbb{R}^3$ with specific structural properties depending on the underlying model. This confirms what has been observed all along that "the Phillips curve emerges not as an unexplained empirical fact, but as a central feature of the solution to a general equilibrium", (Lucas, 1972, p. 122).

To summarize one finds a list of generic structured properties of major macroeconomic variables under rational expectations which must prevail for all stationary or empirical solutions of the dynamical system (4.3.28) along balanced monetary paths. These properties are implications of the structure induced by the equilibrium conditions, by rational expectations, and by the i.i.d. noise process and are not unexplained statistical or random artifacts of an occasional macroeconomic time series.

- Output (real GDP) and real money balances are perfectly correlated and their distributions differ only by a mean shift, i.e. $y^\star \sim m^\star$ and

$$\mathbb{E}\left\{y^\star(\omega)\right\} = \frac{1}{\tilde{c}}\left(g + \mathbb{E}\left\{m^\star(\omega)\right\}\right). \tag{4.4.22}$$

- Output $y^\star$ and employment $L^\star$ are positively correlated showing the typical empirical comovement, but they are not perfectly correlated since

$$y^\star(\omega) = Z(\omega)F(L^\star(\omega)). \tag{4.4.23}$$

- Similarly, the real wage and employment are negatively correlated, but not perfectly since

$$\alpha^\star(\omega) = Z(\omega)F'(L^\star(\omega)). \tag{4.4.24}$$

- How strong the correlations in the last two cases are depends among other things on the curvature of the production function. Nevertheless, the long-run effects of a Hicks neutral productivity shock factors through in a multiplicative way between employment and output and employment and the real wage. This confirms the general intuition that neutral productivity shocks on the micro level filter through to the macro level in a monotonic way. Whether a statistical reversal of these correlations could occur due to other endogenous forces remains to be investigated in specific situations and cannot be determined on a general level.
- The correlation between employment and inflation or real money balances derives from a much more complicated nonlinear structure. Its size and sign (in

(L, θ) space) do not seem to be uniquely defined and reversals seem to be possible. From the definitions above it is apparent that money balances $m^\star$ can be positively or negatively correlated with $\theta^\star(\omega) = m^\star(\omega)/m^\star(\vartheta\omega)$ depending on whether $m^\star\vartheta$ induces a positive or negative time one shift. In other words, the serial correlation (autocorrelation) of real balances matter. As a consequence, employment and inflation (the Phillips curve !) will be positively correlated if (and only if) $m^\star$ is positively correlated with $m^\star \circ \vartheta$. In other words, a positively as well as a negatively sloped Phillips curve are consistent with the main hypotheses of the model: rational expectations and competitive market clearing.

In other words, one finds a set of general well defined stationary macroeconomic relationships between employment and other real economic variables which must prevail under competitive equilibrium and rational expectations. Their determinants and more specific statistical properties can be examined by numerical experiments of computable cases under asymptotic stability for the full nonlinear structure of the model.

4.4.2 Numerical Results: Stationarity and Stability

This section presents the numerical results characterizing the stability and the stationarity of orbits for a specific computable parametric example of the general model. The purpose is twofold. First of all, the numerical exercise demonstrates that it is possible with relatively modest numerical techniques to obtain *exact numerical results* for the general nonlinear economic model without compromising on the qualitative precision by using approximations or linearizations of any of the equations. Second, once the numerical results for a given set of parameters is obtained, a systematic experimental search can be carried out in order to investigate questions of comparative dynamics, statistical properties, policy issues, i.e. of bifurcation features of the solutions of the model.

The central problem of identifying a computable version of the above nonlinear model consists in finding a format which allows to calculate the values at each point in time of the two central objects of the dynamics: the price function and the unbiased predictor for the stochastic case. While the determination of the equilibrium price requires finding only a fixed point of a well defined excess demand function, the calculation of the rational prediction involves computing the inverse of an integral equation. Fortunately, such a solution can again be found through a standard fixed point solution using standard numerical techniques if the production shock is an i.i.d. discrete shock with constant positive probabilities, the case which is used for the following numerical analysis[17].

[17] The results are part of the numerical investigation of the basic model of Chapter 3 in isoelastic form, as introduced in Section 3.2.7. This is joint work with Oliver Claas comprising the investigation of allocative implications in noncompetitive environments, in particular of the bargaining model (as in Böhm & Claas, 2017; Claas, 2017, and in Section 3.5) as well as dynamic and stochas-

For the numerical exercise consider the isoelastic version of the economy analyzed in Section 3.2.7 with a two point production shock $Z \sim \{Z_{\min}, Z_{\max}\}$ with equal probability. This implies an isoelastic random aggregate supply function and a deterministic aggregate demand function with a constant multiplier. For the first set of numerical experiments the values of the respective parameters in consumption, production, and for the government are chosen as in table 4.1. All numerical results are

Table 4.1 Standard parametrization a

$Z_{\min}$	$Z_{\max}$	B	C	c	τ	g	g^*	g^{**}
1.0	1.01	0.6	0.6	0.5	0.7	0.8240	0.8285	0.8328

calculated for the same seed of random numbers in order to compare different orbits appropriately. The discreteness of the production shocks implies that the random dynamical systems (4.3.27) and (4.3.28) becomes an *iterated function system* (IFS), (Barnsley, 1988) or a *Markov switching model*, (see for example Farmer, Waggoner & Zha, 2009a,b). As a consequence, the time series, the attractors as well as the histograms will display some of the typical features of such systems, i.e. phase plot representations are often subsets of multiples of two piece equilibrium manifolds while their invariant marginal distributions may be multimodal, often without densities. Their supports may be fractal sets. Figure 4.19 displays the characteristics of the chosen seed of the noise, part of the time profile and the relative frequency (the histogram). For the first 200 iterations the value for $Z_{\max}$ seems to appear slightly more often than $Z_{\min}$. However, for $T > 1000$ the histogram shows numerically an (almost) equal number of realizations for the two values $\{Z_{\min}, Z_{\max}\}$, indicating that the histogram of the seed approximates an equal probability condition sufficiently.

The convergence of orbits of the random dynamical system (4.3.28) for the stochastic difference equation S to I_3 and to the stationary solution are shown in Figure 4.20, which displays in panel **a** the associated graph of the two time one maps and a phase plot of an orbit starting outside of I_3. Panel **b** shows the time profile of six different orbits, five of which converge to I_3 represented by the large shade interval. Two initial conditions start in I_2 (the small shaded interval) both of which leave this interval in finite time, one diverging the other one converging to I_3.

Figure 4.21 displays the main properties of the random fixed point $q^\star : \Omega \to I_3$ for the stationary solution associated with $Z(\omega)$. Subfigure **a** shows a typical phase plot (an attractor in the space $(q^e_t, q^e_{t+1}) \in I_3 \times I_3$) when there are discrete production shocks. All orbit pairs lie on the disjoint graphs of two nonlinear time one maps associated with the two values of Z. This reveals in particular that the attractor (or supportof the joint distribution) may not be a rectangle in $\mathbb{R}^2$. This also shows that the dynamics of real expectations cannot be approximated well by a one dimensional AR1 system in q^e. Nevertheless, because of stationarity, the two piece

tic issues. I am indebted to him who bore the major burden for programming the computations and who provided the numerical and graphical results shown here.

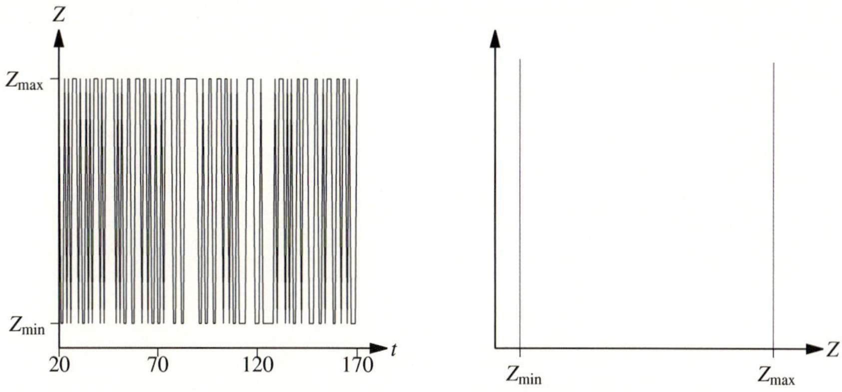

a Time profile of $Z(\omega)$ **b** Histogram of $Z(\omega)$ for $T = 2 \cdot 10^4$

Fig. 4.19 $Z \sim \mathcal{U}\{Z_{\min}, Z_{\max}\}$ with equal probability for $Z_{\min} = 1.0, Z_{\max} = 1.05$

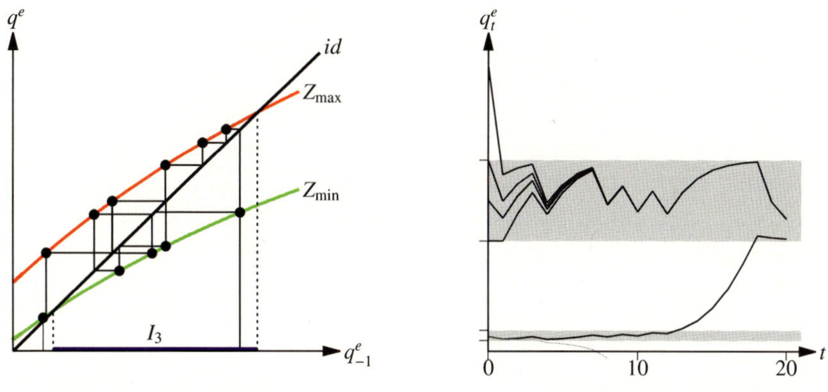

a Convergence to attracting set I_3 **b** Convergence from I_2 and I_3 to $q_3^\star$

Fig. 4.20 Convergence to the random fixed point $q^\star$

attractor implies a well defined and structurally simple autocorrelation such that the two marginal distributions (of the projections onto the two axis) must be identical, as shown in subfigure **c**. The raggedness of the histogram is a typical feature for an IFS, which often does not decrease or become more smooth as the number of iterations becomes large.

In many situations, both fixed points $q_i^\star$, $i = 2, 3$ induce non-degenerate invariant measures $q_i^\star \mathbb{P}$, $i = 2, 3$, whose supports are the full respective intervals I_i, even though the production shocks are concentrated on discrete points and the state space representation indicates attractors as subsets of two disjoint graphs. This non-degeneracy is essentially a consequence of the non-constancy of the time one

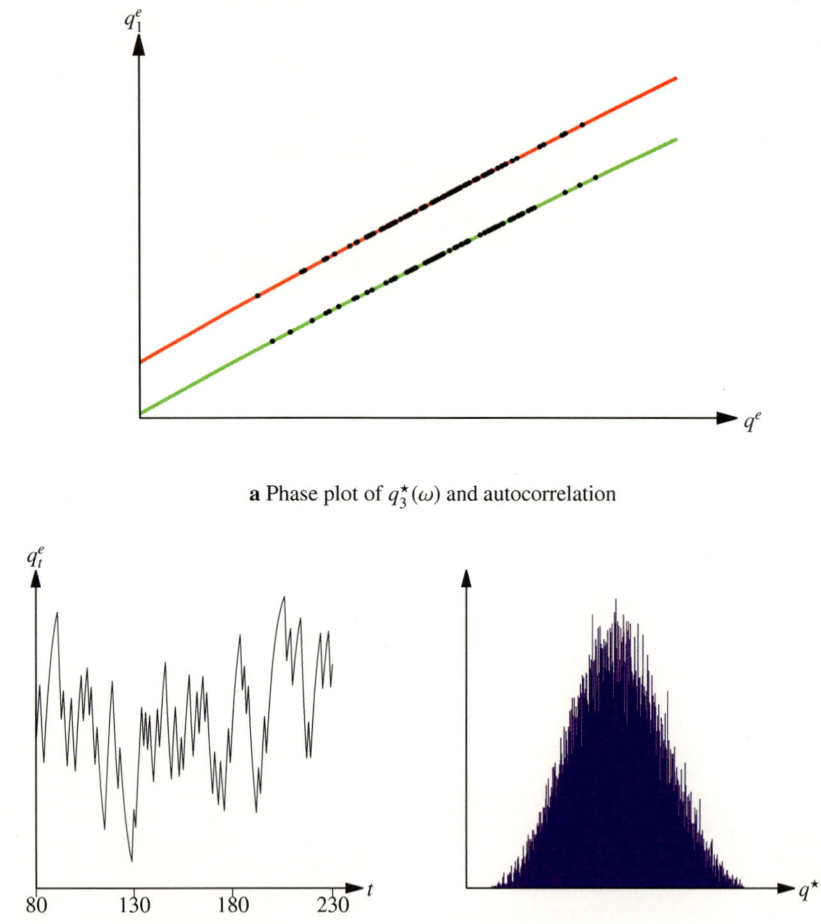

a Phase plot of $q_3^\star(\omega)$ and autocorrelation

b Time profile of $q_3^\star(\omega)$

c Histogram of $q_3^\star(\omega)$; $T = 2 \cdot 10^4$

Fig. 4.21 Stationary solution $q^\star$; $Z_{\min} = 1.0$, $Z_{\max} = 1.01$

map S, preventing finitely discrete stationary solutions to occur under finite discrete shocks. For some parameters of the economy determining the slope of the mapping S and the size of the production shock the invariant measure may have a complicated structure with no density and whose support is a Cantor set with Lebesgue measure zero (see Barnsley, 1988, for examples).

Employment and Output

When Z is a discrete random variable with two values $(Z_{\min}, Z_{\max})$, the marginal distributions of all stationary real variables are influenced in a different way by the discreteness of the shocks as well as by the relationships arising from the equilibrium manifolds under stationarity. Thus, the correlation structure of joint distributions comes from the non-linear mathematical properties of these manifolds. For example, the joint support of stationary employment and stationary output must also consist of two nonempty disjoint subsets of the graphs of the production function associated with the two values. This implies that the employment and output are correlated in a particular way on the two graphs. The support of the stationary measure $L^\star\mathbb{P}$ must satisfy

$$\operatorname{supp}(L^\star\mathbb{P}) \subseteq F^{-1}\left(\frac{1}{Z_{\min}}\operatorname{supp} y^\star\mathbb{P}\right) \cup F^{-1}\left(\frac{1}{Z_{\max}}\operatorname{supp} y^\star\mathbb{P}\right) \tag{4.4.25}$$

Figure 4.22 displays the features of the joint distributionof employment and output (subfigure **a**) together with the time profiles and their histograms which display the typical features of a discrete noise process. The two point distribution of the production shock implies that the joint support of the empirical distribution is a subset of the two graphs of the production function. Subfigure **a** displays the typical comovement of employment and output.

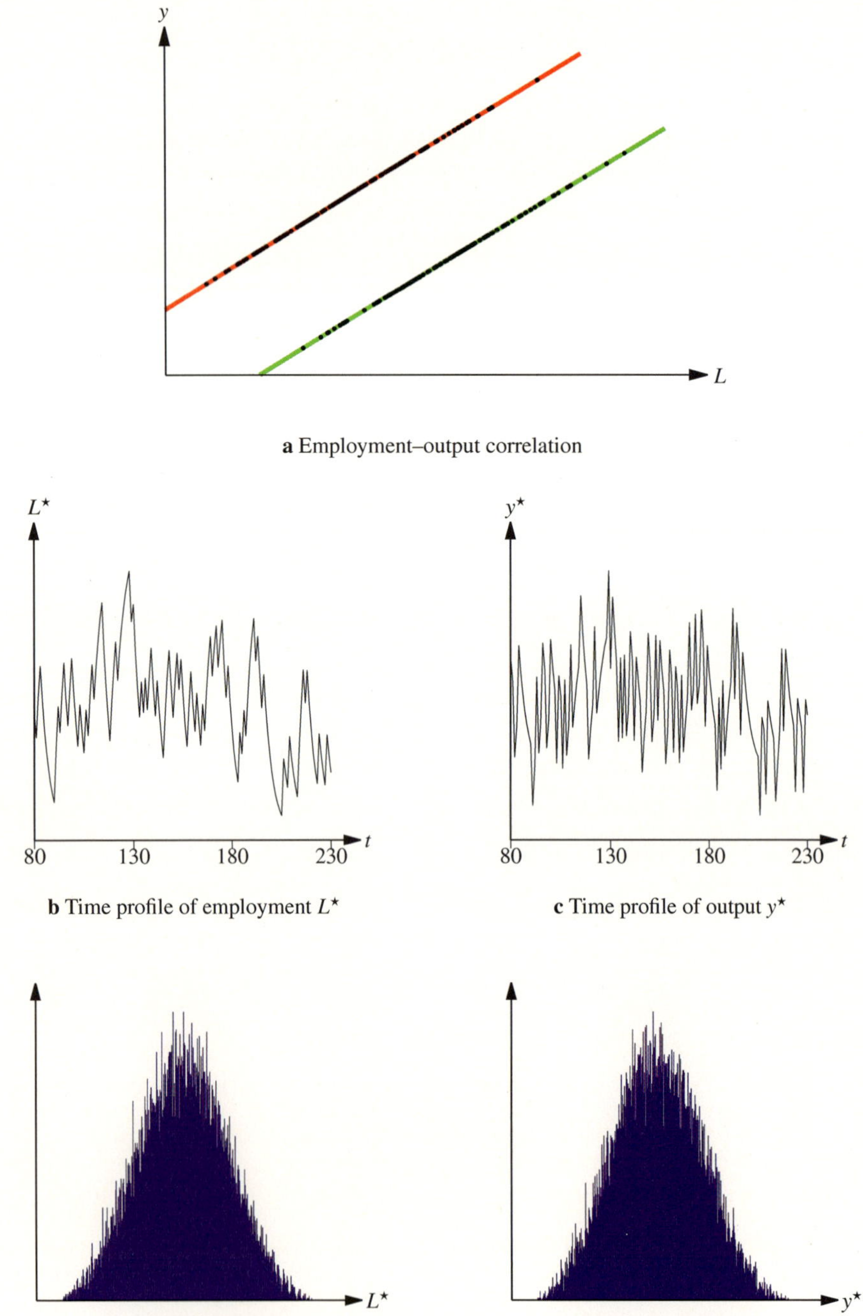

a Employment–output correlation

b Time profile of employment $L^\star$

c Time profile of output $y^\star$

d Stationary employment; $T = 2 \cdot 10^4$

e Stationary output; $T = 2 \cdot 10^4$

Fig. 4.22 Stationary output and employment

Employment and Real Wage

The corresponding diagrams of the statistics between employment and the real wage, Figure 4.23, shows the distinctive properties of an IFS. The two point distribution of the production shock combined with the marginal product rule of profit maximization implies that the joint support in $\mathbb{R}^2_+$ must be a subset of the two associated graphs of the marginal product curves (subfigure **a**). Hicks neutral production shocks induce a negative correlation and not a (positive) empirical comovement of employment and the real wage. However, one finds that the distribution of the real wage is bimodal and not necessarily symmetric, and that its support consists of two (almost) disjoint intervals. The gap in the support arises jointly because of the size of the production shocks and the slope of the marginal product curve. In other words, the long-run dynamics of the real wage fluctuates between the two intervals in a stochastic (non-periodic) way.

Employment and Expected Inflation

A similar phenomenon occurs with respect to the correlation between employment and the expected rate of inflation $\theta^e = p^e/p$. Since the stationary levels of θ^e belong to two small disjoint intervals, its stationary distribution is clearly bimodal and asymmetric, Figure 4.24.

Employment and Inflation

The long-run tradeoff between inflation and employment exhibits the typical Markovian structure in the correlation diagram (the empirical Phillips curve) which is an outcome of the central features of the stationary competitive equilibrium under rational expectations, see Figure 4.25.

Real Money Balances and Inflation

In an analogous way, one obtains the correlation between real balances and inflation as derived in (4.4.20) which also shows the typical Markovian structure in the phase plot (Figure 4.26 **a**). Since output and real balances have the same stationary distribution, the corresponding diagram between $y^\star$ and $\theta^\star$ would show a similar negative correlation. Subfigure **e** in both diagrams also reveals the consequences of the autocorrelation of prices/inflation rates as compared to employment or money balances by an additional nonlinear influence on the shape of the histogram which is less curved (and more triangular).

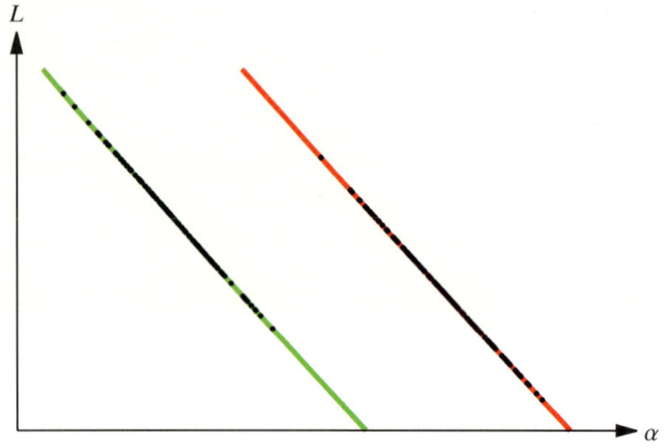

a Real wage–employment tradeoff

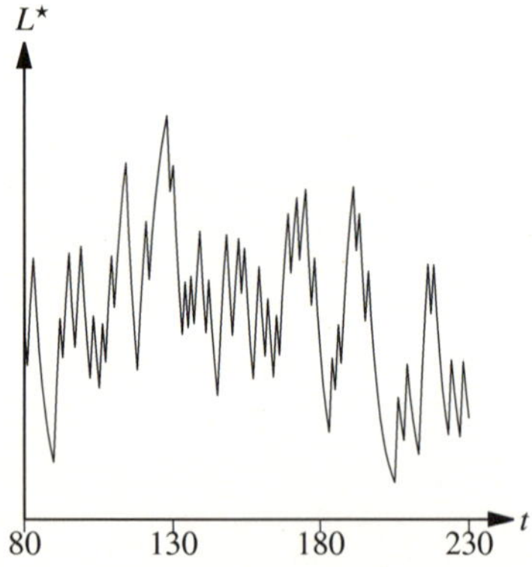

b Stationary employment

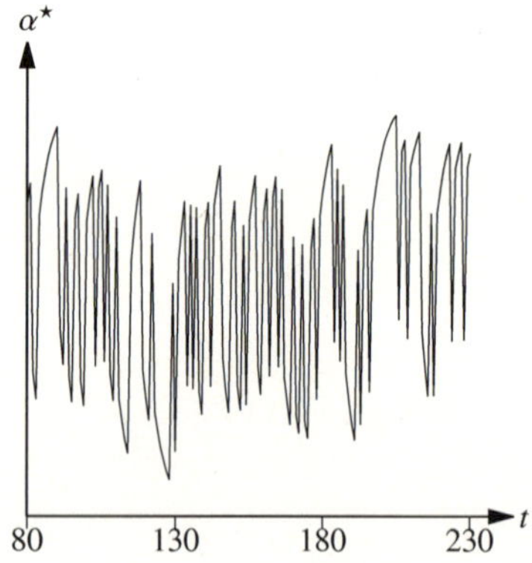

c Stationary real wage

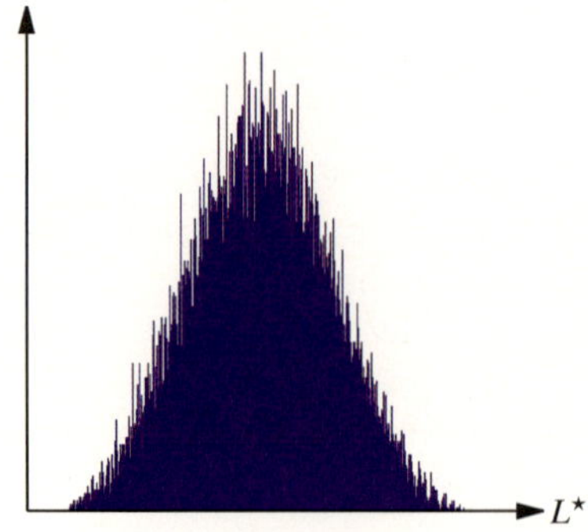

d Stationary employment; $T = 2 \cdot 10^4$

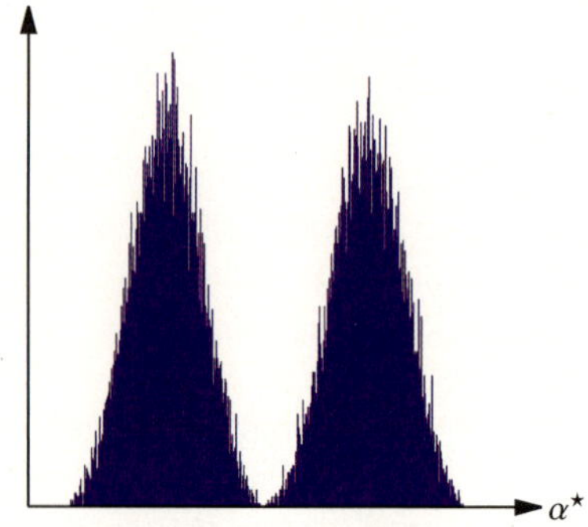

e Stationary real wage; $T = 2 \cdot 10^4$

Fig. 4.23 Stationary employment and real wage

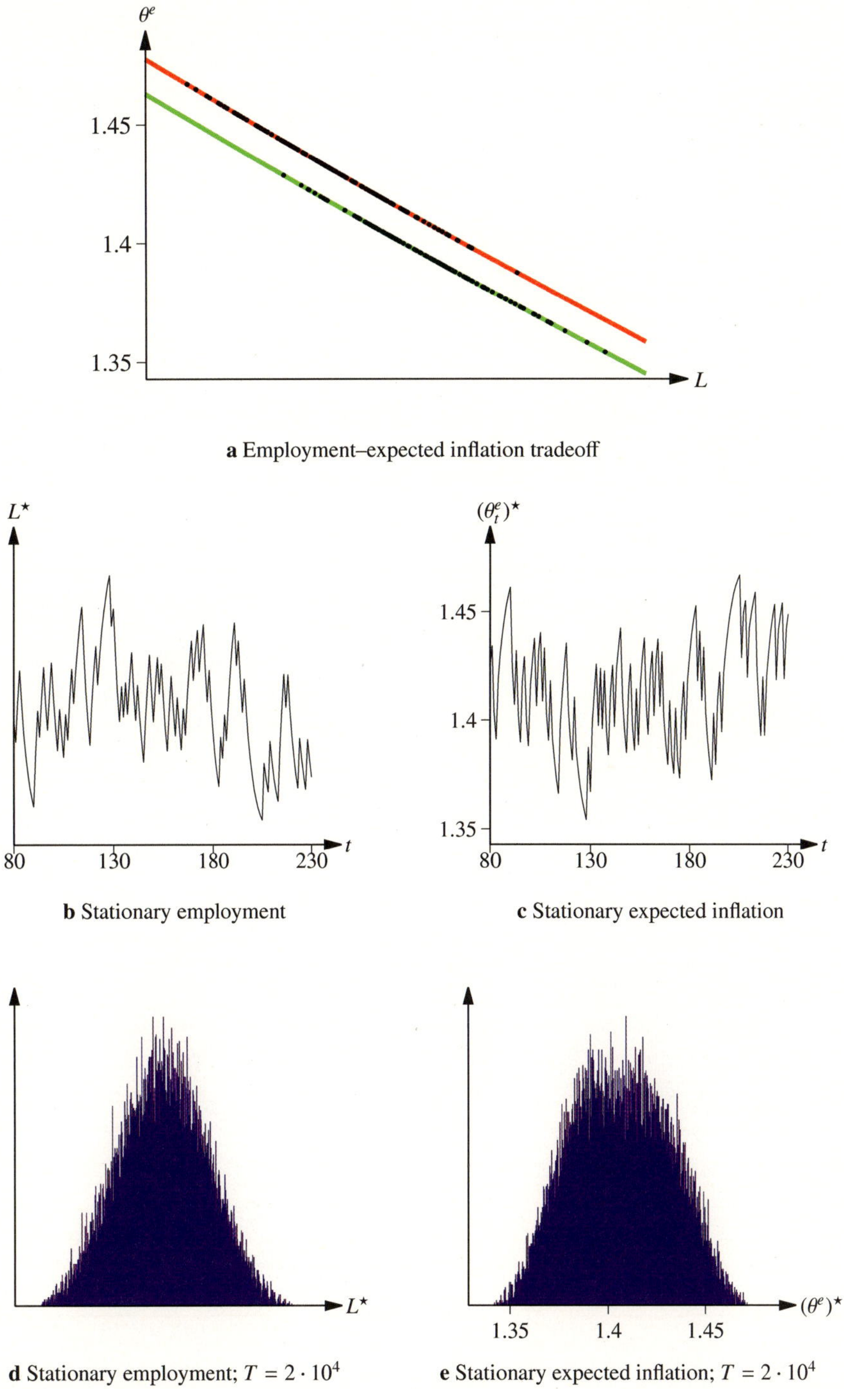

Fig. 4.24 Stationary employment and expected inflation

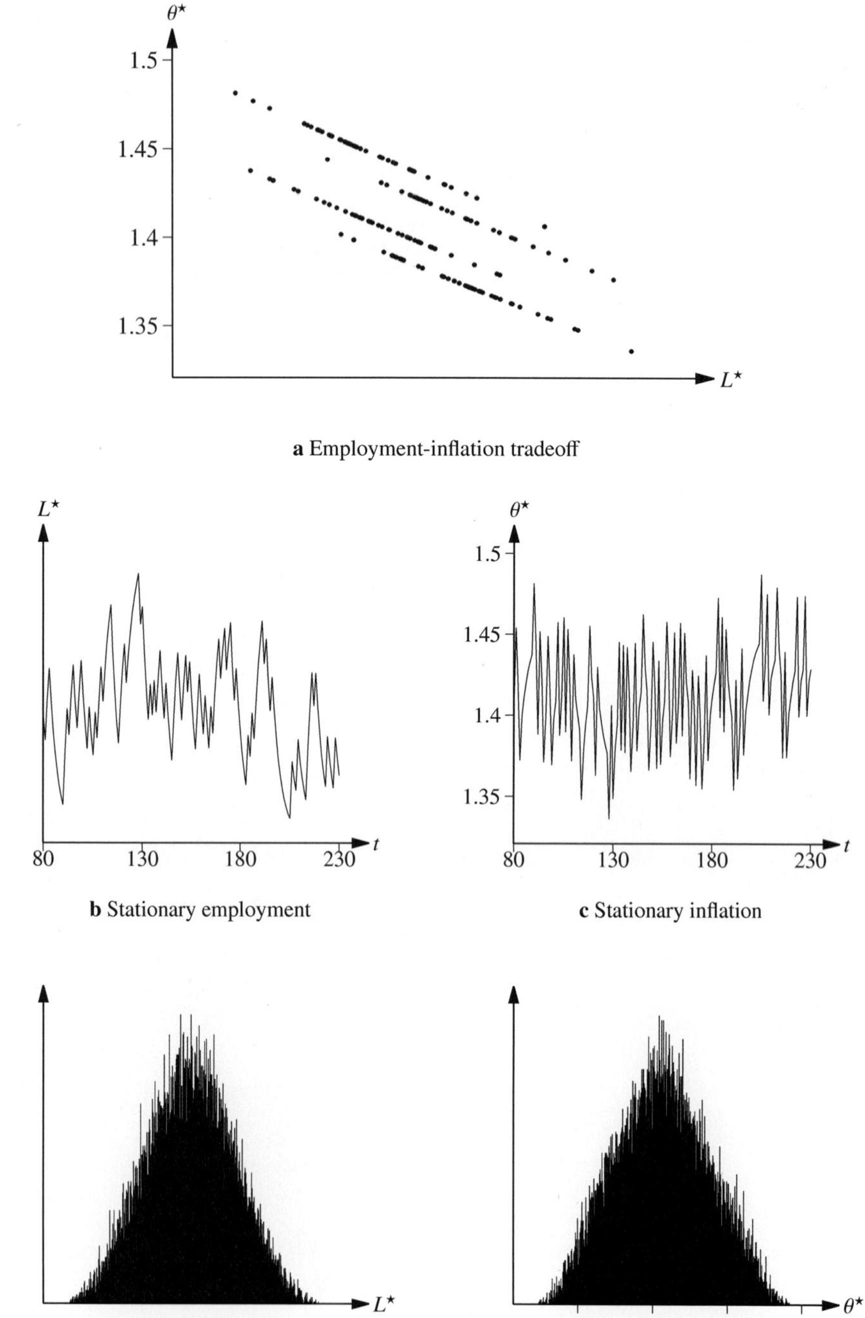

a Employment-inflation tradeoff

b Stationary employment

c Stationary inflation

d Stationary employment; $T = 2 \cdot 10^4$

e Stationary inflation; $T = 2 \cdot 10^4$

Fig. 4.25 Stationary employment and inflation

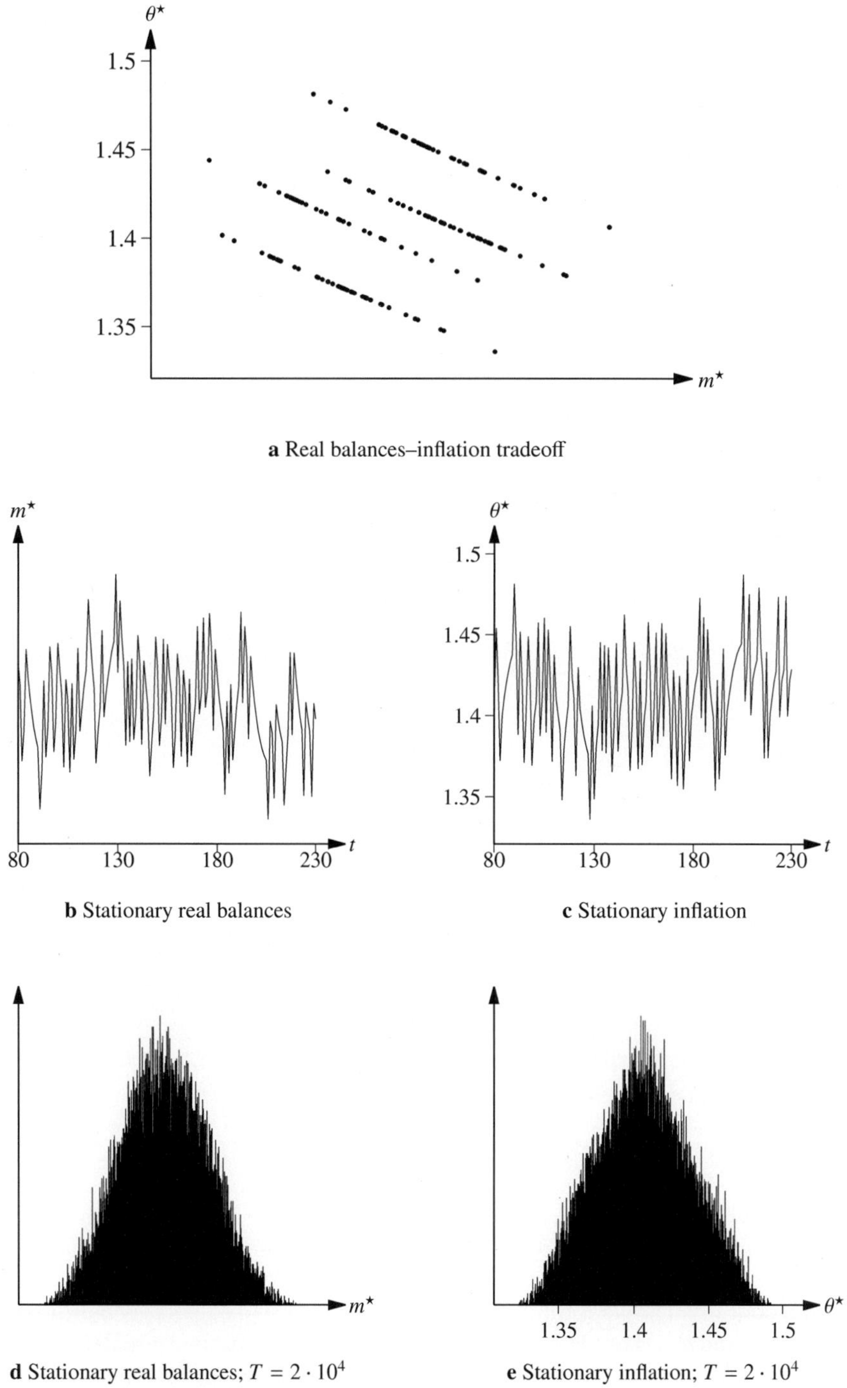

a Real balances–inflation tradeoff

b Stationary real balances

c Stationary inflation

d Stationary real balances; $T = 2 \cdot 10^4$

e Stationary inflation; $T = 2 \cdot 10^4$

Fig. 4.26 Stationary real balances and inflation

4.4.3 Convergence and Stability of Stochastic Balanced Paths

As in the deterministic situation (Section 4.2) convergence of the stochastic intensive form is only a necessary condition for convergence to a balanced random path in the space of money balances and expectations, so that asymptotic convergence to balanced random monetary orbits requires additional restrictions. To avoid writing double time indexes for the remainder, let $p^e_{t-1,t} \equiv p^e_t$. Then, for any $\omega \in \Omega$ and (M_0, p^e_0), the two dimensional random difference equation system (4.3.24)

$$M_{t+1} = M_t\left(\frac{\tilde{c}-\tau^\star}{\tilde{c}}\right)\Big(1 + g\mathcal{P}(1, \psi^*(1, p^e_t/M_t), Z(\vartheta^t\omega))\Big)$$

$$p^e_{t+1} = M_t\,\psi^*(1, p^e_t/M_t)$$

generates an orbit of nominal money balances and expectations $\{(M_t, p^e_t)\}_{t=0}^\infty$ where $M_t = M(t, \omega, (M_0, p^e_0))$ and $p^e_t = p^e(t, \omega, (M_0, p^e_0))$.

Definition 4.4.2 (Balanced random orbits). An orbit

$$\left\{(\bar{M}_t, \bar{p}^e_t)\right\} = \left\{(M(t, \omega, (M_0, p^e_0)), p^e(t, \omega, (M_0, p^e_0))\right\}$$

of (4.3.24), given $\omega \in \Omega$ and (M_0, p^e_0), is called **balanced**, if there exists a random fixed point $q^\star : \Omega \to \mathbb{R}_+$ of (4.3.28)

$$q^e_{t+1} = S(Z_t, q^e_t) := \left(\frac{\tilde{c}}{\tilde{c}-\tau^\star}\right)\frac{\psi^*(q^e_t)}{1 + g\mathcal{P}(1, \psi^*(q^e_t), Z_t)},$$

such that

$$\bar{p}^e_t = \bar{M}_t q^\star(\vartheta^t\omega) \qquad \text{for all } t \geq 0. \tag{4.4.26}$$

Define the distance of an orbit of (4.3.24) to the balanced path associated with $q^\star$ as

$$\Delta_t = \Delta(t, \omega, (M_0, p^e_0)) := p^e(t, \omega, (M_0, p^e_0)) - q^\star(\vartheta^t\omega)M(t, \omega, (M_0, p^e_0)). \tag{4.4.27}$$

A balanced orbit is called **asymptotically stable** if, for all (M_0, p^e_0) in a neighborhood $\mathcal{U}(\bar{M}_0, \bar{p}^e_0, \omega)$,

$$\lim_{t\to\infty} |q^e(t, \omega, p^e_0/M_0) - q^\star(\vartheta^t\omega)| = 0 \quad \text{and} \quad \lim_{t\to\infty} |\Delta(t, \omega, (M_0, p^e_0))| = 0, \quad \mathbb{P}\text{-a.s.}.$$

Figure 4.27 characterizes the convergence issue in the state space of stochastic paths. The two rays $q^\star(\omega)$ and $q^\star(\vartheta\omega)$ describe the one-step movement of the stochastic fixed point moving within the cone of the two blue dotted lines. The one-step move of the orbit in intensive form converging to the random fixed point is indicated by the pair q^e_t and q^e_{t+1}. For (M_t, p^e_t) with distance $\Delta_t \equiv \Delta(t, \omega)$ the two possibile cases of convergence Δ_{t+1} or of divergence with $\Delta_{t+1} > \Delta_{t+1}$ are indicated for the same $\omega \in \Omega$. The diagram visualizes the possibility of convergence or divergence depending on the rate of monetary expansion which is shown to be larger for Δ_{t+1} than for Δ_{t+1} in the one-step description for an arbitrary point (M_t, p_t^e) along the orbit. Theorem 4.4.2 states that convergence (divergence) occurs when the expansionary forces are

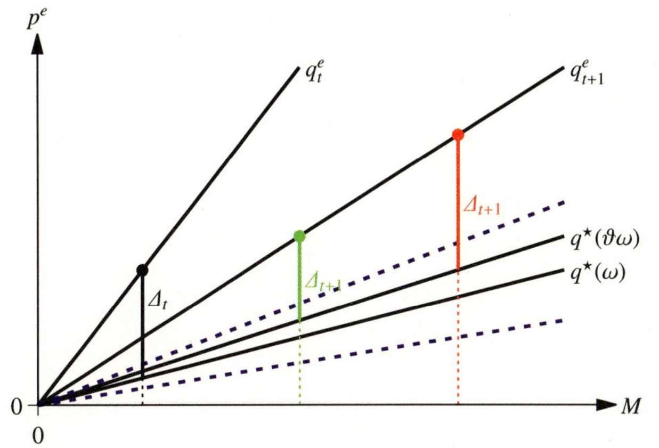

Fig. 4.27 Convergence/divergence to balanced path in (M, p^e)-space

appropriately dominated (or not dominated) *on average* by the rate of contraction of the random fixed point, i.e. keeping the random intensity sufficiently bounded relative to the rate of expansion along the random fixed point.

Theorem 4.4.2.
Let S be differentiable and increasing with respect to q^e and let $q^\star$ be an asymptotically stable random fixed point of

$$q_{t+1}^e = S(\vartheta^t\omega, q_t^e) := \left(\frac{\tilde{c}}{\tilde{c} - \tau^\star}\right) \frac{\psi^*(q_t^e)}{1 + g\mathcal{P}(1, \psi^*(q_t^e), Z(\vartheta^t\omega))}.$$

Then, for almost all $\omega \in \Omega$ and any $q_0^e \in I_3$, $q_0^e \neq q^\star(\omega)$ with $\lim_{t\to\infty} |q^e(t, \omega, q_0^e) - q^\star(\vartheta^t\omega)| = 0$ the distance $\Delta_t := p^e(t, \omega, (M_0, p_0^e)) - q^\star(\vartheta^t\omega)\, M(t, \omega, (M_0, p_0^e))$ satisfies $\mathbb{P}$-a.s.:

$\lim_{t\to\infty} |\Delta_t| = 0$ *if*

$$\mathbb{E}\log(S'(\omega, q^\star(\omega)) + \mathbb{E}\log\left(\frac{\tilde{c} - \tau^\star}{\tilde{c}}\right)(1 + g\mathcal{P}(1, \psi^*(q^\star(\omega)), Z(\omega))) < 0 \tag{4.4.28}$$

$\lim_{t\to\infty} |\Delta_t| = \infty$ *if*

$$\mathbb{E}\log(S'(\omega, q^\star(\omega)) + \mathbb{E}\log\left(\frac{\tilde{c} - \tau^\star}{\tilde{c}}\right)(1 + g\mathcal{P}(1, \psi^*(q^\star(\omega)), Z(\omega))) > 0. \tag{4.4.29}$$

The distance function Δ defines a second random difference equation $\Delta : \Omega \times I_3 \times \mathbb{R} \to \mathbb{R}$, making the pair (S, Δ) a two dimensional random dynamical system $(S, \Delta) : \Omega \times I_3 \times \mathbb{R} \to I_3 \times \mathbb{R}$ with fixed point $(q^\star, 0) : \Omega \to I_3 \times \mathbb{R}$. The proof applies the technique used in Böhm, Pampel & Wenzelburger (2005) for the random growth model.

Proof. From the definition $\Delta_t = M_t(q_t^e - q^\star(\vartheta^t\omega))$ and (4.3.28) one has

$$\begin{aligned}\Delta_{t+1} &= M_{t+1}(q_{t+1}^e - q^\star(\vartheta^{t+1}\omega)) = M_{t+1}(S(\vartheta^t\omega, q_t^e) - S(\vartheta^t\omega, q^\star(\vartheta^t\omega)))\\ &= \frac{M_{t+1}}{M_t}\frac{S(\vartheta^t\omega, q_t^e) - S(\vartheta^t\omega, q^\star(\vartheta^t\omega))}{q_t^e - q^\star(\vartheta^t\omega)}\Delta_t\\ &= \frac{\tilde{c}-\tau^*}{\tilde{c}}\left(1 + g\mathcal{P}(1, \psi^*(q_t^e), Z(\vartheta^t\omega))\right)\frac{S(\vartheta^t\omega, q_t^e) - S(\vartheta^t\omega, q^\star(\vartheta^t\omega))}{q_t^e - q^\star(\vartheta^t\omega)}\Delta_t\end{aligned}$$

implying

$$\frac{\Delta_{t+1}}{\Delta_t} = \frac{\tilde{c}-\tau^*}{\tilde{c}}\left(1 + g\mathcal{P}(1, \psi^*(q_t^e), Z(\vartheta^t\omega))\right)\frac{S(\vartheta^t\omega, q_t^e) - S(\vartheta^t\omega, q^\star(\vartheta^t\omega))}{q_t^e - q^\star(\vartheta^t\omega)}. \quad (4.4.30)$$

Since $\lim_{t\to\infty} |q_t^e - q^\star(\vartheta^t\omega)| = 0$, $\mathbb{P}$-a.s., there exists an $\varepsilon > 0$ sufficiently small and $t_0 = t_0(\varepsilon, \omega) > 0$ sufficiently large such that

$$\left|\mathcal{P}(1, \psi^*(q_t^e), Z(\vartheta^t\omega)) - \mathcal{P}(1, \psi^*(q^\star(\vartheta^t\omega), Z(\vartheta^t\omega))\right| < \varepsilon \quad (4.4.31)$$

$$\left|\frac{S(\vartheta^t\omega, q_t^e) - S(\vartheta^t\omega, q^\star(\vartheta^t\omega))}{q_t^e - q^\star(\vartheta^t\omega)} - S'(\vartheta^t\omega, q^\star(\vartheta^t\omega))\right| < \varepsilon. \quad (4.4.32)$$

By induction we have $\underline{\Delta}_t \le |\Delta_t| \le \overline{\Delta}_t$ for all $t \ge t_0$, for the two linear random dynamical systems

$$\overline{\Delta}_{t+1} = \left(\frac{\tilde{c}-\tau^*}{\tilde{c}}\right)\left[\left(1 + g\mathcal{P}(1, \psi^*(q^\star(\vartheta^t\omega)), Z(\vartheta^t\omega))\right) S'(\vartheta^t\omega, q^\star(\vartheta^t\omega)) + \varepsilon\right]\overline{\Delta}_t \quad (4.4.33)$$

$$\underline{\Delta}_{t+1} = \left(\frac{\tilde{c}-\tau^*}{\tilde{c}}\right)\left[\left(1 + g\mathcal{P}(1, \psi^*(q^\star(\vartheta^t\omega)), Z(\vartheta^t\omega))\right) S'(\vartheta^t\omega, q^\star(\vartheta^t\omega) - \varepsilon)\right]\underline{\Delta}_t \quad (4.4.34)$$

with $\overline{\Delta}_{t_0} = |\Delta_{t_0}|$ and $\underline{\Delta}_{t_0} = |\Delta_{t_0}|$. Therefore, assumption (4.4.28) implies that the upper bound (4.4.33) converges to zero $\mathbb{P}$-a.s. while the condition (4.4.34) implies that the lower bound grows to infinity eventually. Thus, in this case the distance of an orbit to the balanced path diverges under assumption (4.4.29). □

Figure 4.28 displays the convergence features when the balanced orbit associated with $q_3^\star$ is asymptotically stable, a situation which occurs for the parameters given in Table 4.2. The main differences to the values in Table 4.1 consist in a slightly lower government demand and in higher tax rates. This causes lower deficits and thus lower rates of inflation at any one time. For these values – with a small production shock – the time one map S becomes almost linear on I_3, subfigure **a**. Panel **b** and **c**

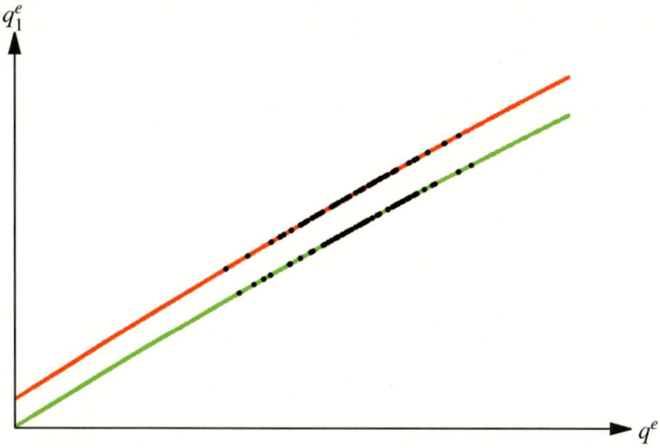

a S becomes almost linear on I_3 for values in Table 4.2

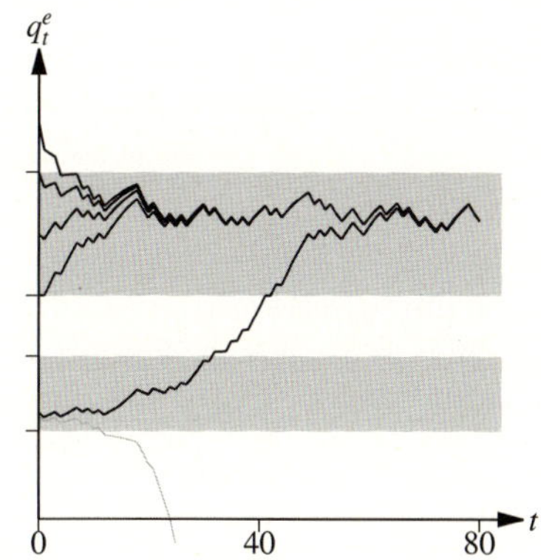

b Convergence from I_2 and I_3 to $q^\star \in I_3$

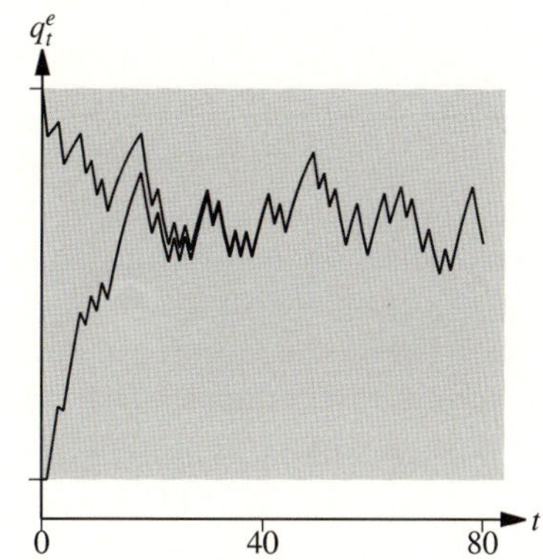

c Convergence in I_3, $t \in [0, 60]$

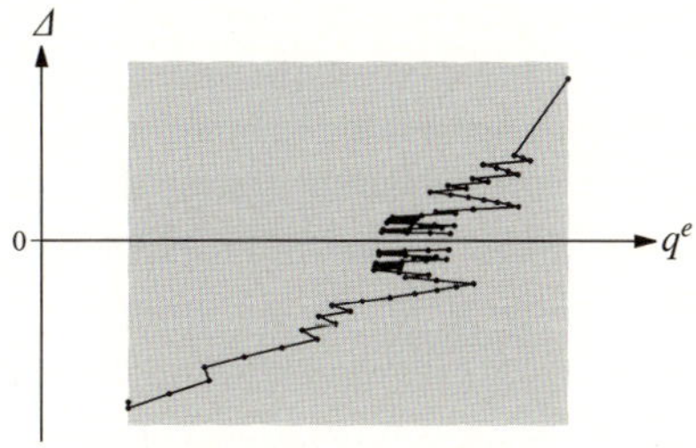

d Convergence in I_3: $t \in [0, 30]$

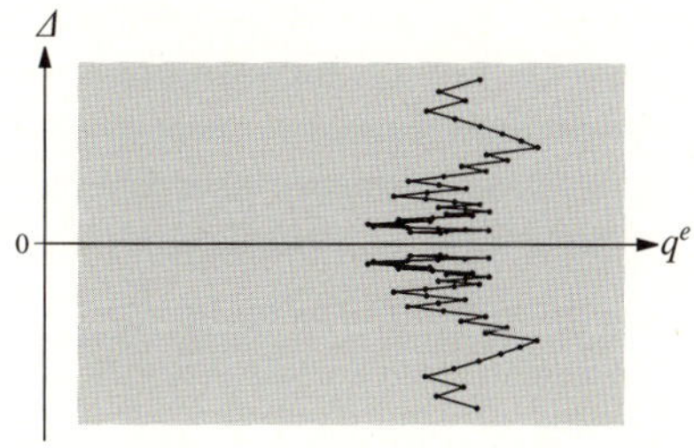

e Symmetry of orbits for $t > 30$

Fig. 4.28 Convergence in (q^e, Δ)-space

Table 4.2 Standard parametrization b

$Z_{\min}$	$Z_{\max}$	B	C	δ	c_s	τ_w	τ_π	g	g^*	g^{**}
1.0	1.01	0.6	0.6	1.0	0.5	0.75	0.75	0.8392		0.8449

show the evolution for six different initial conditions in the space of real expectations (five converging and one diverging) while **d** and **e** display the convergence in $(q^e, \varDelta)$-space for the same ω. Notice the difference in scale between the subfigures **b** and **c**.

Convergence and Growth Rates of Monetary Expansion

In order to understand the conditions (4.4.28) and (4.4.29) for stability/instability it is useful to consider the central equation (4.4.30)

$$\varDelta_{t+1} = \frac{\tilde{c} - \tau^*}{\tilde{c}} \left(1 + g\mathcal{P}(1, \psi^*(q_t^e), Z(\vartheta^t \omega))\right) \cdot \frac{S(\vartheta^t \omega, q_t^e) - S(\vartheta^t \omega, q^\star(\vartheta^t \omega))}{q_t^e - q^\star(\vartheta^t \omega)} \cdot \varDelta_t$$

again which defines a linear random dynamical system in $\varDelta$ whose coefficient consists of a product of two random variables. Since $\lim_{t\to\infty} |q_t^e(t, \omega, q_0^e) - q^\star(\vartheta^t \omega)| = 0, \mathbb{P} - a.s.$ the second term converges to the derivative of S while the first converges to the growth rate of money $\mu_t := M_t/M_{t-1}$ along $q_3^\star$ (see (4.3.24)). Thus, convergence of the distance $\varDelta$ to zero occurs if the growth factor $\mu^\star(\vartheta^t \omega) \cdot S'(q_3^\star(\vartheta^t \omega))$ of the linear system is mean contracting (see Arnold & Crauel, 1992), i.e. if and only if $\mathbb{E}\left(\mu^\star(\vartheta^t \omega) \cdot S'(q_3^\star(\vartheta^t \omega))\right) < 1$, which corresponds to the condition (4.4.28). Since $S'(q_3^\star(\vartheta^t \omega)) < 1, \mathbb{P}-a.s.$, given the assumption for S on I_3, the stability requirement stipulates that the rate of monetary expansion may well be larger than one along the whole orbit of $q^\star$, but it should make the product with S' less than one on average. Figure 4.29 displays the histograms of the rates of monetary expansion in the two cases with their respective means. While money grows at a rate of about 7 percent in the stable case, subfigure **a**, it is about 40 percent in **b** indicating clearly the reason for the instability of the balanced orbit in the case of the parameters of Table 4.1.

4.4.4 The Dynamics of Rational Expectations Equilibria: An Appraisal

The analysis has shown that the balanced random evolution of monetary expanding orbits under rational expectations can be analyzed and described explicitly taking full account of all nonlinearities in the model. The concepts and techniques of the theory of random dynamical systems in the sense of Arnold (1998) provide the

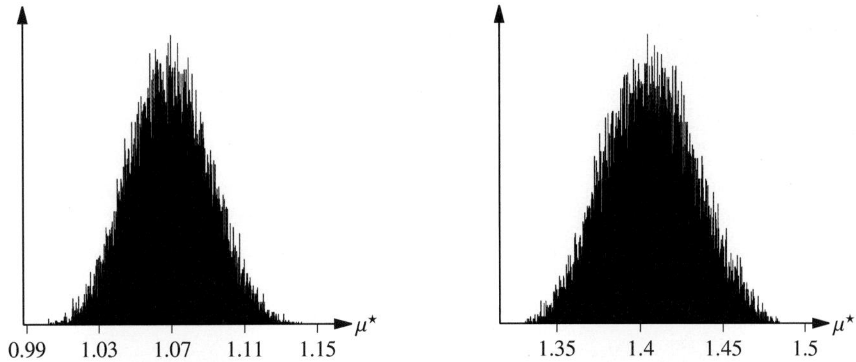

a Money growth stable: $\mathbb{E}\mu^\star = 1.0690$ **b** Money growth unstable: $\mathbb{E}\mu^\star = 1.4063$

Fig. 4.29 Stationary growth rates of money: $T = 2 \cdot 10^4$

necessary methods. These allow to examine all questions of the orbit structure of macroeconomic time series as the primitive and observable object for any recursive nonlinear stochastic macroeconomic model. Combined with numerical techniques this enables investigations of all relevant dynamic and statistical issues demonstrating conditions of convergence or non-convergence and of bifurcation scenarios. These techniques have been applied successfully in other areas of economics such as growth theory and mathematical finance (see Schenk-Hoppé & Schmalfuß, 2001; Böhm & Wenzelburger, 2005; Böhm & Chiarella, 2005).

The mathematical techniques can be applied directly to stochastic orbits of general nonlinear systems. Their relationship to the theory of Markov chains under families of i.i.d. random mappings is well understood offering a unique way to construct an associated Markov chain from a random dynamical system with i.i.d. noise. In other words, the results obtained from an analysis of random orbits can be embedded directly into the more traditional theory of Markov processes for stochastic economic models.

There is no need to suppress the discussion of orbits and analyze only the weaker sequential structure of the associated conditional distributions (as in Bhattacharya & Majumdar, 2004, and by others), which are empirically unobservable. In other words, the time series serve as the identical and essential objects of investigation in the theoretical models as well as in empirical data making an analysis of the development of distributions (in the spirit of the theory of stochastic processes) a natural supplement to the time series approach of the theory of random dynamical systems.

Most importantly, however, from a mathematical point of view there is neither a need to linearize the random difference equations and analyze the stationary orbits of an *approximating* linear macroeconomic model and to draw conclusions about the nonlinear model *only* from properties of the linear system (as done in many applications), nor to analyze the dynamics of moments of conditional distributions (as in Lettau & Uhlig, 2002). Linear approximations allow only a limited characterization of local properties near fixed points which provide little or often misleading information about the behavior of an associated nonlinear system near cycles or (unstable) fixed points (see Böhm, 1999).

The main economic results of this section parallel those of the deterministic situation in a direct fashion showing that the convergence properties of the model in intensity form as well as the one in the state space can be successfully demonstrated in the same way as in the deterministic case (see Section 4.1). For small positive levels of government demand, there exist two expanding random monetary balanced paths, i.e. random fixed points defined on disjoint compact intervals with associated stationary distributions which induce stationary real solutions of the economy with bounded positive real balances, output, employment, and inflation. One finds that the balanced path with high employment and production is always unstable, implying that the associated orbits in the monetary state space are always diverging relative to the balanced path, which makes the stationary balanced solution unobservable empirically. On the other hand, with small enough government demand there are additional conditions under which the balanced path with low employment is asymptotically stable. These conditions are not universally satisfied for all such economies under rational expectations. In particular, as in the deterministic case, given all other parameters of the economy, for large government demand no balanced orbits exist, i.e. balanced monetary expansion under rational expectations is not viable. Thus, there exists a finite threshold level for government consumption beyond which all orbits with rational expectations diverge with excessive monetary expansion outgrowing expectations, i.e. $\lim_{t\to\infty} p_t^e/m_t = 0$ with unbounded real money balances. No real economic variables become stationary and boundary allocations will be obtained. In other words, fiscal parameters determine not only the level of a balanced path but they are also decisive for convergence or divergence.

4.5 Dynamics with Random Aggregate Demand

Models examining the implications of random demand for rational expectations equilibria often use aggregate functions in isoelastic form. Their loglinear forms are often explicitly solvable which provides detailed properties to discuss and compare the different implications for the business cycle, i.e. the differences in the fluctuations of output, prices, and employment and the role of economic policy. On the supply side one often uses the so-called Lucas supply function (originating from Lucas & Rapping, 1969a,b)

$$AS(\theta^e) := AS(1)(\theta^e)^{-\alpha}, \qquad \alpha > 0 \tag{4.5.1}$$

where $\theta^e \equiv p^e_{t+1}/p_t$ is the mean of expected inflation. In Chapter 3 this aggregate functional form was derived from isoelastic functions on the micro level combined with labor market clearing under competitive conditions (see (3.2.53), Section 3.2.7.

To analyze situations with random demand the new classical tradition of the literature uses different 'ad-hoc' functional relations of how random effects enter into the aggregate demand function. These are typically derived from a combinations of versions of their IS-equation and LM-equation which are combined to close the model (see for example Sargent & Wallace, 1975; Taylor, 1977). Using one of their log-linear presentations for the model under discussion here without government debt or bonds, one form could be an isoelastic function as given by

$$AD(M_t, p_t, v_t) := v_t \left(\frac{1}{p_t} \frac{M_t + G_t}{1 - c(1-\tau)} \right)^{\beta}, \qquad \beta > 0, \quad 0 < c \leq 1, \quad 0 < \tau < 1, \tag{4.5.2}$$

where $G_t \geq 0$ denotes nominal government expenditure and v_t is a multiplicative random demand shock.

Let us assume in addition that the government follows a constant deficit rule under a proportional income tax system, as in Section 5.1, with rate $0 < \tau < 1$ allowing expenditures G to exceed tax revenue by a percentage $-1 < \rho < \rho_{\max}$, i.e.

$$G_t = (1+\rho)\tau\, p_t y_t \tag{4.5.3}$$

financing deficits through money creation. Then, aggregate income consistency in any period (the IS equation)

$$Y = M + C + G = M + (1-\tau)cY + (1+\rho)\tau Y$$

induces a proportional income relation

$$Y_t = \frac{M_t}{1 - c(1-\tau) - (1+\rho)\tau}$$

implying that, in any period t, government expenditures are a constant proportion of money balances

$$G_t = \frac{(1+\rho)\tau}{1 - c(1-\tau) - (1+\rho)\tau} M_t \tag{4.5.4}$$

and that money grows at a constant rate ρ_M

$$M_{t+1} = M_t + (G_t - \tau Y_t) = M_t + \underbrace{\left(\frac{\rho\tau}{1 - c(1-\tau) - (1+\rho)\tau} \right)}_{=:\rho_M} M_t .$$

In other words, under the policy constraint, aggregate demand becomes

$$AD(M, p, v) = v_t \left(\frac{M_t}{c(\rho)\, p_t} \right)^{\beta} \tag{4.5.5}$$

where $0 < c(\rho) := 1 - c(1-\tau) - (1+\rho)\tau < 1$ is the inverse of the tax compensated multiplier. Then, commodity market clearing in each period t implies

$$(\theta_t^e)^{-\alpha} = v_t \left(\frac{M_t}{c(\rho)\, p_t} \right)^{\beta} = v_t \left(\frac{1}{c(\rho)} \right)^{\beta} \left(\frac{M_t}{p_t} \right)^{\beta}, \qquad \theta_t^e := \frac{p^e_{t,t+1}}{p_t}. \tag{4.5.6}$$

From this equation one obtains an explicit solution for the random equilibrium price (the random price law) given by

$$p_t = \left(\frac{v_t}{c(\rho)^{\beta}} M_t^{\beta} (p^e_{t,t+1})^{\alpha} \right)^{\frac{1}{\alpha+\beta}}. \tag{4.5.7}$$

Proceeding along the lines of the literature, define endogenous variables in logarithmic form, i.e.

$$\hat{p} := \ln p; \quad \hat{p}^e := \ln p^e; \quad \hat{m} := \ln(M); \quad \hat{v} := \ln(v/c(\rho)^{\beta}).$$

This implies a linear random price law in logarithmic form, (see Figure 4.30)

$$\hat{p}_t = \mathcal{P}(\hat{m}_t, \hat{p}^e_{t,t+1}, \hat{v}_t) := \frac{1}{\alpha+\beta} \left[\beta \hat{m}_t + \alpha \hat{p}^e_{t,t+1} + \hat{v}_t \right] \tag{4.5.8}$$

with additive noise and a linear deterministic monetary equation in log form given by

$$\hat{m}_t = \log(1 + \rho_M) + \hat{m_{t-1}}. \tag{4.5.9}$$

To define rational expectations, let $\mu_t \equiv \mu$ denote the conditional distribution (measure) of $\hat{v}_t$ which is assumed to be independent and identical across time with mean $\mu(\hat{v}) = 0$ and standard deviation $\sigma(\hat{v}) > 0$. Define $\hat{p}^e_{t,t+1}$ as a (log) mean value prediction for the price in $t+1$. $\hat{p}^e_{t,t+1}$ is called a rational or an *unbiased* prediction if the conditional mean expected forecast error in every period is equal to zero, i.e. if

$$\begin{aligned} \mathbb{E}_{t-1}\hat{p}_t - \hat{p}^e_{t-1,t} &= \mathbb{E}_{t-1} \left\{ \mathcal{P}(\hat{m}_t, \hat{p}^e_{t,t+1}, \hat{v}_t) \right\} - \hat{p}^e_{t-1,t} \\ &= \mathbb{E}_{\mu} \left\{ \mathcal{P}(\hat{m}_t, \hat{p}^e_{t,t+1}, \hat{v}_t) \right\} - \hat{p}^e_{t-1,t} \\ &= (\mathbb{E}_{\mu}\mathcal{P})(\hat{m}_t, \hat{p}^e_{t,t+1}) - \hat{p}^e_{t-1,t} \stackrel{!}{=} 0. \end{aligned} \tag{4.5.10}$$

Substituting the price law (4.5.8) into (4.5.10) and taking conditional expectations implies

$$\frac{1}{\alpha+\beta} \left[\beta \hat{m}_t + \alpha \hat{p}^e_{t,t+1} \right] = \hat{p}^e_{t-1,t}. \tag{4.5.11}$$

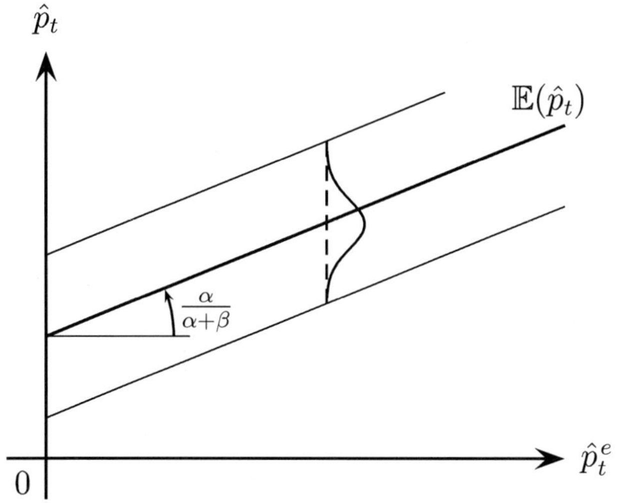

Fig. 4.30 Random log price law $\mathcal{P}(\hat{m}_t, \hat{p}^e_{t,t+1}, \hat{v}_t)$ with demand shocks, $\hat{m}_t$ given

Solving for $\hat{p}^e_{t,t+1}$ one obtains an explicit functional form of the unique unbiased predictor

$$\hat{p}^e_{t,t+1} = \psi_*(\hat{p}^e_{t-1,t}, \hat{m}_t, \mu) := \frac{1}{\alpha}\left[(\alpha+\beta)\hat{p}^e_{t-1,t} - \beta\hat{m}_t\right]. \tag{4.5.12}$$

The unbiased predictor ψ_* is a deterministic function using current (non random) money balances and past (non random) mean predictions $(\hat{p}^e_{t-1,t}, \hat{m}_t) \in \Omega_{t-1}$, as input data from the correct information set. Therefore, given the deterministic equation for money balances, the equations (4.5.8) and (4.5.12) together determine a two dimensional deterministic dynamical system

$$\begin{aligned} \hat{p}^e_{t,t+1} &= \psi_*(\hat{m}_t, \hat{p}^e_{t-1,t}, \mu) := \frac{1}{\alpha}\left[(\alpha+\beta)\hat{p}^e_{t-1,t} - \beta\hat{m}_t\right] \\ \hat{m}_{t+1} &= \log(1+\rho_M) + \hat{m}_t \end{aligned} \tag{4.5.13}$$

in $(\hat{p}^e_{-1}, \hat{m})$ for all measures μ and any size of noise. Any deterministic orbit of this system induces stochastic orbits of prices

$$\hat{p}_t := \mathcal{P}(\hat{m}_t, \psi_*(\hat{p}^e_{t-1,t}, \hat{m}_t), \hat{v}_t) = \hat{p}^e_{t-1,t} + \frac{1}{\alpha+\beta}\hat{v}_t \tag{4.5.14}$$

satisfying rational expectations, i.e. for all t one has

$$(\mathbb{E}_\mu\mathcal{P})(\hat{m}_t, \psi_*(\hat{m}_t, \hat{p}^e_{t-1,t}) = \hat{p}^e_{t-1,t}$$

with $(\hat{m}_t, \hat{p}^e_{t-1,t})$ while the stochastic orbits of (log) output are given by

$$\hat{y}_t := \beta(\hat{m}_t - \hat{p}_t) + \hat{v}_t = \beta\left(\hat{m}_t - \hat{p}^e_{t-1,t}\right) + \frac{\alpha}{\alpha+\beta}\hat{v}_t. \tag{4.5.15}$$

Both equations are linear with additive noise. Most importantly, however, for $\rho_M \neq 0$, the deterministic system (4.5.13) has no steady state and the unbiased expectations dynamics (4.5.12) are globally unstable for each level of $\hat{m}$! Therefore, stationary solutions of (4.5.14) and (4.5.15) do not exist.

Stationary Solutions with Rational Expectations

The literature (for example Taylor, 1977) sometimes discusses the exceptional situation with a balanced budget[18] ($\rho = 0$) implying a zero monetary growth rate

$$\rho_M := \frac{\rho\tau}{1 - c(1-\tau) - (1+\rho)\tau} = 0.$$

This is the only situation when the system (4.5.13) has non-zero fixed points. The *rational expectations process* described by equation (4.5.12) is a linear *deterministic* first order difference equation in $\hat{p}^e$ parameterized in (log) monetary balances $\hat{m}$. For each $\hat{m}$ it has a unique asymptotically *unstable deterministic* steady state $(\hat{p}^e)^* = \hat{m}$ with eigenvalues $\nu_1 = 1 < \nu_2 = \frac{\alpha+\beta}{\alpha}$. Thus, there is a continuum of unstable steady states, one for each $\hat{p} = \hat{m} \in \mathbb{R}$ where each level $\hat{m} = \hat{p}$ induces a stochastic process of prices and output. In other words, there exists a continuum of stationary rational expectations solutions parametrized in $\hat{m}$,

$$\hat{p}^*_t = \hat{m} + \frac{1}{\alpha+\beta}\hat{v}_t \quad \text{and} \quad \hat{y}^*_t = \beta(\hat{m} - \hat{p}^*_t) + \hat{v}_t = \frac{\alpha}{\alpha+\beta}\hat{v}_t. \tag{4.5.16}$$

For each $\hat{m}$, prices and output are perfectly correlated, since (4.5.16) implies

$$\hat{y}^*_t = \alpha\left[\hat{p}^*_t - \hat{m}\right] \tag{4.5.17}$$

with zero cross correlation.

The constant level of rational mean predictions $(\hat{p}^e)^* = \hat{m}$ induces two stationary random variables, log prices and log output, with the following propeties:

- stationary log prices are a random variable with mean $\hat{m}$,
- stationary log output is independent of $\hat{m}$ with zero mean
- and a perfect stochastic tradeoff between log output and log prices,
- where log money balances $\hat{m}$ influence mean log prices but not mean log output.

These properties are interpreted to imply that a systematic monetary policy has transient but no systematic long-run impact on real output. Comparing equations (4.5.15) and (4.5.16), one could indeed infer that random log output depends on the non-random part of money balances and inflation, while the stationary mean does not. Notice, however, that such a conclusion is not consistent with the stationary

---

[18] Compare also the discussion in Section 4.1.4 and Lemma 4.1.2.

characteristics of the solution, since it is made, using properties of the *non stationary* system (4.5.13) which is globally unstable. Thus, any solution or set of data satisfying equation (4.5.15) can never show the stationary features of (4.5.16) for $t \to \infty$.[19]

Some additional remarks could be made regarding the validity of the inferences about properties of the non-stationary paths under rational expectations based on properties of the stationary solution only. The lack of stability of the stationary solution is crucial. Any steady state $(\hat{p}^e)^* = \hat{m}$ of the unbiased predictor (4.5.12) is globally unstable. Therefore, this model predicts divergence with probability one for arbitrary initial conditions $\hat{p}^e_0$ and $\hat{m}_0$, making it difficult (if not impossible) to validate/test the model on empirical data.

The conclusion that systematic changes of policy parameters have only transient but no or specific effects on the stationary solution can only be made under an appropriate analysis where parameters are changed systematically while the noise process remains identical and independent of the parameters. Then transient orbits may be shown to be returning to the levels before the parameters have been changed. This again requires some form of stability/convergence of the solutions which is absent in this model. Therefore, the set of positive conclusions from this model are very restrictive. In essence, a successful analysis about the effectiveness of policy measures in a stochastic model should best be carried out as a bifurcation analysis when the economy performs in an observable i.e. in an attracting stationary environment.

Balanced Monetary Expansion with Rational Expectations

When $\rho_M \neq 0$, nontrivial stationary real solutions can be induced by orbits of balanced monetary expansion or contraction. Converting the system (4.5.13) back to its isoelastic form one obtains a two dimensional system in (p^e_{-1}, M)

$$p^e = \frac{\left(p^e_{-1}\right)^{\frac{\alpha+\beta}{\alpha}}}{M^{\frac{\beta}{\alpha}}} \qquad \text{and} \qquad M_1 = \frac{M}{1+\rho_M}. \tag{4.5.18}$$

Both mappings are homogeneous of degree one. In order to examine existence and stability of balanced paths, consider the dynamics of this system restricted to the unit simplex $\{(p^e_{-1}, m) \in \mathbb{R}^2_+ \mid p^e_{-1} + m = 1\}$ (rather than utilizing the equivalent model in intensive form as before in Section 4.2). This induces one dimensional dynamics given by the time-one map $\mathcal{F} : [0,1] \to [0,1]$

$$p^e = \mathcal{F}\left(p^e_{-1}\right) := \left(p^e_{-1}\right)^{\frac{\alpha+\beta}{\alpha}} \cdot \left(\left(p^e_{-1}\right)^{\frac{\alpha+\beta}{\alpha}} + \frac{1}{1+\rho_m}\left(1 - p^e_{-1}\right)^{\frac{\alpha+\beta}{\alpha}}\right)^{-1}. \tag{4.5.19}$$

[19] One could analyze the backward orbit for $t \to -\infty$, to obtain a partial convergence result to one of the steady states out of the continuum and then calculate the associated stationary solution. In some cases the literature has addressed these issues (see for example Blanchard, 1979).

Lemma 4.5.1.
The mapping $\mathcal{F} : [0,1] \to [0,1]$ (4.5.19) *is continuously differentiable, strictly monotonically increasing, and bijective with three fixed points satisfying*

$$\begin{aligned} \mathcal{F}(0) = 0, \quad \mathcal{F}(\tilde{p}) = \tilde{p}, \quad \mathcal{F}(1) = 1 \\ \mathcal{F}'(0) = 0, \quad \mathcal{F}'(\tilde{p}) > 1, \quad \mathcal{F}'(1) = 0 \end{aligned} \tag{4.5.20}$$

Proof. Rewriting (4.5.19) as

$$\mathcal{F}(x) := \frac{x^a}{x^a + b(1-x)^a} = \frac{1}{1 + b\left(\frac{1-x}{x}\right)^a}, \quad a := \frac{\alpha + \beta}{\alpha} > 1, \quad b := \frac{1}{1 + \rho_M} \tag{4.5.21}$$

one finds

$$\mathcal{F}'(x) = \frac{ab(x(1-x))^{a-1}}{[x^a + b(1-x)^a]^2} > 0 \quad \text{if} \quad x \neq 0, 1.$$

Since $\mathcal{F}(x) = x \neq 0, 1$ if and only if $b = \left(\frac{x}{1-x}\right)^{\frac{1}{a-1}}$, this yields

$$x = \tilde{p} = \frac{b^{a-1}}{1 + b^{a-1}} \qquad \text{and} \qquad \tilde{q} = \frac{\tilde{p}}{1 - \tilde{p}} = b^{a-1} = \left(\frac{1}{1 + \rho_M}\right)^{\frac{\beta}{\alpha}}. \tag{4.5.22}$$

The mapping $\mathcal{F}$ is of sigmoid form, Figure 4.31 **a**, with three fixed points: $\mathcal{F}(0) =$

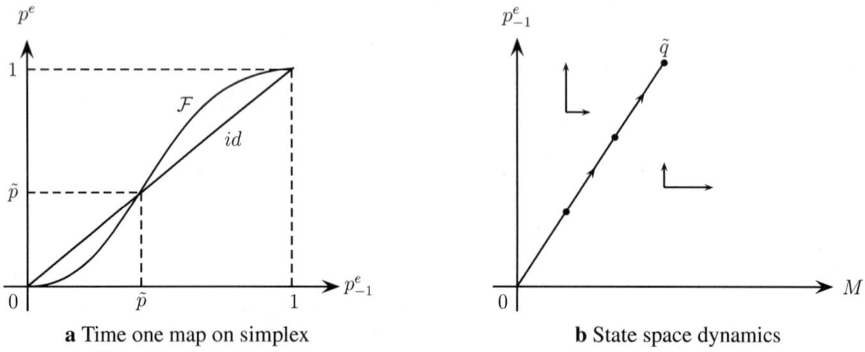

Fig. 4.31 Dynamics with random demand $\rho_M > 0$

$0, \mathcal{F}(1) = 1$, and a strictly positive $\mathcal{F}(\tilde{p}) = \tilde{p}$ which is globally unstable. Therefore, orbits along the balanced path $\tilde{q} = \tilde{p}/(1 - \tilde{p})$ are the only ones which guarantee stationary real allocations where money, price expectations, and mean prices grow at the constant rate ρ_M. All other orbits are non-stationary and diverging with money growing at the constant rate ρ_M while the growth rate of expectations is larger than ρ_M if $(p^e_{-1})_0 > \tilde{q}\, M_0$, or smaller than ρ_M if $(p^e_{-1})_0 < \tilde{q}\, M_0$. □

Chapter 5
Fiscal Policy and the Dynamics of Monetary Equilibrium

In the previous chapters government activity was given by a pair of parameters (g, τ) defining government consumption and a proportional tax rate on income, which were assumed to be constant over time. An important implication of such a stationary government policy was the observation that in each period the government's deficit is endogenous and determined in size and sign at each temporary equilibrium. As a consequence, the amount of fiat money held by consumers as well as the deficit in each period is endogenous leading to explicit endogenous dynamics of the quantity of money induced by the equilibrium conditions on the two markets. In such cases, neither the sign or size of the deficit nor the change of the quantity of money is directly controlled by the government. Chapter 4 derived the major consequences of such policies showing that the two fisal parameters have a decisive influence on the existence, stability, and on the long-run real performance of the macroeconomy under rational expectations.

Some of the consequences of the full endogeneity of the deficit and the money dynamics are undesirable from a policy point of view. This chapter discusses the impact of alternative stationary government fiscal policies designed to control the deficit or the process of money creation, while maintaining the structure of the closed aggregate competitive economy of Chapter 3 when there are no distributional effects, a constant propensity to consume $0 < c < 1$, and a constant proportional income tax with rate $0 < \tau < 1$. One of the main questions to be analyzed will be whether stable balanced orbits with constant real allocations are sustainable in the long-run when the government follows rules which attempt to maintain a constant or stationary deficit over time.

5.1 Deficit Rules and Permanent Inflation/Deflation

The first policy to be analyzed will be a fixed deficit quota rule. In other words, in each period government real demand is adjusted in such a way that the deficit is a constant (positive or negative) fraction $\rho > -1$ of total tax revenue, i.e. government spending in every period satisfies the budget equation

$$pg = (1+\rho)\tau py, \tag{5.1.1}$$

with $0 < \tau < 1$ being the proportional tax rate on overall income. The rate ρ measures the percentage excess of spending over revenue. Negative values correspond to a strict savings policy with an attempt to reduce the outstanding debt/money in the economy. The rule implies in particular that the amount of real government consumption $g = (1+\rho)\tau y$ is endogenous, implying, however, that it is a constant fraction $(1+\rho)\tau$ of real output, in other words, a constant so-called real fiscal share.

Given the deficit rule (5.1.1), income consistency yields

$$py = M + (1+\rho)\tau py + c(1-\tau)py.$$

Thus, one obtains the aggregate demand function given by

$$y = D\left(\frac{M}{p}\right) := \frac{M/p}{1 - [\tau(1+\rho) + c(1-\tau)]} \tag{5.1.2}$$

which is unit elastic with respect to prices. Non-negativity of aggregate income requires that $(1+\rho)\tau + c(1-\tau) < 1$ must hold, which implies an upper bound

$$\rho < \rho_{\max} := \frac{1-\tau}{\tau}(1-c) \tag{5.1.3}$$

for ρ. In other words, under income consistency the deficit quota should be strictly less than the upper bound $\rho_{\max}$ which is a function of the propensity to consume and of the tax rate τ. Notice that with the proportional income tax assumed here, the maximal quota does not depend on the *level* of tax revenue. The maximal deficit quota may well be larger than one if the income tax rate τ is less than one half. This implies in particular that it may be feasible for the government to run a deficit policy where government spending may be more than twice the total tax revenue.

Given ρ, one finds that the quantity of money increases from M in any period to

$$M_1 = M + pg - \tau py = M + \rho\tau py = M\left(1 + \frac{\tau\rho}{(1-c)(1-\tau) - \tau\rho}\right)$$

in the next period, implying a constant growth rate of money equal to

$$\rho_M := \frac{M_1}{M} - 1 = \frac{\tau\rho}{(1-c)(1-\tau) - \tau\rho} = \frac{\rho}{\rho_{\max} - \rho}. \tag{5.1.4}$$

For $1 > \rho_{\max} > \rho$, one has $1 > \rho_{\max} - \rho > 0$, so that $\rho_M > \rho$. Thus, money grows at a constant rate which is larger than the fiscal deficit rule itself and which is an increasing function of the deficit rule.

The analysis of the dynamics under perfect foresight with a fixed deficit rule ρ is now straightforward (see Chapters 3 and 4 for details, in particular Proposition 4.1.1). Given the specific form of the unit elastic aggregate demand function (5.1.2), the price law $p = \mathcal{P}(M, p^e)$ is the unique solution of

$$AS\left(\frac{p^e}{p}\right) = \frac{1}{1 - [\tau(1+\rho) + c(1-\tau)]}\frac{M}{p}.$$

$\mathcal{P}(M, p^e)$ is strictly monotonically increasing in p^e, $\mathcal{P}(M, p^e)/p^e$ strictly decreasing satisfying

$$\lim_{p^e \to 0} \mathcal{P}(M, p^e) = M\mathcal{P}(1,0), \qquad \lim_{p^e \to \infty} \mathcal{P}(M, p^e) = \infty, \qquad \lim_{p^e \to \infty} \mathcal{P}(M, p^e)/p^e = 0.$$

Define the inverse $\mathcal{P}^e$ of the price law by the mapping

$$p^e = \mathcal{P}^e(M, p) := \begin{cases} p\, AS^{-1}\left(\frac{1}{1-[\tau(1+\rho)+c(1-\tau)]}\frac{M}{p}\right) & \text{for } p \geq M\mathcal{P}(1,0) \\ 0 & \text{otherwise} \end{cases}. \tag{5.1.5}$$

Its global properties are given in the next lemma.

Lemma 5.1.1.
If the aggregate supply function is strictly decreasing, then

(a) *the price law $\mathcal{P}(M, p^e)$ and the inverse $\mathcal{P}^e(M, p)$ with respect to p are globally defined,*
(b) *they are both homogeneous of degree one in (M, p) and (M, p^e) respectively,*
(c) *for every $M > 0$, $\mathcal{P}^e(M, p)$ and $\mathcal{P}^e(M, p)/p$ are strictly increasing in p satisfying*

$$\begin{aligned} \lim_{p \to 0} \mathcal{P}^e(M, p) &= \lim_{p \to 0} \frac{\mathcal{P}^e(M, p)}{p} = 0 \\ \lim_{p \to \infty} \mathcal{P}^e(M, p) &= \lim_{p \to \infty} \frac{\mathcal{P}^e(M, p)}{p} = \infty. \end{aligned} \tag{5.1.6}$$

The properties (a) to (c) follow directly from (5.1.5). The inverse $\mathcal{P}^e$ is the unique perfect predictor since $\mathcal{P}(M, \mathcal{P}^e(M, p)) = p$ holds for all $p \geq M\mathcal{P}(1,0)$. If $\mathcal{P}(1,0) = 0$, then the price law is globally invertible and perfect foresight dynamics are globally defined. In other words, the domain of possible perfect foresight dynamics may be restricted, but for states without perfect forecasts expectations are defined to be zero with prices and money balances growing at the constant rate ρ_M.

With these qualifications the dynamics under perfect foresight are described by the system of equations

$$\begin{aligned} p^e &= M\mathcal{P}^e(1, p^e_{-1}/M) \\ M_1 &= (1 + \rho_M)\, M. \end{aligned} \tag{5.1.7}$$

When $\rho_M \neq 0$, the point $(p^e_{-1}, M) = (0,0)$ is the only fixed point. For $\rho_M > 0$, (5.1.7) induces expanding orbits $\{(M_t, p^e_t, p_t)\}$ of money balances, expectations, and prices. In order to understand the dynamics of such monetary expansion under a positive deficit rule ρ, one investigates first the existence of *balanced* paths of monetary growth and of inflation and their stability (see section 4.1).

An orbit $\{(p^e_t, M_t)\}_{t=0}^{\infty}$ of (5.1.7) is called balanced if and only if prices and money grow at the same rate leaving real money balances constant over time, i. e. when $M_{t+1} = (1+\rho_M)M_t$, $p^e_{t+1} = (1+\rho_M)p^e_t$, and $p_{t+1} = (1+\rho_M)p_t$. Such orbits induce a constant level of real money balances $\bar{m} := M_t/p_t$ for all t which imply associated constant allocations in the economy with market clearing. To show existence of balanced paths, define $q^e_{-1} := (p^e_{-1}/M)$. Under perfect foresight $p^e_{-1} = p_t$, so that q^e_{-1} is the inverse of real money balances. Now, consider the one-dimensional dynamical system in q^e_{-1} defined by the time-one map

$$q^e := \frac{p^e}{M_1} = \frac{1}{1+\rho_M}\mathcal{P}^e(1, q^e_{-1}). \tag{5.1.8}$$

Since $\rho_M > -1$ is constant Lemma 5.1.1 implies that this time-one map inherits all the properties of the inverse $\mathcal{P}^e$ of the price law, namely strict monotonicity plus the so-called convex weak Inada conditions[1] (5.1.6). By construction, for $q^e_{-1} \geq \mathcal{P}(1,0)$, predictions are perfect foresight since $\mathcal{P}(M, \mathcal{P}^e(M, p^e_{-1})) = p^e_{-1} = p$. Figure 5.1 displays the qualitative characteristics of the two systems. The system in intensity

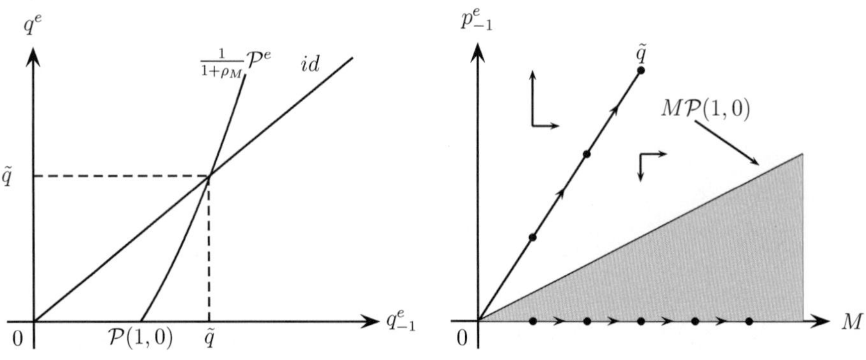

Fig. 5.1 Dynamics with deficit policy $\rho > 0$

form (5.1.8) has exactly two fixed points, zero and a unique positive one $\tilde{q}$ in the domain of the perfect predictor defining a unique positive level $\bar{m} \equiv 1/\bar{p} = 1/\tilde{q}$ of money balances. Moreover, zero is a globally stable fixed point while $\tilde{q}$ is unstable since $\frac{1}{1+\rho_M}\frac{\partial \mathcal{P}^e}{\partial q^e_{-1}}(1, \tilde{q}) > 1$, due to the Inada conditions.

To examine the convergence of an orbit $\{(p_t, M_t)\}$ of (5.1.7) to the balanced path associated with $\bar{p} > 0$ in the domain of perfect foresight, define the distance of an

[1] A mapping $f : \mathbb{R}_+ \to \mathbb{R}_+$ is said to satisfy the convex weak Inada conditions if $g(x) := f(x)/x$ is strictly increasing with $\lim_{x\to 0} g(x) = 0$ and $\lim_{x\to +\infty} g(x) = +\infty$. This is the convex analogue to the concave weak Inada conditions which is often useful in growth theory.

orbit from the balanced path according to (4.2.8) as

$$\Delta_t := p_t - \bar{p}M_t = (\tilde{p}_t - \bar{p})M_t \tag{5.1.9}$$

from which one obtains the recursive equation for the evolution of the distance given by

$$\Delta_{t+1} = (1+\rho_M)\frac{\tilde{p}_{t+1} - \bar{p}}{\tilde{p}_t - \bar{p}}\Delta_t. \tag{5.1.10}$$

Given the monotonicity of the reduced time one map (5.1.8) for $\tilde{p} \equiv q^e_{-1}$, (4.2.8) implies that Δ_t will be positive for all $t \geq 0$ if $\Delta_0 > 0$, while any orbit with a negative deviation $\Delta_0 < 0$ at $t = 0$ implies negative deviation Δ_t for all $t \geq 0$. Therefore, convergence to the balanced path defined by $\bar{p}$ must be monotonic and occurs if and only if $\tilde{p}_t \to \bar{p}$ and $\Delta_t \to 0$ for $t \to \infty$. Therefore, consider the two dimensional dynamical system

$$\begin{aligned} \tilde{p}_{t+1} &= \frac{\mathcal{P}^e(1, \tilde{p}_t)}{1+\rho_M} \\ \Delta_{t+1} &= (1+\rho_M)\frac{\dfrac{\mathcal{P}^e(1, \tilde{p}_t)}{1+\rho_M} - \bar{p}}{\tilde{p}_t - \bar{p}}\Delta_t. \end{aligned} \tag{5.1.11}$$

Its non zero steady state $(\bar{p}, 0)$ has the two associated real eigenvalues

$$\lambda_1 = \frac{1}{1+\rho_M}\frac{\partial \mathcal{P}^e}{\partial q^e}(1, \bar{p}) \quad \text{and} \quad \lambda_2 = \frac{\partial \mathcal{P}^e}{\partial q^e}(1, \bar{p}), \tag{5.1.12}$$

which are both positive with $\lambda_1 > 1$. Asymptotic convergence of the system (5.1.11) to $(\bar{p}, 0)$, however, requires $\max(\lambda_1, \lambda_2) < 1$, which is clearly violated. Therefore, the balanced path associated with $\bar{p}$ is asymptotically unstable. In addition, applying Theorem 4.2.3 one can show that $\lim_{t\to\infty} |\Delta_t| = \infty$ for all orbits with $p_0/M_0 \neq \bar{p}$. In other words, balanced monetary growth with perfect foresight under a fixed positive deficit rule is globally unstable.

As a consequence, if the government follows a fixed deficit rule $\rho > 0$, the economy can achieve only three possible types of behavior in the long run whose characteristics depend on the initial condition, see Figure 5.1. Let $\bar{m} \equiv 1/\bar{p} \equiv 1/\bar{q}$ denote the level of real money balances associated with the positive steady state $\bar{p}$ of the reduced system (5.1.8).

(a) If the economy starts at a level of money balances (M_0, p_0) such that $M_0/p_0 = \bar{m}$, balanced monetary growth with perfect foresight is possible along the path $\bar{q}$. The real allocation in the economy remains constant over time while prices and the quantity of money grow at the same rate ρ_M. This is, however, a zero probability event.

(b) If $M_0/p_0 < \bar{m}$, then prices grow at a faster rate than the quantity of money, which implies

$$\lim_{t\to\infty} \frac{M_t}{p_t} = 0.$$

Therefore, output and employment tend to zero as well, since

$$\lim_{t\to\infty} y_t = \lim_{t\to\infty} D(M_t/p_t) = 0.$$

Thus, in the limit money has no value inducing all trade to converge to zero activity.

(c) Conversely, if $M_0/p_0 > \bar{m}$, money balances grow at a faster rate than prices implying that real money balances grow beyond bound and orbits will leave the region of perfect foresight (the gray shaded region in Figure 5.1) in finite time. Under the predictor $\mathcal{P}^e$ predictions drop to zero in one step and prices become $p_t = M_t\mathcal{P}(1,0)$ guaranteeing that prices and money grow at the constant rate ρ_M associated with constant real balances $1/\mathcal{P}(1,0)$, constant output and employment at their maximal level. Therefore, if the production function is bounded above or if labor supply is bounded from above, then $\mathcal{P}(1,0)$ is positive. Aggregate supply becomes bounded from above as well and output grows to its upper bound. Such orbits do not have perfect foresight, but the levels of output and employment are constant after some finite time.

Deficit Rules with Random Production

It seems not surprising that the property of global divergence occurs also in the more general situation under rational expectations when production is random. Consider the situation where the production technology is subject to a recurring multiplicative (Hicks neutral) positive productivity shock $Z \in [Z_{\min}, Z_{\max}]$ at the beginning of every period, as analyzed in Section 4.3. It was shown there that the random aggregate supply function is of the form $AS(Z, p^e/p)$ which is decreasing in (p^e/p) and increasing in Z with $E_{AS}(Z) > 1$. p^e is now the mean predicted future price.

Combined with the aggregate demand function (5.1.2) under the deficit rule, commodity market clearing becomes

$$\frac{M/p}{1-[\tau(1+\rho)+c(1-\tau)]} = AS(Z, p^e/p) \tag{5.1.13}$$

which implies directly the following lemma (see also Lemma 5.1.1).

Lemma 5.1.2.
Let the assumptions of Proposition 4.3.1 be satisfied.

(a) *For every $(M, p^e) \gg 0$ and $Z \in [Z_{min}, Z_{max}]$, there exists a unique positive temporary equilibrium price $p > 0$ solving equation* (5.1.13) *inducing a random price law $\mathcal{P} : \mathbb{R}^2_+ \times [Z_{min}, Z_{max}] \to \mathbb{R}_+$.*

(b) *For any conditional measure μ on $[Z_{min}, Z_{max}]$ the mean price law*

$$(\mathbb{E}_\mu \mathcal{P})(M, p^e) := \mathbb{E}_\mu \{\mathcal{P}(M, p^e, \cdot)\} := \int \mathcal{P}(M, p^e, Z) d\mu(Z) \tag{5.1.14}$$

is a continuous function, homogeneous of degree one and increasing strictly in (M, p^e) *satisfying*

$$\lim_{p^e \to \infty} (\mathbb{E}_\mu \mathcal{P})(M, p^e) = +\infty \quad \textit{and} \quad \lim_{p^e \to 0} (\mathbb{E}_\mu \mathcal{P})(M, p^e) = M\,(\mathbb{E}_\mu \mathcal{P})(1, 0) \geq 0$$

$$\lim_{p^e \to \infty} \frac{(\mathbb{E}_\mu \mathcal{P})(M, p^e)}{p^e} = 0 \quad \textit{and} \quad \lim_{p^e \to 0} \frac{(\mathbb{E}_\mu \mathcal{P})(M, p^e)}{p^e} = \infty \tag{5.1.15}$$

The properties of the random dynamical system inducing orbits under rational expectations can now be obtained following the same procedure as in Section 4.3. Given the conditional measure μ of the production shock Z in any period, a prediction p^e is unbiased for μ at (M, p^e_{-1}) if and only if the conditional forecast error is equal to zero, i.e. if p^e solves

$$(\mathbb{E}_\mu \mathcal{P})(M, p^e) := \mathbb{E}_\mu\{\mathcal{P}(M, p^e, \cdot)\} := \int \mathcal{P}(M, p^e, Z) d\mu(Z) \stackrel{!}{=} p^e_{-1}. \tag{5.1.16}$$

It follows from Lemma 5.1.2 that the mean price law $(\mathbb{E}_\mu \mathcal{P})$ has a unique inverse with respect to p^e for $p^e \geq M\,(\mathbb{E}_\mu \mathcal{P})(1, 0)$, which is strictly increasing in p^e for any M and μ. Therefore, there exists a globally defined predictor which is unbiased on that domain given by

$$\psi^*(M, p^e_{-1}, \mu) := \begin{cases} (\mathbb{E}_\mu \mathcal{P})^{-1}(M, p^e_{-1}) & \text{if } p^e_{-1} \geq M\,(\mathbb{E}_\mu \mathcal{P})(1, 0) \\ 0 & \text{otherwise} \end{cases} \tag{5.1.17}$$

solving $(\mathbb{E}_\mu \mathcal{P})(M, \psi^*(M, p^e_{-1}, \mu)) = p^e_{-1}$. The mapping ψ^* is homogeneous of degree one and strictly increasing in (M, p^e_{-1}), while $\psi^*(M, p^e_{-1})/p^e_{-1}$ is strictly increasing in (p^e_{-1}) inheriting further properties from (5.1.15) of the price law for $p^e_{-1} \geq M\,(\mathbb{E}_\mu \mathcal{P})(1, 0)$, i.e.

$$\begin{aligned} \lim_{p^e_{-1} \to 0} \psi^*(M, p^e_{-1}, \mu) &= \lim_{p^e_{-1} \to 0} \frac{\psi^*(M, p^e_{-1}, \mu)}{p^e_{-1}} = 0 \\ \lim_{p^e_{-1} \to \infty} \psi^*(M, p^e_{-1}, \mu) &= \lim_{p^e_{-1} \to \infty} \frac{\psi^*(M, p^e_{-1}, \mu)}{p^e_{-1}} = \infty. \end{aligned} \tag{5.1.18}$$

As a consequence, the dynamical system with unbiased predictions is given by

$$\begin{aligned} p^e &= \psi^*(M, p^e_{-1}, \mu) = M\psi^*\left(1, \frac{p^e_{-1}}{M}, \mu\right) \\ M_1 &= (1 + \rho_M)M. \end{aligned} \tag{5.1.19}$$

If the production shocks are i.i.d. the system is *deterministic* being driven exclusively by the predictor and structurally identical to the case without production shocks. Nevertheless, the deterministic (real) expectations induce a stochastic process of random prices, wages, output, and employment in all cases of a non-degenerate shock while nominal money balances grow at the constant rate ρ_M.

If $q_M \neq 0$, the system (5.1.19) has no non zero fixed point and it behaves the same way as (5.1.7). For μ constant, it induces a one dimensional system in $q^e_{-1} := p^e_{-1}/M$ whose fixed points define balanced paths of expected real money balances. Let

$$q^e = \mathcal{F}(q^e_{-1}, \mu) := \frac{1}{1+\rho_M} \psi^*(1, q^e_{-1}, \mu) \tag{5.1.20}$$

define the time one map in intensity form which is strictly increasing satisfying the convex Inada conditions (see Lemma 5.1.2). It has exactly two fixed points, one equal to zero and a positive one $\tilde{q} > 0$ with $\mathcal{F}'(\tilde{q}) > 1$. Thus, the balanced path associated with $\tilde{q}$ is unstable and the distance $|\Delta_t|$ of all non-balanced orbits with rational expectations $\left\{(p^e_{t-1}, M_t)\right\}$ from the balanced path $\tilde{q}$ diverges to infinity. Orbits starting from $q^e_0 > \tilde{q}$ have rational expectations and diverge to ∞ with zero output in the limit (a non-monetary state). Orbits starting from $q^e_0 < \tilde{q}$ loose the rational expectations property and reach $q^e = 0$ in finite time. However, they induce a positive balanced path of prices and money growing at the rate ρ_M with real balances equal to $1/\mathcal{P}(1, 0)$ and constant output and employment (see Figure 5.1). Therefore, $q^e_{-1} = 0$ is the only attracting balanced path with basin of attraction given by $q^e_{-1} < \tilde{q}$.

When the production shocks are more general implying a random conditional mean process of the production shock, unbiased predictions will turn the difference equation (5.1.18) for the predictor ψ^* into a stochastic one induced by a process $\{\mu_t\}$. Still, the balanced path with $q^e = 0$ remains the only attracting stationary solution, while the positive one turns into an unstable random fixed point. Therefore, for all initial conditions in the random basin of attraction of the zero random fixed point, real money balances (and therefore output and employment) follow a stationary solution $M/p = 1/\mathcal{P}(1, 0, Z)$ (after some finite time) induced by the random fixed point $q^e = 0$ and the random price law. Nominal money balances grow in a deterministic fashion at the rate ρ_M. Therefore, random balanced expansion or contraction under rational expectations with positive output and employment is a zero probability event. All orbits outside the basin of attraction of zero diverge with probability one almost surely. Thus, all (non zero) orbits under rational expectations must be non stationary and diverging.

5.2 Random Deficits and Monetary Expansion

Given the findings about the nonstationary consequences of a fixed deterministic fiscal rule, a government may consider a stochastic stationary rule where deficits and surpluses occur in a random way with a stationary distribution, for example with a mean deficit of $\rho \neq 0$ over time. Within the modeling structure above, assume now that the government chooses its deficit quota in each period in a random way at the beginning of each period. For simplicity, assume that the quota is i.i.d. across time and uniformly distributed with mean $\bar{\rho}$, for example $\rho \sim \mathcal{U}[\bar{\rho} - \epsilon, \bar{\rho} + \epsilon]$. The aggregate demand function becomes random

$$D(M/p, \rho) = \frac{1}{1 - [\tau(1+\rho) + c(1-\tau)]} \frac{M}{p}$$

with a random multiplier. It remains unit elastic in the current price. Market clearing implies a random price law $\mathcal{P}(M, p^e, \rho)$ whose mean law for a given conditional measure μ in an arbitrary period is

$$(\mathbb{E}_\mu \mathcal{P})(M, p^e) := \int \mathcal{P}(M, p^e, \rho)\mu(d\rho) \tag{5.2.1}$$

Since $\mathcal{P}(M, p^e, \rho(\omega))$ has an explicit inverse with respect to p^e (on the range of AS) given by

$$p^e = pAS^{-1}\left[\frac{M/p}{1 - [\tau(1+\rho(\omega)) + c(1-\tau)]}\right] \tag{5.2.2}$$

which is monotonically increasing in ρ, it preserves all the properties of the price law of the situation with a fixed deterministic policy rule. Therefore, $(\mathbb{E}_\mu \mathcal{P})(M, p^e)$ inherits the same properties of the mean price law as under random production, i.e. homogeneity, monotonicity, and the weak Inada conditions. Therefore, Lemma 5.1.2 implies that there exists a unique globally defined unbiased predictor $\psi^\star(M, p^e_{-1}, \mu)$ preserving the convex weak Inada conditions of (5.1.18). Money balances adjust in a random fashion with the growth factor

$$1 + \rho_M(\omega) := \frac{(1-c)(1-\tau)}{(1-c)(1-\tau) - \tau\rho(\omega)}. \tag{5.2.3}$$

Thus, the dynamics of the economy under a random policy rule are governed by a homogeneous two-dimensional random dynamical system

$$\begin{aligned} p^e &= \psi^*(M, p^e_{-1}, \mu) = M\psi^*\left(1, \frac{p^e_{-1}}{M}, \mu\right) \\ M_1 &= (1 + \rho_M(\omega))M. \end{aligned} \tag{5.2.4}$$

The unbiased predictor ψ^* is a deterministic function (assuming an i.i.d. fiscal policy with constant measure). Exploiting the homogeneity, this induces the one-dimensional system in intensity form with state variable $q^e_{-1} := p^e_{-1}/M$ given by

$$q^e = \frac{\psi^*\left(1, q^e_{-1}, \mu\right)}{1 + \rho_M(\omega)}. \tag{5.2.5}$$

From Lemma 5.1.2 and the Inada conditions of ψ^* it follows that the system 5.2.5 has two random fixed points, a positive one which is asymptotically unstable and zero which is asymptotically stable with non-degenerate positive basin of attraction. Thus, associated with the unstable random fixed point, there exist positive stochastic balanced orbits which are unobservable empirically. Therefore, as in all previous cases, balanced expansion with stationary positive levels of employment, output, and real money balances is a zero probability event.

Concluding, we find that a permanent expansionary or contractionary fiscal deficit policy with $\rho(\omega) \neq 0$ has decisive and undesirable real effects with probability one. In an intertemporal general equilibrium model with money as the store of value, with optimizing agents and competitive market clearing at all times, balanced expansion with positive stationary output and employment under rational expectations or perfect foresight is a zero probability event.

In other words, a government policy choosing a constant/stationary fiscal deficit rule in such an economy in the best of all logical worlds of economic rationality while letting market forces do the intertemporal adjustments fails to implement the desired stationary equilibrium by itself with probability one. In that sense there is no stabilizing invisible hand. Moreover, experimental or empirical time series data which could confirm the properties of such a monetary model would most likely fail to exist.

Chapter 6
The Keynesian Model with Money

Following a reappraisal by Clower (1965, 1967) and Leijonhufvud (1968) of attempts to analyze unemployment phenomena in the Keynesian spirit in monetary environments Barro & Grossman (1971), Benassy (1975b), Drèze (1975), Younès (1975), Malinvaud (1977), and others presented pioneering work integrating disequilibrium trading principles into the Hicksian monetary intertemporal framework. This induced a reconsideration of the Keynesian criticism of equilibrium theory as being an inadequate tool to describe unemployment situations in aggregative economies. Within two decades a considerable number of publications presented extensions and justifications for this approach attempting to provide microeconomic foundations to Keynesian macroeconomics with unemployment configurations. The surge of publications with the joint perspective of microeconomic foundations to potential macroeconomic applications was designed to describe a modeling framework to characterize stationary disequilibrium configurations with typical permanent price and wage rigidities.

The contribution of that literature consists in developing general principles and concepts of trading under situations without market-clearing *and* defining associated allocative equilibrium concepts. Following the original contributions by Drèze (1975); Younès (1975); Barro & Grossman (1976) general approaches were presented as a new macroeconomic theory for disequilibrium allocations with price rigidities (as in Benassy, 1983, 1986; Malinvaud, 1980; Böhm, 1980, 1989). These are defined as a system of prices and constraints which guarantee market clearing under an extended concept of excess effective demand and quantity constraints as an extension of the standard equilibrium in the sense of Arrow and Debreu (see Drèze, 1991, 2001; Younès, 1975). By construction these so-called fixed-price equilibria are a static equilibrium concept within an arbitrary temporary situation of an economy detached from dynamic considerations. Therefore, as in the theory of temporary equilibrium with market clearing, this literature does not provide a description of the *intertemporal* adjustments of prices and constraints between any two periods as a descriptive part of the temporary configuration in an economy.

The realization of fixed-price equilibria requires *and* is perceived of as a virtual adjustment mechanism à la tâtonnement (duly recognized by researchers in the area and expressed by Drèze, 1991, 2001). Its equilibrium points are not the limit point of a dynamic adjustment mechanism of prices and constraints of the Hicksian model, but only a zero of an excess demand system extended to prices *and* constraints (see Drèze, 1991, 2001). Therefore, prices and constraints satisfy static rigidity conditions or consistency. They do not describe rigidity or persistence of prices over time.

This literature did not receive general acceptance as a foundation for stationary configurations of a dynamic macroeconomic theory as originally intended by its proponents. The critics claim that it fails to provide sufficient reason why prices are necessarily rigid to explain for example a lasting or a permanent cause for unemployment. The critique is justified since the allocative description of equilibria *at fixed prices* is an extended static equilibrium concept detached from dynamic considerations in the Hicksian sense of intertemporal adjustments.

In order to overcome the critique it is conceptually meaningful to apply the rationing methods and use the trading principles to model a dynamic adjustment of prices – but remove the consistency conditions (and the tâtonnement requirement) of the short-run equilibrium concept based on relations between rigid prices and constraints. In other words, the rationing is assumed to occur at *temporarily inflexible* prices which are wrong (measured by excess demand criteria), i.e. *temporary price rigidity* prevails in a state of an intertemporal economy. Then, *dynamic price rigidity* is said to prevail if it is a stationary state of the economy with price adjustment, i.e. a fixed point of the dynamical system with total rigidity, or an invariant set (or an attractor) with restricted rigidity, but not the zero of an extended modified static excess demand function based on a tâtonnement adjustment in a virtual quantity setting. Chapters 6 to 8 provide the associated extensions of the model of Chapter 3 without imposing the equilibrium conditions on constraints showing that the trading principles allow inferences on dynamic, i.e. intertemporal rigidity when combined with dynamic price adjustments.

6.1 Disequilibrium: Trading when Markets Do Not Clear

Invoking the rationing principles of the literature of fixed-price equilibria in situations when equilibria on the commodity market or the labor market do not exist implies a natural extension of the equilibrium model of Chapter 3 to analyze consistent income determination and feasible trading in markets. But it can also be applied successfully when prices and wages are not sufficiently flexible in the given period to obtain simultaneous equilibrium in the two markets. In particular, when both of them are fixed *at the beginning of each period* trades and incomes have to be determined under disequilibrium conditions *before* prices and wages readjust at the beginning of next period.

In order to describe feasible allocations with market evaluations in disequilibrium under general economic principles it is necessary to define trading rules which imply feasible allocations under income consistency at *all* prices and wages in every period. Formally, this means that a generic state of the economy in an arbitrary period now consists of a list of (at least) four variables (p, w, M, p^e) while the objective of investigation is to define and characterize two mappings $\mathcal{Y} : \mathbb{R}^4_+ \to \mathbb{R}_+$ and $\mathcal{L} : \mathbb{R}^4_+ \to \mathbb{R}_+$ from this space which describe the actual levels of output and of employment in any period. Under structural stationarity of the economy these mappings will be the same to be applied in every period. They are the time-invariant objects to be studied in a stationary environment. Therefore, time series of such economies are restricted to the graph of these two mappings, i.e. observable orbits are contained in subsets of the associated six-dimensional space. Similar restrictions apply for models with random perturbations where the observable time series are contained in subsets of the random graph of the mappings, as shown in Sections 4.3 to 4.5.

The principles of trading at arbitrary prices together combined with the techniques for the model under market clearing (Chapters 3 and 4) provide a generic basis for a dynamic macroeconomic model when markets do not necessarily clear at all times. There are two basic principles characterizing the trading rules under disequilibrium in a market environment with price taking agents which seem to be generally accepted. They form the basis of the literature on temporary equilibria with quantity constraints:

(a) Agents should never be forced to trade more than they actually propose to trade in any market, even when there might be additional constraints in other markets. Thus, trading in disequilibrium should always be *voluntary* for any agent taking into account possible cross effects arising between markets or from multiple restrictions.
(b) If rationing occurs in a market, only the larger quantity of the desired supply or demand should be restricted. In other words, the short side in a market always determines the size of actual trading, leaving no room for additional desirable trading in a market. This principle is referred to as the *short-side rule*.

These two principles imply for the macroeconomic model of Chapter 3 that the level of *traded output* in any period is determined by the minimum of supply, demand, and of total capacity in the economy, i.e. actual output is defined as[1]

$$y_t = \mathcal{Y}\left(p_t, w_t, M_t, p^e_{t,t+1}\right) \tag{6.1.1}$$

$$:= \min\left\{D\left(\frac{M_t}{p_t}, \frac{p^e_{t,t+1}}{p_t}, g, \tau\right), n_f F\left(h\left(\frac{w_t}{p_t}\right)\right), n_f F\left(\frac{1}{n_f} N\left(\frac{w_t}{p_t}, \frac{p^e_{t,t+1}}{p_t}\right)\right)\right\}$$

with associated actual level of employment

$$L_t = n_f F^{-1}\left(\frac{1}{n_f}\mathcal{Y}\left(p_t, w_t, M_t, p^e_{t,t+1}\right)\right) =: \mathcal{L}\left(p_t, w_t, M_t, p^e_{t,t+1}\right). \tag{6.1.2}$$

[1] The same letters $\mathcal{Y}$ and $\mathcal{L}$ are used for the output and employment maps in the disequilibrium situations as in Chapter 3 for the output and employment maps in temporary equilibrium. Since the domains of the maps are different, the usage should be clear from the context, and no confusion should arise.

The trading rule (6.1.1) assumes a uniform rationing mechanism of allocating shortages among firms. Since they are assumed to use an identical production function, their labor demand and commodity supply under competitive prices and wages are identical. Uniform rationing seems natural, but other principles could be applied as well[2]. This aggregate formulation encompasses also many models with heterogeneous consumers, such as workers with heterogeneous productivities, firm owners with different propensities to consume, with group specific taxation, or with multiple or heterogeneous firms, when effects from the income distribution do not influence aggregate demand and supply.

With the determination of real trades in the two markets, the associated nominal levels are uniquely determined simultaneously: total income[3] $p\mathcal{Y}(p, w, M, p^e)$, labor income $w\mathcal{L}(p, w, M, p^e)$ and profits $p\mathcal{Y}(p, w, M, p^e) - w\mathcal{L}(p, w, M, p^e)$, as well as the government deficit as the difference of government spending minus tax revenue $pg - \tau p\mathcal{Y}(p, w, M, p^e)$. In other words, given the pair of government parameters (g, τ), the list (p, w, M, p^e) of prices, wages, money balances, and price expectations defines the four dimensional state space of the economy and the two mappings $\mathcal{Y} : \mathbb{R}^4_+ \to \mathbb{R}_+$ and $\mathcal{L} : \mathbb{R}^4_+ \to \mathbb{R}_+$ yield for *every* state of the economy uniquely all real and nominal quantities.

It is important to analyze these two mapping in detail from a global perspective to show the role of prices and wages, of money balances, and of expectations for the determination of output and employment, and the possible types of disequilibrium regimes. Given the simplifying assumptions that firms are identical and that they do not hold inventory, it suffices to analyze the mapping $\mathcal{Y}$ since the level of employment is a monotonic transformation of output. Consider an arbitrary price–wage pair $(p, w) \neq (p^*, w^*)$ where $p^* = \mathcal{P}(M, p^e)$ and $w^* = \mathcal{W}(M, p^e)$ are the so-called Walrasian levels under market clearing. From equation (6.1.1) one observes that there are three distinct generic disequilibrium situations which may occur when at most one of the three functions is determining the minimum (first discussed by Barro & Grossman, 1971; Benassy, 1975a,b, 1982, 1983; Malinvaud, 1977) who also introduced the terminology used here.

(1) When output is determined by aggregate demand with aggregate capacity and supply by producers being larger, is referred to as a *Keynesian unemployment* disequilibrium, denoted **K**. The level of employment is lower than labor supply *and* desired output is larger than aggregate demand. In this case, the supply side is rationed on both markets.
(2) When producer supply is determining the level of output and employment with labor supply *and* aggregate demand for commodities being larger, this is called a *classical unemployment* situation, denoted **C**. In this case, the production sector is determining the transaction levels in both markets while the consumption

[2] When there are inventories or multiple markets in the economy the application of the two principles imply a more involved description of feasible levels of output and employment than the simple minimum rule here.

[3] To avoid unnecessary notation for the temporary analysis, let $p^e \equiv p^e_{t,t+1}$.

sector experiences rationing on both markets. In other words, there is demand rationing on the commodity market *and* supply rationing on the labor market.

(3) When output and employment are determined by the capacity constraint, producers are demand rationed on the labor market *and* consumers are demand rationed on the commodity market. This case has been labeled as a situation of *repressed inflation* denoted **I**.

Given the continuity of the three functions in definition (6.1.1), $\mathcal{Y}$ itself is continuous as well and the three disequilibrium regimes define open subsets in the state space $\mathbb{R}^4_{++}$. It is customary to refer to these subsets as the respective *regions* of Keynesian unemployment, of classical unemployment, and of repressed inflation respectively. The dividing boundaries between the different regions are closed sets (lines or surfaces) where one of the inequalities changes to an equality. Table 6.1 displays the three (four) mutually disjoint cases, exhibiting also the corresponding boundaries where rationing switches from the demand side to the supply side.

Table 6.1 Disequilibrium regimes

		Labor Market	
		Supply Rationing	Demand Rationing
Commodity Market	Supply Rationing	**K**	$\mathbf{K} \cap \mathbf{I}$
	Demand Rationing	**C**	**I**

Let (g, τ) be fixed. For given $(M, p^e) \gg 0$ the three different disequilibrium regimes have a natural representation as sections in the two dimensional space $\mathbb{R}^2_+$ of prices and wages as the graphs and contours of the mappings $\mathcal{Y}$ or $\mathcal{L}$. The boundaries of the respective regions must be identical to parts of the three equilibrium curves introduced in Chapter 3. Specifically, the boundary between classical and Keynesian employment, denoted $\mathbf{K} \cap \mathbf{C}$ must be part of the commodity market clearing curve CE, the boundary $\mathbf{K} \cap \mathbf{I}$ is a subset of the price-wage frontier, and the boundary between $\mathbf{C} \cap \mathbf{I}$ must be a subset of the labor market clearing curve LE (see Section 3.2.6). Figure 6.1 depicts the graph of the mapping $\mathcal{Y}$ and the associated output/employment contours in (p, w)−space.

When demand is homogeneous of degree zero in money balances and prices, output and employment are homogeneous of degree zero in (p, w, M, p^e). Thus, output can also be considered as a function of three variables only, the real wage $\alpha := w/p$, real money balances $m := M/p$, and the expected rate of inflation $\theta^e := p^e/p$. The

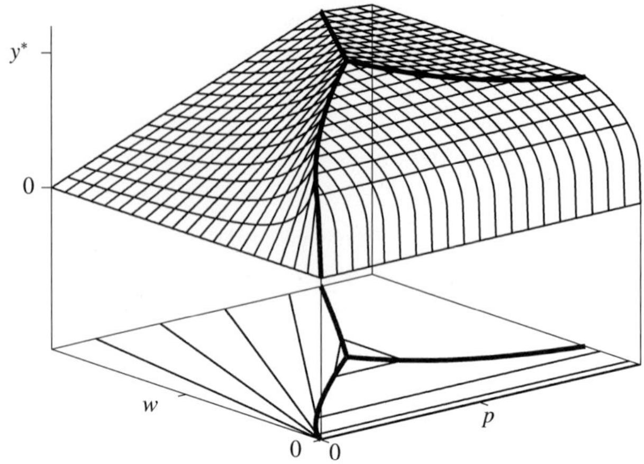

Fig. 6.1 Output surface and contours in price-wage space; (M, p^e) given

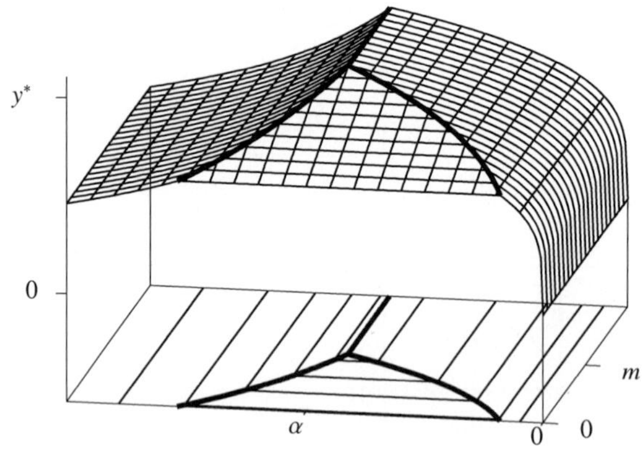

Fig. 6.2 Output surface and contours in the space of real money balances and real wage; θ^e given

geometric representation of the associated graph and the contours of $\mathcal{Y}(m, \alpha, \theta^e)$ in (m, α)−space for fixed θ^e are shown in Figure 6.2. Observe that the real wage is the decisive determinant in classical unemployment and in inflationary states. It plays no role in Keynesian states, while real money balances play the crucial role in the latter two. Maximal output and employment are obtained at a level of the real wage $\alpha^* = \mathcal{W}(M, p^e) / \mathcal{P}(M, p^e)$ corresponding to the temporary Walrasian equilibrium $(p^*, w^*) = (\mathcal{P}(M, p^e), \mathcal{W}(M, p^e))$.

6.2 Feasible States under Disequilibrium Trading

Consider now in more detail an economy with the structural features of Chapter 3 when agents in both markets are price takers and when intertemporal preferences over consumption of workers and share holders are homothetic for cohorts of overlapping generations living two periods satisfying Assumptions 3.1.1 and 3.2.1. Although the different parts of the model were discussed in detail in Chapter 3 it is useful to us retrace some of them here[4]. In a first step, it will be assumed for the worker-consumers that the disutility of work v is constant implying that aggregate labor supply is constant and equal to $L_{\max} > 0$. Thus, full capacity output is constant and there is no expectations effect on the allocation or on savings from the labor supply side. While this implies that the aggregate supply function is constant (and therefore no longer globally invertible!), the assumption still guarantees, together with the other properties, that there is a unique temporary equilibrium. This specialization is not an essential restriction for the discussion of disequilibrium situations, but it will make some of the qualitative properties of allocations and of the disequilibrium dynamics of Chapter 7 more transparent. Chapter 8 reintroduces the endogeneity of labor supply and shows the ensuing effects on the dynamics.

In a second step, given that intertemporal preferences are homothetic, this allows that propensities to consume out of income depend on expected future prices. This reintroduces an expectations effect on the demand side generalizing property (iv) of Assumption 3.2.1. As a consequence the propensity to consume out of current income becomes a function of the expected inflation rate. Thus, consistency of aggregate expenditure and aggregate income means

$$p_t y_t = M_t + c\left(\theta_t^e\right)(1-\tau)\, p_t y_t + p_t g,$$

i.e. any level of aggregate income with consistent feasible output y_t has to be equal to the sum of total expenditures quantities by the government, the young and old consumers at the given price level p_t.

Solving for y_t, with $m_t = M_t/p_t$ denoting real balances, yields the aggregate demand function $D\left(m_t, \theta_t^e\right)$ under income consistency, given by

$$y_t^d = D\left(m_t, \theta_t^e\right) := \frac{m_t + g}{1 - c\left(\theta_t^e\right)(1-\tau)}, \tag{6.2.1}$$

which has the slightly more general form as in Assumption 3.2.1. It is homogeneous of degree zero in money balances, prices, and price expectations. In this case, the intertemporal equilibrium structure with an expectations feedback in the price law reappears via the propensity to consume[5]. Since the emphasis of this chapter and the

[4] The repetition may seem redundant, but it will make this chapter self-contained. It will allow to develop one additional version of the macroeconomic model from a simple microeconomic structure of consumers. Producers and the government will be treated identically as before.

[5] This guarantees the same homogeneity properties of the price law as in Chapter 3. The details for the perfect foresight dynamics under market clearing are different from those in Chapter 4. Their qualitative properties and consequences for the dynamics are not discussed here.

next one is on the consequences of temporary price and wage rigidities on the allocations and on the dynamics, the expectations effects will only play a secondary role. The question of perfect foresight under disequilibrium will not be treated. Therefore, where necessary adaptive predictors rather than perfect predictors will be used.

In an attempt to keep the structure otherwise to an essential minimum to obtain a workable forward recursive model for the remainder of this chapter, the model of Chapter 3 will be chosen with a single producer ($n_f = 1$) and a single member of each two-period-lived consumer, worker and shareholder. In each period t, the production of output is instantaneous from labor without the possibility of inventory holding. The production function $F : \mathbb{R}_+ \to \mathbb{R}_+$ is strictly concave and satisfies the Inada conditions implying the global invertibility properties of supply and demand by the producer under profit maximizing behavior. Formally, for each pair of prices and wages $(p, w) \gg 0$, labor demand and commodity supply by the producer is defined as

$$z^* := h\left(\frac{w}{p}\right) = \arg\max_z pF(z) - wz \qquad \text{and} \qquad y^* := F\left(h\left(\frac{w}{p}\right)\right)$$

as given in (3.1.1). As before, the government desires to purchase a quantity $g \geq 0$ of the commodity in each period at the market price $p \geq 0$, and raises revenue by levying a uniform proportional income tax on young consumers' income at the rate τ, $0 \leq \tau \leq 1$.

6.2.1 The Role of Prices, Wages, and Money Balances

Applying the two principles of voluntary trading and of the short side rule for the determination of feasible output and employment (as defined in (6.1.1) and (6.1.2)) to the specific case with constant labor supply one obtains an output level y_t

$$y_t = \mathcal{Y}\left(p_t, w_t, M_t, p^e_{t,t+1}\right) = \min\left\{D\left(\frac{M_t}{p_t}, \frac{p^e_{t,t+1}}{p_t}\right), F\left(h\left(\frac{w_t}{p_t}\right)\right), F(L_{\max})\right\},$$

and an employment level L_t

$$L_t = \mathcal{L}\left(p_t, w_t, M_t, p^e_{t,t+1}\right) = F^{-1}\left(\mathcal{Y}\left(p_t, w_t, M_t, p^e_{t,t+1}\right)\right).$$

Thus, given government parameters (g, τ), for any period t and any list of data $(p_t, w_t, M_t, p^e_{t,t+1})$, a feasible state of trades at the macroeconomic level is a pair (y_t, L_t) of aggregate output and employment determined uniquely as the minimum of aggregate demand, producer notional supply, and capacity output for any list of current prices and wages, expected prices, and money holdings of the old consumer. It is straightforward to verify the statements made in the following theorem.

Theorem 6.2.1.
Given the assumptions on production and on preferences one has:

$$\forall\,(p_t, w_t, M_t, p^e_{t,t+1}) \gg 0, \quad \forall\, g \geq 0, \quad \forall\, 0 \leq \tau \leq 1:$$

(a) there exists a unique positive pair (y_t, L_t) *of feasible levels of output and employment given by the two functions* $\mathcal{Y}$ *and* $\mathcal{L}$,

(b) the type of temporary feasible state is either

C – *classical unemployment*
K – *Keynesian unemployment*
I – *repressed inflation*

or one of the 4 boundary cases,

(c) the functions $\mathcal{Y}$ *and* $\mathcal{L}$ *are continuous and homogeneous of degree zero in* $(p_t, w_t, M_t, p^e_{t,t+1})$. *Hence, using the homogeneity property, one may write*

$$y_t = \mathcal{Y}(\alpha_t, m_t, \theta^e_t) \equiv \mathcal{Y}(1, \alpha_t, m_t, \theta^e_t)$$

$$L_t = \mathcal{L}(\alpha_t, m_t, \theta^e_t) \equiv \mathcal{L}(1, \alpha_t, m_t, \theta^e_t)\,.$$

The theorem states that, given the government parameters (g, τ), temporary feasible states are uniquely defined for each pair (p, w) of prices and wages and for any level of aggregate money balances and expectations (M, p^e) which were the state variables for the model with market clearing. Since prices and wages are given at the beginning for each period, a state of the economy is now a list $(p, w, M, p^e) \in \mathbb{R}^4_+$ which is the state space of the economy.

In order to understand the properties of the maps of output and employment it is useful to analyze their properties with respect to prices and wages in detail for given (M, p^e). Since the minimum rule defines feasible output as a pointwise minimum of three differentiable functions which have partially different domains it follows that the two maps will not be differentiable on the boundaries of the regimes. Figure

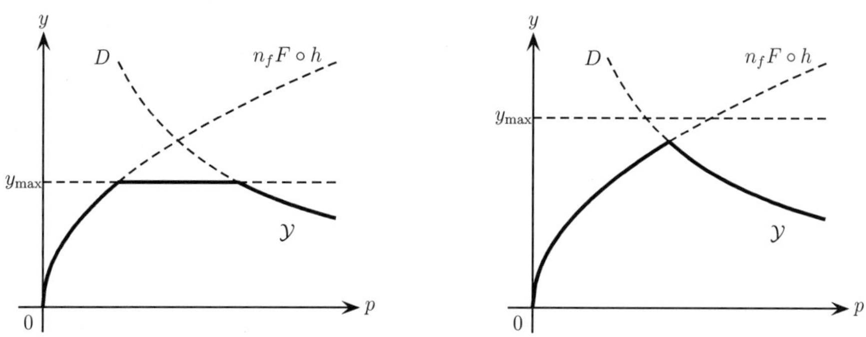

a Low wage – low y_{max} **b** High wage – high y_{max}

Fig. 6.3 The minimum rule determining output: (M, p^e) and L_{max} given

6.3 displays typical qualitative features of the output map for the case of constant labor supply $L_{\max}$ portraying a section of the graph of $\mathcal{Y}$ in (p, y)–space for given values of (w, M, p^e). This identifies the sources for the occurrence of the types **C, K, I** of disequilibrium state and the reasons for the occurrence of the typical non-differentiability of $\mathcal{Y}$. It indicates also that the local properties of the two mappings must depend on the disequilibrium type which prevails at each state $(p, w, M, p^e,)$.

The local comparative statics properties of the two mappings are easily established for any given state of the economy, taking proper account of the function which induces the minimum and which identifies the regime.

$$\begin{array}{llll} \dfrac{\partial \mathcal{L}}{\partial p} > 0 & \text{and} & \dfrac{\partial \mathcal{L}}{\partial w} < 0 & \text{for } (p, w, M, p^e) \in \mathbf{C} \\ \dfrac{\partial \mathcal{L}}{\partial p} < 0 & \text{and} & \dfrac{\partial \mathcal{L}}{\partial w} = 0 & \text{for } (p, w, M, p^e) \in \mathbf{K} \\ \dfrac{\partial \mathcal{L}}{\partial p} = 0 & \text{and} & \dfrac{\partial \mathcal{L}}{\partial w} = 0 & \text{for } (p, w, M, p^e) \in \mathbf{I} \end{array} \tag{6.2.2}$$

These show that the effect of parametric changes of prices and wages are specific to the regimes. In a similar fashion one can derive the effects with respect to money balances and expectations, as well as government demand g and the tax rate τ.

$$\begin{array}{llll} \dfrac{\partial \mathcal{L}}{\partial M} = 0 & \text{and} & \dfrac{\partial \mathcal{L}}{\partial p^e} < 0 & \text{for } (p, w, M, p^e) \in \mathbf{C} \\ \dfrac{\partial \mathcal{L}}{\partial M} > 0 & \text{and} & \operatorname{sgn} \dfrac{\partial \mathcal{L}}{\partial p^e} = \operatorname{sgn} c'(\theta^e) & \text{for } (p, w, M, p^e) \in \mathbf{K} \\ \dfrac{\partial \mathcal{L}}{\partial M} = 0 & \text{and} & \dfrac{\partial \mathcal{L}}{\partial p^e} = 0 & \text{for } (p, w, M, p^e) \in \mathbf{I} \end{array} \tag{6.2.3}$$

To obtain the global characteristics of the mappings $\mathcal{Y}$ and $\mathcal{L}$, the boundaries between the three regimes **C,K,I** have to be established which are defined by the solutions of pairwise equalities of two functions of (6.1.1) while the third is not binding. In other words,

$$\begin{array}{lll} (p, w, M, p^e) \in \mathbf{C} \cap \mathbf{K} & \text{if and only if} & D(p, M, p^e, g, \tau) = F(h(w/p)) < F(L_{\max}) \\ (p, w, M, p^e) \in \mathbf{C} \cap \mathbf{I} & \text{if and only if} & F(h(w/p) = F(L_{\max}) < D(p, M, p^e, g, \tau) \\ (p, w, M, p^e) \in \mathbf{I} \cap \mathbf{K} & \text{if and only if} & D(p, M, p^e, g, \tau) = F(L_{\max}) < F(h(w/p)) \end{array} \tag{6.2.4}$$

defines the boundaries in $\mathbb{R}^4_+$ between the open disequilibrium regimes. It is customary to describe the properties of the output and employment functions geometrically either as functions of prices and wages (p, w) for given (M, p^e), or, exploiting the homogeneity of the mappings, as a function of real money balances and real wages (m, α) for a given expected rate of inflation θ^e, since aggregate demand is

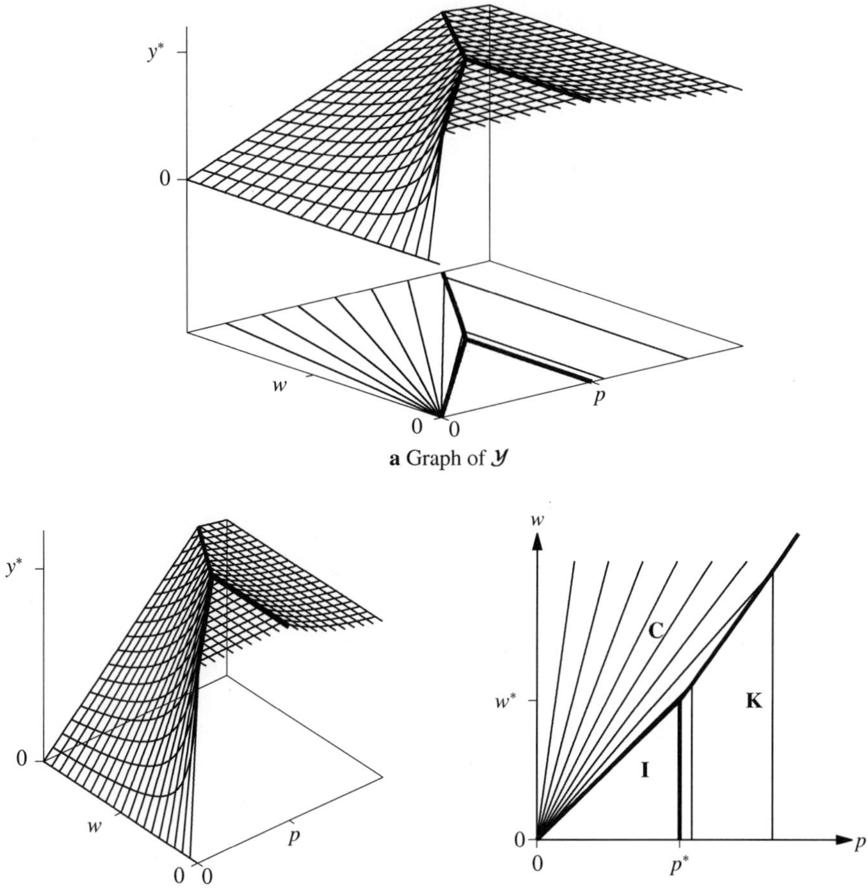

a Graph of $\mathcal{Y}$

b Graph of $\mathcal{Y}$

c Contours of $\mathcal{Y}$

Fig. 6.4 Surface and contours of output map $\mathcal{Y}$ with L_{max}; (M, p^e) given

homogeneous of degree zero in (M, p, p^e). Figure 6.4 portrays the graph of $\mathcal{Y}$ and the associated contour map in (p, w)−space. The point (p^*, w^*) denotes the market clearing situation. For all wage rates $w > 0$, a sufficiently high commodity price always induces a Keynesian unemployment regime and a price low enough induces a classical regime. On the other hand, the inflationary regime occurs only for $(p, w) \ll (p^*, w^*)$. Maximal output and employment is obtained in the triangular region $(0, p^*, w^*)$ in (p, w)−space. The iso-employment/iso-output contours for $L < L_{max}$ are those curves with constant real wages in the classical region **C** and their continuous connection to the vertical lines in the Keynesian region **K**. Similar to Figure 6.4 one obtains the graph of $\mathcal{Y}$ as a surface in $\mathbb{R}^3_+$ for given θ^e and the associated contours in (m, α)−space as shown in Figure 6.5. The two contour plots confirm the comparative statics effects of (p, w) and (m, α) respectively.

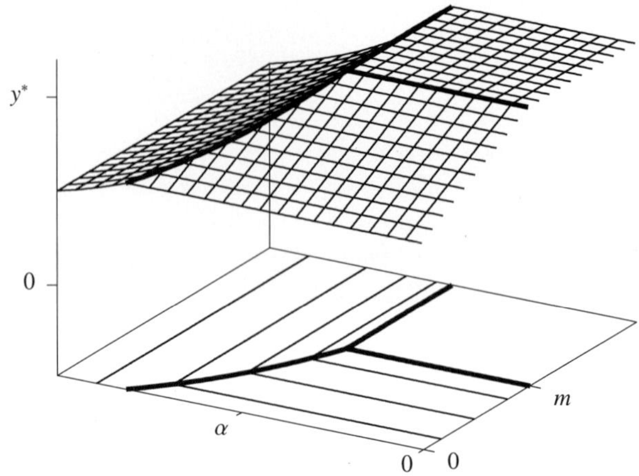

Fig. 6.5 Surface and contours of output map $\mathcal{Y}$ with $L_{\max}$ and θ^e given

6.2.2 The Role of Government Parameters

The comparative statics effects of changes of government demand g and of the tax rate τ are obtained in a similar fashion. It is obvious that their signs must be regime specific. Since labor supply is constant and equal to $L_{\max}$, fiscal parameters enter only in the aggregate demand function. Thus, they cannot have an effect on aggregate employment under classical employment and repressed inflation. Given the assumption under demand rationing, an increase of government demand induces a crowding out effect on private consumption in the classical and the inflationary regime. In the Keynesian regime, however, one observes the well-known positive effect on employment and output from additional fiscal consumption demand and the inverse effect through the tax multiplier on output from an increase of the tax rate τ.

$$\begin{aligned}
&\frac{\partial \mathcal{L}}{\partial g} = 0 \quad &&\text{and} \quad &&\frac{\partial \mathcal{L}}{\partial \tau} = 0 \quad &&\text{for } (p, w, M, p^e) \in \mathbf{C}\\
&\frac{\partial \mathcal{L}}{\partial g} > 0 \quad &&\text{and} \quad &&\frac{\partial \mathcal{L}}{\partial \tau} < 0 \quad &&\text{for } (p, w, M, p^e) \in \mathbf{K}\\
&\frac{\partial \mathcal{L}}{\partial g} = 0 \quad &&\text{and} \quad &&\frac{\partial \mathcal{L}}{\partial \tau} = 0 \quad &&\text{for } (p, w, M, p^e) \in \mathbf{I}
\end{aligned} \tag{6.2.5}$$

To derive the global impact observe first that an increase of government demand with $L_{\max}$ given implies a proportional increase of the wage and the price level under market clearing (p^*, w^*). This in turn implies that for every wage rate w the commodity market clearing price must increase which induces a uniform shift of the boundary $(\mathbf{C} \cap \mathbf{K}) \cup (\mathbf{I} \cap \mathbf{K})$ to the right in (p, w)–space. In other words, both regions

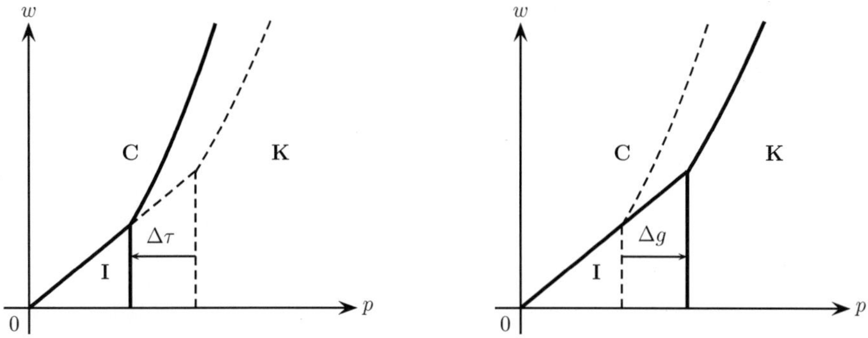

a An increase of the tax rate **b** An increase of government demand

Fig. 6.6 Role of government parameters on regimes; $L_{\max}$, (M, p^e)

of classical unemployment **C** and **I** of repressed inflation increase, see Figure 6.6 **b**. In other words, a band of Keynesian states turn into classical or inflationary states.

A uniform increase of the income tax rate τ under constant exogenous labor supply has no labor market effect, but it induces a reduction of the consumption multiplier. Therefore, for every w, this implies a price decrease under commodity market clearing which means a uniform shift of the boundary $(\mathbf{C}\cap\mathbf{K})\cup(\mathbf{I}\cap\mathbf{K})$ to the left in (p, w)−space, see Figure 6.6 **a**. Thus, the region of Keynesian unemployment increases while the other two shrink. The same type of displacement of the frontier of the Keynesian region can be shown to appear in the alternative presentation in the space of real balances and real wages.

6.2.3 Endogenous Labor Supply

The properties of local comparative statics do not change substantially in the more general case when labor supply by young consumers is a function of the real wage and of price expectations, as assumed in Assumption 3.2.1. Let aggregate labor supply be given by $L^S = N(\alpha, \theta^e)$ which is increasing in the real wage and decreasing in expected inflation. Aggregate capacity is simply given by the available labor supply in every period. If, in addition, aggregate consumption demand can still be written as a function of real net income, with a propensity to consume being a function of expected inflation only, then, except for some additional non-linearity, the previous comparative-statics analysis is of a very similar form. At any list of prices, wages, and price expectations (p, w, p^e) feasible states in the economy are given by

$$
\begin{aligned}
y_t &= \mathcal{Y}\left(p_t, w_t, M_t, p^e_{t,t+1}\right) \\
&= \min\left\{D\left(\frac{M_t}{p_t}, \frac{p^e_{t,t+1}}{p_t}\right), n_f F\left(h\left(\frac{w_t}{p_t}\right)\right), n_f F\left(\frac{1}{n_f} N\left(\frac{w_t}{p_t}, \frac{p^e_{t,t+1}}{p_t}\right)\right)\right\}
\end{aligned}
$$

and associated employment level

$$
L_t = n_f F^{-1}\left(\frac{1}{n_f}\mathcal{Y}\left(p_t, w_t, M_t, p^e_{t,t+1}\right)\right) = \mathcal{L}\left(p_t, w_t, M_t, p^e_{t,t+1}\right).
$$

The real wage is the decisive determinant in classical unemployment and in inflationary states. It plays no role in Keynesian states while real money balances play the crucial role in the latter two. Maximal output and employment are obtained at a

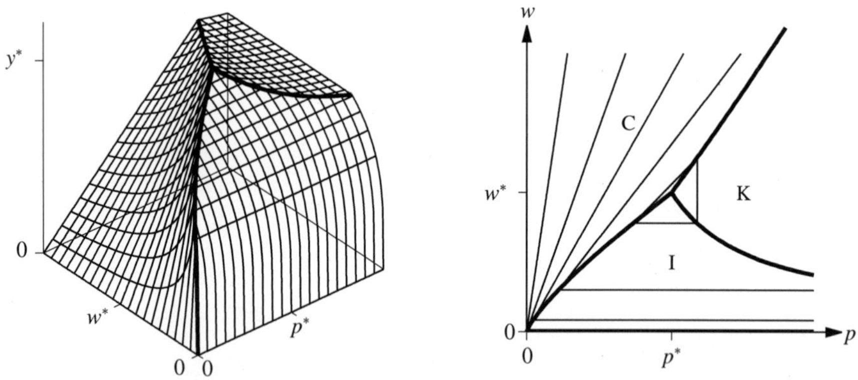

a Surface of $\mathcal{Y}$ **b** Contours of $\mathcal{Y}$

Fig. 6.7 Output surface and contours with endogenous labor: (p, w)−space

level of the real wage corresponding to the Walrasian equilibrium. Figures 6.7 and 6.8 show the geometric properties of the output map in the two possible representation in (p, w)−space and in (m, α)−space. Notice that the boundaries of the regimes in (p, w)−space coincide partially with the *LE* curve, the *CE* curve, and the capacity frontier defined in Section 3.2.

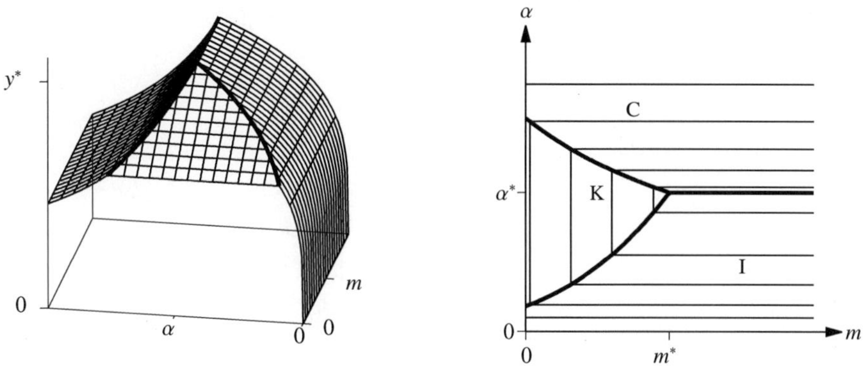

a Surface of $\mathcal{Y}$ **b** Contours of $\mathcal{Y}$

Fig. 6.8 Output surface and contours with endogenous labor: (m,α)–space

Changes of the fiscal parameters on the partition of the state space under endogenous labor supply are quite distinct from the previous case and worth analyzing since they cause well determined movements/shifts of the boundaries of each regime. An increase in government consumption consists of an additive shift in aggregate demand, so that the real wage at the Walrasian equilibrium remains unchanged implying also that the boundary $\mathbf{C} \cap \mathbf{I}$ remains unchanged. However, the boundaries $\mathbf{K} \cap \mathbf{C}$ and $\mathbf{K} \cap \mathbf{I}$ must be shifted accordingly leading to an enlargement of the regions **C** and **I**. In other words, some Keynesian states become classical or inflationary. This property points out that changes of fiscal parameters may lead to regime switching at constant prices and wages, see **b** in Figures 6.9 and 6.10. Parametric changes of

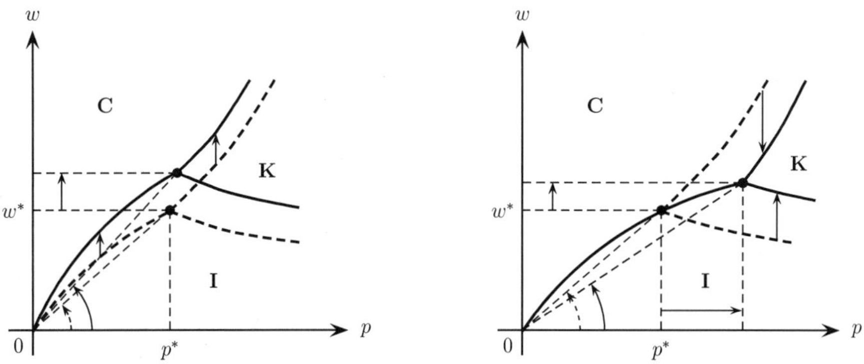

a An increase of the tax rate **b** An increase of government demand

Fig. 6.9 Role of government parameters on regimes: (p,w)–space; (M,p^e) given

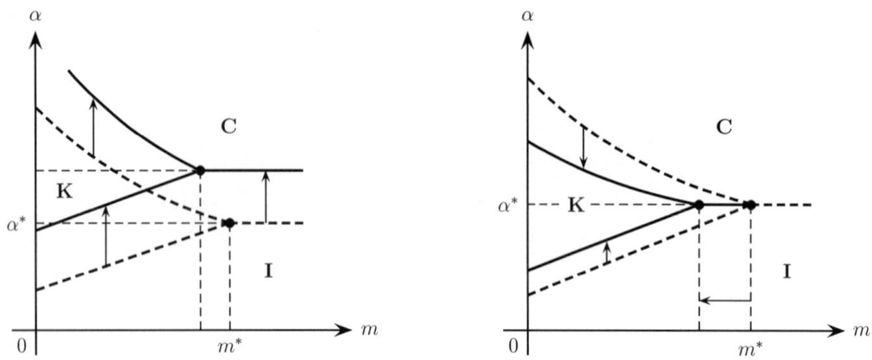

Fig. 6.10 Role of government parameters on regimes: (m, α)–space; θ^e given

the tax rate τ induce a simultaneous influence on the demand multiplier and on the labor supply. This implies a negative effect on the Walrasian equilibrium real wage causing an upward shift of the equilibrium *LE* curve or the boundary $\mathbf{C} \cap \mathbf{I}$. As a consequence, the classical region becomes smaller with an increase and displacement of the regions **K** and **I**.

6.3 The IS-LM Model Revisited

The common IS-LM model of the literature assuming price and wage rigidities in one way or another (as for example in Abel & Bernanke, 2005; Blanchard, 2003; Romer, 2005, or in other text books) can be written directly in the form of the prototype model here. What needs to be done to reveal the formal relationship is to make prices and wages explicit in the IS-LM model *and* to define feasibility in markets under the prevailing rigidities by extending/adding the output/supply side of the economy under the given labor supply conditions to the demand determined income equations.

Consider the following version of the standard aggregative model in the Keynesian tradition with nominal income determination $Y \equiv p\,y$ given by

$$\begin{aligned} p\,y &= C((1-\tau)p\,y) + p\,i(r) + p\,g \\ \bar{M} &= M(p\,y, r). \end{aligned} \tag{6.3.1}$$

The function $M(py, r)$ denotes the demand for money, $\bar{M}$ the money supply, and $C((1-\tau)Y)$ is aggregate private consumption expenditures as a function of nominal income, $i(r)$ is real investment demand with the usual properties. These two fundamental equations in (6.3.1) determine income consistent aggregate commodity demand under simultaneous market clearing in the money/bond market. Thus, under the usual assumptions of the Keynesian model

$$\begin{aligned} M_r < 0, \qquad & i' < 0 \\ M_Y > 0, \qquad & C_Y < \frac{C((1-\tau)Y)}{(1-\tau)Y} < 1 \end{aligned} \tag{6.3.2}$$

for each given $(p, \bar{M})$, a unique solution of (6.3.1) defines two functions: an income consistent aggregate demand function $y^d = D(p, \bar{M})$ and an equilibrium rate of interest $r(p, \bar{M})$.

The income-expenditure equation in (6.3.1) is independent of the level of nominal assets (money or bonds) held by the private sector. This means essentially that asset holders exert no consumption demand from their wealth. Therefore, the money held by the private sector enters the aggregate demand function for commodities only through the equilibrium condition in the asset market[6]. Applying the Implicit Function Theorem to (6.3.1) using (6.3.2) one obtains

$$0 > \frac{\partial D}{\partial p}(p, \bar{M}) = -\frac{y}{p} \cdot \frac{M_r(\frac{C}{Y} - C_Y) + pi' M_Y}{M_r(1 - C_Y) + pi' M_Y} > -\frac{D(p, \bar{M})}{p} \tag{6.3.3}$$

for the properties of the demand function. Thus, the induced income consistent aggregate demand function is downward sloping with an elasticity greater than minus one, i.e. with the same qualitative properties as those found in Chapter 3. In the case (6.3.1) presented here, $D(p, \bar{M})$ contains no explicit expectations feedback and no distinction is made with respect to the distribution of factor incomes. In addition, if the demand function for money is homogeneous of degree one in income, then aggregate commodity demand is not necessarily homogeneous of degree zero in $(p, \bar{M})$ and a function real balances $(\bar{M}/p)$ unless consumption expenditure (Y) is linear in income as well.

In this formulation the level of income $Y \equiv pD(\bar{M}, p)$ as a function of the price level is exclusively demand determined, asking only for income consistency and money/asset market consistency. Higher prices imply higher nominal income since

$$\frac{\partial Y}{\partial p} = \frac{\partial}{\partial p}(pD(\bar{M}, p)) = D(\bar{M}, p)\,(1 + E_D(p)) > 0,$$

but are associated with lower demand levels $y = D(p, \bar{M})$. Moreover, for any given price level p the first equation in (6.3.1) cannot be interpreted as an equilibrium of the commodity market[7]. In summary, if these additional conditions hold commodity

[6] When the monetary policy follows a so-called Taylor rule setting the real interest rate rather than the level of the money supply, money holdings are endogenous and not an argument of the aggregate demand function (see Romer, 2000). Then, however, the income consistent aggregate demand function depends on the interest rate and on expected inflation.

[7] Nevertheless, this is done frequently in the literature. $Y \equiv pD(p, \bar{M})$ induces equality of total income and planned expenditures at an arbitrary price level p, i.e. guaranteeing income-expenditure consistency and not equality of supply and demand in the output market.

demand implied by the IS-LM model satisfies Assumption 3.2.1 and would be an equivalent description of aggregate demand conditions as in Chapter 3. As a consequence, an equilibrium analysis with labor market clearing, the so-called IS-LM-AS model in the sense of Romer (2000), would be isomorphic to the AS-AD equilibrium version above fully described by an equilibrium price law and its associated properties.

Nothing can be deduced from the IS-LM model about real output and employment without the equality to an associated aggregate supply function. Expenditure consistency alone ignores the real sector of the economy completely, i.e. the supply side/production and labor market conditions in several ways. At any given pair of prices and wages $(p, w) \gg 0$:

(a) $y^d = D(p, \bar{M})$ may be larger or smaller than total capacity output in the given period;

(b) $y^d = D(p, \bar{M})$ may be larger or smaller than the quantity producers would like to sell on the commodity market;

(c) the amount of labor required to produce $y^d = D(p, \bar{M})$ may be larger or smaller than the amount of labor supplied by consumers.

Therefore, adding to the IS-LM model a specified aggregate supply function AS defined as an equilibrium relation for the labor market implies that an *equilibrium version* of the IS-LM model with aggregate supply equal to aggregate demand is isomorphic to the standard AS-AD Model under market clearing as shown in Chapter 3. In the general case, the aggregate supply function AS may well be one derived under noncompetitive conditions (as in Section 3.3), with wage bargaining (Section 3.5), or efficiency wages (Section 3.7.3), or in other forms of labor market conditions (for example as those discussed in Romer, 2005; Blanchard, 2003). In all cases, temporary equilibria would be of the type discussed with market clearing inducing a description given by a price law and a wage law inheriting the implications from the generic model of temporary equilibrium of Chapter 3.

If prices and wages are fixed or equilibrium levels for the associated IS-LM-AS model do not exist, conclusions on trading at disequilibrium have to derived appropriately. Adding the supply side/the production sector at *given prices and wages* (p, w) induces a consistent version of the IS-LM model under fixed prices and wages which is structurally equivalent to the one derived in the previous section as extension of the equilibrium model from Chapter 3. To see this, assume for simplicity a labor market with constant inelastic supply $L_{\max} > 0$. Then, one obtains a capacity constrained aggregate supply relation (assuming $n_f = 1$ for simplicity)[8]

$$y^s = \widetilde{AS}(p, w, L_{\max}) := \min\left\{F\left(h\left(\frac{w}{p}\right)\right), F(L_{\max})\right\}. \tag{6.3.4}$$

If prices and wages are fixed within the period, the principles of voluntary trading and of one sided rationing imply that feasible transactions on each market are de-

[8] This function is not an aggregate supply function AS in the sense of Chapter 3.

termined by the minimum rule. Thus, given money balances $\bar{M}$, prices and wages (p, w), actual output and employment are determined by

$$y = \widetilde{\mathcal{Y}}(p, w, \bar{M}) := \min\left\{D(p, \bar{M}), y^s\right\} = \min\left\{D(p, \bar{M}), F\left(h\left(\frac{w}{p}\right)\right), F(L_{\max})\right\} \tag{6.3.5}$$

and by

$$L = \widetilde{\mathcal{L}}(p, w, \bar{M}) := F^{-1}\left(\widetilde{\mathcal{Y}}(p, w, \bar{M})\right) = F^{-1}\left(\min\left\{D(p, \bar{M}), F\left(h\left(\frac{w}{p}\right)\right), F(L_{\max})\right\}\right), \tag{6.3.6}$$

which are functions identical to the ones derived from the minimum principle underlying (6.1.1). Their qualitative properties are equivalent to those shown in Figures 6.1 and 6.2. Thus, completing the IS-LM model with fixed prices and wages in a consistent way by adding the supply side of the economy *and* by applying the same principles of trading in situations of excess supply *and* excess demand leads to a general consistent theory of the determination of income *and* of feasible allocations in disequilibrium.

As a consequence, (6.3.1) together with (6.3.5) define the feasible temporary structure of a general macroeconomic model with fixed prices and wages for which the determination of the short run characteristics of employment and output can be analyzed in the same way as in the extension of the AS-AD model. In other words, adding an aggregate supply function completes the IS-LM Model as an equilibrium model for the determination of a pair (p, w) of equilibrium values *and* feasible levels of output and employment.

The determination of income postulated in the IS-LM model for fixed prices and wages is only a *partial* and *incomplete* description of a closed monetary economy with undefined feasible trades in the output and the labor market. Its completion with consistency for the markets for labor and output makes the IS-LM-AS an equilibrium version of an AS-AD model of Chapter 3. With fixed prices and wages it induces a model allowing for disequilibrium trading whose feasible states are isomorphic to those of the extended AS-AD Model (see also Benassy, 1983). Compared to the model of Chapter 3 with money only, the IS-LM model (6.3.1) is interpreted to possess a more extended asset structure with government bonds and money. This plays no role for the temporary allocative analysis at this stage. It becomes relevant as soon as the evolution of asset positions in the economy are analyzed. This requires a specification of the demand effects from asset holders, of the intertemporal budget conditions of the government and of asset holders, or of policy rules by a central bank inducing changes of the income consistent aggregate demand function.

6.4 Keynes and Hicks: A Synthesis

The analysis of the determination of allocations and incomes under disequilibrium trading at given prices and wages in an arbitrary period has led to a significant qualitative extension and generalization of trading scenarios compared to the equilibrium analysis of Chapter 3. It was shown that three[9] distinct regimes of non-clearing configurations between the labor and the output market occur generically. They are uniquely determined with disequilibrium features characterized by typical spillover effects between the labor and the commodity market. Generically, disequilibrium

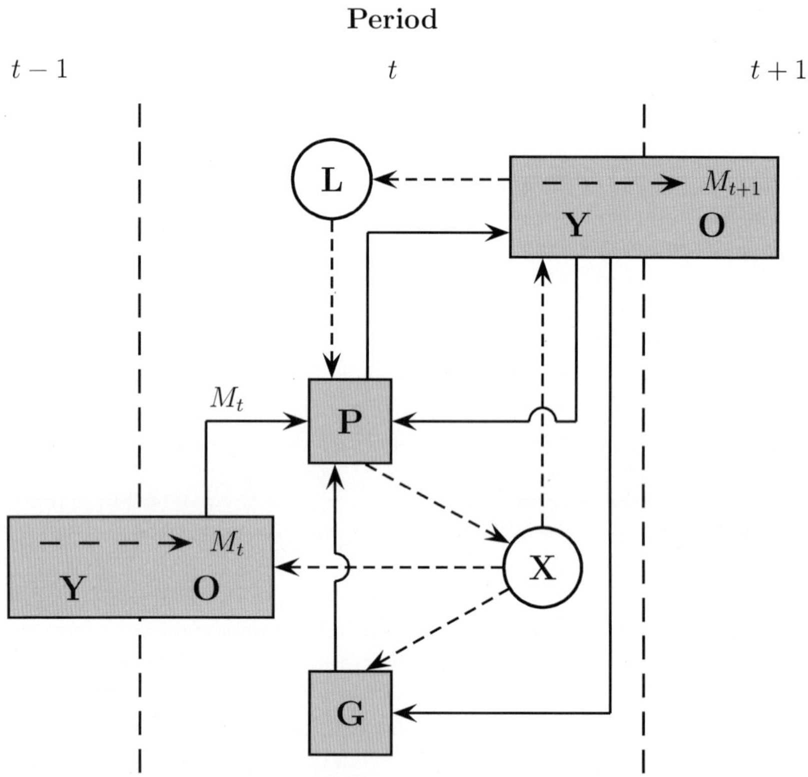

Fig. 6.11 Time flow diagram of the model

trading always involves *either* trading on the same short side of both markets imposing binding constraints on two separate sectors (i.e. simultaneous supply or demand rationing on both markets), *or* supply and demand rationing occurs being imposed

[9] The occurrence of only three regimes is due to the asymmetry in the intertemporal modeling for consumers versus producers. If the behavior of producers were modeled in an intertemporal setting allowing inventory holding four disequilibrium configurations would occur in general.

on the agents of one sector alone (consumers) with no binding constraint for the production sector. The resulting trading environment is one where on each market the aggregate outcome is well-defined and obtainable in each period. Applying the rationing principles (6.1.1) only does not imply that the associated feasible trades satisfy additional conditions as typically imposed in so-called equilibria with quantity rationing.

The two trading principles underlying voluntariness and the short-side rule seem natural and somewhat undisputed in an environment when prices and wages are taken parametrically by most agents. More extensive and more detailed rationing mechanisms can be introduced in different ways into the model for each market, in particular under heterogeneity of agents. As long as these are operating recursively their outcome in each market is well-defined and the recursive nature of the mappings is preserved.

The implications of trading and budget consistency combined with the short-side rule implies that the temporary states are still consistently described by the interaction of the agents on markets as given by the flow diagram of Figure 3.1 repeated here to underline the structural equivalence of the equilibrium and the disequilibrium model. The circles identify the two markets for labor **L** and the consumption good **X**. With the intertemporal embedding of consumers into the framework of overlapping generations, the temporary structure of trading becomes stationary remaining unchanged over time. Trade/exchange between agents imply that the real flows through markets (indicated by the dashed lines) balance to zero. Income consistency and the closedness of the expenditure system implies that monetary/expenditure flows between sectors (corresponding to the solid arrows between the square boxes labeled **P**, **G**, and **Y O**) must balance as well.

The disequilibrium scenarios depend in a unique way on the state variables of the economy, on money balances and expectations, and on the prevailing prices and wages which are temporarily fixed. Therefore, for a consistent description of the dynamic development of the economy it will be necessary to describe the adjustment of prices and wages in addition to the changes of expectations and of money balances in order to preserve the forward recursive structure.

Given the basic Assumption 3.2.1 and a pair of fiscal parameters (g, τ) the extension of the equilibrium model of Chapter 3 to the situation with given prices and wages in each period implied an extended state space with $(p, w, M, p^e) \in \mathbb{R}^4_+$ (rather than $\mathbb{R}^2_+$ for the model with market clearing) and two allocative mappings $\mathcal{Y} : \mathbb{R}^4_+ \to \mathbb{R}_+$ and $\mathcal{L} : \mathbb{R}^4_+ \to \mathbb{R}_+$ determining actual trades in the two markets. Therefore, if the structural assumptions determining these mappings are part of the stationary description of the macroeconomic environment the two mappings will be constant over time. As a consequence the set of observable monetary and real states of such an economy are exclusively determined by these two maps. It must be a subset of the graph of the mapping $(\mathcal{Y}, \mathcal{L}) : \mathbb{R}^4_+ \to \mathbb{R}^2_+$, a subset of $\mathbb{R}^6_+$. In other words, from an empirical point of view examining observable time series of a stationary economy an understanding of the properties of the graph of this map must become the goal of an analysis since it is the unique *time invariant* object to be studied under any parametric approach.

Chapter 7
Dynamics in Disequilibrium – Endogenous Business Cycles

The presentation of the model in Chapter 6 showed that the dimension of the state space of the monetary economy in disequilibrium for which the dynamics needs to be defined increased from two to four adding the vector of prices and wages (p, w) to the pair of money balances and expected prices (M, p^e). With $\mathbb{R}^4_+$ being the state space four mappings have to be defined to describe the evolution of the economy under the disequilibrium mechanism embodied in the two mapping $(\mathcal{Y}, \mathcal{L})$: in addition to money balances and price expectations the adjustment of prices and wages have to be given[1]. Therefore, in order to describe time series of such economies (evolving on the graph of $(\mathcal{Y}, \mathcal{L})$) requires a formal description of the mappings governing the changes of prices and wages as well as of money balances and expectations, each of which are mappings from $\mathbb{R}^4_+$ to $\mathbb{R}_+$. If these are time independent the resulting dynamical system will be autonomous and the observable time series will be contained in subsets of the graph of the mappings $(\mathcal{Y}, \mathcal{L})$, a piecewise smooth surface in $\mathbb{R}^6_+$. Thus, from a structural point of view the space of observable time series for the extended disequilibrium model is the *same* as for the equilibrium model (see Section 4.1), except the associated mappings are different. Some of them play different logical roles.

Two simplifications and reductions of dimension follow from the homogeneity and the fact that the current model does not allow inventory holding. First, both mappings $\mathcal{Y} : \mathbb{R}^4_+ \to \mathbb{R}_+$ and $\mathcal{L} : \mathbb{R}^4_+ \to \mathbb{R}_+$ are homogeneous of degree zero in (p, w, M, p^e) so that they can be reduced to $\mathcal{Y} : \mathbb{R}^3_+ \to \mathbb{R}_+$, $(\alpha, m, \theta^e) \mapsto \mathcal{Y}(\alpha, m, \theta^e) = y$ with $\alpha = w/p$, $m = M/p$, $\theta^e = p^e_1/p$. Second, since there is no inventory holding $F \circ \mathcal{L} = \mathcal{Y}$. i.e. $\mathcal{Y}$ is a monotonic tansformation of $\mathcal{L}$. Therefore, knowing the properties of $\mathcal{Y}$ is sufficient to know $\mathcal{L}$. Thus, any orbit $\left\{(p_t, w_t, M_t, p^e_{t,t+1})\right\} \in \mathbb{R}^4_+$ of nominal values induces a sequence of quadruples $\{(\alpha_t, m_t, \theta^e_t, y_t = \mathcal{Y}(\alpha_t, m_t, \theta^e_t))\}$ of real variables defining output and all other parts of the allocation in disequilibrium at any one time. Therefore, in the end it suffices to

[1] Stationary states of the model with an exogenously fixed labor supply were discussed in Böhm (1989). A first dynamic analysis was presented in Böhm (1978b, 1993) with a first numerical analysis in Böhm, Lohmann & Lorenz (1994).

describe and understand the dynamics of the nominal system in $\mathbb{R}_+^4$ and the orbits in real space on the graph of $\mathcal{Y} \subset \mathbb{R}_+^4$ which contains the bounded balanced paths[2].

7.1 Money Balances and the Government Budget

To complete the description of the dynamic features of the economy, the evolution of money balances and of expectations has to be defined. Since markets do not clear in general, there exist inter market spill overs which are regime specific and which have an impact on private savings by consumers. As a consequence, the evolution of money balances, as a result of private savings is no longer uniquely related to the government budget in the same way as under market clearing.

When there are two agents trading in a market only, the application of the minimum rule implies a unique outcome for the allocation and for the evolution of money balances and the government deficit. Therefore, in the model here with only one producer and one worker/consumer there is a unique outcome in all rationing situations of the labor market and those with supply side rationing. However, when there is insufficient supply in the goods market, demand rationing occurs, so that three agents in the market are affected: young consumers, old consumers, and the government. This may cause forced savings of young consumers since they may not be able to spend as much on current consumption. If old consumers are not able to spend all of their wealth, then a rule has to be found determining who receives the unspent part of the existing money balances. Finally, if the demand by the government is rationed, then the budget deficit will clearly be influenced by the rationing mechanism.

The state of the economy being given by the list $(p_t, w_t, M_t, p_{t+1}^e, g, \tau)$, let the aggregate levels of output and employment be determined by the minimum rule through the functions $\mathcal{L}$ and $\mathcal{Y}$. Then, in order to obtain a determinate outcome of savings and of the deficit, it is necessary to define a specific rationing rule for the commodity market. It will be assumed here that under demand rationing a hierarchical rationing mechanism will be applied: the government is served first, then the old consumer, and then the young consumer. This implies that actual consumption of the young consumer in period t for given output $y_t = \mathcal{Y}(p_t, w_t, M_t, p_{t,t+1}^e, g, \tau)$ is given by

$$x_t = \max\left\{0, y_t - g - \frac{M_t}{p_t}\right\}. \tag{7.1.1}$$

As a consequence, actual savings of the young, i.e. initial money balances at the beginning of the next period are

[2] This contrasts with the situation under market clearing in Chapter 4 given by two dimensional dynamics of prices and expectations and the equilibrium set, a subset of $\mathbb{R}_+^2$, defined by the unit contour of the price law (3.2.40).

$$\begin{aligned} M_{t+1} &= (1-\tau)\,p_t y_t \;-\; p_t x_t \\ &= p_t \left[\min\left\{ y_t, \frac{M_t}{p_t} + g \right\} - \tau\, y_t \right], \qquad y_t = \mathcal{Y}(\alpha_t, m_t, \theta^e_{t,t+1}, g, \tau) \qquad (7.1.2) \\ &=: p_t\, \mathcal{M}\,(\alpha_t, m_t, \theta^e_{t,t+1}, g, \tau). \end{aligned}$$

which defines the difference equation for the money dynamics. If there is no demand rationing of old consumers, i.e. if $M_t/p_t \leq y_t - g$, the change in the money stock is equal to the government deficit and the equation corresponds to the usual government budget constraint. Otherwise young agents are demand rationed to zero and savings is equal to total net income. Thus, possible amounts of money unspent by old consumers are confiscated/destroyed by the central bank and are not distributed to young consumers as transfers[3].

7.2 Adaptive Expectations

For a complete description of the dynamics of the economy, two features still need to be described: the expectations formation by the consumers and the price and wage adjustment for each constellation of disequilibrium. For an analysis of the temporary structure of the model expected prices are essentially a parameter for the economy at time t. However, for a full dynamic analysis an explicit description has to be given of how young consumers make a particular forecast. In contrast to the analysis under market clearing in Chapter 4 with perfect foresight, here it is assumed that expectations by consumers follow an adaptive procedure, based on some fundamental principles of time series analysis common and acceptable to most forecasting principles. Any precise form of such a forecasting rule may be subject to revisions or criticisms, depending on the level of rationality one would like to impose as a modeler.

It is obvious that the form will have an influence on the dynamics of the system when there is an expectations feed back. The particular form chosen here assumes only some basic general principles and there is no claim of generality or empirical validity. Moreover, since the structural results of the next section do not depend on these features, there is no additional need to defend or justify the general principles. For the dynamic analysis of the next section it will be assumed that the prices expected by consumers depend on observed prices over some finite past and that all generations use the same expectations function (predictor) with an identical length of memory. Let $T = 0, 1, 2, 3, \ldots$ denote the number of periods before t for which generation t considers past prices to be relevant for their forecast. Then, the expected price $p^e_{t,t+1}$ for period $t+1$ will be a homogeneous function of the vector of past and current prices $(p_t, p_{t-1}, p_{t-2}, \cdots, p_{t-T})$ of the following general form:

---

[3] Other mechanisms than (7.1.1) like equal rationing or equal proportional rationing could be used here as well. They would induce different consequences for actual savings and the development of money balances.

$$p^e_{t,t+1} = p_t\, \Psi\left(\frac{p_t}{p_{t-1}}, \cdots, \frac{p_{t+1-T}}{p_{t-T}}\right).$$

Homogeneity implies that the forecasting rule is consistent with inflationary or deflationary observation, a feature commonly used in also in the computation of price indexes. Using the homogeneity one finds that the forecast for the inflation rate can be written as a function of past observed rates of inflation

$$\theta^e_t = \Psi(\theta_{t-1}, \cdots, \theta_{t-T}),$$

in other words, generations of consumers make a point forecast of the inflation rate from t to $t+1$ on the basis of the past $T = 0, 1, 2, \ldots$. inflation rates. Apart from continuity and consistency of a steady state property of inflation/deflation confirmation, no further specific assumption will be imposed on Ψ.

Assumption 7.2.1 *The expectations function* $\Psi : \mathbb{R}^T_{++} \to \mathbb{R}_{++}$

$$p^e_{t,t+1} = p_t\, \Psi\left(\frac{p_t}{p_{t-1}}, \cdots, \frac{p_{t+1-T}}{p_{t-T}}\right) \quad T \geq 1 \tag{7.2.1}$$

is continuous and satisfies $\Psi(\theta, \ldots, \theta) = \theta, \quad \forall\, \theta > 0,$ *and* $\Psi \equiv 1$ *if* $T = 0$.

This formulation encompasses most forecasting rules commonly used in adaptive econometric time series models as well as in financial market models of trend forecasting, trend reversals etc..

Notice that the elements described so far - namely the government budget constraint, the expectations function, and the allocation function $\mathcal{Y}$ - are sufficient information to define steady states with perfect foresight, in spite of the fact that a specific price and wage adjustment process has not been defined. In general, sequences of temporary feasible states would be given by sequences of prices, wages, money balances, and price expectations. Due to the homogeneity of the allocation function, a three dimensional state space suffices with the coordinates for real wages, real money balances, and expected inflation rates.

7.3 Price and Wage Adjustment – The Law of Supply and Demand

For a complete description of the dynamic evolution of the economy, the wage and price adjustment mechanisms have to be defined. In a competitive economy, when there is no specific agent setting the price in the market, it is natural to assume that prices and wages respond to the disequilibrium situation in each market. The general principle is straightforward. Supply (demand) rationing in any particular period leaves a potential seller (buyer) unsatisfied, who then would be willing to sell for a lower (buy for a higher) price, leading to lower (higher) prices in the respective market in the next period. Such a rule corresponds most closely to those applied in all tâtonnement models, which is used in non-tâtonnement models as well. In this context it is immaterial whether this rule is implemented by an anonymous auctioneer, the government, or whether the producer and the worker follow it. Such a rule is essentially myopic. It is defined using exclusively information provided by the economy at date t through the disequilibrium configuration on each market. Hence, it does not take any past history of states or prices nor any expectations for the future into account.

It is useful (see Böhm, 1989, 1993) to divide the adjustment mechanism into two parts, namely a function measuring the strength of the disequilibrium and a function determining the magnitude of the actual price and wage adjustment. Let (s^c, s^ℓ) denote a pair of numbers between minus one and plus one, measuring the disequilibrium for the commodity and for the labor market respectively. They are best interpreted to indicate percentages of unsatisfied desired transactions, but other monotonic transformations could be used as well. Negative numbers mean supply rationing, positive numbers mean demand rationing. Thus, for example, $s^\ell = -0.3$ could indicate an unemployment rate of 30 per cent. These measures depend on the disequilibrium state in every period, i.e. on the three variables (α, m, θ^e). Therefore, the two associated signaling functions σ^c and σ^ℓ generating the measures are defined on the state space $\mathbb{R}^3_+$ of the economy.

$$\sigma^c : \mathbb{R}^3_{++} \to [-1, +1]; \qquad (\alpha, m, \theta^e) \mapsto \sigma^c(\alpha, m, \theta^e) = s^c$$

$$\sigma^\ell : \mathbb{R}^3_{++} \to [-1, +1]; \qquad (\alpha, m, \theta^e) \mapsto \sigma^\ell(\alpha, m, \theta^e) = s^\ell$$

The signaling functions will be called consistent with the disequilibrium state if the following assumption is satisfied.

Assumption 7.3.1 (Disequilibrium Signaling)

$$\sigma^c(\alpha, m, \theta^e) = \begin{cases} > 0 & (\alpha, m, \theta^e) \in int\,(\boldsymbol{C} \cup \boldsymbol{I}) \\ < 0 & (\alpha, m, \theta^e) \in int\,\boldsymbol{K} \\ = 0 & (\alpha, m, \theta^e) \in \boldsymbol{K} \cap \boldsymbol{C}, \end{cases}$$

$$\sigma^\ell(\alpha, m, \theta^e) = \begin{cases} > 0 & (\alpha, m, \theta^e) \in int\,\boldsymbol{I} \\ < 0 & (\alpha, m, \theta^e) \in int\,(\boldsymbol{C} \cup \boldsymbol{K}) \\ = 0 & (\alpha, m, \theta^e) \in \boldsymbol{C} \cap \boldsymbol{I}, \end{cases} \tag{7.3.1}$$

$$(\alpha, m, \theta^e) \in (\boldsymbol{K} \cap \boldsymbol{I}) \setminus \boldsymbol{WE} \Rightarrow \begin{cases} \sigma^\ell(\alpha, m, \theta^e) \geq 0 \geq \sigma^c(\alpha, m, \theta^e) \\ \text{and} \\ \sigma^\ell(\alpha, m, \theta^e) > \sigma^c(\alpha, m, \theta^e) \end{cases}$$

σ^ℓ, σ^c are continuous except on $\boldsymbol{K} \cap \boldsymbol{I}$.

The first two sets of conditions present the natural notion of disequilibrium, i.e. indicating actual rationing if and only if the measure is nonzero. The third condition indicates that the adjustment on the state space $\mathbb{R}^3_+$ must be discontinuous at $(\alpha, m, \theta^e) \notin$ **WE**. The property of sign preservation for the given signal together with

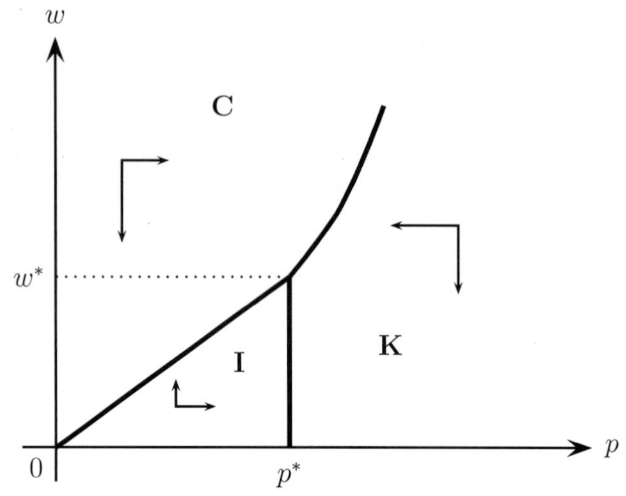

Fig. 7.1 Consistent price and wage adjustment with constant labor supply

continuity would imply that both signals would have to be zero on $\mathbf{K} \cap \mathbf{I}$, in spite of the fact that disequilibrium in both markets prevails on the boundary $\mathbf{K} \cap \mathbf{I} \setminus \mathbf{WE}$, violating sign consistency (see Table 6.1). In other words, consistency of the signal and of the adjustment according to the Law of Supply and Demand is inconsistent with continuity[4]. Therefore, in order to avoid that the adjustment mechanism for the economy settles artificially at a non-Walrasian equilibrium when at least one market signal is clearly different from zero, joint consistency requires that at least one of the market signaling functions should be discontinuous. Thus, at least one of the disequilibrium signals must be nonzero on $\mathbf{K} \cap \mathbf{I} \setminus \mathbf{WE}$ if the adjustment function is to be continuous with respect to the signal.

The price and wage adjustment functions now map the disequilibrium signals of any period in a sign preserving fashion into rates of price or wage change as discussed above. A rule with such features will be referred to as the Law of Supply and Demand.

Assumption 7.3.2 (The Law of Supply and Demand) *A price adjustment rule* $\mathcal{p}$ *and a wage adjustment rule* $\mathcal{w}$

$$\begin{aligned} \mathcal{p} : [-1, +1] \to (-1, +\infty), \qquad & p_{t+1} = p_t\Big(1 + \mathcal{p}(s_t^c)\Big) \\ \mathcal{w} : [-1, +1] \to (-1, +\infty), \qquad & w_{t+1} = w_t\Big(1 + \mathcal{w}(s_t^\ell)\Big) \end{aligned} \tag{7.3.2}$$

[4] This feature results from the structural asymmetry of the model which has *intertemporal* consumption decisions but *atemporal* production decisions. The discontinuity does not arise in a model where the decision problem in both sectors is a true intertemporal one, as for example with the possibility of inventory holding for producers.

are said to correspond to the ***Law of Supply and Demand*** *if they are continuous, strictly monotonically increasing and satisfy* $p(0) = w(0) = 0$[5].

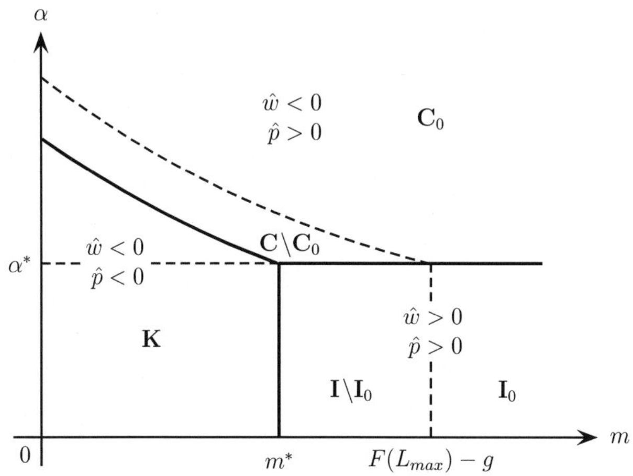

Fig. 7.2 Consistent price and wage adjustment principle

A price and wage adjustment rule satisfying Assumptions 7.3.1 and 7.3.2 will be called consistent. With these specifications the description of the dynamic system is now complete. There are essentially two equivalent functional forms which can be used to describe the dynamic process. One form considers the expected inflation rate θ_t^e as the third state variable whereas the other one considers the actual inflation rate θ_t. In either case the system of difference equations is of order T from $\mathbb{R}_+^3$ to itself. If $T = 0$, or in some other special cases where expectations do not matter, the dynamic system reduces to a first order difference equation from $\mathbb{R}_+^2$ to itself. Here, the formulation with expected inflation rates is given. Figure 7.2 indicates the qualitative features of the consistent price and wage adjustment principle in the temporary state space (m, α).

Figure 7.3 displays the recursive feedback structure of the economy. Square boxes represent the functions with possible multiple inputs but with a unique output, while circles identify the corresponding variables determined by the functions. The diagram emphasizes the forward recursive flow character which is a necessary requirement. There are no intermediate loops between the functions, requiring the determination of a solution or a fixed point[6] which is the essential property in order to obtain an explicitly defined dynamical system.

As a consequence one now obtains the complete description of the dynamical system given by four homogeneous difference equations (7.1.2), (7.2.1), (7.3.2)

---

[5] The two adjustment functions p and w using lower case letters should not be confused with the upper case calligraphic letters $\mathcal{P}$ and $\mathcal{W}$ used for the price law and the wage law respectively in Chapter 3.

[6] Some early qualitative results are given in Böhm (1978a) and Benassy (1984).

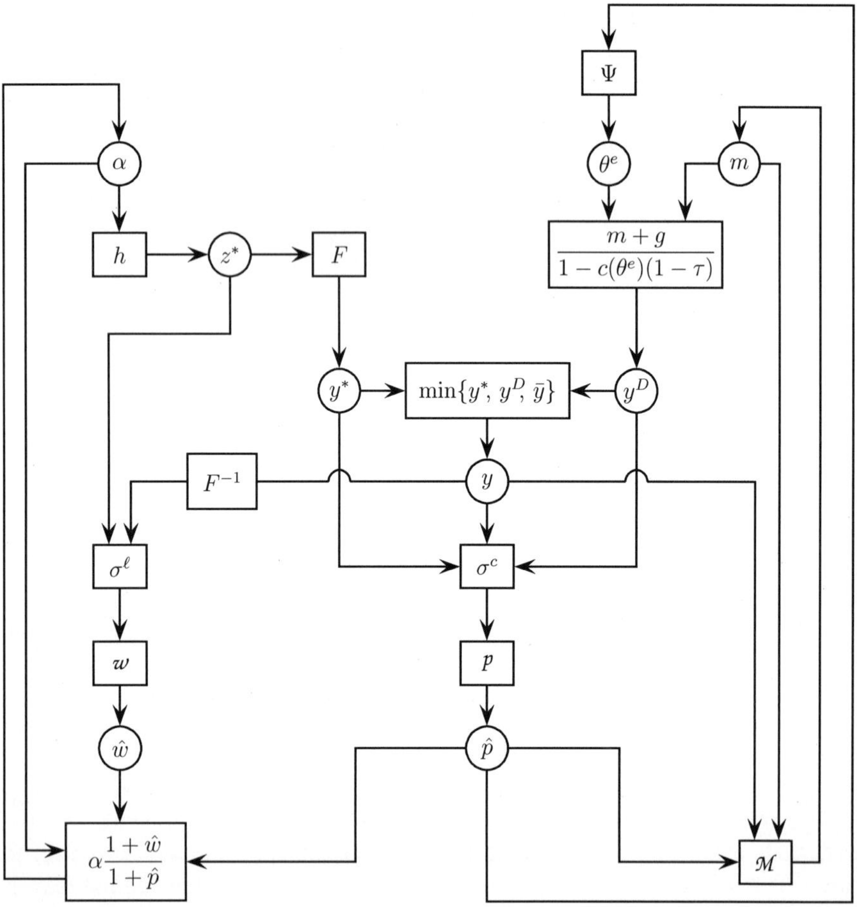

Fig. 7.3 Feedback structure of the complete real model

together with (7.3.1) which describe the time-one shift of the four state variables $(M_t, p_t, w_t, p^e_{t,t+1})$ as a delay system of order $4 + T$. Using the homogeneity one obtains the associated $3 + T$-dimensional system[7]

[7] As in Chapter 4 one could proceed here with an analysis of the full dimensional monetary system to discover that stationary solutions will be rare for the homogeneous system, as is the case under market clearing. The reasons for nonexistence of fixed points will be structurally the same, but the non-generic sets of parameters for which they exist will be more complex.

$$\alpha_{t+1} = \alpha_t \frac{1 + w(s_t^\ell)}{1 + p(s_t^c)}, \qquad s_t^j = \sigma^j(\alpha_t, m_t, \theta_t^e), \quad j = c, \ell,$$

$$m_{t+1} = \frac{\min\{y_t, m_t + g\} - \tau y_t}{1 + p(s_t^c)}, \quad y_t = \mathcal{Y}(\alpha_t, m_t, \theta_t^e) \tag{7.3.3}$$

$$\theta_{t+1}^e = \Psi(\theta_t, \dots, \theta_{t-T+1}), \qquad \theta_t = 1 + p(s_t^c).$$

in $(\alpha, m, \theta^e, \theta_{-1}, \dots, \theta_{-T}) \in \mathbb{R}_+^{3+T}$ whose fixed points define the balanced paths of the homogeneous system. A fixed point $(\alpha, m, \theta) \gg 0$ induces constant allocations with perfect foresight respecting the Law of Supply and Demand. The forward iteration of the mapping (7.3.3) generates orbits (i.e. paths of real wages, money balances, and expectations) with consistent price and wage adjustment, which in turn induce a sequences of levels of output and employment under disequilibrium with their respective characteristics. In other words, the features of persistent nonstationary orbits describe the induced features of *endogenously generated* business cycles which are caused by the nonlinearities of the mappings.

7.4 Existence and Uniqueness of Stationary Real States

Fixed points of the dynamical system (7.3.3) are the stationary solutions or the steady states of the economy in real terms which define balanced paths of the monetary economy[8]. It is straightforward to verify the following properties of consistent steady states. If (α, m, θ) is consistent and stationary, then

(a) $\theta^e = \theta$,

(b) $(\alpha, m, \theta) \notin \mathbf{C}$,

(c) $\theta > 1$ if and only if $(\alpha, m, \theta) \in \mathbf{I}$,

$\theta = 1$ if and only if $(\alpha, m, \theta) \in \mathbf{WE}$,

$\theta < 1$ if and only if $(\alpha, m, \theta) \in \mathbf{K}$.

In other words, classical states can never persist in the long run. However, expectations are fulfilled, i.e. consistent steady states have the perfect foresight property. In addition, Keynesian unemployment states must be deflationary, whereas states of repressed inflation must have a positive inflation rate. Thus, money balances, prices, and wages decrease at constant rates in Keynesian states while they increase at constant rates in inflationary states. Moreover, as the following theorem indicates, the type of steady state for this prototype economy strictly depends on the production characteristics of the economy *and* on the relative sizes of government demand to taxes, but neither on the adjustment speeds on any of the markets nor on consumer characteristics.

[8] The two terms steady states and stationary states will be used here equivalently.

Theorem 7.4.1 (Type Uniqueness of Stationary States).
Let the basic Assumption 3.2.1 be satisfied and assume that labor supply is constant and given by $L_{\max} > 0$. *Then:*

$$\begin{aligned} (\alpha, m, \theta) \in \mathbf{K} \quad &\Leftrightarrow \quad g < \tau F(L_{\max}) \\ (\alpha, m, \theta) \in \mathbf{WE} \quad &\Leftrightarrow \quad g = \tau F(L_{\max}). \\ (\alpha, m, \theta) \in \mathbf{I} \quad &\Leftrightarrow \quad g > \tau F(L_{\max}) \end{aligned} \tag{7.4.1}$$

If c is independent of expectations, then there exists a unique steady state (α, m, θ).

Proof. Let $(\alpha, m, \theta) \gg 0$ denote a steady state. Then:

$$w(\sigma^{\ell}(\alpha, m, \theta)) = p(\sigma^{c}(\alpha, m, \theta)) = \theta - 1 \tag{7.4.2}$$

$$m\theta = \min\{y, m + g\} - \tau y \tag{7.4.3}$$

$$y = \min\left\{\frac{m + g}{1 - c(\theta)(1 - \tau)}, F(h(\alpha)), F(L_{\max})\right\} \tag{7.4.4}$$

(a) a. Let $0 \ll (\alpha, m, \theta) \in \mathbf{K}$.
Then: $\theta < 1$, $y < y_{\max}$ and $m\theta = m + g - \tau y$. Therefore,

$$0 > m(\theta - 1) = g - \tau y > g - \tau y_{\max}.$$

b. Conversely, let $g < \tau y_{\max}$ and assume that $\theta \geq 1$. Then:

$$\begin{aligned} m \leq m\theta &= \min\{y_{\max}, m + g\} - \tau y_{\max} \\ &\leq m + g - \tau y_{\max} < m. \end{aligned}$$

Therefore, $\theta < 1$ and $y < y_{\max}$ Hence, $(\alpha, m, \theta) \in \mathbf{K}$.

(b) a. Let $0 \ll (\alpha, m, \theta) \in \mathbf{I}$. Then:
$\theta > 1$, $y = y_{\max}$ and $m\theta = \min\{y_{\max}, m + g\} - \tau y_{\max}$. Therefore, if $g \leq \tau y_{\max}$, this would imply

$$\begin{aligned} m < m\theta &= \min\{y_{\max}, m + g\} - \tau y_{\max} \\ &\leq m + g - \tau y_{\max} \leq m, \end{aligned}$$

which is a contradiction. Hence, $g > \tau y_{\max}$.

b. Conversely, let $g > \tau y_{\max}$ and assume that $\theta \leq 1$. Then

$$\begin{aligned} m \geq m\theta &= \min\{y, m + g\} - \tau y \\ &= m + g - \tau y > m + \tau y_{\max} - \tau y \geq m \end{aligned}$$

which is a contradiction. Hence, $0 \ll (\alpha, m, \theta) \in \mathbf{I}$

(c) The case $0 \ll (\alpha, m, \theta) \in \mathbf{WE} \quad \Leftrightarrow \quad g = \tau y_{\max}$ follows. □

Multiple stationary states may occur in more general cases when the expectations effect in consumption is nonzero or when labor supply is endogenous.

7.5 Real Business Cycles with Constant Labor Supply

In spite of the fact that the uniqueness of steady states and their qualitative properties can be derived in general, the dynamic properties of the system seem intractable analytically on a general level. Even for the simplest case with no expectations memory ($T = 0$), the resulting planar system of a first order difference equation has a degree of complexity, for which closed form solutions cannot be obtained without further assumptions. Moreover, even in such cases, properties of global dynamics can only be derived using numerical techniques. Therefore, a class of parametric specifications of the assumptions for consumer characteristics, expectations formation, technology, and price and wage adjustments, denoted **C**, **E**, **F**, **P**, is chosen for the remaining part of this chapter. For these an explicit time one map is obtained, for which the numerical simulation results are shown.

Assumption 7.5.1 (Assumption C) *For $\delta > 0$ and $-\infty < \rho < 1$:*

$$u(x_t, x_{t+1}) = \begin{cases} \dfrac{1}{\rho}\left(x_t^{\rho} + \delta\, x_{t+1}^{\rho}\right) & \rho \neq 0 \\[2ex] \ln\, x_t + \delta\, \ln\, x_{t+1} & \rho = 0. \end{cases}$$

Assumption **C** describes the standard CES intertemporal utility function with substitution parameter ρ and time discount parameter δ, which is a special case of Assumption 3.1.1. The Cobb-Douglas case, i.e. $\rho = 0$, implies that current demand is independent of expected prices. In this situation the dynamics are independent of expectations and the system is two-dimensional and of first order regardless of the length of memory T and of the expectations function Ψ.

Assumption 7.5.2 (Assumption E:) *For $\tilde{T} = 0, 1, 2, 3, \ldots$ and $T = \min\{t-1, \tilde{T}\}$:*

$$\theta_t^e = \frac{p_{t+1}^e}{p_t} = \begin{cases} \dfrac{1}{T}\displaystyle\sum_{k=1}^{T} \theta_{t-k} & T > 0 \\[2ex] 1 & T = 0. \end{cases}$$

This expectations function uses a simple unweighted averaging procedure which is assumed here for simplicity[9]. Any other more sophisticated econometric forecasting method could be applied as well. One of the issues still unresolved in this context is

[9] It is clear that the form of the forecasting rule has an influence on the dynamics of the system because of the expectations feedback operating through the propensity to consume. Since the impact of the forecasting rule on the dynamics will not be examined here the simplest averaging procedure may suffice (see, however, Lorenz & Lohmann, 1996).

whether perfect prediction is possible*and* under which conditions the perfect foresight orbits are stable.

Assumption 7.5.3 (Assumption F) *The production function is of the isoelastic form*

$$y_t = \frac{A}{B} z_t^B, \qquad A > 0, \qquad \textit{and} \qquad 1 > B > 0.$$

Assumption 7.5.4 (Assumption P) *For* $0 < \gamma < 1\,,\, 0 < \kappa < 1$*:*

$$p_{t+1} = p_t \begin{cases} 1 + \gamma \dfrac{y_t^D - y_t}{y_t^D}, & y_t^D > y_t \\[2ex] 1 + \kappa \dfrac{y_t - y_t^*}{y_t^*}, & \textit{otherwise}, \end{cases}$$

and for $0 < \mu < 1\,,\, 0 < \lambda < 1$*:*

$$w_{t+1} = w_t \begin{cases} 1 + \lambda \dfrac{L_t - L_{\max}}{L_{\max}}, & L_{\max} > L_t \\[2ex] 1 + \mu \dfrac{z_t^* - L_t}{z_t^*}, & \textit{otherwise}. \end{cases}$$

Here the usual effective excess demand percentages are chosen as disequilibrium measures and the adjustment functions $\hat{p}$ and $\hat{w}$ are linear with constant parametrically chosen adjustment coefficients.

Apart from the two government parameters (g, τ), the Assumptions **C**, **E**, **F**, **P**, require the numerical specification of ten parameter values: $(\delta, \rho, T, L_{\max})$ for the consumer, (A, B) for the producer, and the adjustment coefficients $(\gamma, \kappa, \lambda, \mu)$ for the two markets. $L_{\max}$ and A are essentially scaling parameters which can be set equal to one without loosing any structural information. Then the real wage α^* which clears the labor market is also equal to one as well for all other possible parameter values. This leaves eight essential parameters plus (g, τ) to be chosen. Figure 7.4 indicates the role of the fiscal parameters for the appearance of stationary states when the scaling parameters are set to $A = L_{\max} = 1$. In this case $y_{\max} = F(L_{\max}) = 1/B$ and the condition for the unique type of the stationary state (Theorem 7.4.1) shows that Keynesian states occur if and only if $Bg < \tau$. This divides the parameter space (g, τ) into the two regions of Keynesian **K** and inflationary **I** states by the line $\tau = Bg$, along which stationary Walrasian equilibria occur.

7.5.1 Keynesian Stationary States

Let $g < \tau y_{\max}$. Then $(\alpha, m, \theta) \gg 0$ is a Keynesian steady state with consistent price and wage adjustment if and only if it is a solution of the following equations:

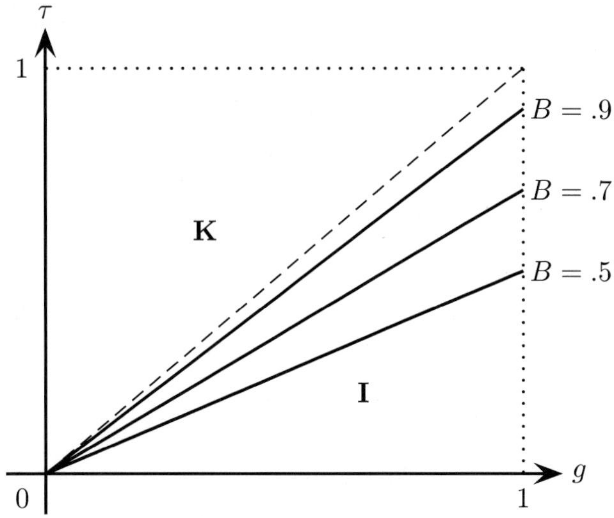

Fig. 7.4 The role of fiscal parameters on types of stationary states; $A = L_{\max} = 1$

$$m\theta = m + g - \tau y \tag{7.5.1}$$

$$\theta = 1 + \lambda\left(\frac{L}{L_{\max}} - 1\right) \tag{7.5.2}$$

$$\theta = 1 + \kappa\left(\frac{y}{y^*(\alpha)} - 1\right) \tag{7.5.3}$$

such that

$$y = \frac{m+g}{1 - c(\theta)(1-\tau)} \tag{7.5.4}$$

$$L = F^{-1}\left(\frac{m+g}{1 - c(\theta)(1-\tau)}\right) \tag{7.5.5}$$

with $L < L_{\max}$, $0 < \theta < 1$. When there is no expectations feed back in demand, i.e. when $c(\theta) \equiv c$ is independent of expectations, the dynamical system is two dimensional in the space of real money balances and the real wage (m, α), so that the stationary solutions can be represented in the usual two-dimensional diagram.

Consider first the set of stationary money balances $\Delta m = 0$, which must be a solution of equations (7.5.1), (7.5.2), and (7.5.4). Equations (7.5.1) and (7.5.4) yield a unique demand consistent relationship between stationary inflation θ and stationary real balances m:

$$\theta = \frac{m+g}{m}\frac{(1-c)(1-\tau)}{1-c(1-\tau)}. \tag{7.5.6}$$

Its graph corresponds to the disequilibrium analogue of the equilibrium set under

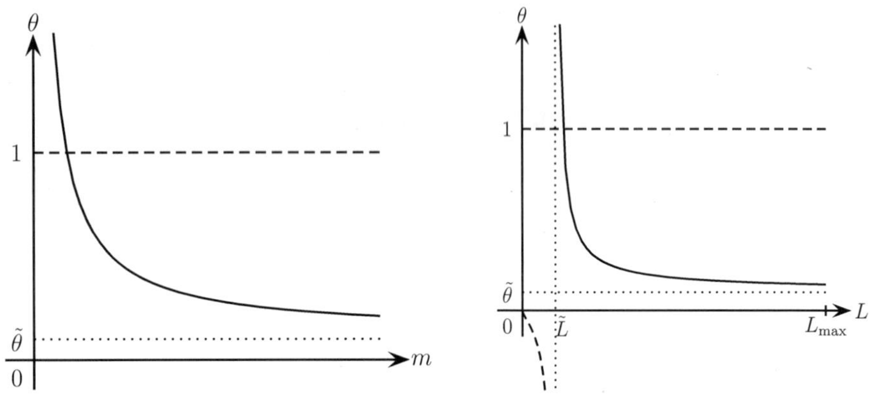

a Real money balances vs. inflation: (7.5.6) **b** Employment vs. inflation: (7.5.7)

Fig. 7.5 Tradeoffs for Keynesian stationary states

market clearing of Chapter 3 describing the time-invariant set of observable stationary pairs of real money balances and inflation rates (see Figure 7.5 **a**) without requiring consistency properties with respect to a price and wage adjustment mechanism.

Using the parametrization in the employment level L (7.5.5) rather than in money balances reveals that there exists a second relationship between stationary inflation and employment

$$\theta = \frac{(1-c)(1-\tau)}{1-c(1-\tau)} \frac{(1-c(1-\tau)F(L)}{(1-c(1-\tau)F(L)-g}, \tag{7.5.7}$$

also independent of the real wage α, describing the unique tradeoff between employment and inflation, a stationary Phillips curve (see Figure 7.5 **b**). For $g > 0$, this function has a pole at a positive employment level $\tilde{L} > 0$ and is a downward sloping relationship between employment and inflation, i.e. the associated long-run Phillips curve (measured wih respect to unemployment) is an *upward sloping* function with a positive upper bound for the rate of unemployment and a lower positive bound on deflation.

In order to find the stationary level of real balances under consistent price adjustment (i.e. with money and prices contracting at the same rate $0 < \theta < 1$) equations (7.5.2) and (7.5.5) imply a second monotonically increasing relationship

$$\theta = 1 + \lambda \left[\frac{F^{-1}\left(\dfrac{m+g}{1-c(1-\tau)} \right)}{L_{\max}} - 1 \right] \tag{7.5.8}$$

between inflation and real money balances. Both are independent of the real wage α. Therefore, (7.5.6) and (7.5.8) imply a unique solution $(\bar{m}, \bar{\theta})$ as stationary levels under consistent price adjustment. Thus, $\Delta m = 0$ is a vertical line in (m, α) space at the level $0 < \bar{m} < m^\star$ with $\bar{m} \to m^\star$ as $g \to \tau y_{\max}$ (see Figure 7.6 **a**). In other words, given the rationing mechanism on the supply side of the commodity market consistent price adjustment alone determines the level of stationary money balances $\bar{m}$. Therefore, output $\bar{y}$, employment $\bar{L} < L_{\max}$, and the unemployment rate

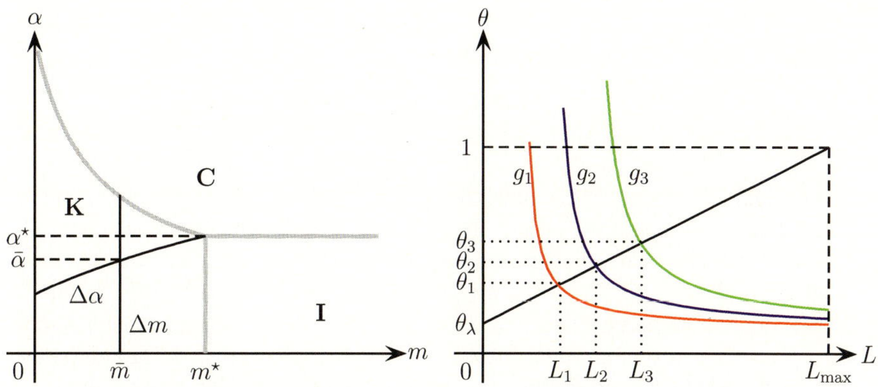

a Keynesian stationary state $(\bar{m}, \bar{\alpha})$ **b** Role of $g_1 < g_2 < g_3 < \tau y_{\max}$

Fig. 7.6 Keynesian stationary states for $g < \tau y_{\max}$, $c(\theta) \equiv c$

$(L_{\max} - \bar{L})/L_{\max}$ in a Keynesian situation are independent of the real wage.

To determine the stationary real wage under consistent wage adjustment the nominal wage has to adjust appropriately given the rationing mechanism on the labor market. Therefore, consider the set of possible stationary real wages $\Delta\alpha = 0$ defined by equations (7.5.2), (7.5.3), (7.5.4), and (7.5.5). The first two imply

$$\lambda\left(\frac{L}{L_{\max}} - 1\right) = \kappa\left(\frac{F(L)}{y^\star(\alpha)} - 1\right) \tag{7.5.9}$$

which yields[10]

$$y^\star(\alpha) = \frac{F(L)}{\frac{\lambda}{\kappa}\left(\frac{L}{L_{\max}} - 1\right) + 1} = \frac{\kappa L_{\max} F(L)}{\lambda(L - L_{\max}) + \kappa L_{\max}} \tag{7.5.10}$$

Solving for α as a function of m using (7.5.4) and (7.5.5) implies the solution

[10] To guarantee a positive solution under consistency the adjustment speed κ on the commodity market should not be too small relative to λ, for example $\kappa \geq \lambda$ suffices; see Theorem B.3.1.

$$\bar{\alpha} = h^{-1}\left(F^{-1}\left[\frac{\kappa L_{\max} F(\bar{L})}{\lambda\left(\bar{L} - L_{\max}\right) + \kappa L_{\max}}\right]\right) \quad \text{with} \quad \bar{L} = F^{-1}\left(\frac{\bar{m} + g}{1 - c(1-\tau)}\right). \quad (7.5.11)$$

Any solution $(\bar{L}, \bar{\theta})$ with $0 < \bar{L} < L_{\max}$ and $(1 - c(1-\tau))F(\bar{L}) > g$ is unique and implies $0 < \bar{\theta} < 1$. Therefore, one obtains a unique steady state $(\bar{\alpha}, \bar{m}, \bar{\theta})$ for any $0 < g < \tau y_{\max}$ and adjustment coefficients $0 < \lambda \geq \kappa < 1$ (see Figure 7.6 **a**) which is in fact a Keynesian stationary state.

In addition, (7.5.6) and (7.5.8) imply that the long-run level of employment $\bar{L}$ and the deflation factor $\bar{\theta}$ are strictly increasing in government demand g (and strictly decreasing in the tax rate τ). In other words, the two long-run employment multipliers are positive (negative) (see Figure 7.6 **b**).

$$\frac{\partial \bar{L}}{\partial g} > 0, \quad \frac{\partial \bar{\theta}}{\partial g} > 0, \quad \frac{\partial \bar{\alpha}}{\partial g} > 0; \qquad \frac{\partial \bar{L}}{\partial \tau} < 0, \quad \frac{\partial \bar{\theta}}{\partial \tau} < 0, \quad \frac{\partial \bar{\alpha}}{\partial \tau} < 0. \quad (7.5.12)$$

Equation (7.5.12) describes the stationary effects of employment, inflation, and of the real wage for parametric policy changes under consistent price and wage adjustment. Concerning the effect on the stationary real wage, one finds that the term in square brackets of (7.5.11) is a decreasing function of employment, so that α is increasing in real money balances under $\Delta\alpha = 0$ for $\kappa \geq \lambda$. Thus, one also obtains a positive government multiplier on the long-run real wage. In summary, the commodity demand conditions and the speed of the price adjustment determine the stationary level of employment, of money balances, and of the rate of deflation while the speed of the wage adjustment on the labor market then determines the stationary real wage.

7.5.2 *Inflationary Stationary States*

The analysis of inflationary stationary states is somewhat simpler than in the Keynesian case, since the level of employment is always fixed at $L_{\max}$, so that inflation rates and money balances depend essentially on the price and wage adjustments responding to excess demand.

Let $g > \tau y_{\max}$. Applying the same procedure as before to determine a stationary state (m, α) one obtains two equations which are independent of the real wage α, while the adjustment equation on wages depends on the excess demand for labor which is a function of the real wage. Specifically, one has

$$\theta = \frac{1}{m} \min \left[(1-\tau) y_{\max}, m + g - \tau y_{\max} \right] \tag{7.5.13}$$

$$\theta = 1 + \gamma \left(1 - y_{\max} \frac{1 - c(1-\tau)}{m+g} \right) \tag{7.5.14}$$

$$\theta = 1 + \mu \left(1 - \frac{L_{\max}}{z^*(\alpha)} \right). \tag{7.5.15}$$

The first is strictly decreasing in m as the minimum of two strictly decreasing func-

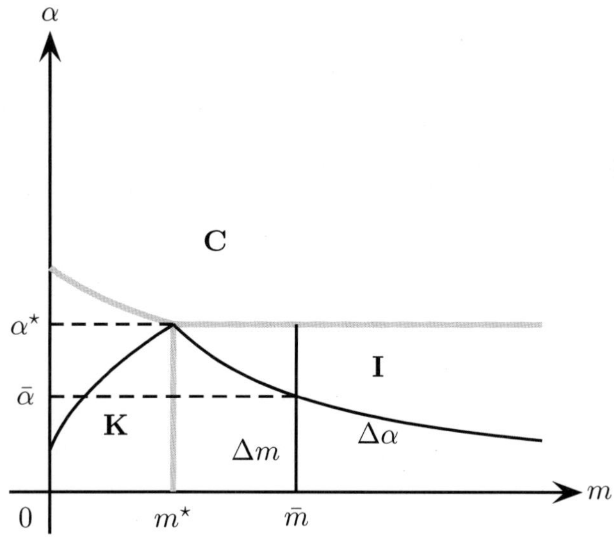

Fig. 7.7 Inflationary stationary states $(\bar{m}, \bar{\alpha})$ for $c(\theta) \equiv c$

tions while the second one is strictly increasing. Thus, there exists a level of stationary money balances $\bar{m} > m^\star$ and an associated inflation rate $\bar{\theta} > 1$ which are independent of α. Equations (7.5.14) and (7.5.15) define the set of solutions for $\Delta\alpha = 0$. Rearranging terms and solving for α one also obtains an explicit solution

$$\alpha = h^{-1} \left(\frac{L_{\max}}{1 + \frac{\gamma}{\mu} \left(y_{\max} \frac{1-c(1-\tau)}{m+g} - 1 \right)} \right) \tag{7.5.16}$$

which is a decreasing function in m for $m \geq m^\star$. Therefore, there exists a unique stationary solution $(\bar{m}, \bar{\alpha})$ (see Figure 7.7).

7.5.3 Asymptotic Stability

Given Theorem 7.4.1 on the type-uniqueness of fixed points of the dynamical system (7.3.3) generic stationary states are contained in open sets of the state space and the dynamics in the neighborhood are defined by smooth mappings specific to the regions **K** and **I**. Therefore, a local stability analysis can be carried out using standard methods of dynamical systems theory. Note that this cannot be done for the non-generic case of the Walrasian equilibrium with the adjustment mechanism of the law of supply and demand.

In order to provide a detailed structural analysis of the generic dominant dynamic features of the general model and in an effort to keep the analysis as tractable as possible the analytical properties will only be examined for the situation with fixed labor supply and when there is no expectations feed back in consumption ($\rho = 0$). In this case the propensity to consume $c = 1/(1 + \delta)$ is constant and the dynamical system 7.3.3 is two-dimensional. In the extended numerical analysis the general case with an expectations feed back and with endogenous labor supply will be examined. The two dimensional system $\mathcal{F} : \mathbb{R}^2_+ \to \mathbb{R}^2_+$, $(\alpha_t, m_t) \mapsto \mathcal{F}(\alpha_t, m_t) = (\alpha_{t+1}, m_{t+1})$ under these restrictions satisfying conditions **C**, **F**, and **P** is given by the set of equations (7.5.17) to (7.5.20)

$$
\begin{aligned}
\alpha_{t+1} &= \alpha_t \frac{1 + w(s^\ell(\alpha_t, m_t))}{1 + p(s^c(\alpha_t, m_t))} \\
m_{t+1} &= \frac{\min((1-\tau)y_t, m_t + g - \tau y_t)}{1 + p(s^c(\alpha_t, m_t))}
\end{aligned}
\tag{7.5.17}
$$

such that for $0 < \gamma < 1$, $0 < \kappa < 1$:

$$
p(s^c(\alpha_t, m_t)) = \begin{cases} \gamma \dfrac{y_t^D - y_t}{y_t^D} & y_t^D > y_t \\ \kappa \dfrac{y_t - y_t^*}{y_t^*} & \text{otherwise,} \end{cases}
\tag{7.5.18}
$$

and for $0 < \mu < 1, 0 < \lambda < 1$:

$$
w(s^\ell(\alpha_t, m_t)) = \begin{cases} \lambda \dfrac{L_t - L_{\max}}{L_{\max}} & L_{\max} > L_t \\ \mu \dfrac{z_t^* - L_t}{z_t^*} & \text{otherwise.} \end{cases}
\tag{7.5.19}
$$

where $y_t = \min\left(y_t^D, y_t^*, y_{\max}\right)$ and

$$y_t^D = \frac{m_t + g}{1 - c(1-\tau)}, \quad y_{\max} = \frac{A}{B}(L_{\max})^B, \quad y_t^* = \frac{A}{B}\left(\frac{\alpha_t}{A}\right)^{\frac{B}{B-1}}, \quad z_t^* = \left(\frac{\alpha_t}{A}\right)^{\frac{1}{B-1}}. \tag{7.5.20}$$

Then one has the following result originally shown by Kaas (1995) for the more general model. A proof for the isoelastic case is given in Appendix B.3 employing Theorem B.3.1 which provides further properties.

Theorem 7.5.1.

*Assume the parametric specifications given above for Assumptions **C**, **F**, **P**. Then, for every pair of government parameters* $(g, \tau) \neq 0$*:*

(a) there exists a unique steady state $(\alpha, m) \gg 0$ *such that*

1. (α, m) *is a Keynesian steady state* $\iff g < \tau y_{\max}$
2. (α, m) *is an inflationary steady state* $\iff g > \tau y_{\max}$
3. (α, m) *is a Walrasian steady state* $\iff g = \tau y_{\max}$;

(b) for $g \neq \tau y_{\max}$ *the steady state* (α, m) *is asymptotically stable for all adjustment factors* $0 < \gamma, \kappa, \lambda, \mu < 1$ *if* $0 < B \leq 1/2$;

(c) if (α, m) *is an inflationary steady state, then both eigenvalues* ν_1, ν_2 *are real and*

1. $-1 < \max(\nu_1, \nu_2) < 1$
2. $\min(\nu_1, \nu_2) < -1 \implies \mu > \bar{\mu} \geq 2(1-B)$ *and* $B > 1/2$;

(d) if (α, m) *is a Keynesian steady state, then*

1. (ν_1, ν_2) *complex* $\implies |\nu_i| < 1$
2. (ν_1, ν_2) *real* $\implies -1 < \max(\nu_1, \nu_2) < 1$
3. $\min(\nu_1, \nu_2) < -1 \implies \kappa > \bar{\kappa} \geq \lambda$ *and* $B > 1/2$.

Theorem 7.5.1 indicates that the dynamic behavior in the isoelastic case is determined by the elasticity of production B and its interplay with the adjustment parameters on the two markets. If $0 < B \leq 1/2$ steady states whether Keynesian or inflationary are globally stable regardless of the adjustment parameters. Type uniqueness implies that Keynesian balanced paths appear with this configuration more frequent than inflationary under the respective ranges of the fiscal parameters (g, τ). The case $B = 1/2$ corresponds to a quadratic cost function (linear marginal costs).

Statements (c) and (d) indicate that in either region a steady state (α, m) can undergo at most a period-doubling bifurcation when one of the real roots has modulus less than minus one. The instability in the inflationary region is caused by an overshooting property of the wage adjustment caused by a combination of the associated adjustment speed μ and by a high value of the production elasticity $0 < B < 1$ with respect to labor. Since the elasticity of the demand for labor demand equals $-1/(B-1)$ the condition $\mu > 2(1-B) \iff \mu/(1-B) > 2$ states that the product of the two elasticities of wage adjustment and of wage response of labor demand must be larger than two. This translates into a linear negative tradeoff between adjustment speed and production elasticity portrayed in Figure 7.8 **b**.

When the steady state is Keynesian the period-doubling bifurcation requires large values of the adjustment parameter κ (downward flexibility) on the commodity market, causing the overshooting property via the commodity market. The necessary relation to the elasticity of labor cannot be estimated as directly as in the inflationary situation. The dependence is highly nonlinear. In Figure 7.8 **a** the stability boundary has been drawn qualitatively for a situation as $\kappa > \bar{\kappa} \approx 2(1 - B)$. Thus, in both cases, the two curves drawn should not be interpreted as the bifurcation curves. These lie above and to the right of the dashed line.

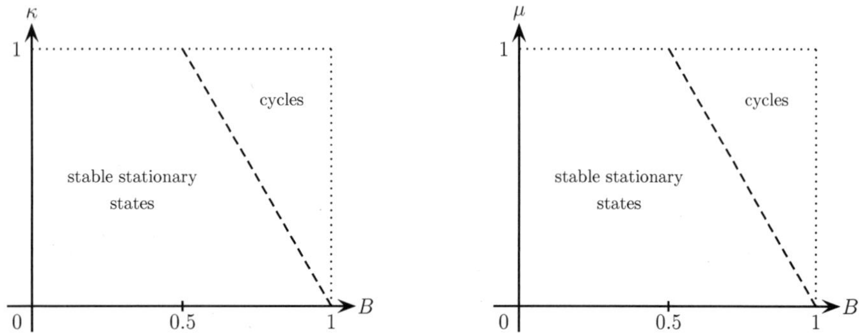

a Keynesian steady states $g < \tau y_{\max}$ **b** Inflationary steady states for $g > \tau y_{\max}$

Fig. 7.8 Regions of stability and of attracting cycles

After the period doubling the global behavior for a given labor elasticity and adjustment speeds is strongly influenced by the government parameters, causing a wide range of cycles and complex behavior as government demand and the tax rate take on different values. More complex bifurcation scenarios as well as multiple steady states may occur with an expectations feed back in consumption, i.e. when $\rho \neq 0$ (see Kaas, 1995), or when expectations effects are induced through endogenous labor supply. These are also confirmed by the numerical experiments in the next section.

It is one of the surprising findings of the following numerical analysis, that in some cases the period doubling bifurcation occurs quite near the necessary condition. For example, the numerical analysis shows that a period-doubling bifurcation occurs for some pairs of the fiscal parameters (g, τ) near the Walrasian steady state already for $B = 0.7$ and adjustment speeds $\lambda = \mu = \gamma = \kappa = 0.6$, or when $B = 0.8$, one observes period doubling bifurcations for adjustment speeds as low as $\lambda = \mu = \gamma = \kappa = 0.4$. This also occurs under endogenous labor supply (see Section 7.6) where period doubling occurs at lower adjustment speeds.

7.5.4 Endogenous Cycles and Complexity

Given Theorem 7.5.1 one may expect that the system will indeed generate endogenous business cycles for a large open set of reasonable parameter values, since the necessary condition for a period doubling bifurcation is easily satisfied. In spite of the fact, that the proof of the above theorem is fairly straightforward, it seems very difficult to obtain stronger analytical results. As a consequence a structural numerical analysis will be used to seek confirmation of the analytical properties, but mainly to provide further detailed insight into the global dynamic behavior and to understand the role of the parameters for the dynamic development of such economies[11]. First, however, consider the set of parameters

$$A = L_{\max} = 1,\ \delta = 1,\ \rho = 0,\ B = 0.5,\ g = 1,\ \tau = 0.5,\ \gamma = \kappa = \lambda = \mu = 0.5.$$

The values for $A = L_{\max} = 1$ are scaling parameters which have no influence on the dynamic features of the model except the location of the steady states and of the boundaries of the regimes. Choosing the consumption parameters $\rho = 0$ and $\delta = 1$, defines a benchmark case, since with the Cobb-Douglas utility function with equal time preference implies a constant propensity to consume equal to 0.5 with no expectations feed back. The next six values are just the midpoints of their allowable intervals. Given these, $g = 1$ yields a Walrasian steady state with half of total output consumed by the government. All numerical experiments with these values show convergence to the Walrasian steady state for any initial conditions (p_1, w_1, M_1). Notice that theses parameter values violate the necessary condition for the period doubling bifurcation. The same is true for parameter values 'close by' with any expectations memory between zero and twenty. No cycles were found. Thus, the steady state seems globally stable for all values near the benchmark case and the condition for period doubling seems to be 'tight'.

In contrast, when the adjustment speeds are simultaneously increased to 0.6 and the elasticity of production is increased to 0.9, a rich spectrum of endogenous finite and complex cycles is observed. For a discussion of situations with endogenous cycles, the following set of parameter values can be considered as a benchmark serving as a reference point for situations with endogenous cycles. The first set of

Table 7.1 Standard set of parameters used

A	$L_{\max}$	δ	ρ	B	g	τ	γ	κ	λ	μ
1.0	1.0	1.0	0.0	0.9	—	0.25	0.6	0.6	0.6	0.6

diagrams show some of the typical situations with finite cycles. They depict the

[11] The results are calculated using the C++-based software package MACRODYN (see Böhm & Schenk-Hoppé, 1998; Böhm, 2003).

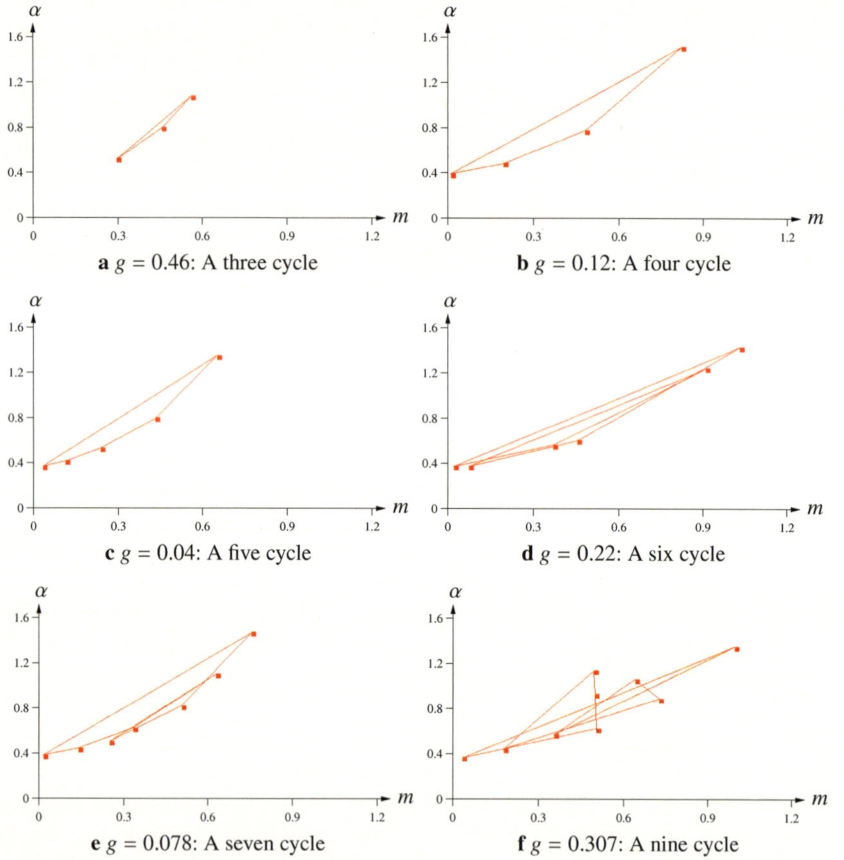

a $g = 0.46$: A three cycle

b $g = 0.12$: A four cycle

c $g = 0.04$: A five cycle

d $g = 0.22$: A six cycle

e $g = 0.078$: A seven cycle

f $g = 0.307$: A nine cycle

Fig. 7.9 Stable finite cycles: $B = 0.9$, $\tau = 0.25$, $\gamma = \kappa = \lambda = \mu = 0.6$

time paths for the employment level and for the two state variables (α_t, m_t), plus the attractor in state space (Figures 7.9 and 7.10). All cycles exhibit a counterclockwise orientation. The pattern of regime switching is quite different between them, as is the location of the steady state. Notice, however, that the parameter sets differ only in the level of government demand, g which induces a large variety of stable cycles of 'almost any' order.

If government demand is chosen to be $g = 0.28$ together with $\tau = 0.25$ one obtains a complex non-cyclical time series, as portrayed in Figure 7.11 for two different time windows. The value $g = 0.28$ is approximately equal to the number $g = \tau y_{\max} = 10/36 = 0.2\overline{7}$ which implies an associated Walrasian equilibrium as steady state with $(\bar{m}, \bar{\alpha})$ with $\bar{\alpha} = 1$ and

$$\bar{m} = \frac{1 - c(1 - \tau)}{B} - \frac{\tau}{B} = \frac{(1 - c)(1 - \tau)}{B} = 0.3375.$$

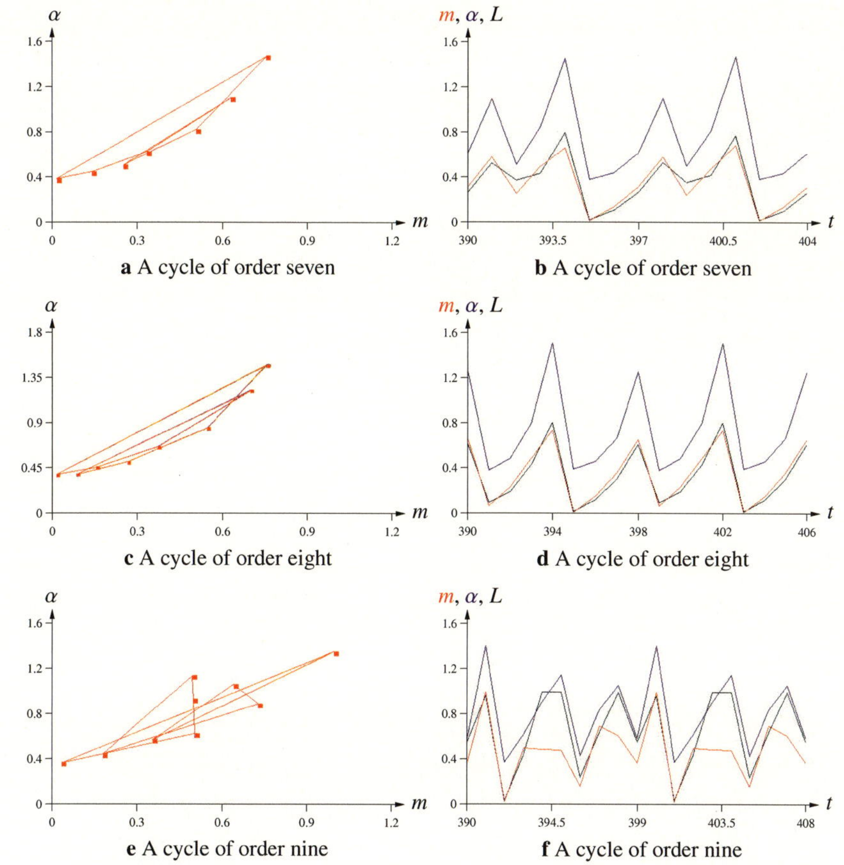

Fig. 7.10 Comovement of real wages, money balances, and employment in finite cycles

The time series show quite strikingly that the behavior is aperiodic, i.e. they do not show signs of regular periodicity. In addition, they reveal that the movement in the state space displays regional temporary persistence for time intervals of different length *and* non-periodic recurrence to regions. Similar non-periodic orbits can be shown to exist for all pairs (g, τ) satisfying $0.9g \approx \tau$. For $t \in [300, 400]$ high real wages correlate with low levels of money balances, indicating that the orbit moves predominantly in the Keynesian region. For $t \in [800, 1000]$, high money balances correlate with low real wages indicating that the orbit is predominantly in the inflationary region, features which can also be deduced from the shape of the attractor, see Figure 7.12 **a**.

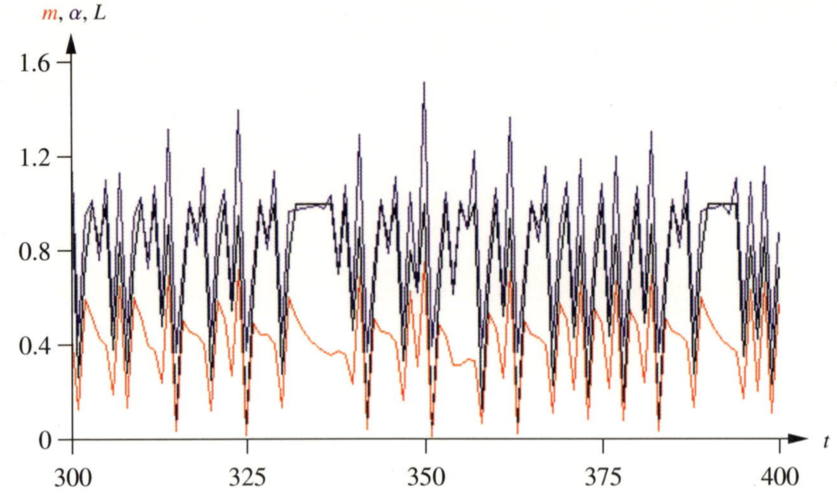

a Time window one of a complex time series

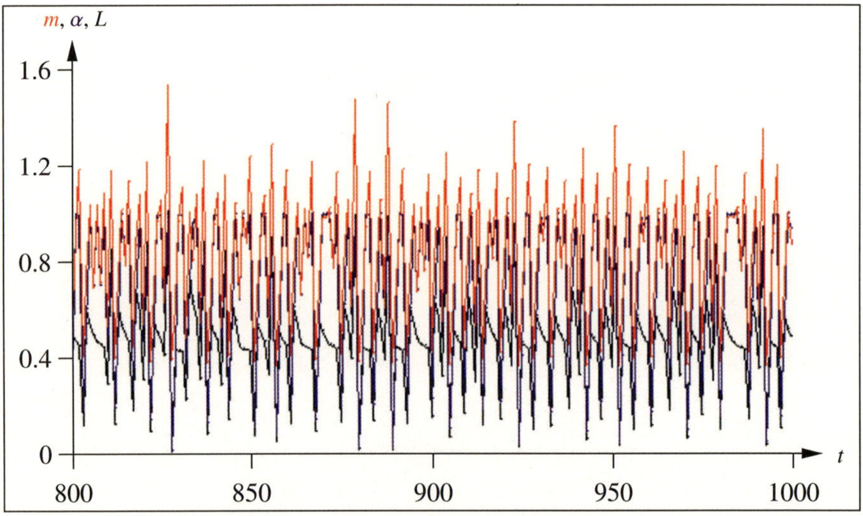

b Time window two of a complex time series

Fig. 7.11 Two time windows of a complex orbit; $g = 0.28$, $B = 0.9$, $\tau = 0.25$, $\gamma = \kappa = \lambda = \mu = 0.6$

Figure 7.12 displays the attractor in state space (panels **a** and **b**) for a time series of length $T = 10^5$ as well as some of the tradeoffs between inflation and employment. The complexity is also revealed by the histograms for the real wage, real money balances, and employment level of Subfigures **c**, **d**, and **h** showing high frequencies in the respective intervals.

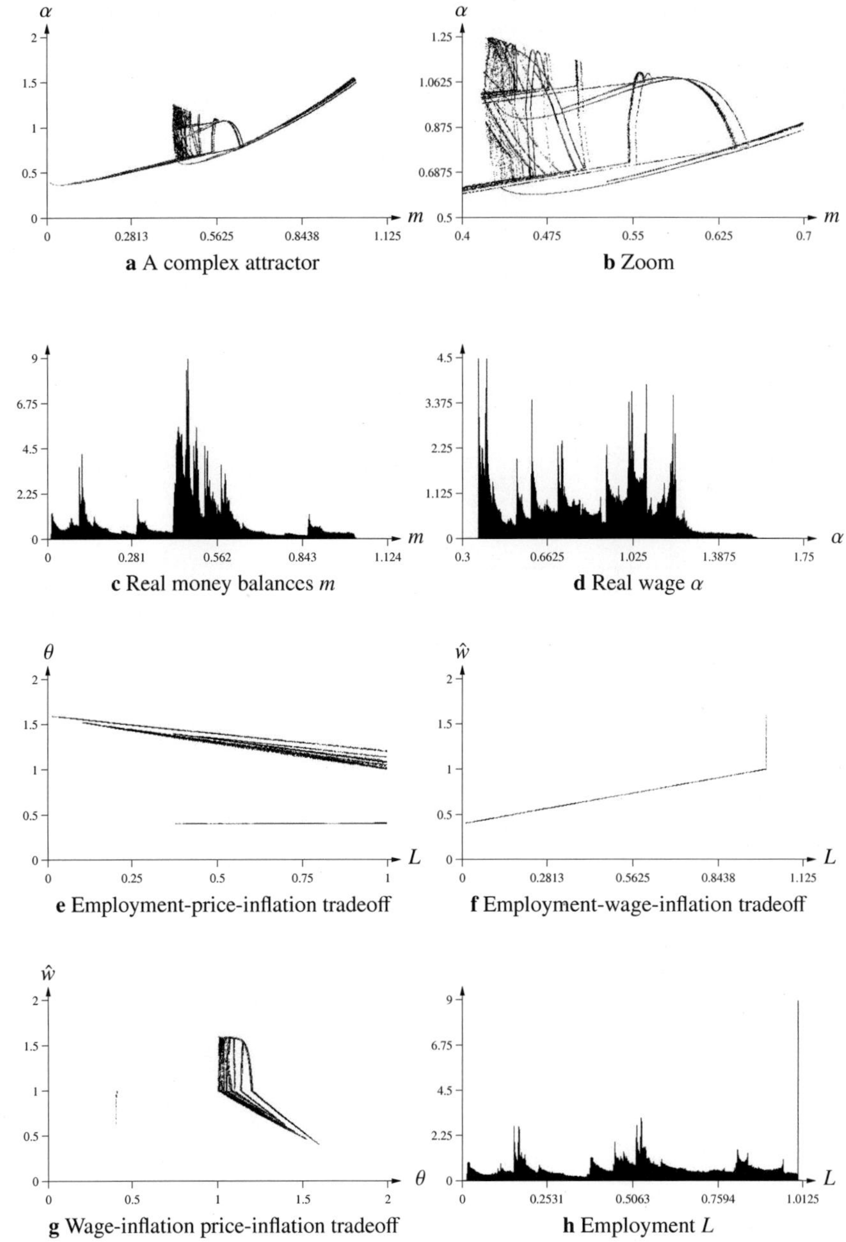

Fig. 7.12 Attractor and statistics of a chaotic orbit for $g = 0.28$, $B = 0.9$, $\tau = 0.25$, $\gamma = \kappa = \lambda = \mu = 0.6$; $T = 10^5$

Figures 7.13 and 7.14 present two bifurcation diagrams with government demand g and the tax rate τ as parameters keeping the elasticity of production at $B = 0.9$. The last 90% of a time series of length $T = 2000$ are shown where the colors indicate the relative frequencies (using the color coding from a topographical map: low=deep blue, medium=green, brown=high). In both diagrams the same level of the elasticity of production $B = 0.9$ is used as for the set of diagrams showing attracting cycles of different orders in Figures 7.9 and 7.10. Thus, their respective periodicity windows can be identified in Figure 7.13. These show that the fiscal parameters have a decisive influence on the type of the long-run behavior of the economy. The qualitative features of the analysis can be summarized as follows:

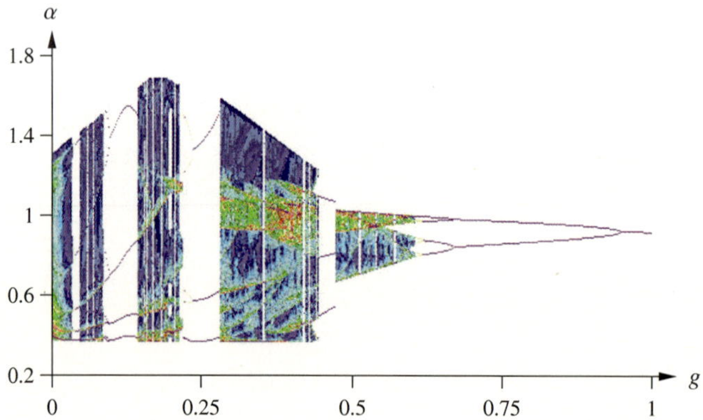

Fig. 7.13 g Bifurcation: $\tau = 0.25$, $B = 0.9$, $\gamma = \kappa = \lambda = \mu = 0.6$

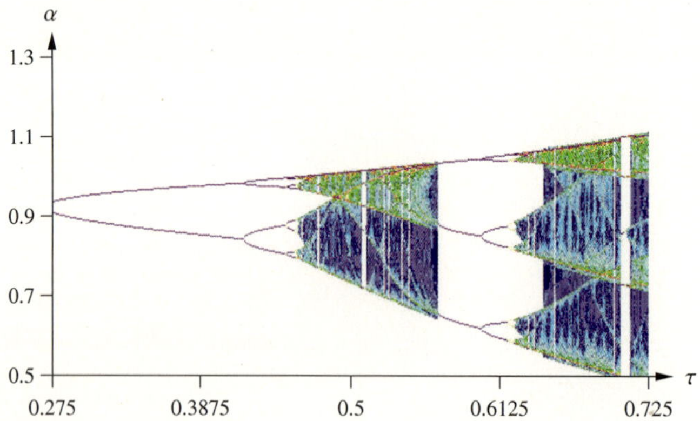

Fig. 7.14 τ Bifurcation: $g = 0.2\bar{7}$, $B = 0.9$, $\gamma = \kappa = \lambda = \mu = 0.6$

(a) There are large open sets of parameter values for which the unique steady state whether Keynesian of inflationary is stable and others for which it is unstable.

There seem to be no immediate heuristic or economically intuitive arguments which make stability or instability more plausible for certain sets than others, except those discussed in the context of Theorem 7.5.1.

(b) There are large sets of values for some parameters such that there exist stable cycles of order 2, 3, 4, 5, 7, 9, and higher, as well as their multiples, depending on the values of the remaining parameters. For example, for a set of $(\delta, \rho, \tau, B, \gamma, \kappa, \lambda, \mu)$ different values of (g, τ) generate the above mentioned cycles. Thus, an economist may conclude, that, other things being equal, different economic policies "cause" business cycles of different order. Figures 7.9 and 7.10 show different cases when the parameter g is changed only while all other values remain constant.

(c) Continuous changes of some parameters generate typical bifurcation phenomena of period doubling and the usual routes to chaos known from one dimensional systems with a period three cycle (see Figures 7.13 and 7.14) (see Li & Yorke, 1975). This is to be expected knowing that there exist stable cycles of different order depending on the values of one parameter, other things being equal. Hence, for small changes of some parameters the stability of a cycle of order k is lost and the system changes to a stable cycle of order $2k$ or $k/2$, as the case may be.

(d) For open sets of parameter values, no stable cycles exist and long iterated sequences show all features of irregular, possibly chaotic orbits. The associated trajectories induce complex attractors which are independent of initial conditions in most cases.

These numerical experiments reveal that all of the four dynamic features listed above can be obtained for systems with $\rho = 0$, i.e. for the simplest two dimensional macroeconomic system where expectations and memory do not matter. On the other hand no specific features have been detected in experiments with values for $\rho \neq 0$ and $\tau > 0$ which were not reproducible for a model with $\rho = 0$. Therefore, all further results reported here deal primarily with the case of $\rho = 0$ only. It is an open question to what extent expectations and memory in demand matter in this model, especially given the sensitivity of many standard macroeconomic models to those two elements[12]. Some numerical evidence of the role of expectations under endogenous labor supply is provided in Section 7.6. Economically the experiments confirm and extend the properties given by Theorem 7.5.1.

- High price and wage adjustment coefficients lead to instability in conjunction with a high elasticity of production B, causing an overshooting property of the wage dynamics.
- Therefore, the curvature of the production function plays a major role for instability, cycles, and possibly complex behavior.

[12] The role of the length of a perceived memory are ambiguous and not transparent. Some numerical results are reported in Lorenz & Lohmann (1996) which often contradict perfect foresight and are in contrast to early analytical findings; compare Grandmont & Laroque (1986).

7.5.5 Coexisting Attracting Cycles and Sensitivity on Initial Conditions

In addition to the possibility of high order stable cycles and complex orbits, the model also exhibits some cases with coexisting asymptotically stable cycles of different order. Thus, there is sensitivity of orbits with respect to initial conditions. Moreover, the associated basins of attraction exhibit a complex/fractal structure.

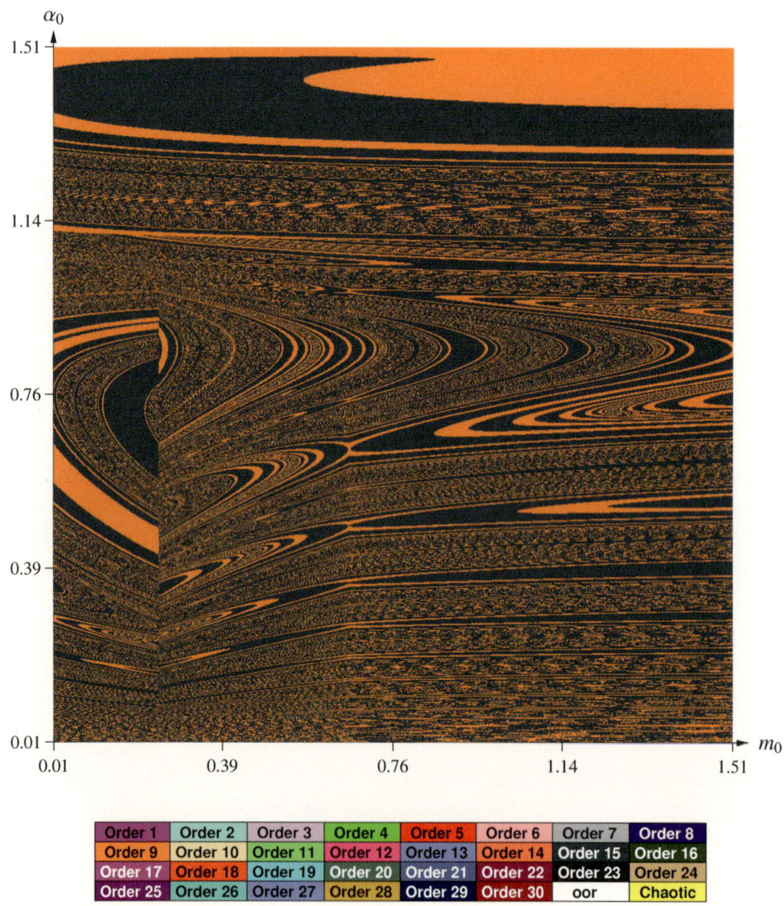

Fig. 7.15 Basins of attraction of coexisting cycles of order 9 and 15

Figure 7.15 displays the basins of attraction of two coexisting cycles of order 9 and order 15, which occur for the valus of parameters given in Table 7.2. The basins

Table 7.2 Parameters for coexisting cycles

A	L_{max}	δ	ρ	B	g	τ	γ	κ	λ	μ
1.0	1.0	1	0.0	0.9	0.45	0.25	0.7	0.7	0.7	0.7

for the two coexisting cycles have been calculated as a perodicity diagram (a cyclogram) for the space of initial conditions (m_0, α_0). They consist of nested self similar sets with a partial fractal structure. The diagram also shows clearly the boundary between the Keynesian and the inflationary region as a vertical line.

Figure 7.16 characterizes the sensitivity on initial conditions over the real wage α_0 as parameter in a bifurcation diagram for fixed real money balances m_0, display-

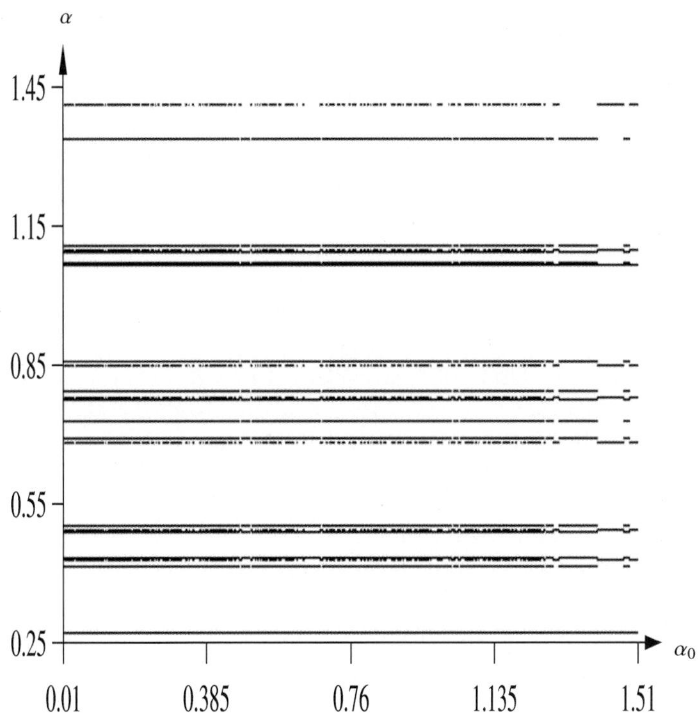

Fig. 7.16 Sensitivity on initial conditions of cycles of order 9 and 15

ing the discontinuities which occur when α_0 moves across the boundaries of the basins of the cycles.

7.5.6 Fiscal Policy and the Business Cycle

The bifurcation diagrams 7.13 and 7.14 reveal that the two fiscal parameters, the government demand g and the income tax rate τ, have a decisive impact on the dynamic behavior. They suggest that for given values of the adjustment speeds and the production parameter, it could be up to the government to decide whether the steady state is asymptotically stable or whether cycles occur. This possibility to control or determine the global behavior, however, depends in a monotonic way on the labor elasticity B. Figures 7.17 **a** to **f** display a sequence of periodicity diagrams (or bifurcation diagrams with the two parameters (g, τ)) displaying for each fixed vector of parameters $(g, \tau, B, \gamma, \kappa, \lambda, \mu)$ the long-run characteristic of an orbit for the fixed parameters by assigning a unique color according to the code given in subfigure **g**, i.e. a cycle of order $k \equiv$ color of order k or non-periodic≡yellow. A border line in parameter space between any two solid colors indicates the occurrence of an associated bifurcation as the parameter values (g, τ) cross the line. In particular, the line between the areas magenta and cyan (subfigures **a** to **e**) describe the bifurcation line where the period doubling occurs in the two regions[13].

The six diagrams display the respective periodicity features for different values of B, showing the qualitative behavior of the economy in the long run for all values of (g, τ) while holding all other parameters fixed. The line in the parameter space (g, τ) given by the equation $\tau = Bg$ marks the boundary between Keynesian and inflationary steady states since $A = L_{\max} = 1$.

The color magenta marks the region of government parameters where the associated steady state is globally stable. For all diagrams the adjustment speeds have been set uniformly to $\gamma = \kappa = \lambda = \mu = 0.6$. It follows from Theorem 7.5.1 that the critical values for the period doubling bifurcations are $\mu \geq 2(1 - B)$ or $\kappa \geq \bar{\kappa} \approx 2(1 - B)$. Since $\mu = 0.6 = 2(1 - B) = 2(1 - 0.7)$ no cycles exist for $B < 0.7$ and the steady of both types are globally stable. However, for $\lambda = \mu = 0.6$ with $B = 0.7$ and $Bg > \tau$ the first period-doubling bifurcation occurs in the inflationary region indicated by the line generated by the change of color from magenta to cyan[14]. Similarly, for $Bg < \tau$ and $0.6 = \kappa = \lambda = 2(1 - B) = 2(1 - 0.7)$ a loss of stability occurs in the Keynesian region, see Subfigure 7.17 **a**.

For larger B the diagrams portray the occurrence of further period doubling bifurcations including a large area of parameters with a stable cycle of order three for $B \geq 0.85$. As B increases, the regions of magenta become smaller and move

---

[13] The algorithm of MACRODYN uses a box counting procedure developed by Lohmann & Wenzelburger (1996) which identifies the order of the cycle or non-periodicity

[14] Since the condition for the bifurcation in the inflationary region is a linear one, numerically the bifurcation line appears as a piecewise linear relationship between g and τ for which the inflation rate is constant, see also Theorem B.3.1. This does not hold in the Keynesian region.

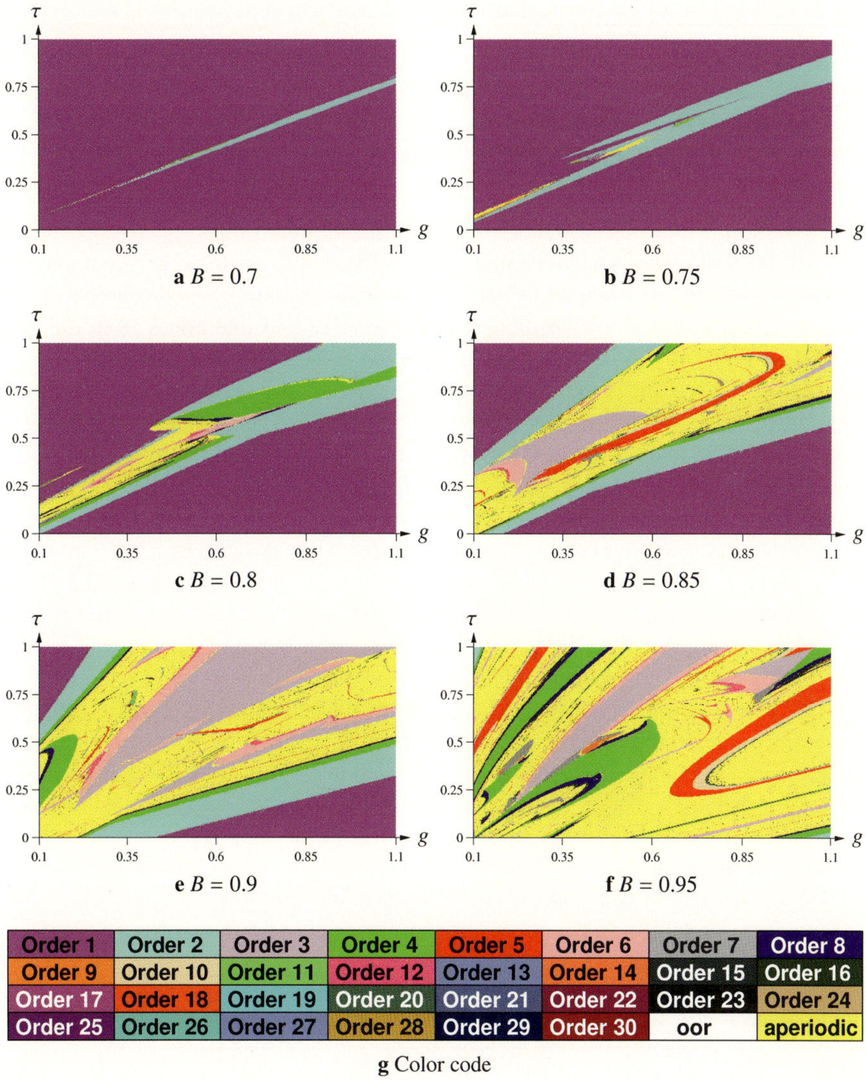

Fig. 7.17 Destabilizing impact of increasing labor elasticity; $\gamma = \kappa = \lambda = \mu = 0.6$

toward the extreme points $(g, \tau) = (1, 0)$, the extreme inflationary steady state or to $(g, \tau) = (0, 1)$, the extreme Keynesian steady state. For $B = 0.95$ no value of (g, τ) induces a stable steady state any more showing that the government looses such a control over the stability of the steady state completely. Thus, the economy exhibits endogenous cycles for any stationary government policy (g, τ). As a consequence, a stationary fiscal policy can asymptotic convergence. A stabilizing policy, if it existed, would have to be a non-stationary rule which would have to use structural information of the model as well as time series data.

7.6 Real Business Cycles with Endogenous Labor Supply

The assumption of constant labor supply within the analysis so far may seem restrictive for two reasons to determine whether the results on the role of parameters for the occurrence of business cycles as well as of the stabilizing/destabilizing effects of fiscal policies hold in the general case. The first one arises from the fact that under constant labor supply the expectations effect from the consumer side is eliminated. This reduces the dimension of the dynamical system to two which allows a geometric representation in the full state space of real balances and the real wage. Simultaneously, without the expectations effect the partition of the state space into the three regimes (as subsets of the state space $\mathbb{R}^2_+$) remains unchanged for any given set of parameters. Thus, the steady states and the type of switching between the regimes in cycles can be identified directly from the attractor in $\mathbb{R}^2_+$. When labor supply is endogenous both features are no longer true since the regimes are defined as open subsets in $\mathbb{R}^3_+$. Second, with constant labor supply actual employment (resp. actual output) in the inflationary regime is always constant and equal to $L_{\max}$ (resp. $y_{\max}$). This makes employment effects trivial while making all other real effects a simple redistribution in consumption between old and young consumers and the government. Therefore, it is of interest to examine whether the features of the real business cycles change in substantial ways when labor supply is endogenous. This allows for real cross market effects and endogenous levels of employment under demand rationing implying different characteristics of the long run behavior and the structure of bifurcations.

The parameters of the model chosen for the extension with isoelastic labor supply are given in Table 7.3. For programming reasons the parameter for the disutility

Table 7.3 Parameters for endogenous labor supply

A	δ	ρ	B	D	g	τ	γ	κ	λ	μ
1.0	1	0.0	0.7 – 0.9	0.2 – 0.99	0.5	0.35	0.6	0.6	0.6	0.6

of labor in Section 3.2.7 had to be renamed here to D from C where $1 + C$ was the elasticity of the disutility of labor implying an elasticity of the labor supply function of $1/C$. Therefore, within the numerical analysis of the remaining sections of this chapter $D \neq 0$ means that labor supply is endogenous with elasticity $1/D$ corresponding also to the elasticity of the aggregate labor supply function of Assumption 3.2.1. The case of $D = 0$ corresponds to the situation with $L_{\max}$ constant[15].

The determination of the impact of a positive elasticity of labor supply on the results of the previous section has been carried out through a systematic numerical search and rerun of the results for values of $D \in (0, 1)$. These indicate that there are no main qualitative structural breaks or discontinuities occurring for $D > 0$ versus $D = 0$. From a general point of view, however, two general observations can be made. The first one indicates that a positive elasticity induces a smoothing effect in bifurcations. The second qualitative observation is that the elasticity by itself is somewhat secondary and does not lead to major changes alone. However, its impact arises typically in conjunction with specific ranges of the elasticity of production.

7.6.1 Fiscal Policy and the Role of Endogenous Labor Supply

The numerical experiments reveal that the conditions for the determination of the type of the steady states and the local conditions for stability and the occurrence of a period-doubling bifurcation in the three dimensional case are essentially identical to the ones with constant labor supply. For all values of $D \in (0.0, 1.0)$ the equilibrium types are globally determined as in Theorem 7.4.1 and the local bifurcations are as given in Theorem 7.5.1. In other words, the government parameters (g, τ) determine the type of the steady state (via a modified result on the type uniqueness, possibly with coexistence of multiple steady states) while the elasticity of the production function in conjunction with the adjustment speeds of prices and wages determine the occurrence of the bifurcation. Figure 7.18 displays the characteristics of the real business cycles for six different values $D = 0.2, 0.5, 0.8, 0.9, 0.95, 0.99$ of the elasticity of labor supply and an elasticity of production $B = 0.7$.

As can be seen from the diagrams **a** to **f** there exists a narrow monotonic band of parameters with $Bg \approx \tau y_{\max}$ which divides the space (g, τ) into asymptotically stable Keynesian (above) and inflationary states (below). This band of values (g, τ) is relatively insensitive to changes of the parameter D. Thus, there exists a critical line with positive slope in the parameter space (g, τ) along which for each fixed $D > 0$ Walrasian steady states exist.

[15] Confusion with D as the symbol for the income consistent aggregate demand function in Chapters 3 to 6 should not arise.

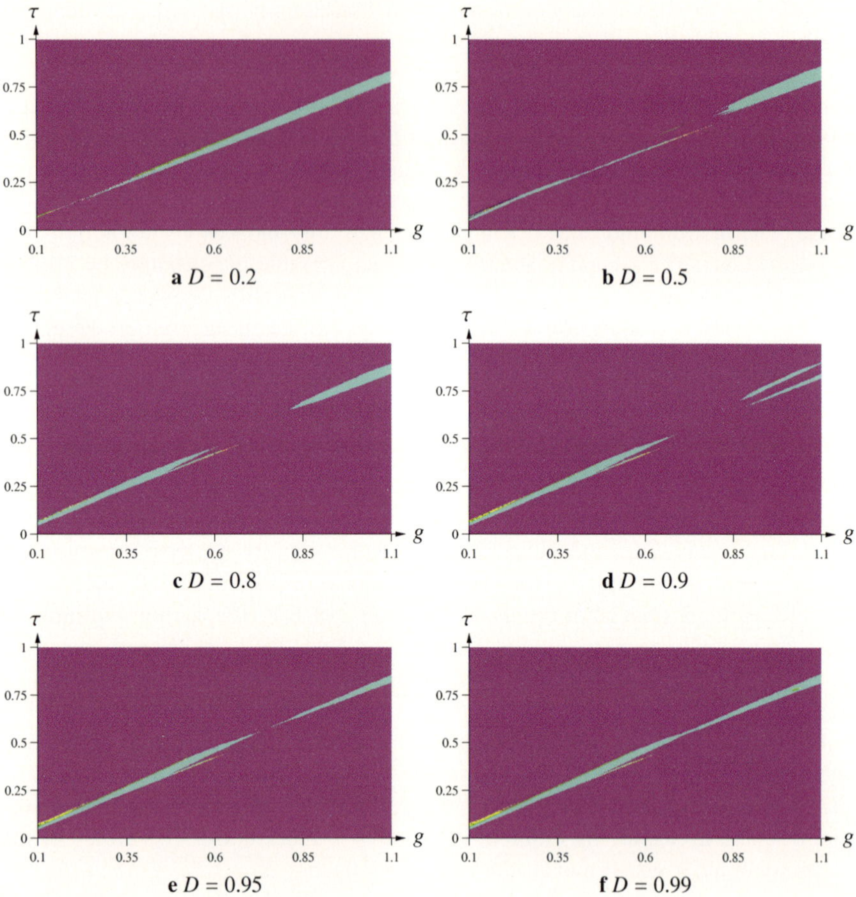

Fig. 7.18 Stabilizing fiscal policies for $B = 0.7$

Since the Walrasian steady states are asymptotically unstable they cannot be observed through forward iteration, but they could be determined numerically as the associated contour of the mapping $\mathcal{Y}$. For values below (above) inflationary (Keynesian) hyperbolic and asymptotically attracting steady states prevail. Thus, the dynamic characteristics of the economy for all values of the fiscal parameters correspond to the case with constant labor supply as in Figure 7.17 **a**.

As the value of the elasticity of production is increased to $B = 0.8$ all periodicity diagrams for $D \in \{0.0, 0.2, 0.3, 0.4, 0.5, 0.6, 0.7, 0.8, 0.9, 0.99\}$ maintain the same geometric structure. Six of them are shown in Figure 7.19. In particular, the bifurcation lines for the occurrence of period-doubling from steady state to a period-two cycle in both regimes are very similar to those under constant labor supply as in Figure 7.17 **b**.

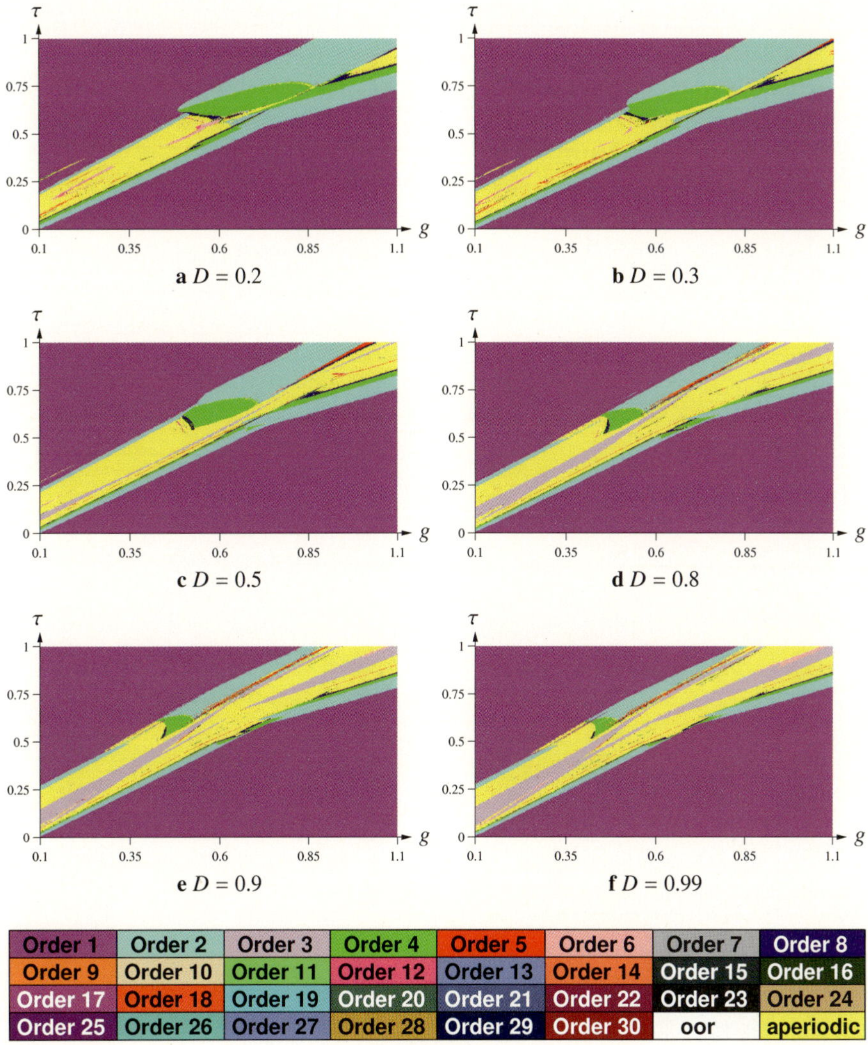

g Color code

Fig. 7.19 Stabilizing fiscal policies for $B = 0.8$

Finally, Figure 7.20 provides the comparison with the situation under constant labor supply, Figure 7.17 **f**. This shows that the effectiveness of a stabilizing fiscal policy is essentially driven by the elasticity of the production function B (as in the case with constant labor supply) and does not depend on the elasticity of the labor supply function. As a consequence all one-parameter bifurcations with fixed positive elasticity of labor supply plotting one of the parameters of the economy with constant labor supply against one of the two state variables, the real wage or real

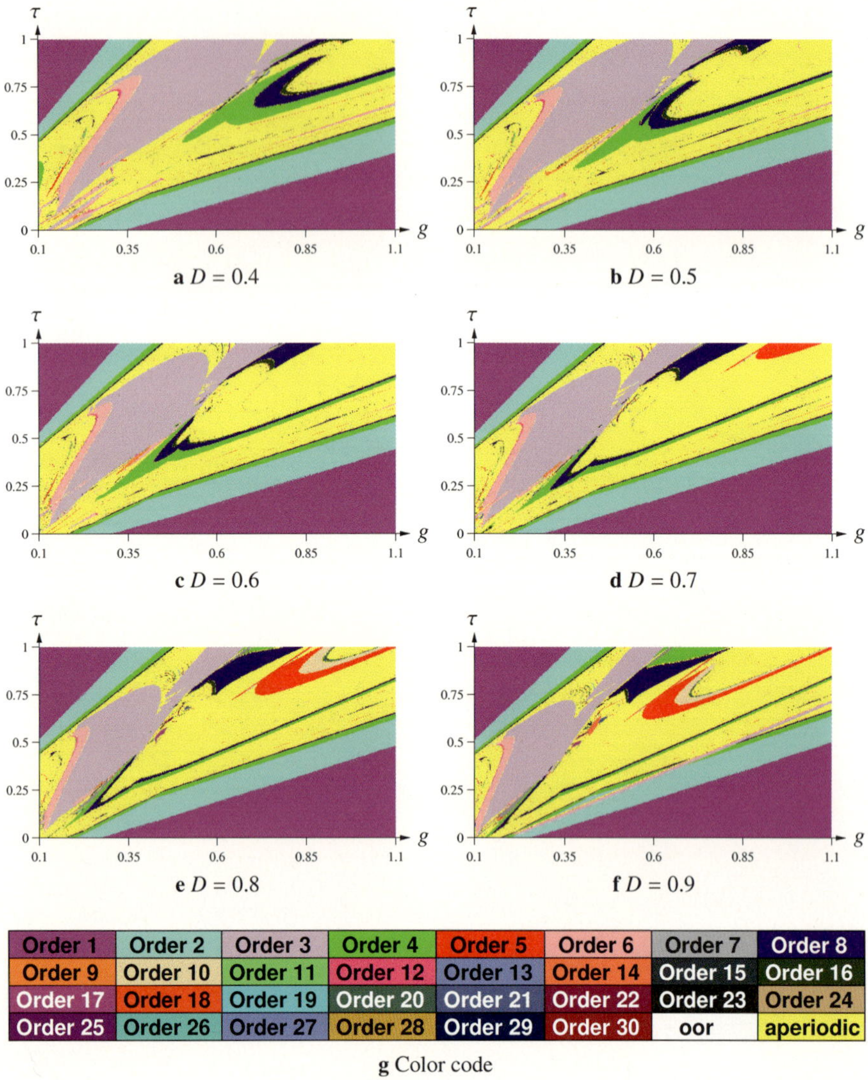

Fig. 7.20 Stabilizing fiscal policies for $B = 0.9$

money balances, would be qualitatively indistinguishable from the the case with oconstant labor supply. In particular, the bifurcation diagrams of Figures 7.13 and 7.14 with respect to government demand g and the tax rate τ would show the same cascades of periodic windows and of irregular cycles. Thus, the significance of the elasticity of endogenous labor supply by itself seems to be of a second order for the occurrence and the nature of the real business cycles of the prototype economy.

7.6.2 The Joint Role of Elasticities in Production and Labor Supply

A closer examination of the periodicity diagrams of Figure 7.20 reveals additional effects resulting from *joint* changes of the parameters B and D. While the bifurcation curves from steady state to a cycle of order two remain unchanged as D inreases, the areas and the period-doubling bifurcation lines for cycles from order four to order eight (green-to-blue) change substantially in subfigures **a** to **f**. At the same time, period-doubling windows for cycles of order five to order ten (red-to-beige) appear in subfigures **d** to **f** which are not present in **a** to **c**. This indicates the presence of joint effects between the two elasticities which cannot be distinguished from single bifurcation diagrams. In order to investigate these joint effects, con-

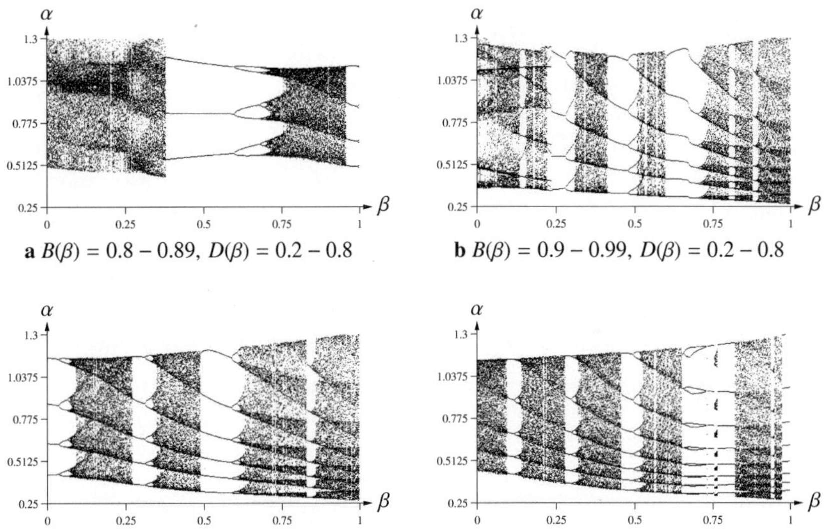

a $B(\beta) = 0.8 - 0.89$, $D(\beta) = 0.2 - 0.8$ **b** $B(\beta) = 0.9 - 0.99$, $D(\beta) = 0.2 - 0.8$

c $B(\beta) = 0.9 - 0.99$, $D(\beta) = 0.5 - 0.8$ **d** $B(\beta) = 0.9 - 0.99$, $D(\beta) = 0.9 - 0.99$

Fig. 7.21 Role of B and D on cycles; $\lambda = \mu = \gamma = \kappa = 0.6$

sider a simultaneous change of the parameters along the diagonal of a box of size $[B_{\min}, B_{\max}] \times [D_{\min}, D_{\max}]$, i.e. convex combinations of pairs of parameters, defined by

$$B(\beta) := \beta B_{\min} + (1-\beta)B_{\max}, \quad D(\beta) := \beta D_{\min} + (1-\beta)D_{\max}, \quad 0 \leq \beta \leq 1. \tag{7.6.1}$$

For a production elasticity $B > 0.8$ there exist a noticeable joint effect from a simultaneous change of D, see Figure 7.21. After each window with a period doubling of order three a cascade of regular windows of order four, five, six, etc. appear as β increases from zero to one. This corresponds to a so-called period-adding bifurcation which appears frequently in one dimensional piece-wise linear systems with one discontinuity (see Avrutin & Sushko, 2013). In other words, there are open sets of parameters $(B, D) \in (0, 1)^2 \subset \mathbb{R}^2_+$ for which regular attracting cycles of the corresponding stepwise increasing order and their period-doubling cascades occur. Thus, attracting cycles of many different finite orders exist for ranges of the elasticities which seem to be related to unknown structural features hidden in the model. No economic interpretation for these properties are known.

7.7 Keynesian Real Business Cycles Revisited

This chapter has provided a characterization of the features of the real dynamics[16] for the extended disequilibrium version of the basic model of Chapter 3 proposed in Chapter 6. It represents a consistent completion of a standard Keynesian macroeconomic model of a closed monetary economy built along microeconomic principles (such as those stipulated by Assumption 3.2.1). The time-one maps of each of the state variables are typically nonlinear under these assumptions. They inherit their main properties from the fact that the disequilibrium trading is specific to each regime which cannot be described by standard smooth functions as in most tâtonnement models. Therefore, the time-one map of the economic model is a nonlinear piecewise smooth system which in turn implies that the dynamics exhibit a variety of additional features which were not present in the analysis of the dynamics of temporary equilibria as in Chapter 4. As a consequence the dynamics in typical and interesting cases are characterized by several generic structural properties.

Switching between different regimes occurs along asymptotic orbits, necessarily so for any business cycles, a feature which arises typically in piecewise smooth systems.

The change of money balances is endogenous since real government demand is assumed to be given and expenditures are financed by a proportional income tax plus through money. This fully integrates fiscal and monetary policy in a consistent way, but it also implies that money balances change endogenously specific to regimes.

The price and wage adjustments respond sluggishly to excess demand signals on the two markets. Obviously, these are specific to each regime. Thus, the corresponding adjustment mechanisms are described by mappings which are specific to each regime as well.

Adaptive expectations of a simple statistical nature were chosen here, mainly because their role for the allocative features of the two-dimensional prototype model could be neglected. A variety of different heuristic, statistical principles, or learning

[16] The homogeneity of the system induces dynamics in $\mathbb{R}^4_+$ of nominal values with an unbounded range of the mappings as seen in Chapter 4. Their dynamic behavior has not been analyzed for the disequilibrium model here requiring an adaptation of the methods of Section 4.2 to investigate stability and convergence to expanding or contracting cycles.

rules could have been chosen as well. Their impact on the dynamics could have been examined also (as in Lorenz & Lohmann, 1996). Their effectiveness relative to the dynamics under perfect foresight was not the subject of this chapter.

As a consequence, the dynamics of prices and wages, of expectations, and of money balances induce changes of output, consumption, and employment over time. This gives rise to a consistent forward recursive model within which standard dynamic issues can be examined, e.g. whether the real stationary states are Walrasian equilibria, whether they are stable and monotonically converging, or whether there exist fluctuations in the medium run and in the long run. The dynamic evolution is defined and generated from fundamental principles of national income accounting, of feasibility, and of basic elements of temporary equilibrium analysis. Since identical rules and mechanisms are chosen in each period the model is stationary. The mathematical description of the evolution is given by a time-invariant mapping, making it an autonomous dynamical system on $\mathbb{R}^3_+$ with a large forward-invariant compact set.

In spite of its economic simplicity (no exogenous randomness, no inventories, no capital formation, no effects from differential consumption etc.), the structure induces a relatively complex nonlinear dynamic model which endogenously generates cycles of high order as well as irregular behavior. Moreover, it is shown that the structural causes of business cycles are embodied in the characteristics of the fundamental building blocks of an economy with optimizing agents and government activities, market feasibility, and the adjustment rules applied for prices and wages. This disproves a common supposition that the Keynesian model cannot be used for the analysis of business cycle phenomena. In contrast, the results here show that disequilibrium models of the Keynesian type, whether in this simple or in more elaborate versions, provide a valid theoretical framework for studies of the causes and features of the business cycle. In particular, the model provides an explicit framework within which the role of an active government policy on the steady states, their stability, and on the business cycle can be studied. The fact that the parametric version of the economy uses commonly accepted standard isoelastic functional forms underlines the importance of the general exercise. Given the list of finitely many parameters, one obtains a large open set of parametric economies for which *all* important dynamic issues can be studied using numerical techniques. This creates an experimental laboratory for the investigation of complex dynamic systems of higher dimensional spaces whose qualitative properties could not be obtained differently.

This chapter supplies an explicit view of the main dynamic features of the basic model. Many different issues and features could be investigated for the same model. One of the basic findings indicates that a stationary fiscal policy becomes a crucial bifurcation parameter which interacts with the parameters of the technology in a particular predictable way. Further investigations should be designed to determine whether there are others and what their relationship is. These findings contrast with most of the common characterizations of the role of fiscal policy shown in standard references (as in Blinder & Solow, 1973; Tobin & Buiter, 1976; Turnovsky, 1977, 1995). Often they are based on incomplete dynamic descriptions, restrictions imposing stationarity, or tâtonnement arguments. A more elaborate comparison identifying the differences in the modeling would be required to distinguish the sources for the different conclusions.

The role of the expectations feed back in consumption and in labor supply has been largely neglected. The question whether perfect foresight is possible has been left unanswered. The issue seems to be of a secondary nature when cycles occur. Since it is known that adaptive expectations themselves may be the source of cycles while under market clearing the rational expectations orbits are often unstable, the role of expectations formation in relation to policy as a source of business cycles might be more important (see Evans & Honkapohja, 2001). A comparison of results and methods with those used in Chapter 4 may lead to the appropriate answers.

Answering the stability issue will most likely also provide an explanation to the more general question why expectations in the Keynesian model seem to play such a secondary role, a fact which contrasts sharply with a large part of macroeconomic folklore based on the market clearing theory as analyzed in Chapters 3 and 4. Most importantly, the impact of a random environment for the disequilibrium analysis has not been discussed so far. Additional modifications of the results on the business cycles and their long-run characteristics are shown to exist when productivity shocks or fiscal demand shocks occur, which are analyzed in Chapter 8.

Chapter 8
Disequilibrium Dynamics with Random Perturbations

An analysis of the role of random perturbations for the dynamics of business cycles in macroeconomic models still represents one of the most challenging tasks in economic theory. The approach taken in Chapter 4 for equilibrium dynamics introduced some new and useful concepts which allowed a *dynamic* analysis of *stochastic* phenomena, i.e. the possibility to provide an investigation of stability and convergence properties of random time series as well as to exhibit their time-invariant statistical properties induced by the stationary orbits. They allow a complete analysis of random data under their transient as well as their stationary or time-invariant behavior within the same mathematical framework. The central issue here will be to apply these methods to the disequilibrium setting in order to characterize and show the main features of the cyclical nature of real and monetary phenomena resulting from the interplay of the time-one map under disequilibrium and random perturbations.

The primary purpose of the next sections is to examine the effect of random productivity on the behavior of the real sector of the nonlinear monetary dynamical system with sluggish price and wage adjustment (as in Chapter 7) and to compare and contrast the results with the methodology and findings of current theories of nonlinear stochastic business cycle models. It seems best and most convincing in situations when the associated deterministic dynamics are stable converging to a steady state rather than in situations when attracting cyclical or complex behavior occurs. In this case the question whether a linearization could be appropriate to describe the long-run stochastic dynamics might best be posed and answered.

In the literature, for most non-linear difference equations a linear approximation of the difference equations near the statistical mean of deterministic fixed points combined with an assumption of *additive* perturbations is used (for example Lettau & Uhlig, 2002, or many contributions to the literature) to characterize the statistical properties of the stationary solution of the nonlinear attracting random system. The justification of the validity of such an approximation is based on the assumption that the limiting behavior of the nonlinear random system under small enough perturbations can be approximated by (is approximately similar to) the one generated by small random displacements of the deterministic fixed points.

The intuition supporting this assumption asserts the existence of a *convexification effect* induced by the random mixing of fixed points which is assumed to occur *and* which includes/intersects sufficiently the limiting set (the attractor and support of the measure) of the stationary solution of the true nonlinear system. In addition, the moments of the true and the approximating distributions are claimed to be close to each other so that they can serve as empirical estimates of the true distribution.

Unfortunately, in general, the mixing of the fixed points does not generate a convexification which is related in a systematic way to the support of the actual stationary solution of a nonlinear random difference equation. Thus, it is questionable whether such an approximation provides reliable statistical characteristics and it seems difficult to identify structural conditions under which such an approximation may be appropriate.

The origin of this deviation lies in the fact that the long-run stochastic behavior consists always of a joint interaction of the *dynamical mapping* **and** of the *random perturbation*. This is true even in situations when the perturbation is additive and small. At best the first moment (the statistical mean) of the two invariant distributions are approximately equal. In general, however, the limiting behavior of the nonlinear system with small noise often has no resemblance with the convexifying effect caused by the randomization of a set of deterministic fixed points (see Böhm, 1999, 2001). It will be shown for some simple examples that the convexifying effect is not present implying that the true dynamics are quite different from a linear approximation near steady states and associated randomization.

8.1 Random Productivity with Constant Labor Supply

8.1.1 Hicks Neutral Productivity Shocks $A(\omega)$

Consider first a production shock which changes the productivity via a random scale effect as in Section 4.3 (see Böhm, 2001). The parameter A is assumed to follow a stationary random process with values in some strictly positive interval. As a consequence, the evolution of the real variables of the economy whether in permanent equilibrium or in disequilibrium is described by a two dimensional stochastic difference equation (a random dynamical system with real noise in the sense of Arnold, 1998). Since the perturbation is assumed to be i.i.d. the dynamical system becomes a Markov process.

The long-run development is well understood for systems consisting of random attracting affine mappings. In this case the long-run behavior is determined by a stable random fixed point, the random analogue of a fixed point or stationary state of a deterministic system (see Böhm, 2006, for an application to the multiplier-accelerator model of Samuelson, 1939). Since the system here is nonlinear and only piece-wise differentiable, these results are not directly applicable. However, a careful numerical analysis of the model here provides reliable qualitative characteristics indicative enough of the long-run invariant behavior of the system.

Due to the multiplicative form of the production shock the parametric influences of changes of A can be traced out analytically for the set of induced deterministic dynamical system. Notice first that the level of full employment output $y_{\max}$, the real wage α^*, and real money balances m^* at temporary equilibrium are linear functions of the production shock A

$$y_{\max} = \frac{A}{B}(L_{\max})^B \quad \Longrightarrow \quad \alpha^* = A(L_{\max})^{B-1}, \quad m^* = \frac{\tilde{c}A}{B}(L_{\max})^B - g \tag{8.1.1}$$

with $\tilde{c} := 1 - c(1 - \tau)$, implying that the random equilibrium set of the economy for all A is a subset of a straight line with *positive* slope in the state space,

$$\alpha^* = \frac{B}{\tilde{c}\, L_{\max}}(m^* + g), \tag{8.1.2}$$

which is independent of A. Therefore, the random displacements of the Walrasian equilibrium of the economy $(m^*, \alpha^*)(A(\omega)) \in \mathbb{R}^2_+$ occur on a one-dimensional *deterministic* set in the state space. As a consequence, the random equilibrium dynamics under productivity shocks and rational expectations induces orbits of real balances, real wages, and of output which are perfectly correlated, and which are generated by linear equations with multiplicative noise. Thus, their time series are linear images of the noise process only. Moreover, the money balances, nominal GDP, and the price level follow random walks. Second, under the disequilibrium adjustment, due to the type uniqueness of the stationary state, changes in A induce a nonlinear displacement of the steady state *with* possibly a change of the disequilibrium type.

Discrete Productivity Shocks: An Example

Assume as a first case that $A(\omega)$ takes on three distinct values $0 < A_1 = 0.99 < A_2 = 1 < A_3 = 1.01$ with equal probability $1/3$ while the remaining parameters are as in Table 8.1. Each of the three parameter configurations have an associated

Table 8.1 Values of parameters for discrete random scale effect $A(\omega)$

$c = 10/13$	$L_{\max} = 1$	$B = 0.7$	$g = 0.5$	$\tau = 0.35$
$\tilde{c} = 0.5$	$\gamma = 0.6$	$\kappa = 0.6$	$\lambda = 0.6$	$\mu = 0.6$

stationary state, where (m_1, α_1) is of the inflationary type **I**, (m_2, α_2) is a Walrasian equilibrium **WE**, and (m_3, α_3) is of the Keynesian type **K** where **WE** is the mean of the three equilibria. In addition, $B = 0.7$ and $\lambda = \mu = \gamma = \kappa = 0.6$ implies $2(1 - B) = 0.6 = \lambda = \mu$. Thus, Keynesian and inflationary states associated with different values of A are asymptotically stable in the disequilibrium dynamics (see Theorem 7.5.1).

The resulting random dynamical system consists of an i.i.d. mixing of three mappings whose stationary states are asymptotically attracting. According to the general intuition for random dynamical systems, the associated random orbits are converging if the perturbation is small. The example assumes a variation/deviation from the Walrasian level A_2 of $\pm 1\%$. However, the results shown are robust in a neighborhood of the parameter values chosen, in particular for $0 < A_2 - A_1 < \epsilon$ when the difference is smaller. For arbitrary initial conditions the dynamics converge globally to an invariant set of the type analyzed below, showing also convergence to a well defined invariant distribution with computable moments and quantiles in all experiments.

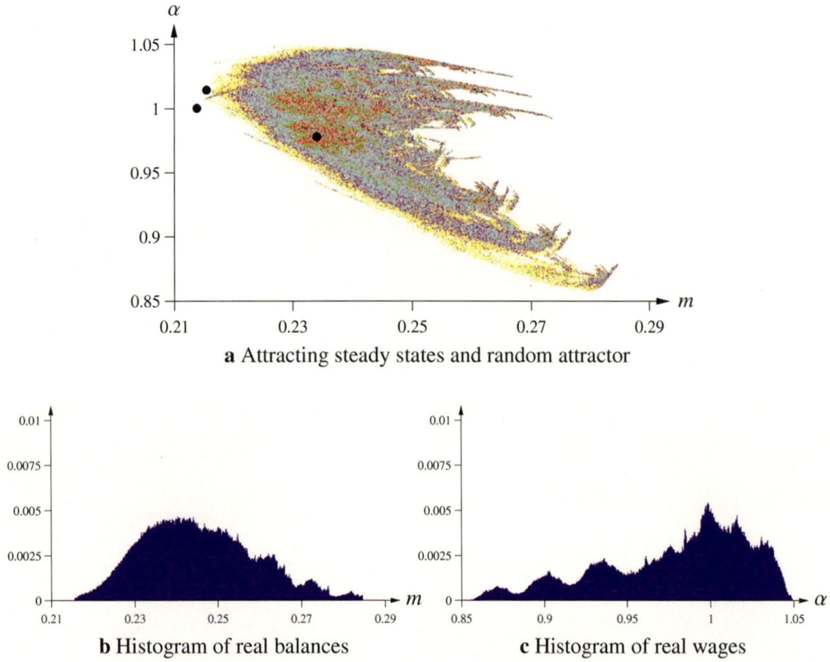

Fig. 8.1 Hicks neutral productivity shock: $A(\omega)$ uniform i.i.d. ~ $\{0.99, 1, 1.01\}$; $T = 10^6$

Figure 8.1 panel **a** shows the attractor (the support of the invariant distribution) of the long-run development of the economy in state space with the three steady states of type **WE**, **K**, **I** (from left to right) inserted as black dots. The three fixed points (α_i, m_i), $i = 1, 2, 3$ define a nondegenerate triangle in the state space where the Walrasian equilibrium is clearly is not the mixing center under the probabilities 1/3 each. The plot shows the values of the state variables (m_t, α_t) for the last 90 % of a series of length $T = 10^6$ iterations with increasing density/frequency from low=yellow-to-light-red-to-cyan-to-green-to-dark red=high.

The attractor is of a specific nonlinear form not related or similar to a rectangle or parallelogram which would be typical for an attractor of an AR-1 system with additive noise. Most importantly, only the inflationary steady state with the lowest real wage α_3 associated with $(\alpha_3, m_3) \in \mathbf{I}$ lies fully inside the attractor, while $(\alpha_2, m_2) \in \mathbf{WE}$ is outside the attractor and the state $(\alpha_1, m_1) \in \mathbf{K}$ is on the border of the attractor. Nevertheless, the triangle $(\alpha_1, m_1) - (\alpha_2, m_2) - (\alpha_3, m_3)$ intersects the attractor, but its barycenter lies near the boundary of the attractor.

Given equal probability for all three values (α_i, m_i), $i = 1, 2, 3$, the Walrasian steady state is the mean of the three associated Walrasian equilibria, but clearly not the mean of the three steady states. Thus, the attractor is clearly not contained in the convex hull of the three associated Walrasian states (which lie on a straight line with the Walrasian one as the midpoint). It is also not a subset of the convex hull of the three associated stationary states (which form a triangle) and therefore clearly not a "noisy" steady state surrounding the barycenter of the triangle.

Surely, neither the convexified set of the Walrasian equilibria (α_i, m_i), $i = 1, 2, 3$, nor the convexified triangle of the three disequilibrium steady states can be reasonably considered an acceptable approximation of the true invariant set. An analysis hypothesizing that the long-run sample path is a mixing of the three attracting steady states systematically leads to wrong conclusions about the generating process from any data sample. The true long-run development cannot be described even approximately as a "noisy steady state" (a mixture) of the three Walrasian equilibria or of the stationary states. This observation, which is known for stationary solutions of stochastic difference equations, confirms one of the fundamental insights into the properties of stable long-run behavior in random dynamical systems with nonlinearities or non additive noise: even under the strongest contraction properties of the family of mappings and the simplest i.i.d. noise process, their *dynamic* randomization leads to attractors and invariant distributions which bear little resemblance with a *randomization of static fixed points*. The structural insight underlines the importance of the mixing of different attracting *processes* which play the important role in the long-run dynamics determining a genuinely different invariant distribution. From an economic point of view, the conceptual fallacy using a 'steady state approximation' in a dynamic economy is caused by ignoring the importance of the structural features of the adjustment process. Thus, a time series analysis of an orbit of a random dynamical system which estimates the model using an approximation via fixed points will typically fail to obtain unbiased results.

Panels **b** and **c** of Figure 8.1 display the marginal densities of the invariant distribution. In Table 8.2 some standard statistical measures (mean, variance, skewness, and excess kurtosis)[1] are calculated. These exhibit features contradicting normality as well as uniformity of the underlying limiting process which would be the case qualitatively for an AR-system with additive noise.

Finally, Figure 8.2 shows part of the sample path of employment after subtracting a transient phase and the long-run density, which has a long-run mean = 0.974771, a variance = 0.0016634, and standard deviation = 0.0407848.

---

[1] For definitions and recursive calculations of moments see Appendix A.

Table 8.2 Moments for $A(\omega)$ uniform i.i.d. $\sim \{0.99, 1, 1.01\}$; $T = 10^6$

	mean	variance	standard deviation	skewness	kurtosis
real balances	0.245631	0.000183807	0.0135576	0.390719	0.424401
real wage	0.978802	0.00194908	0.0441483	-0.692988	-0.37539

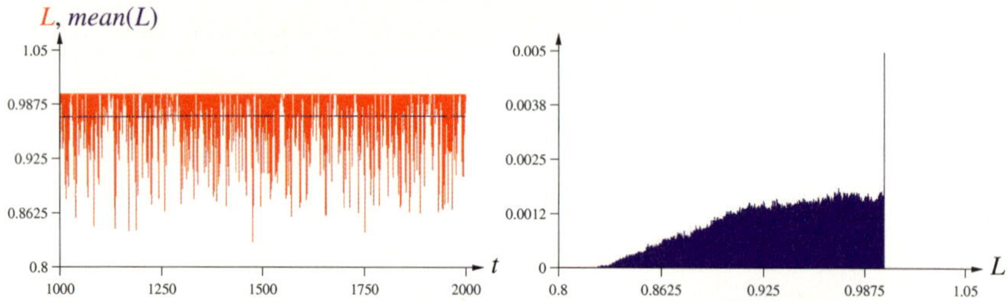

a Sample path **b** Density/histogram

Fig. 8.2 Employment under Hicks neutral productivity shock: $A(\omega)$ uniform i.i.d. $\sim \{0.99, 1, 1.01\}$; $T = 10^6$

To complete the qualitative analysis with discrete shocks, additional statistical/qualitative properties of the stationary solution are shown describing the trade-offs between employment and the real wage and between employment and price inflation. Figure 8.3 panel **a** shows a clear positive correlation between the real wage

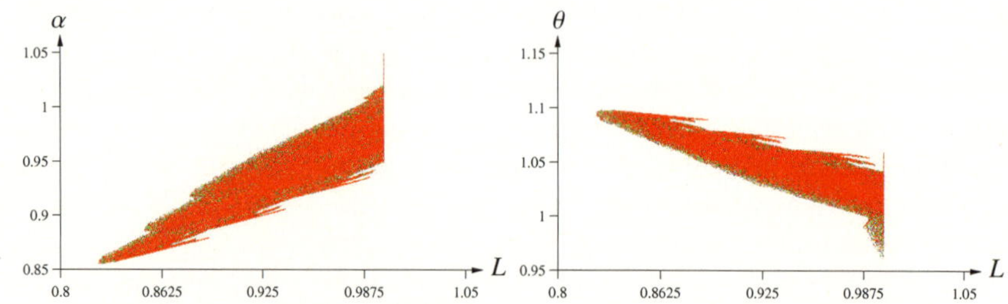

a Real wage-employment correlation **b** Inflation-employment correlation

Fig. 8.3 Hicks neutral productivity shock: $A(\omega)$ uniform i.i.d. $\sim \{0.99, 1, 1.01\}$; $T = 10^6$

and employment. i.e. high levels of employment are related to high real wages empirically, a relationship opposite to the labor demand function of the producer. Similarly, there is a clear negative correlation between price inflation and employment shown in panel **b**. Thus, statistically one observes a negatively sloped Phillips curve, a result opposite to comparative statics finding when comparing Keynesian steady states. Together these results indicate that random perturbations in a dynamic setting often induce long-run properties which do not coincide qualitatively with the comparative statics effects derivable for the associated deterministic models. In other words, the observable statistical correlations between variables which are not state variables describe features of the time-invariant mappings determining these variables which are state dependent, but these are without a delay effect or dynamic functional relationship from the dynamical system. Therefore, for models within this class the tradeoff between employment and inflation is a static/stationary relationship whose properties derive from the dynamics in the state space and not from a *dynamic relation* between employment adjustments and responses to expected inflation rates.

Smooth Productivity Shocks

It is known that dynamical systems with discrete perturbations may generate attractors which are Cantor sets whose associated invariant distributions have no densities (occurring for so-called *iterated function systems*, see Barnsley, 1988). Therefore, it is sometimes argued that random difference equations with smooth perturbations behave quite differently generating more appropriate statistical properties for an empirical/econometric analysis. For the present model this is not the case. The long-run behavior of the current Keynesian disequilibrium model remains essentially the same if the production shocks are changed from a discrete uniform distribution over the three points $\{0.99, 1, 1.01\}$ to a uniform i.i.d. distribution over the interval $[0.99, 1.01]$ with the same mean.

Figure 8.4 displays the attractor and the statistical properties of the state variables and the employment characteristics in the uniform smooth case. There is, as there should be, a noticeable smoothing effect of the attractor which induces a smoothing of the distributions of all other random variables. Notice, however, that the location of the attractor and its general shape does not change much. Moreover, except for the smoothing effect, the distributions are very similar, a fact also reflected in the similarity of the observed moments, see Table 8.3. Moreover, the statistical correlations between employment, the real wage and the inflation rate are essentially identical to the ones in the discrete case. Thus, observable time series induced by multiplicative production shocks, whether discrete or smooth, are indistinguishable for sample paths except for the smoothing effect in the distributions.

Table 8.3 Moments for $A(\omega)$ uniform i.i.d. $\sim [0.99, 1.01]$; $T = 10^6$

	mean	variance	standard deviation	skewness	kurtosis
real balances	0.245631	0.000183807	0.0135576	0.390719	0.424401
real wage	0.98149	0.00149129	0.0386172	-0.667982	-0.493506
employment	0.977788	0.0012058	0.0347247		

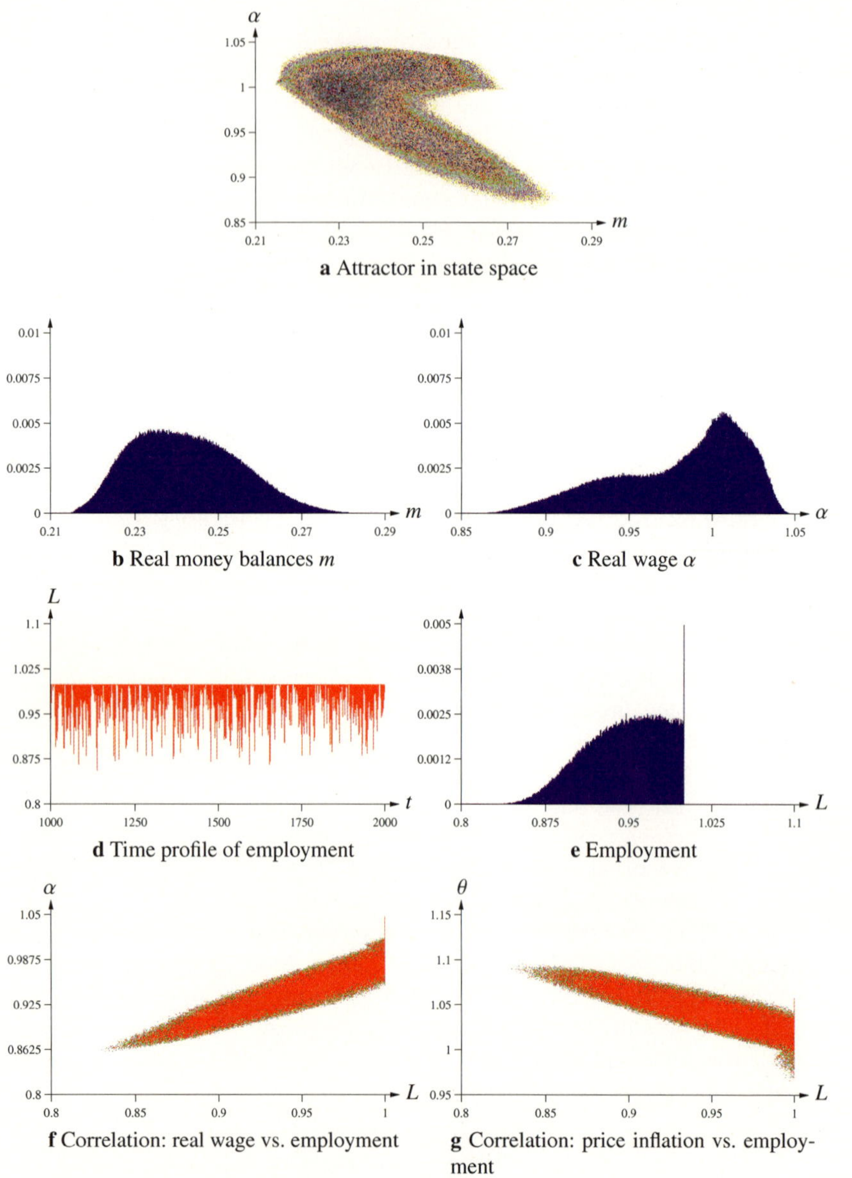

Fig. 8.4 Hicks neutral productivity shock: $A(\omega)$ uniform i.i.d. $\sim [0.99, 1.01]$; $T = 10^6$

The Role of Parameters: Stochastic Bifurcations

The previous experiments were directed toward understanding the role of noise near the Walrasian steady state given by $B = 0.7$ and $g = 0.5, \tau = 0.35$ with values of $A \in [0.99, 1.01]$ (Table 8.1) while keeping the levels of the two government parameters (g,τ) fixed. However, as shown in detail for the deterministic case in Section 7.5.6, the fixed level of government demand has a decisive influence on the type of the business cycle, i.e. on the periodicity and the distribution in the state space. Similar effects have to be expected with random productivity, whether they are operating on the scale or on the elasticity of the technology.

In the noisy case it is not useful to consider periodicity diagrams in parameter space (as for example Figure 7.17) since stationary random solutions for stochastic difference equations do not correspond to cycles of finite order. Instead, colors can be used to indicate the *ranges* of frequencies occurring for the stationary distribution in the state space, as in Figure 8.4 **a**. However, when the noise process is scaled continuously within the model the numerical analysis sometimes reveals situations of particular finite integer order of the (marginal) distributions of some of the variables which have their counterpart in the deterministic case when the noise becomes small and reduces to zero.

In order to study the influence of a *deterministic* parameter on the stationary evolution of the model when a second parameter undergoes permanent randomization, it is useful to carry out a bifurcation analysis with the restriction that for each level of the deterministic parameter the experiment is executed for the same initial condition of the state variables in the economy *and* for the same seed of the noise path of the randomized parameter. In this case, a bifurcation diagram portrays the isolated influence of the parameters on the long-run development of the state variables.

The Role of the Elasticity of Production

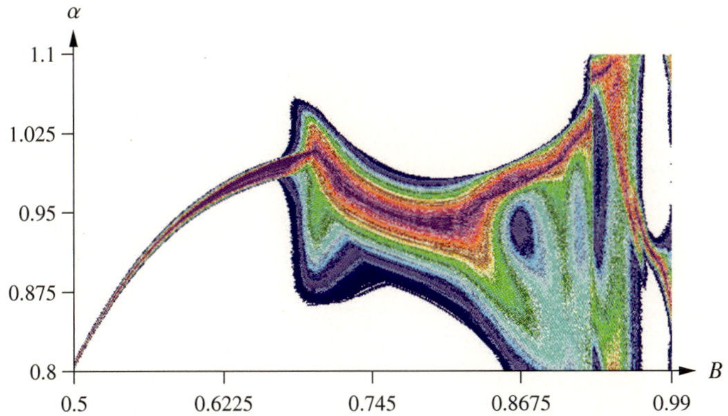

Fig. 8.5 Role of the production elasticity B under Hicks neutral productivity shock: $A(\omega)$ uniform $\sim [0.99, 1.01]$

Figure 8.5 portrays the effect of different levels of the production elasticity on the long-run behavior of the real wage. It follows from the Stability Theorem 7.5.1 and the type-uniqueness that, for each value of A, the steady state is of the Keynesian type and is locally attracting for all $B < 0.7$. In addition, for the given size of the small noise the attraction coefficients are stronger as B is further away from the deterministic bifurcation point and from the boundary of the Keynesian region. Thus, locally, no regime switching occurs for low values of B and stochastic movements will be confined to a neighborhood of the set of parametric Keynesian states in that region alone. As B increases and approaches the critical value 0.7 the support of the stationary solution expands and the overshooting property of the wage mechanism induces regime switching, additional random fluctuations, and a widening of the support of the invariant distribution, as shown by the diagram. The color coding chosen corresponds to the common height indicator of a topographical map: from high=purple>red>yellow>green>cyan>light blue>dark blue=low. Thus, the widening of the attractor is a clear indication of frequent stochastic regime switching primarily between **K** and **I**. Observe that the fluctuations of the real wage are noticeably stronger than the fluctuations of $\pm 1\%$ of $A(\omega)$ which is equal to the level of the random Walrasian real wage, since $L_{\max} = 1$. As the elasticity of production becomes large fluctuations widen considerably (more than 10% above and below the Walrasian level), multimodal distributions occur, and orbits remain predominantly in the inflationary regime. These issues will not be pursued here (see Sections 8.2.1 and 8.2.3 below).

The Role of Fiscal Policy under Productivity Shocks

An execution of the stochastic bifurcation analysis to the policy parameters is straightforward and reveals additional structural features of the effects of policy in a stochastic environment. Figure 8.6 shows the implications of different levels of

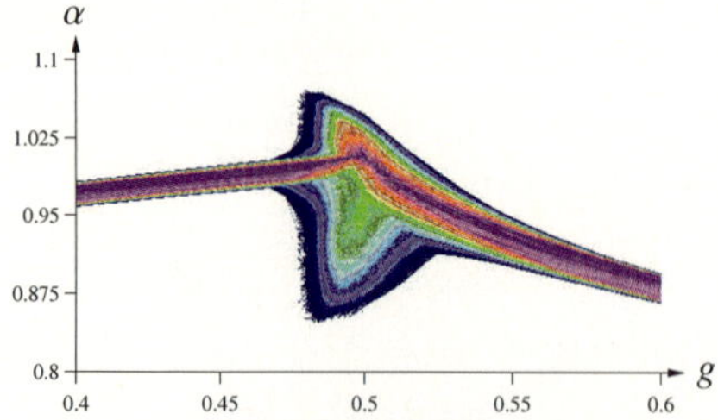

a Local role of g with $B = 0.7$

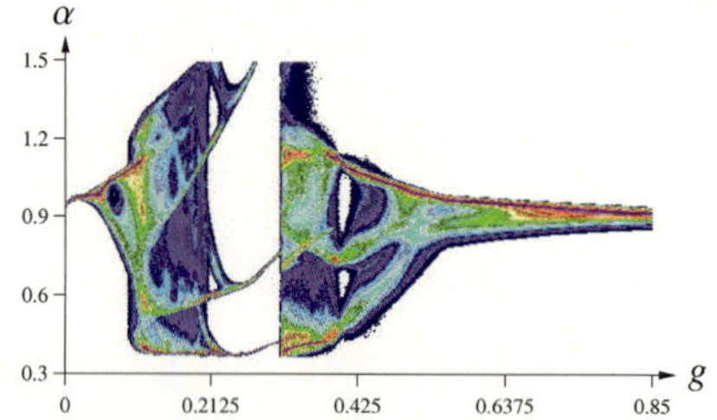

b Global role of g with $B = 0.9$

Fig. 8.6 Role of government demand under Hicks neutral productivity shock: $A(\omega)$ uniform $\sim [0.99, 1.01]$; $\tau = 0.35$

government demand and the interaction with the elasticity of production. Subfigure **a** characterizes the relatively strong local attraction property for a low value of the production elasticity B. It identifies the regimes switching situations in an interval near the Walrasian equilibria for values of $g \approx 0.5$, as strongly attracting Keynesian states for g to the left of the interval and strongly attracting inflationary states to the right.

However, for $B = 0.9$ the global bifurcation scenario changes substantially and the Subfigure **b** could be compared with corresponding Figure 7.13 in the deterministic case, although the two tax rates are not the same, $\tau = 0.35$ in the stochastic and $\tau = 0.25$ in the deterministic case. The difference has little significance for the comparison to be carried out.

First, since $A(\omega) \in [0.99, 1.01]$, the full employment real wage equals $\alpha^*(\omega) = A(\omega)$ since $L_{\max} = 1$ which gives $y_{\max}(\omega) = A(\omega)/B$. Second, the condition for a Walrasian equilibrium is $g = \tau y_{\max}$ which occurs in the deterministic case at $g = 0.25/0.9 = 0.2\bar{7}$ and for the stochastic case in the range $g \in [0.31185, 0.392\bar{7}]$. For government demand larger (smaller) than these values steady states are inflationary (Keynesian) respectively.

Comparing the two bifurcation diagrams there are significant similarities and differences. In particular, the economy is stabilized in the inflationary region in both cases for high values of government demand and real wages below the Walrasian wage. For lower values of g employment fluctuations persist in both cases with two significant intervals of government demand where the frequency is of order three. There are two parametric windows of g with cycles of order three in the deterministic case and also two windows for the parameter g with a stochastic periodicity of the same order for the real wage fluctuating between three disjoint invariant sets. This implies that the support of the invariant (marginal) measure consists of three disjoint intervals for the real wage and there exist three separate invariant measures which are deterministic fixed points of the same order. In other words, the data could be interpreted as a 'noisy cycle' of deterministic(!) order three. Thus, qualitatively, the level of government demand on the whole range displays a long-run random behavior revealing a dynamic complexity similar to the deterministic bifurcation diagram, i.e. a similar spectrum of random cycles appears with a large region of a period-three scenario in the middle. In general, however, such a conclusion cannot be drawn for the, stationary development of orbits if the (projection of the) support of the invariant distribution consists of finitely many disjoint subsets. The existence of a support consisting of disconnected subsets does not necessarily mean that the corresponding state variable fluctuates with a deterministic period between the disjoint random parts of the attractor.

Thus, often for the parametrized economy the attracting deterministic cycles caused by different levels of government demand reappear under small noise for some levels of g inducing stationary distributions centered around or near the attracting deterministic cycle. In contrast, however, for low values of government demand the stochastic model shows convergence and stabilization to a stationary solution concentrated on a narrow interval of the real wage while in the deterministic case cyclical behavior with large fluctuations persist for all low levels of government demand. This shows that such a local similarity of the bifurcation with small and without noise does not always hold[2].

[2] This fact is known in the literature, see for example Arnold & Wihstutz (1982); Arnold & Boxler (1991, 1992); Lasota & Mackey (1994) who also provide examples where typical features of a

8.1.2 Random Elasticity of Production: Perturbation of Curvature $B(\omega)$

The multiplicative random shock in the previous section is a pure scale effect and corresponds to a so-called Hicks neutral productivity shock implying that the marginal productivity of both factors and their shares in output are effected in the same way. Thus the relative shares are constant. However, for a constant scale factor A, an increase in the elasticity for labor not only implies an increase in labor productivity, but it also induces an increase in labor share and a decrease of output assigned to capital. In other words, random changes of the elasticity of labor causes asymmetric productivity changes on output and on output shares. It is an interesting issue whether such production shocks have important consequences for the long-run development of the economy which are different from those induced by pure scale effects.

Consider the following scenario with fixed parameters, chosen as in the previous case and repeated in Table 8.4, but now with a random elasticity B distributed on the

Table 8.4 Values of parameters for $B(\omega)$

$c = 10/13$	$L_{\max} = 1$	$A = 1$	$g = 0.5$	$\tau = 0.35$
$\tilde{c} = 0.5$	$\gamma = 0.6$	$\kappa = 0.6$	$\lambda = 0.6$	$\mu = 0.6$

three discrete values $0 < B_1 = 0.69 < B_2 = 0.7 < B_3 = 0.71$ with equal probability of $1/3$, or a uniform distribution over the interval $[0.69, 0.71]$ with mean $\mathbb{E}B(\omega) = 0.7$.

In order to make the same comparison here under random elasticity between the dynamics under market clearing versus the disequilibrium dynamics, equations (8.1.1) and (8.1.2) imply that the competitive real wage is unaffected by the noise of $B(\omega)$ because of $A = L_{\max} = 1$. Thus,

$$\alpha^* = \frac{A}{L_{\max}} (L_{\max})^B, \quad m^* = \frac{\tilde{c}A}{B} (L_{\max})^B - g, \qquad \tilde{c} := 1 - c(1-\tau) \tag{8.1.3}$$

implies that the real wage is constant under all values of $B(\omega)$ and real money balances $m^*(\omega)$ take all the randomness. This means that the random equilibrium set of the economy for all B is a subset of a *horizontal* line in the state space at $\alpha^* = 1$.

Therefore, in the discrete case, one finds that there exist three associated stationary states $(\alpha_i, m_i), i = 1, 2, 3$, satisfying $\alpha_1, \alpha_3 < \alpha_2 = 1$ and $m_1 < m^*(B_1)$, $m_2 = m^*(B_2)$, and $m_3 > m^*(B_3)$ such that and $L_3 = L_2 = L_{\max} = 1 > L_1$ with

$(m_1, \alpha_1) \in \mathbf{K}$ is of the Keynesian type with $\theta_1 < 1$,

---

deterministic bifurcation are destroyed by randomization (also Arnold, Bleckert & Schenk-Hoppé, 1999). In some cases this may occur even with additive noise (see Crauel & Flandoli, 1998).

$(m_2, \alpha_2) \in \mathbf{WE}$ is a Walrasian equilibrium with $\theta_2 = 1$,

$(m_3, \alpha_3) \in \mathbf{I}$ is of the inflationary type with $\theta_3 > 1$.

The three states induce a nontrivial triangle in state space, so that the existence of a convexification effect relative to a linearization near steady states can be compared again with the true converging behavior of the dynamics under disequilibrium.

The numerical results under the parameters chosen in Table 8.4 are shown in Figure 8.7. One finds that the overshooting feature of the wage dynamics is reinforced

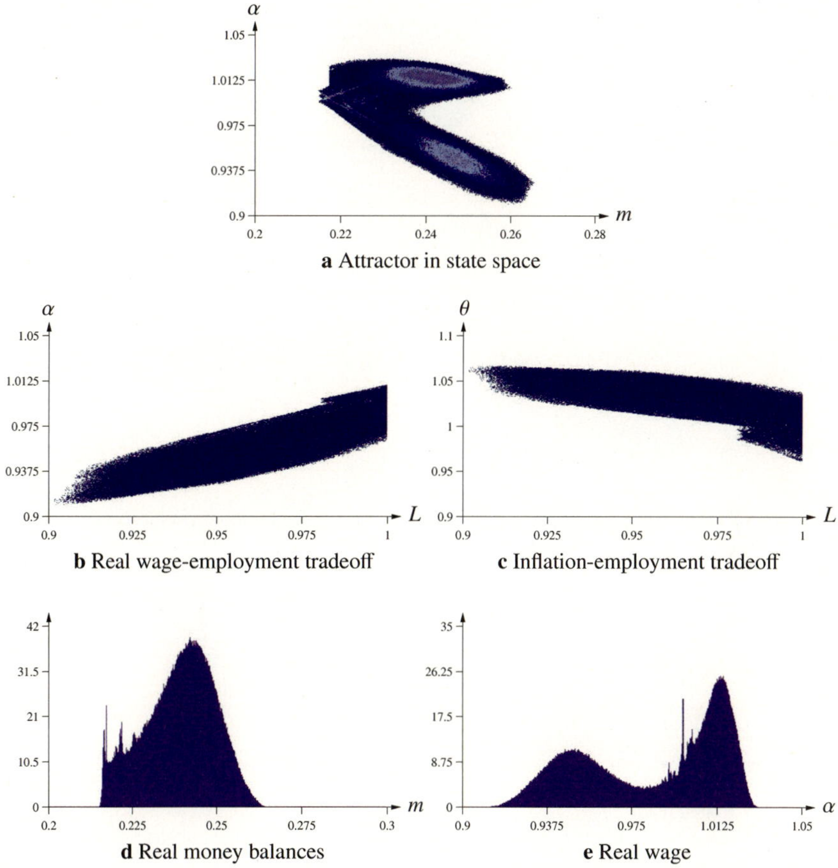

Fig. 8.7 Random production elasticity: $B(\omega)$ uniform i.i.d. $\sim \{0.69, 0.7, 0.71\}$; $T = 10^6$

with random elasticity, partly because of the relatively high level of the adjustment speed. Observe that the condition for asymptotic stability of the steady states is $2(1 - B) > \lambda, \mu$ which holds only for the **WE** and **K** states, but is violated for the **I** state since $2(1 - B_3) = 0.58 < 0.6 = \mu$. (The mean eigenvalue still indicates stability.) As the attractor shows (Figure 8.7 panel **a**) the overshooting property of the real wage adjustment mechanism induces the same splitting of the attractor in the direction of the real wage as under the random scale effect which eventually induces the bimodality of the distribution of the real wage (panel **e**) with a high frequency in inflationary states and low in Keynesian and classical states. In contrast, the random elasticity makes both correlations of employment with the real wage and the rate of inflation still positive but significantly less so than under random scale effects (panel **b** and **c**). Inflation and employment are still negatively correlated, i.e. the Phillips curve is relatively flat but still increasing.

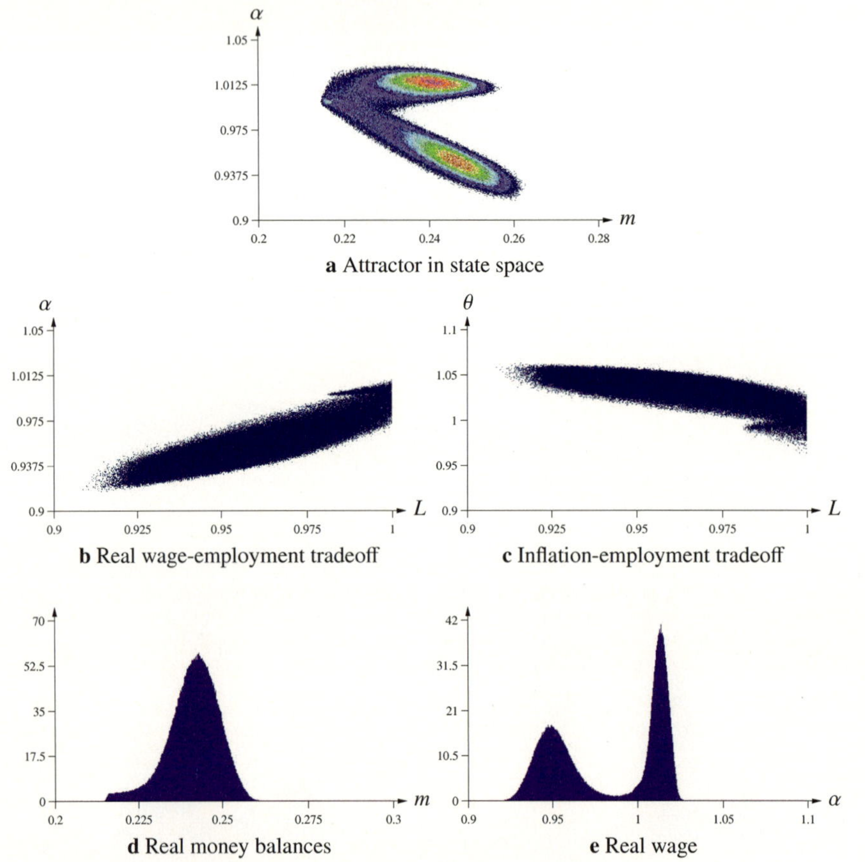

a Attractor in state space

b Real wage-employment tradeoff

c Inflation-employment tradeoff

d Real money balances

e Real wage

Fig. 8.8 Random production elasticity: $B(\omega)$ uniform i.i.d. ~ $[0.69, 0.71]$

The Role of Government Demand with Random Curvature

Finally, one could ask for the sensitivity of these results with respect to government demand and whether the global effects are significantly different from those under a multiplicative Hicks neutral production shock. Figure 8.9 portrays these effects for an interval of levels of government demand g on the real wage (the last 90% of an orbit of length $T = 10^6$) with small noise on the elasticity of production $B(\omega)$ with mean $\mathbb{E}\{B(\omega)\} = 0.9$ with values of all other parameters as shown in Table 8.5. These are similar to the ones used in the deterministic case for the corresponding

Table 8.5 Values of parameters for random elasticity $B(\omega)$

$c = 10/13$	$L_{\max} = 1$	$B \sim$ i.i.d.$[0.89, 0.91]$	$g \in [0.0, 0.9]$	$\tau = 0.35$
$\tilde{c} = 0.5$	$\gamma = 0.6$	$\kappa = 0.6$	$\lambda = 0.6$	$\mu = 0.6$

Figure 7.13, although the two tax rates are not the same. First, since $A = L_{\max} = 1$, the full employment real wage equals $\alpha^* = 1$ for all g and B. Second, the condition for a Walrasian equilibrium is $Bg = \tau$ which occurs in the deterministic case at $g = 0.25/0.9 = 0.27\bar{7}$ and for the stochastic case in the range $g \in [0.385, 0.393]$. For government demand larger (smaller) than these values steady states are inflationary (Keynesian) respectively.

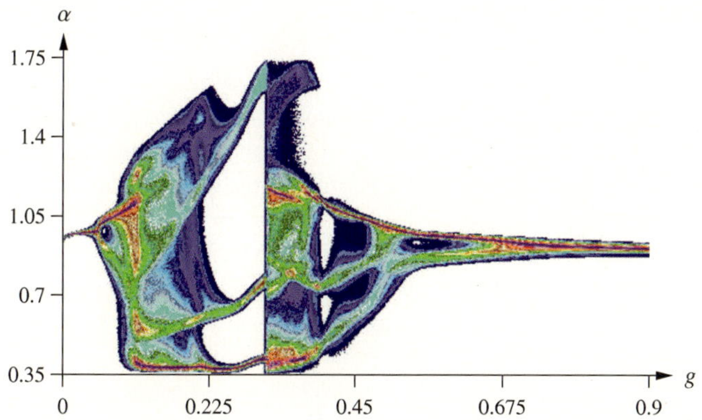

Fig. 8.9 Role of government demand under random production elasticity: $B(\omega)$ uniform i.i.d. $\sim [0.89, 0.91]$

One finds that the qualitative features of the stochastic bifurcation diagram under comparable shocks are quite similar. The global role of different levels of government demand induce a similar separation of the effects of g: a stabilizing global effect for large values of government demand in the inflationary region and a more cyclical fluctuations in the Keynesian region for low values, while there exists an intermittent range, including the interval of values supporting Walrasian equilibria, where fluctuations and regime switching occur, exhibiting also two windows of frequencies of order three. Thus, in both random scenarios, the changes of g dominate the fluctuations in size making the range of the random changes of the parameter to appear as small.

8.2 Stationary Business Cycles with Constant Labor Supply

For a more detailed analysis of the stationary characteristics of the real variables (real money balances, real wages, employment, and inflation) over the business cycle under random perturbations it is informative to examine to what extent the main parameters influence the *observable* real aggregates and to establish their typical correlations. The previous numerical analysis was directed toward understanding when and how the economy deviates from Walrasian stationary states and switches between the regimes under disequilibrium trading guided by the range of values of parameters identifiable from Theorems 7.4.1 and 7.5.1. However, these also identify the ranges of the parameters for which stochastic business cycles converge and *remain* in each of the regimes under a given fiscal policy. Thus, it is possible to compare the stationary behavior of the real business cycle under the two specific regimes, Keynesian **K** or inflationary **I** and establish the differences.

8.2.1 Stationary Keynesian States under Random Productivity

Hicks Neutral Production Shocks $A(\omega)$

Consider first the situation with random productivity $A(\omega)$ inducing a Keynesian business cycle with unemployment. Parametrically, the steady states are of the Keynesian type if $Bg < \tau A$. They are asymptotically attracting for $2(1 - B) > \mu, \kappa$. Therefore, let the fiscal policy be determined by $g = 0.5$ and $\tau = 0.35$ and assume the adjustments being set to $\lambda = \kappa = \gamma = \mu = 0.6$. Then, for low values of B all states are asymptotically of the Keynesian type without switching and they are locally attracting for an appropriately chosen random perturbation of $A(\omega)$. This implies asymptotically attracting stationary behavior of all orbits to a random fixed point $(m^*(\omega), \alpha^*(\omega))$, i.e. the existence of two stationary random variables $(m^*, \alpha^*) : \Omega \to \mathbb{R}^2_+$ describing the stationary evolution along the business cycle in the state space $\mathbb{R}^2_+$ inducing a unique invariant measure for all values including $B = 0.7$.

Figure 8.10 shows the result for $B \in \{0.3, 0.4, 0.5, 0.6, 0.7\}$ plotting in each case the last 90% of a stochastic orbit of length $T = 10^6$ while keeping the value ratio of $\mathbb{E}y_{\max} = \mathbb{E}\{A(\omega)\}/B$ in a comparable range. The diagram shows clear but different correlation patterns of the two state variables depending on the elasticity of the production function. The attractor undergoes a definite change in the correlation between real balances and the real wage being negative for $B < 0.5$ and positive for $B > 0.5$. Notice also that stationary second moments are highest when $B = 0.5$. This is partly due to the fact that the labor demand function is unit elastic for $B = 0.5$ implying a constant wage bill. But an exact causal explanation of the zreo correlation from the elasticity condition is difficult.

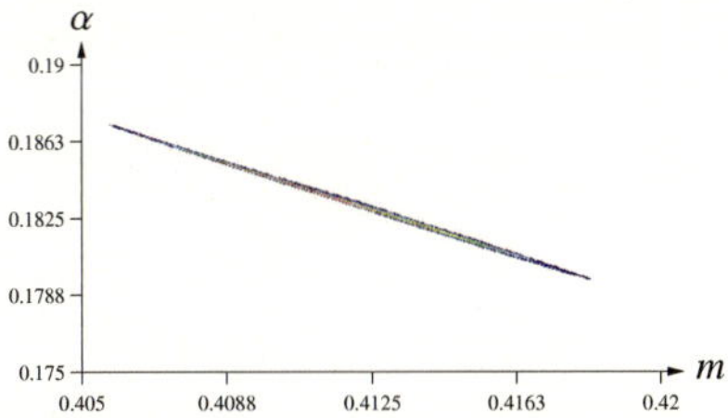

a $A(\omega)$ uniform i.i.d. $\sim$ $[0.69, 0.71]$; B = 0.3

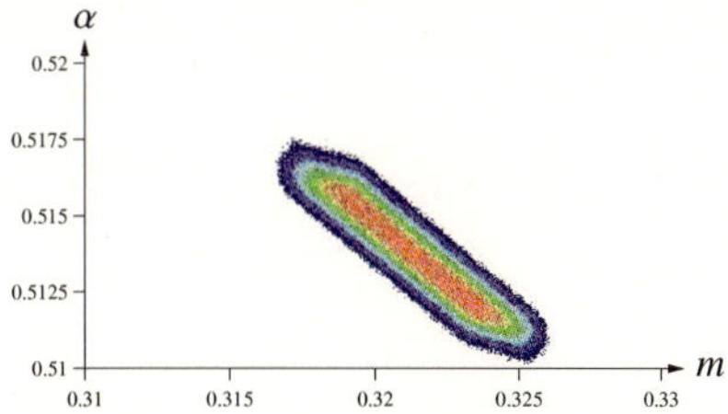

b $A(\omega)$ uniform i.i.d. $\sim$ $[0.79, 0.81]$; B = 0.4

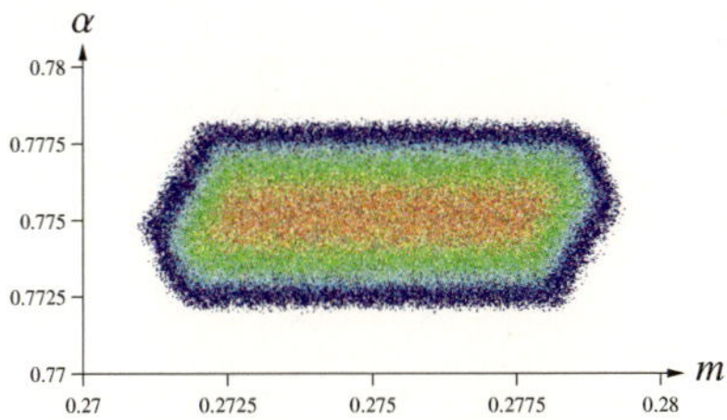

c $A(\omega)$ uniform i.i.d. $\sim$ $[0.89, 0.91]$; B = 0.5

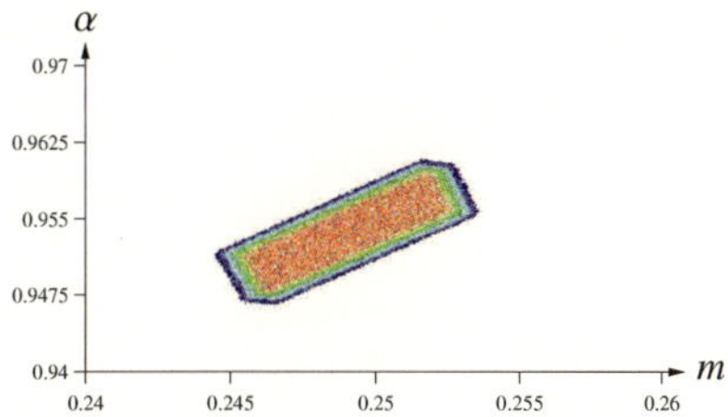

d $A(\omega)$ uniform i.i.d. $\sim$ $[0.99, 1.01]$; B = 0.6

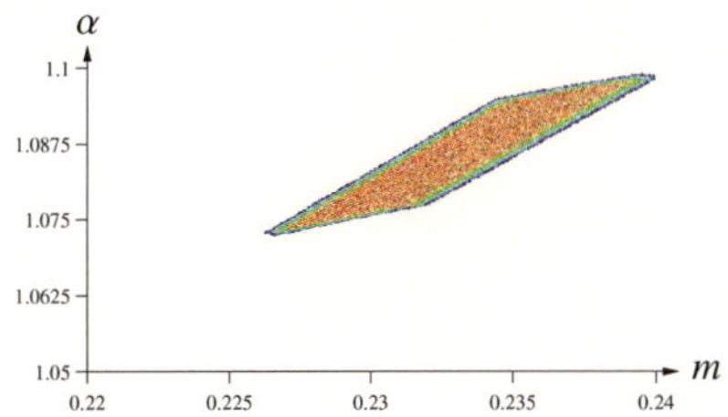

e $A(\omega)$ uniform i.i.d. $\sim$ $[1.09, 1.11]$; B = 0.7

Fig. 8.10 Keynesian attractor for $A(\omega)$ with $B \in \{0.3, 0.4, 0.5, 0.6, 0.7\}$, $g = 0.5$, $\tau = 0.35$, $T = 10^6$

Figure 8.11 displays the stationary money-inflation contour under disequilibrium which exhibits a clear negative correlation in the data for all values of $B \in \{0.3, 0.4, 0.5, 0.6, 0.7\}$. This is the attracting invariant set on the random graph of the price adjustment function which corresponds to the attracting set on the random unit price contour under market clearing in Section 4.3. For $B > 0.7$ regime switching occurs and the monotonic correlation disappears. Finally, Figure 8.12 displays the stationary inflation-employment tradeoff, the long-run Phillips curve which has a clear positive correlation between employment and inflation for all values of $B \in \{0.3, 0.4, 0.5, 0.6, 0.7\}$. In other words, higher levels of employment appear together with higher levels of inflation (i.e. lower levels of deflation!) in the time series. Numerically, the correlation seems to be higher with less variance as the elasticity of production increases.

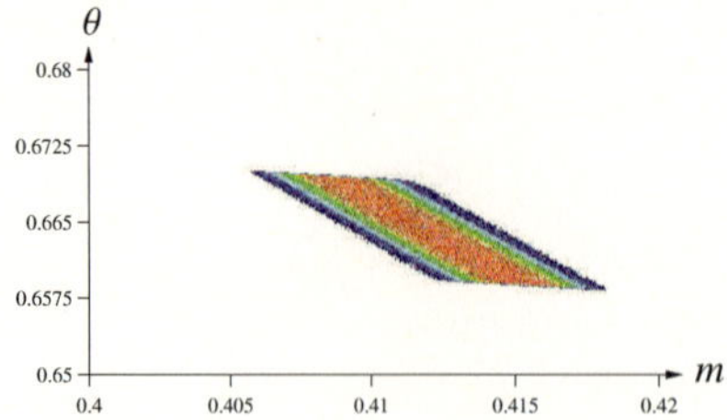

a $A(\omega)$ uniform i.i.d. $\sim$ $[0.69, 0.71]$; B = 0.3

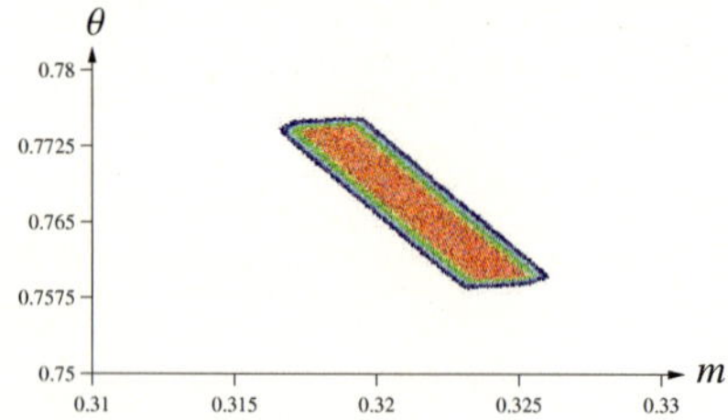

b $A(\omega)$ uniform i.i.d. $\sim$ $[0.79, 0.81]$; B = 0.4

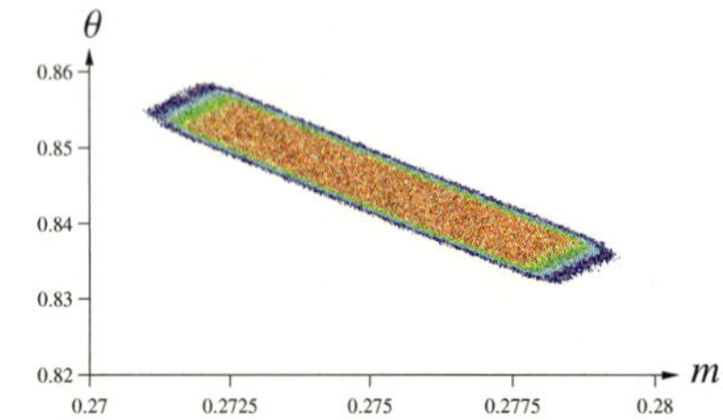

c $A(\omega)$ uniform i.i.d. $\sim$ $[0.89, 0.91]$; B = 0.5

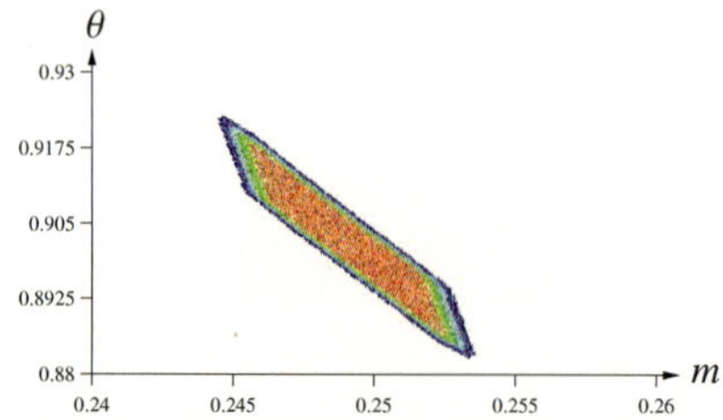

d $A(\omega)$ uniform i.i.d. $\sim$ $[0.99, 1.01]$; B = 0.6

Fig. 8.11 Keynesian money-price contours under $A(\omega)$; $B \in \{0.3, 0.4, 0.5, 0.6\}$, $g = 0.5$, $\tau = 0.35$, $T = 10^6$

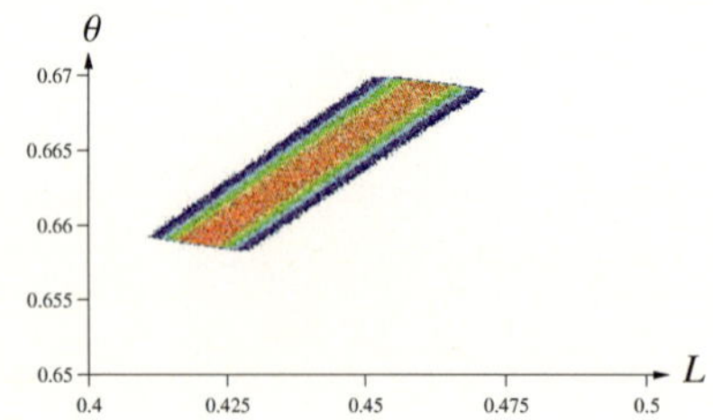

a $A(\omega)$ uniform i.i.d. $\sim$ $[0.69, 0.71]$; B = 0.3

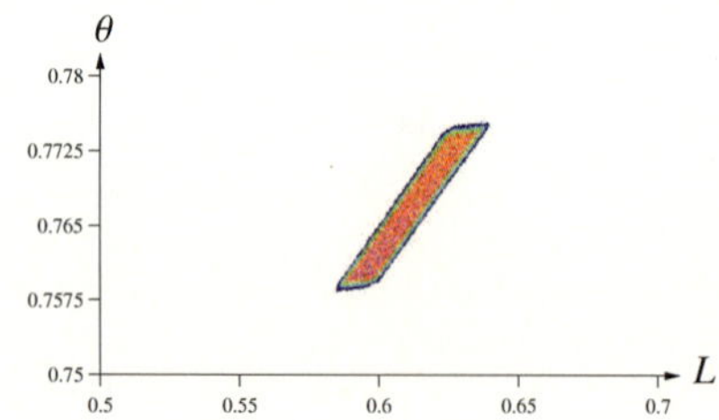

b $A(\omega)$ uniform i.i.d. $\sim$ $[0.79, 0.81]$; B = 0.4

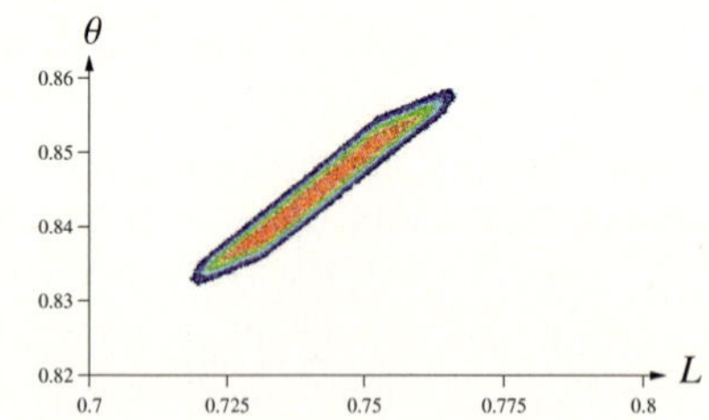

c $A(\omega)$ uniform i.i.d. $\sim$ $[0.89, 0.91]$; B = 0.5

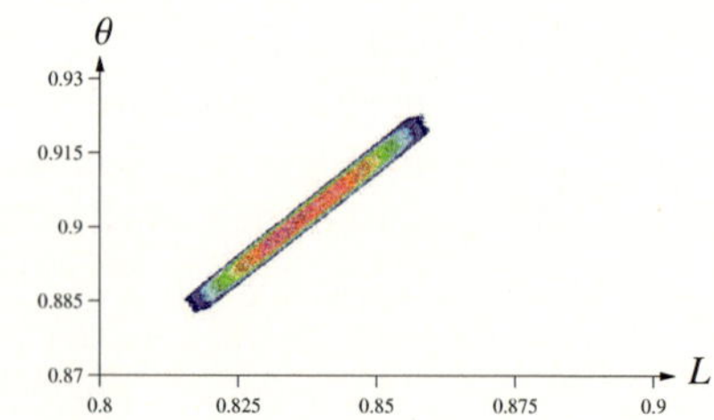

d $A(\omega)$ uniform i.i.d. $\sim$ $[0.99, 1.01]$; B = 0.6

Fig. 8.12 Keynesian Phillips curves under $A(\omega)$ for $B \in \{0.3, 0.4, 0.5, 0.6\}$, $g = 0.5$, $\tau = 0.35$, $T = 10^6$

Random Elasticities of Production $B(\omega)$

The next set of diagrams presents the corresponding stationary features of the real business cycles for a list of values for the parameter $A = \{0.7, 0.8, 0.9, 1.0\}$ while the elasticity of production B undergoes randomization in such a away that the attract-

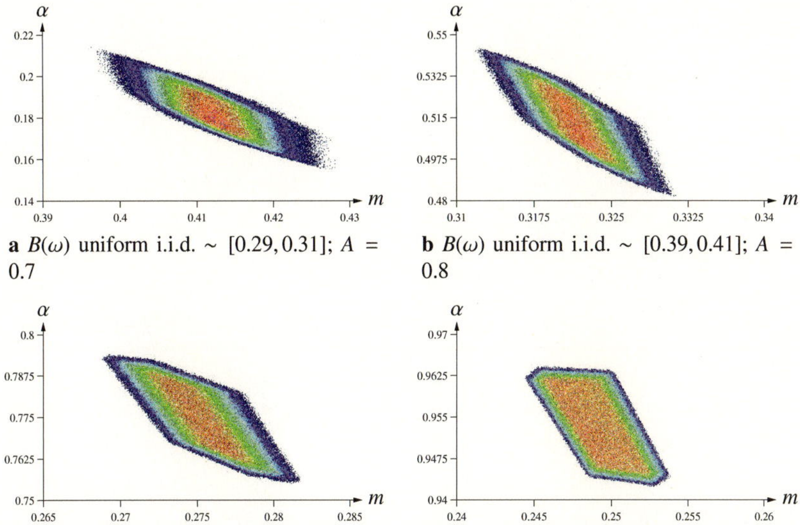

a $B(\omega)$ uniform i.i.d. $\sim$ $[0.29, 0.31]$; $A = 0.7$

b $B(\omega)$ uniform i.i.d. $\sim$ $[0.39, 0.41]$; $A = 0.8$

c $B(\omega)$ uniform i.i.d. $\sim$ $[0.49, 0.51]$; $A = 0.9$

d $B(\omega)$ uniform i.i.d. $\sim$ $[0.59, 0.61]$; $A = 1.0$

Fig. 8.13 Keynesian attractor under $B(\omega)$ for $A \in \{0.7, 0.8, 0.9, 1.0\}$, $g = 0.5$, $\tau = 0.35$, $T = 10^6$

ing set is fully Keynesian. In contrast to the case with random scale of production the attractor has a clear negative correlation for all values of A *even* for the situations when $\mathbb{E}\{B(\omega)\} = 0.5$ which was the critical value of B at which the reversal of the correlation of the attractor occurs.

Figure 8.13 shows the numerical outcomes for the first four values displaying a clear negative correlation between the two state variables with a unimodal distribution. For $\mathbb{E}\{B(\omega)\} > 0.7$ regime switching occurs again including classical and inflationary states along the orbits.

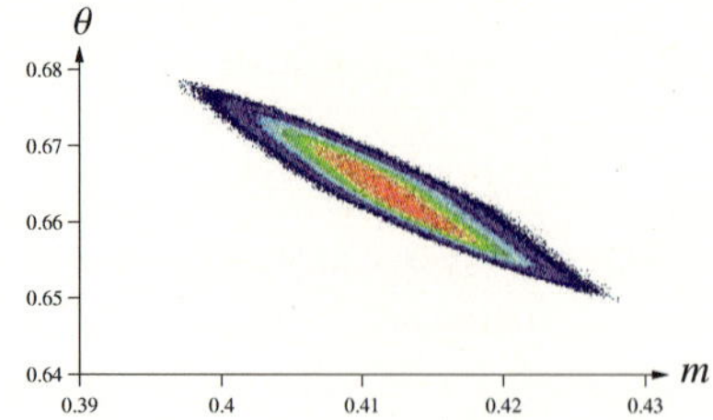

a $B(\omega)$ uniform i.i.d. $\sim$ $[0.29, 0.31]$; A = 0.7

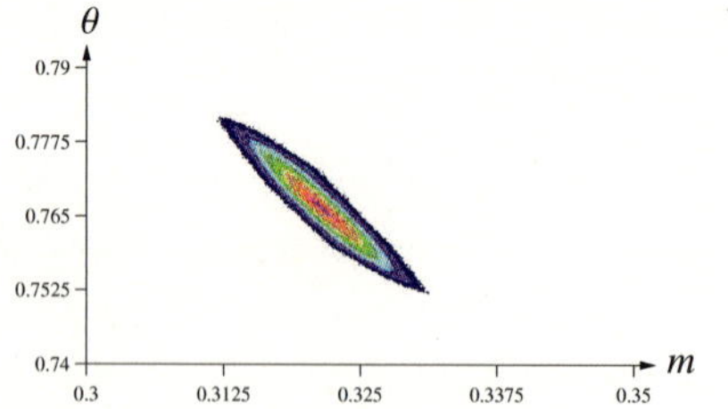

b $B(\omega)$ uniform i.i.d. $\sim$ $[0.39, 0.41]$; A = 0.8

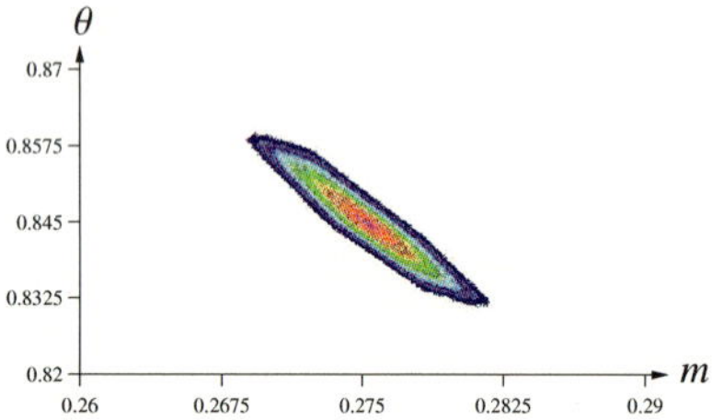

c $B(\omega)$ uniform i.i.d. $\sim$ $[0.49, 0.51]$; A = 0.9

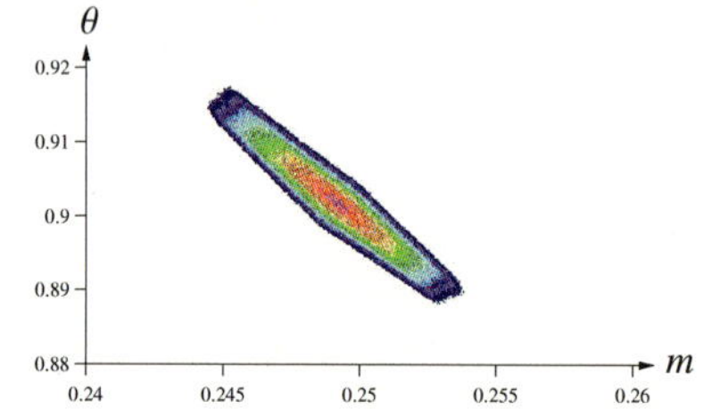

d $B(\omega)$ uniform i.i.d. $\sim$ $[0.59, 0.61]$; A = 1.0

Fig. 8.14 Keynesian money-price contours for $B(\omega)$ for $A \in \{0.7, 0.8, 0.9, 1.0\}$, $g = 0.5$, $\tau = 0.35$, $T = 10^6$

Figure 8.15, finally, shows the Phillips curves under different values of the scale parameter A. In contrast to the situation with random scale it shows a weak positive tradeoff appears which seems to be higher and more prominent with higer value of the mean $\mathbb{E}\{B(\omega)\}$ of the elasticity. In addition, the attracting set displays a high variance of inflation and relatively low variation of employment. The sensitivity analysis reveals that the type of randomness in productivity (scale versus curvature) has a decisive impact on the tradeoff. As in the cases before, the weak tradeoff continue to exist in the Keynesian regime for all values up to $\mathbb{E}\{B(\omega)\} = 0.7$. For larger values regime switching occurs for the stationary solution and the tradeoff is no longer monotonic.

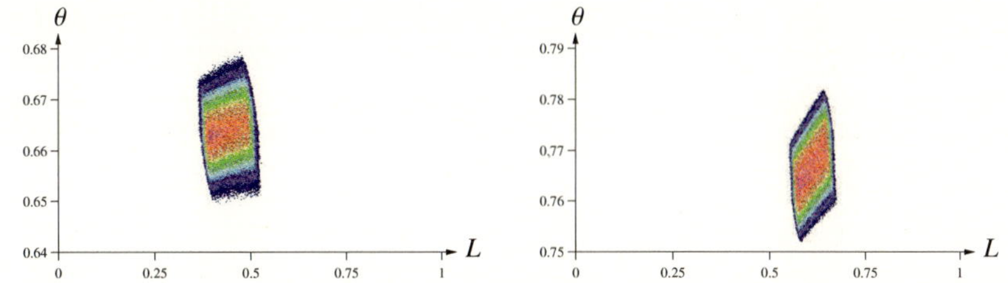

a $B(\omega)$ uniform i.i.d. $\sim$ $[0.29, 0.31]$; $A = 0.7$

b $B(\omega)$ uniform i.i.d. $\sim$ $[0.39, 0.41]$; $A = 0.8$

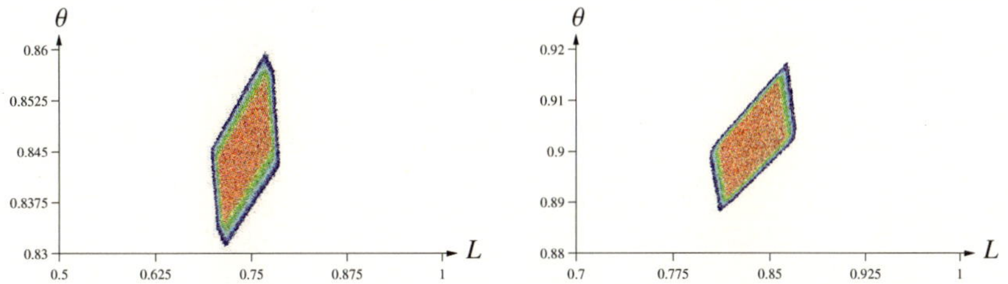

c $B(\omega)$ uniform i.i.d. $\sim$ $[0.49, 0.51]$; $A = 0.9$

d $B(\omega)$ uniform i.i.d. $\sim$ $[0.59, 0.61]$; $A = 1.0$

Fig. 8.15 Keynesian Phillips curves under $B(\omega)$ for $A \in \{0.7, 0.8, 0.9, 1.0\}$, $g = 0.5$, $\tau = 0.35$, $T = 10^6$

8.2.2 Stationary Keynesian States with Random Demand

So far the sensitivity analysis concentrated on the role of random of effects from the supply side of the economy, in particular from the technology. Given the specific form of the aggregate demand function random demand effects may arise from the level of government demand, from the tax rate, or from the propensity to consume, as long as no additional expectations effects are introduced. Since all of the parame-

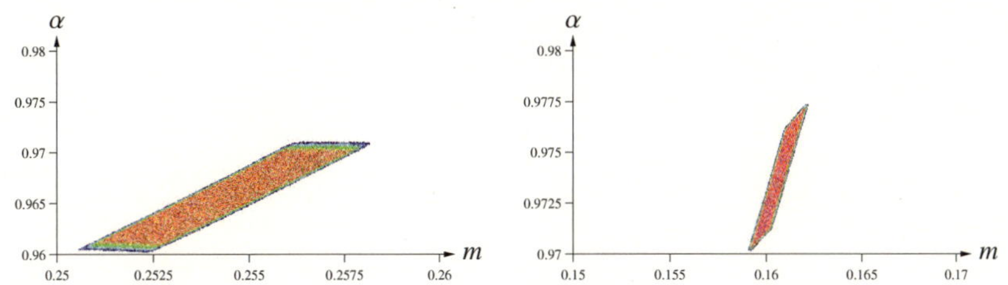

a $g(\omega)$ uniform i.i.d. $\sim$ $[0.19, 0.21], \tau = 0.2$

b $g(\omega)$ uniform i.i.d. $\sim$ $[0.59, 0.61], \tau = 0.5$

Fig. 8.16 Keynesian attractors under $g(\omega)$ for $\tau \in \{0.2, 0.5\}$, $B = 0.7, A = 1.0$, $T = 10^6$

ters enter the demand function in a linear form with an influence on the multiplier or on the additive shift one might expect similar qualitative results under stationarity for all three perturbations. In fact, a detailed numerical study of randomizations of the three parameters (g, τ, c) shows that small smooth i.i.d. randomizations induce the same qualitative features of the three attracting sets analyzed in this section, the attractor in state space, the money-price contour, and the Phillips curve. All of them show positive correlations between the paired variables with some similarity to those of linear random difference equations when the remaining parameters are kept at levels inducing asymptotically attracting Keynesian states.

Figures 8.16 to 8.18 show the results with two selections of numerical output for each attracting set of the anlysis with random government demand at the given tax rates. The principle properties, in particular the *positive* correlation of all pairs

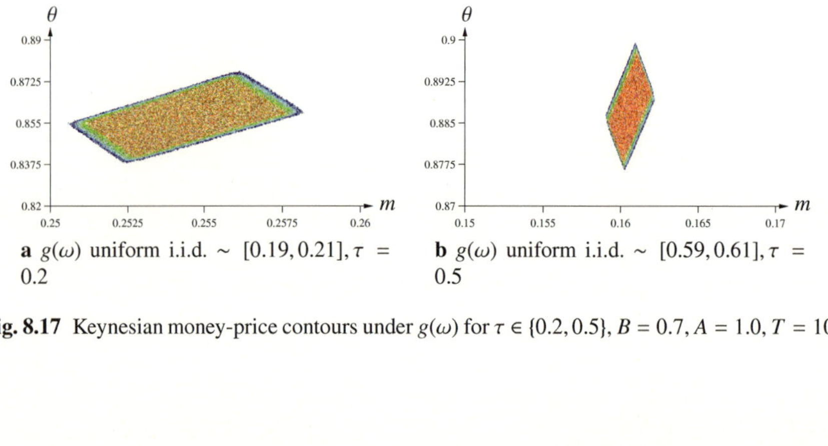

a $g(\omega)$ uniform i.i.d. $\sim$ $[0.19, 0.21], \tau = 0.2$ **b** $g(\omega)$ uniform i.i.d. $\sim$ $[0.59, 0.61], \tau = 0.5$

Fig. 8.17 Keynesian money-price contours under $g(\omega)$ for $\tau \in \{0.2, 0.5\}$, $B = 0.7, A = 1.0, T = 10^6$

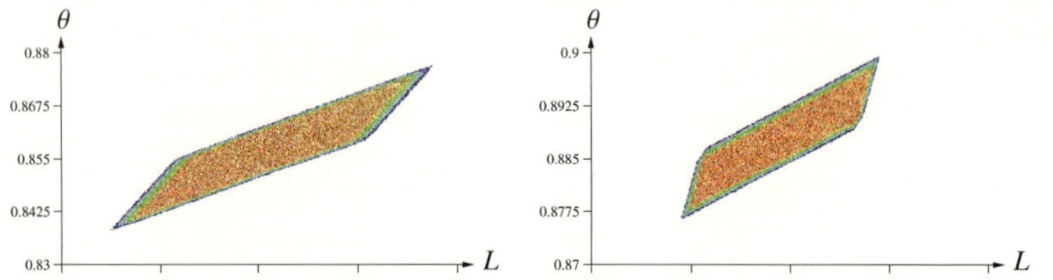

a $g(\omega)$ uniform i.i.d. $\sim$ $[0.19, 0.21], \tau = 0.2$ **b** $g(\omega)$ uniform i.i.d. $\sim$ $[0.59, 0.61], \tau = 0.5$

Fig. 8.18 Keynesian Phillips curves under $g(\omega)$ for $\tau \in \{0.2, 0.5\}$, $B = 0.7, A = 1.0, T = 10^6$

of variables persist in the experiments, only varying in the degree and the domain. When parameters fall outside the Keynesian domain, regime switching occurs and parts of the limiting orbits enter in the classical and in the inflationary regime for all sufficiently long orbits, here, as before, chosen to be $T = 10^6$.

8.2.3 Stationary Inflationary States under Random Productivity

Consider again first the situation with random productivity $A(\omega)$ inducing a stationary real business cycle in the inflationary regime with full employment $L_{\max}$. Parametrically, the steady states are of the infaltionary type if $Bg > \tau A$. They are asymptotically attracting for $2(1-B) > \lambda, \gamma$. In order to keep the results between the two stationary outcomes comparable with those of the previous section assume identical values for the fiscal policy to be determined by $g = 0.5$ and $\tau = 0.35$ and, in addition, assume the adjustments being set to $\lambda = \kappa = \gamma = \mu = 0.6$. Then, for high values of B all states are asymptotically inflationary without switching and locally attracting for appropriately chosen random perturbations of $A(\omega)$. As in the Keynesian case this implies asymptotically attracting behavior of all orbits to an

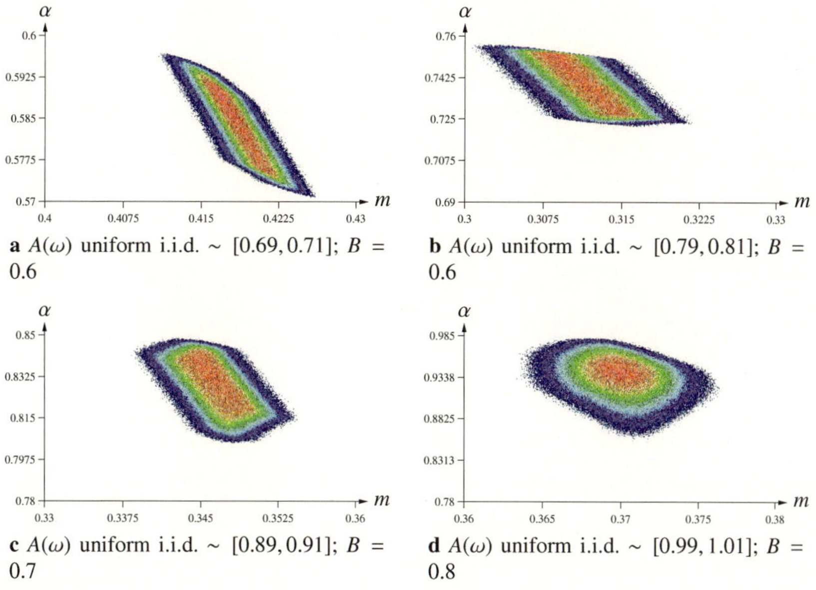

a $A(\omega)$ uniform i.i.d. $\sim [0.69, 0.71]$; $B = 0.6$

b $A(\omega)$ uniform i.i.d. $\sim [0.79, 0.81]$; $B = 0.6$

c $A(\omega)$ uniform i.i.d. $\sim [0.89, 0.91]$; $B = 0.7$

d $A(\omega)$ uniform i.i.d. $\sim [0.99, 1.01]$; $B = 0.8$

Fig. 8.19 Inflationary attractor under $A(\omega)$ for $B \in \{0.5, 0.6, 0.7, 0.8\}$, $g = 0.5$, $\tau = 0.35$, $T = 10^6$

inflationary random fixed point $(m^*(\omega), \alpha^*(\omega))$, i.e. the existence of two stationary random variables $(m^*, \alpha^*) : \Omega \to \mathbb{R}^2_+$ describing the evolution along the business cycle in the state space $\mathbb{R}^2_+$ with an associated unique invariant measure for all values including $B = 0.5$ for appropriately adjusted $A(\omega)$. Figure 8.19 displays for different attractors in state space with varying but typical negative correlation between real money balances and real wages. This contrasts sharply with the attractor in the Keynesian case (Figure 8.10) where the sign of the correlation depends monotonically on B showing a positive (negative) correlation for values of $B > 0.5$ ($B < 0.5$). Thus, surely and as to be expected, the properties of stationary real business cycles in the two disjoint regimes posess distinctly different features. In other words, a time series analysis would be able to identify these differences from the data and could provide the appropriate information and estimates which characterize the difference of the cycles for any two identical parametric economies under the influence of different unobservable production shocks.

Figure 8.20 shows the money-inflation tradeoff to be negative in the inflationary regime which is again the opposit to the Keynesian case (Figure 8.11). The geometric form of the attracting set is a parallelogram which indicates that there is strong

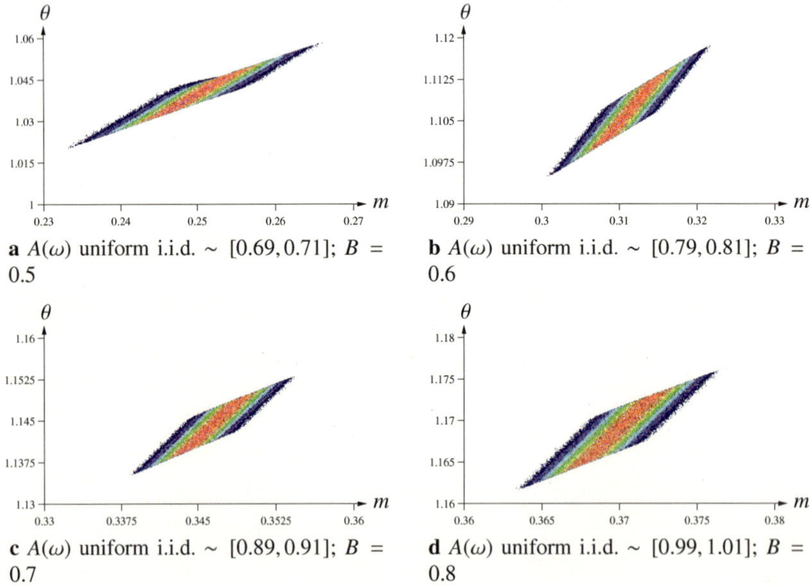

a $A(\omega)$ uniform i.i.d. $\sim$ $[0.69, 0.71]$; $B = 0.5$

b $A(\omega)$ uniform i.i.d. $\sim$ $[0.79, 0.81]$; $B = 0.6$

c $A(\omega)$ uniform i.i.d. $\sim$ $[0.89, 0.91]$; $B = 0.7$

d $A(\omega)$ uniform i.i.d. $\sim$ $[0.99, 1.01]$; $B = 0.8$

Fig. 8.20 Inflationary money-price contours under $A(\omega)$ for $B \in \{0.5, 0.6, 0.7, 0.8\}$, $g = 0.5$, $\tau = 0.35$, $T = 10^6$

linearity in the inflation mapping. This phenomenon is typical in affine systems with additive i.i.d perturbations (as in AR(1) systems) which suggests that here the dynamic relationship between money balances and the inflation rate could be written as an AR(1) subsystem with additive noise in the space $\mathbb{R}^3_+$.

The effects of a random elasticity $B(\omega)$ on attracting inflationary stationary business cycles are given in Figures 8.21 and 8.22. Both confirm the orientation of the correlation of the state space attractor and of the money-price contours of the random productivity shock $A(\omega)$ with a surprising apparent linearity in seven of the eight numerical experiments. Why this linearity feature disappears in the attractor for $A = 1$ and $\mathbb{E}\{B(\omega)\} = 0.8$ (Figure 8.21 **d**) is unclear.

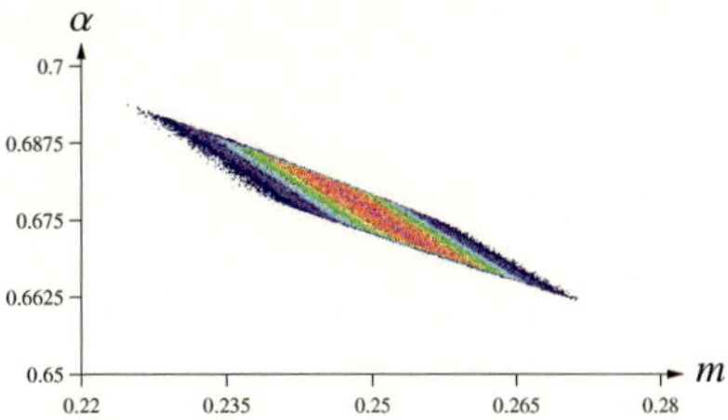

a $B(\omega)$ uniform i.i.d. $\sim$ $[0.49, 0.51]$; A = 0.7

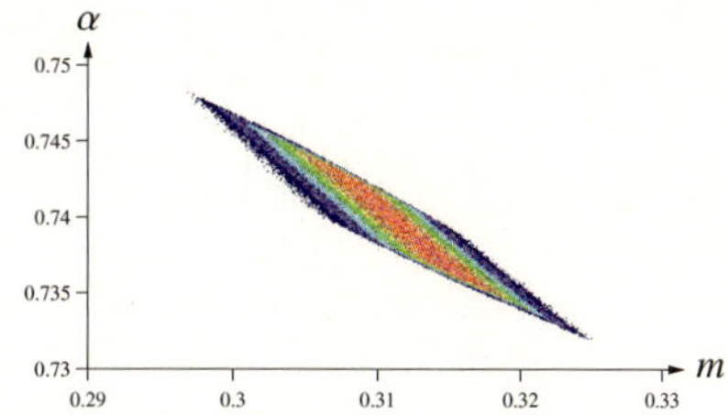

b $B(\omega)$ uniform i.i.d. $\sim$ $[0.59, 0.61]$; A = 0.8

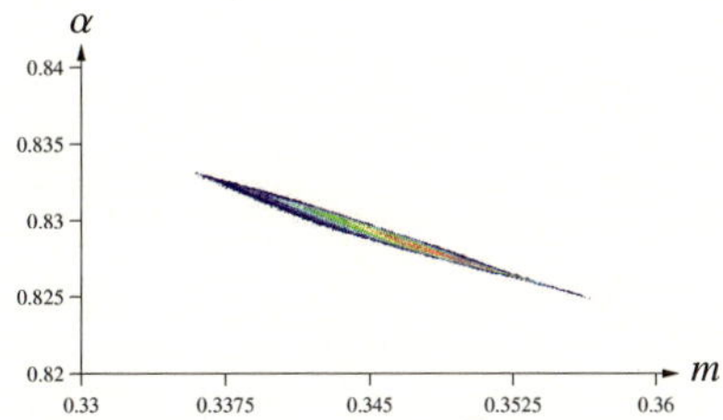

c $B(\omega)$ uniform i.i.d. $\sim$ $[0.69, 0.71]$; A = 0.9

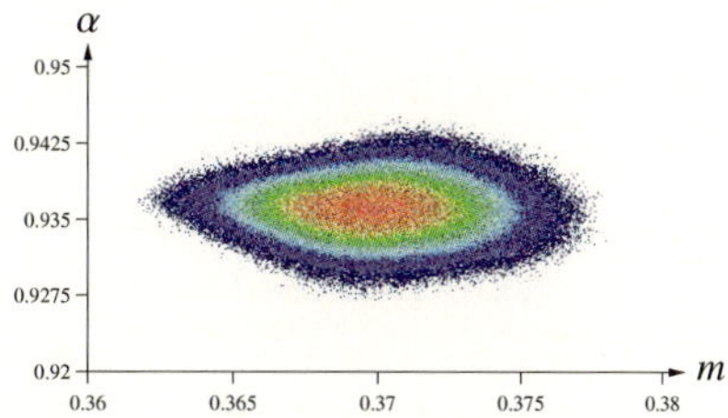

d $B(\omega)$ uniform i.i.d. $\sim$ $[0.79, 0.81]$; A = 1.0

Fig. 8.21 Inflationary attractor under $B(\omega)$ for $A \in \{0.7, 0.8, 0.9, 1.0\}$, $g = 0.5$, $\tau = 0.35$, $T = 10^6$

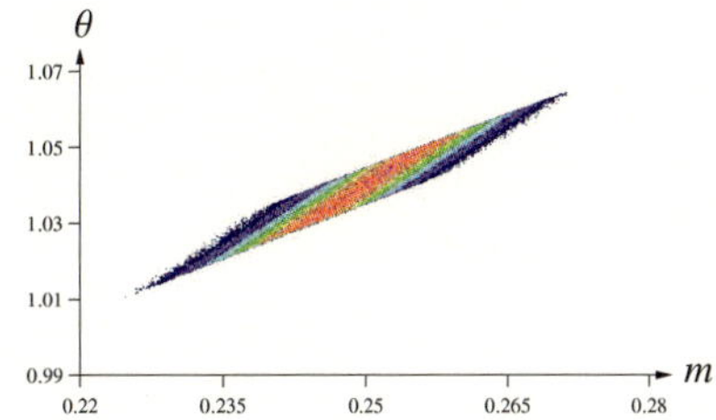

a $B(\omega)$ uniform i.i.d. $\sim$ $[0.49, 0.51]$; A = 0.7

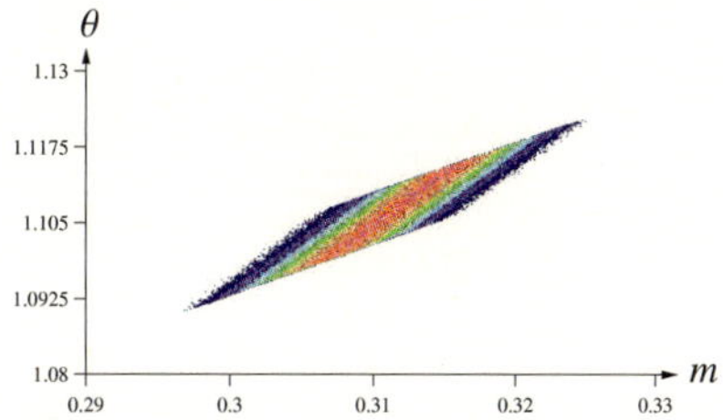

b $B(\omega)$ uniform i.i.d. $\sim$ $[0.59, 0.61]$; A = 0.8

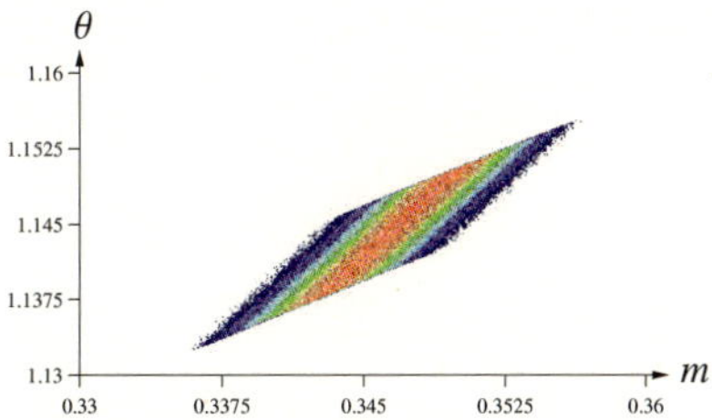

c $B(\omega)$ uniform i.i.d. $\sim$ $[0.69, 0.71]$; A = 0.9

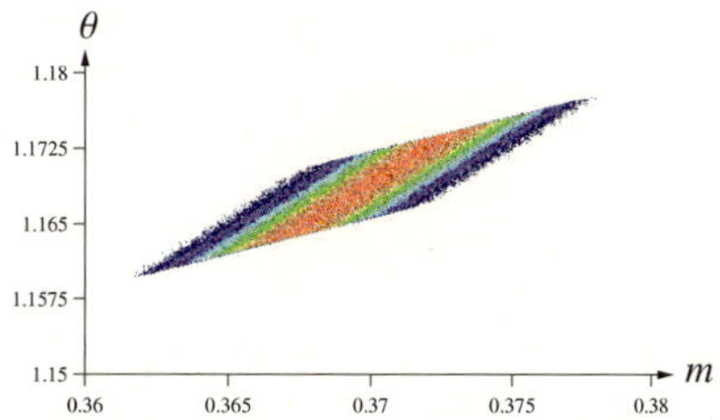

d $B(\omega)$ uniform i.i.d. $\sim$ $[0.79, 0.81]$; A = 1.0

Fig. 8.22 Inflationary money-price contours under $B(\omega)$ for $A \in \{0.7, 0.8, 0.9, 1.0\}$, $g = 0.5$, $\tau = 0.35$, $T = 10^6$

An investigation of effects from random demand do not seem to reveal additional insights which are significant and different. Since the level of output/employment in the inflationary region is constant random demand induces a one-to-one redistribution of output between the government and the private sector, i.e. a perfect correlation between real balances and inflation. This implies a negative correlation (almost linear) for the attractor in the state space. When labor supply is endogenous this changes and additional spillover effects occur and attractors show new results (see Section 8.3).

8.3 Random Productivity with Endogenous Labor Supply

This section contains selected numerical results for the basic model in isoelastic form with endogenous labor supply[3] as derived in Section 3.2.7. As in all previous experiments without stochastic effects it is of interest whether the characteristics of the real business cycles change in substantial ways when labor supply is not constant but endogenous inducing more cross market effects which may change the long-run properties of the cycle. In the following experiments the same three levels $D = 0.2, 0.5, 0.7$ of the parameter are used for the supply elasticity rather than running a bifurcation analysis with continuous changes which would limit the graphical presentation to considering one variable at a time only. Since the state space of the real business cycles is two-dimensional most of the stationary correlation between any two pairs of variables are describable by the associated attracting invariant set. The remaining parameters are kept at the corresponding levels of the previous experiments as shown in Table 8.6. The comparison is carried out for each specific

Table 8.6 Values of parameters under endogenous labor supply and random productivities

$c = 10/13$	$D \in \{0.2, 0.5, 0.7\}$	$g = 0.5$	$\tau = 0.35$
$\gamma = 0.6$	$\kappa = 0.6$	$\lambda = 0.6$	$\mu = 0.6$

randomization with the same random seed (one each for $A(\omega)$ and for $B(\omega)$) to make the results for different values of the parameters comparable in a meaningful way. Each iteration is of length $T = 10^6$ and the diagrams of the attractors plot the last 90% of pairs of an orbit.

[3] The elasticity of the labor supply function was parametrized as $1/C$ where $1+C$ was the elasticity of the function of the disutility of labor, see Section 3.2.7For programming reasons the parameter had to be renamed from C to D. Thus, for the remaining sections of this chapter $D \neq 0$ means that labor supply is endogenous with elasticity $1/D$. This corresponds also to the elasticity of the aggregate labor supply function given in Assumption 3.2.1. The case of $D = 0$ corresponds to the situation with $L_{\max}$ constant.

8.3.1 Employment and the Inflation Tradeoff with $A(\omega)$

The Phillips Curve with Keynesian Unemployment: $D = 0.2, 0.5, 0.7$

The first experiment examines the implications for the stationary Phillips curve resulting from the interaction of the labor supply parameter D and the elasticity of production under a uniform Hicks neutral production shocks $A(\omega)$. As in the case with fixed labor supply (compare Figure 8.12) stationary states for low values of the elasticity of production B are Keynesian states. Figure 8.23, subfigures **a**, **c**, **e** shows no noticeable effect from the labor supply elasticity *and* the same clear positive correlation between employment and inflation as in the case with constant labor supply. Subfigures **b**, **d**, **f** show the distribution (counting measure) within the three regimes

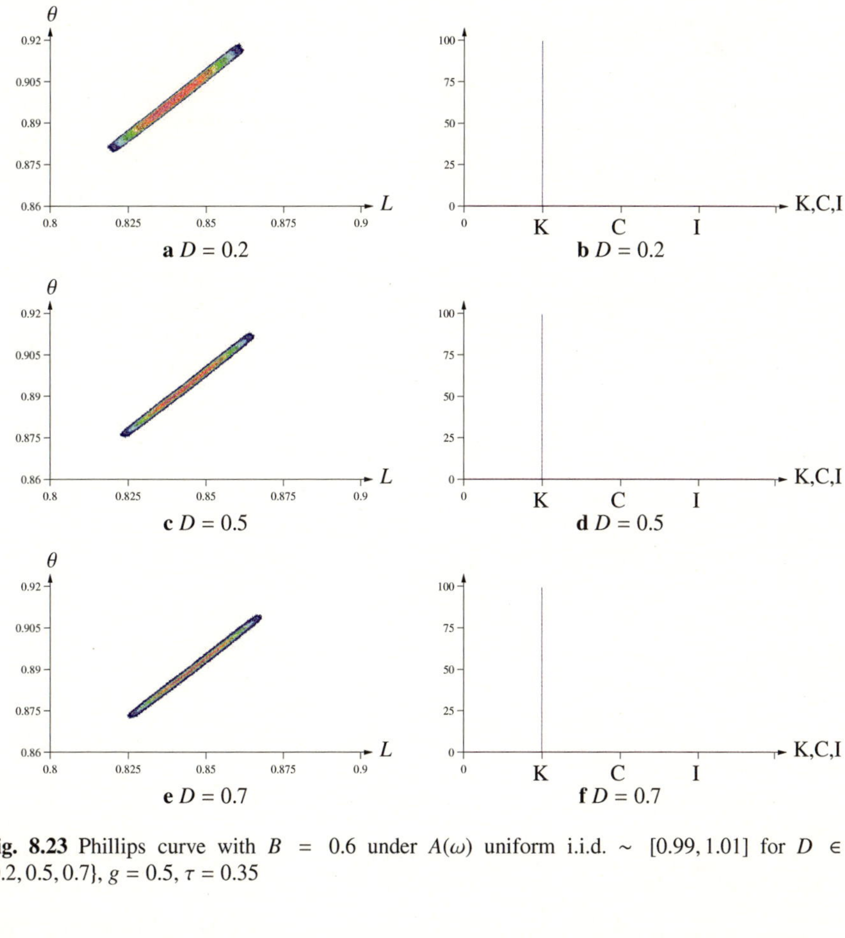

Fig. 8.23 Phillips curve with $B = 0.6$ under $A(\omega)$ uniform i.i.d. $\sim [0.99, 1.01]$ for $D \in \{0.2, 0.5, 0.7\}$, $g = 0.5$, $\tau = 0.35$

K, **C**, and **I**, confirming that the orbit remains in the Keynesian regime for all three levels of D. In other words, at $B = 0.6$, the variations of D neither change the type of the attractor nor the correlation compared to the case with constant labor supply, a property which is also confirmed for higher values of D.

The Phillips Curve with Regime Switching C – I for $D = 0.2, 0.5, 0.7$

As the elasticity of production increases to $B = 0.7$, the stationary orbit no longer remains in the Keynesian regime but switches between classical and inflationary states. Simultaneously, the correlation changes from positive to negative, Figure

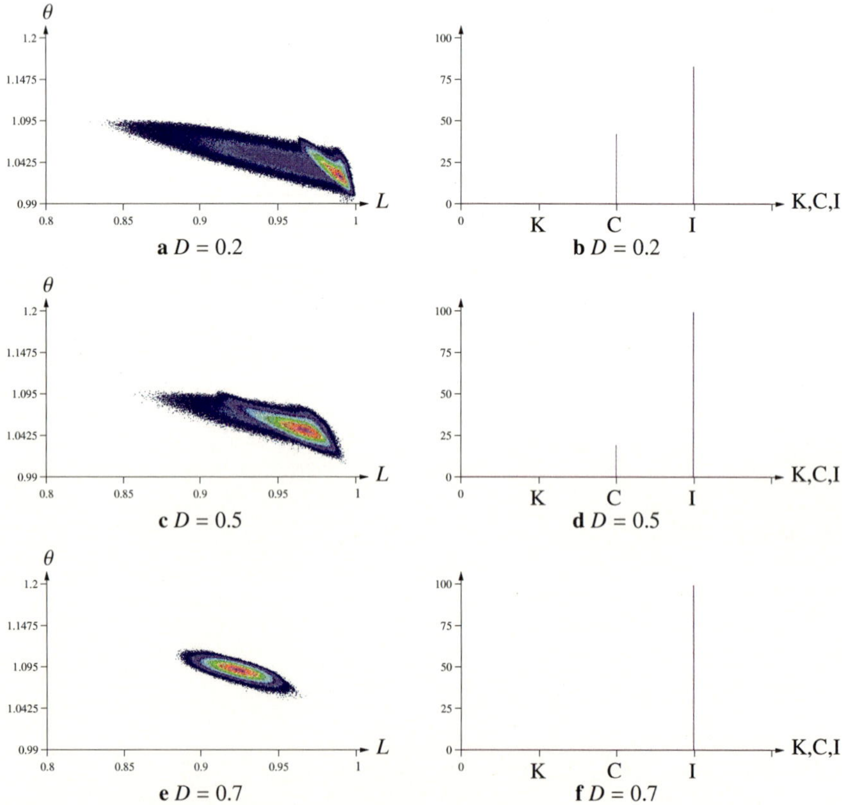

Fig. 8.24 Phillips curve with $B = 0.7$ under $A(\omega)$ uniform i.i.d. $\sim [0.99, 1.01]$ for $D \in \{0.2, 0.5, 0.7\}$, $g = 0.5$, $\tau = 0.35$

8.24, while loosing its typical property (as in Figure 8.23) corresponding to the form of an attractor which often appears in linear stochastic models. An increase of the parameeter D enforces the switching and ultimately the change to an all inflationary Phillips curve. In addition, the form of the attracting set reveals rotations in the form of the attractor resulting typically from complex roots in affine difference equations.

An Inverted Phillips Curve in Regime I for $D = 0.2, 0.5, 0.7$

With further increases of the elasticity of production to $B = 0.8$, the attractors remain in the inflationary regime for all B as they should. The sign of the correlation continues to prevail and the attractors become less circular (i.e. with less rotation), Figure 8.25 shows the changes for $B = 0.8$ which have an almost linear property for $D = 0.7$. Further experiments with larger values for D confirm this tendency.

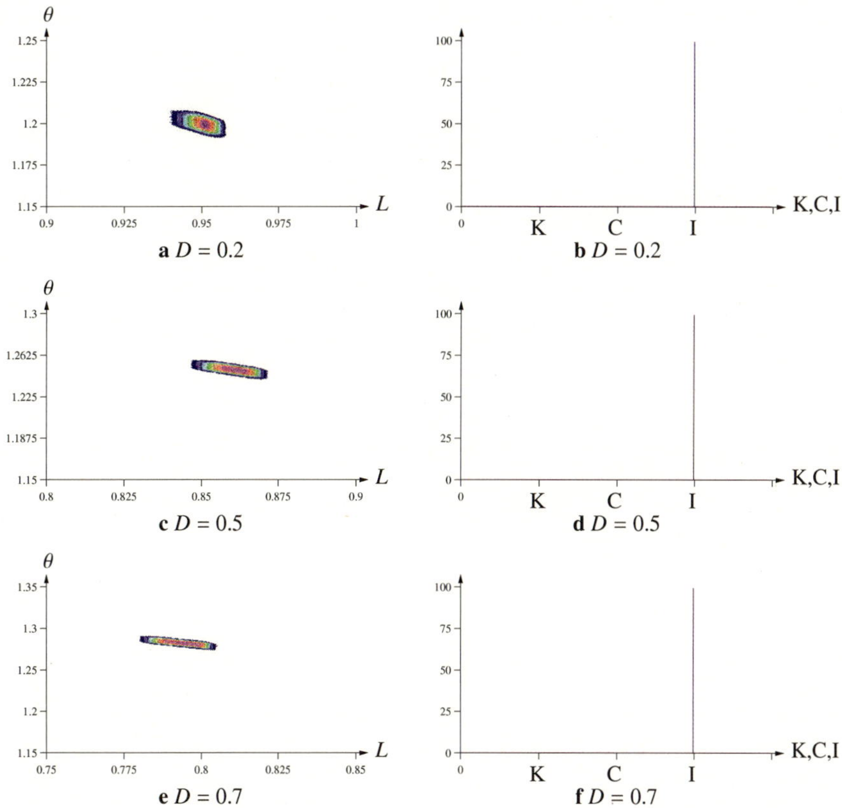

Fig. 8.25 Phillips curve with $B = 0.8$ under $A(\omega)$ uniform i.i.d. $\sim [0.99, 1.01]$ for $D \in \{0.2, 0.5, 0.7\}$, $g = 0.5$, $\tau = 0.35$

For $B = 0.9$ the Phillips curve (and the associated state space attractor) change their characteristic, Figure 8.26 loosing the linearity but maintaining and reinforcing the negative correlation as D increases.

In summary, when productivity shocks are Hicks neutral, one concludes that the endogeneity of labor supply as sutch does not change the long-run positive tradeoff between inflation and employment in Keynesian regimes as shown for the case with constant labor supply, see Figure 8.12. The correlation is positive in these cases, i.e.

deflation rates closer to one in absolute value are correlated with higher levels of employment. The sign of the correlation, i.e. of the long-run tradeoff between inflation

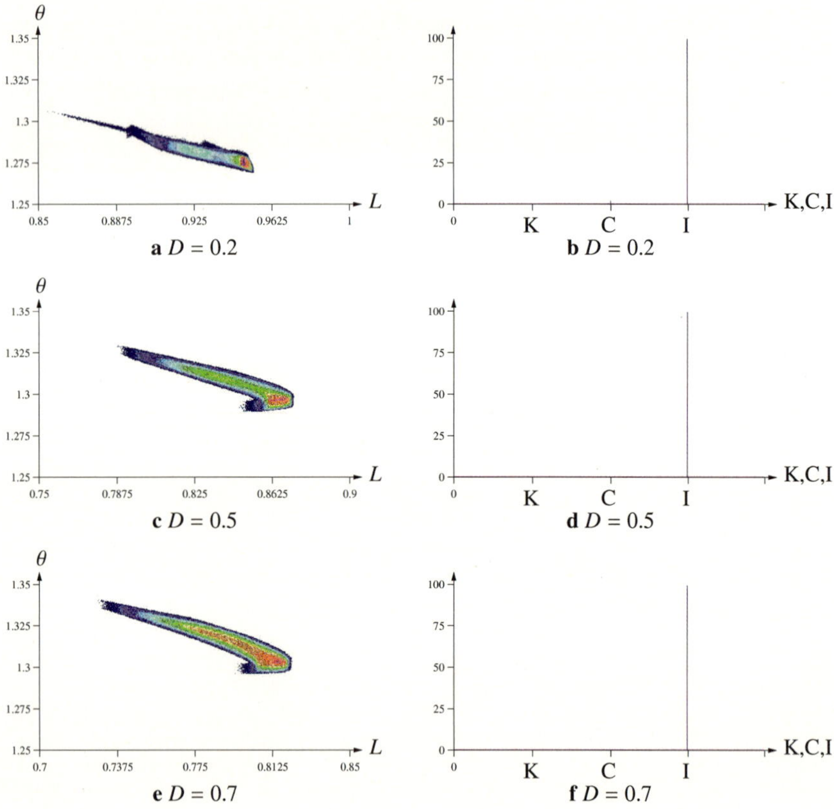

Fig. 8.26 Phillips curve with $B = 0.9$ under $A(\omega)$ uniform i.i.d. $\sim [0.99, 1.01]$ for $D \in \{0.2, 0.5, 0.7\}$, $g = 0.5$, $\tau = 0.35$

and employment with Hicks neutral production shocks depends **primarily** on the elasticity of production B which also induces the switching effect between regimes. However, the elasticity of labor supply seems to smoothen the distribution effect on the attractor as D increases, i.e. as the elasticity $1/D$ becomes smaller. Thus, with higher elasticities of production Keynesian regimes disappear and demand rationing occurs inducing stationary inflationary regimes which exhibit an inverse correlation between employment and inflation.

8.3.2 The Phillips Curve with Random Elasticity $B(\omega)$

When the production scale remains constant at $A = 1$ and the elasticity of production is random the role of endogenous labor remains secondary but not totally unimportant. When Keynesian stationary cycles occur, i.e. with low $B(\omega)$, the Phillips curve under random curvature exhibits the same positive correlation and independence from the labor the supply elasticity as in the Hicksian case, but with noticeably higher variance of inflation rates, compare Figure 8.27 and Figure 8.23. In the nu-

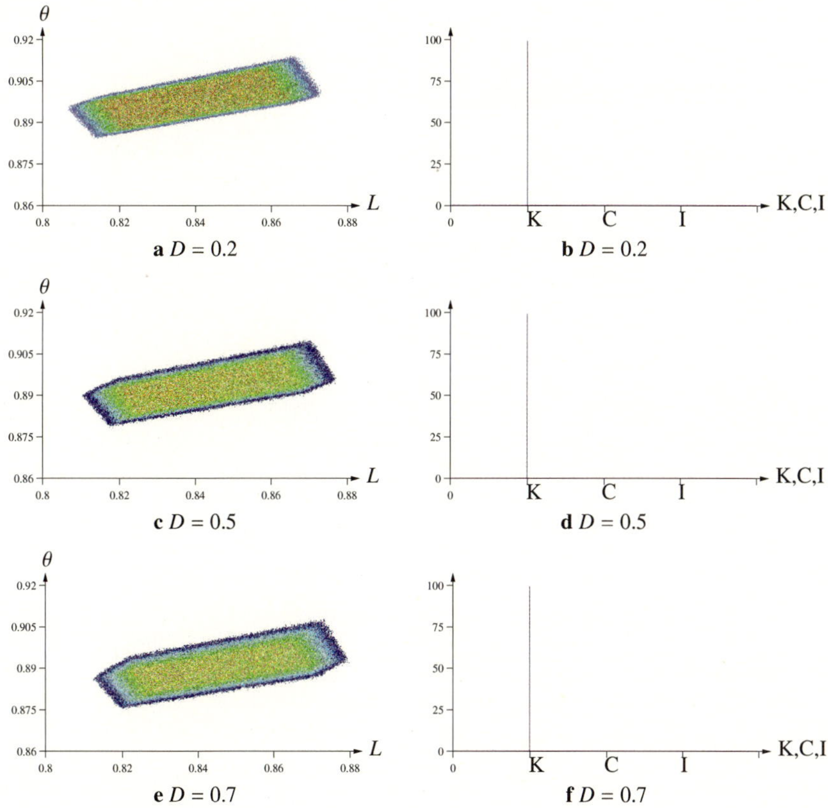

Fig. 8.27 Phillips curve under $B(\omega)$ uniform i.i.d. $\sim$ $[0.59, 0.61]$ for $D \in \{0.2, 0.5, 0.7\}$, $g = 0.5$, $\tau = 0.35$, $T = 5 \cdot 10^5$

merical experiment shown the two parameter configurations are such that $B = 0.6$ with $\mathbb{E}\{A(\omega)\} = 1.0$ leads to a geometrically indistinguishable attractor compared to $A = 1.0$ and $\mathbb{E}\{B(\omega)\} = 0.6$. Thus, when production shocks are typically not observable, the statistical features of the Phillips curve would be indistinguishable for the two different shocks, making an identification from the data difficult.

Figures 8.27 to 8.30 show the qualitative changes for $D = 0.2, 0.5, 0.7$ as $\mathbb{E}\{B(\omega)\}$ increases from 0.6 to 0.9. The experiments are again for $g = 0.5$, $\tau = 0.35$ with $T = 10^6$.

As the mean of an i.i.d. shock of the elasticity of production increases switching from Keynesian to inflationary stationary states occur which simultaneously changes the positive correlation between employment and inflation to a negative one while the elasticity of labor supply smoothens out the attractor and the Phillips curves but also reinforces the switching to inflationary states.

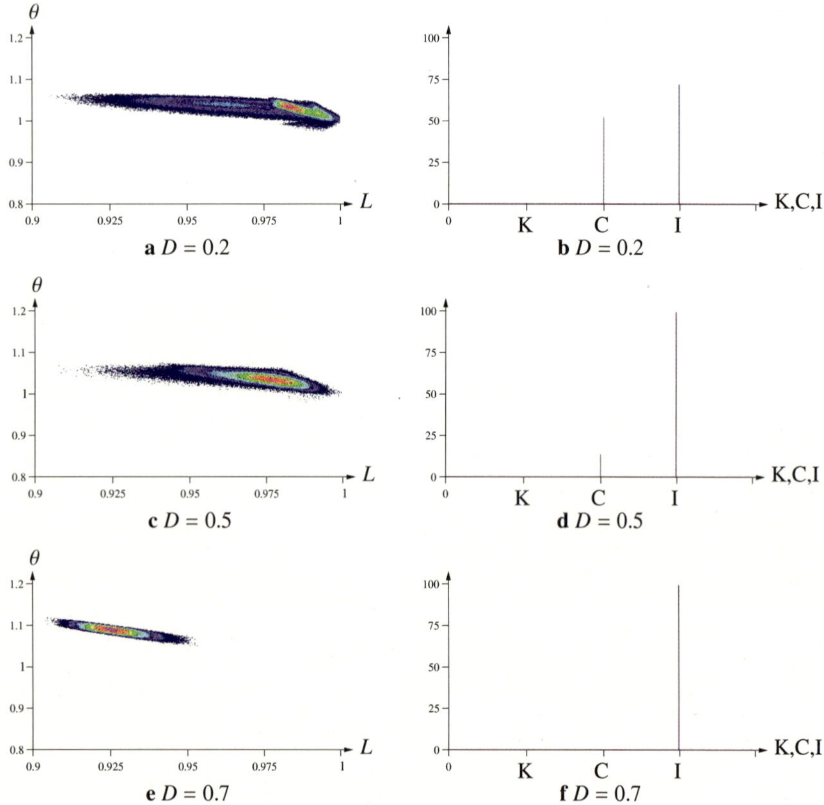

Fig. 8.28 Phillips curve under $B(\omega)$ uniform i.i.d. $\sim [0.69, 0.71]$ for $D \in \{0.2, 0.5, 0.7\}$, $g = 0.5$, $\tau = 0.35$, $T = 5 \cdot 10^5$

For $\mathbb{E}\{B(\omega)\} = 0.8$ no more regime switching occurs and a weak negative correlation persists, but the attractor exhibits properties with additional rotation in the data, see Figure 8.29.

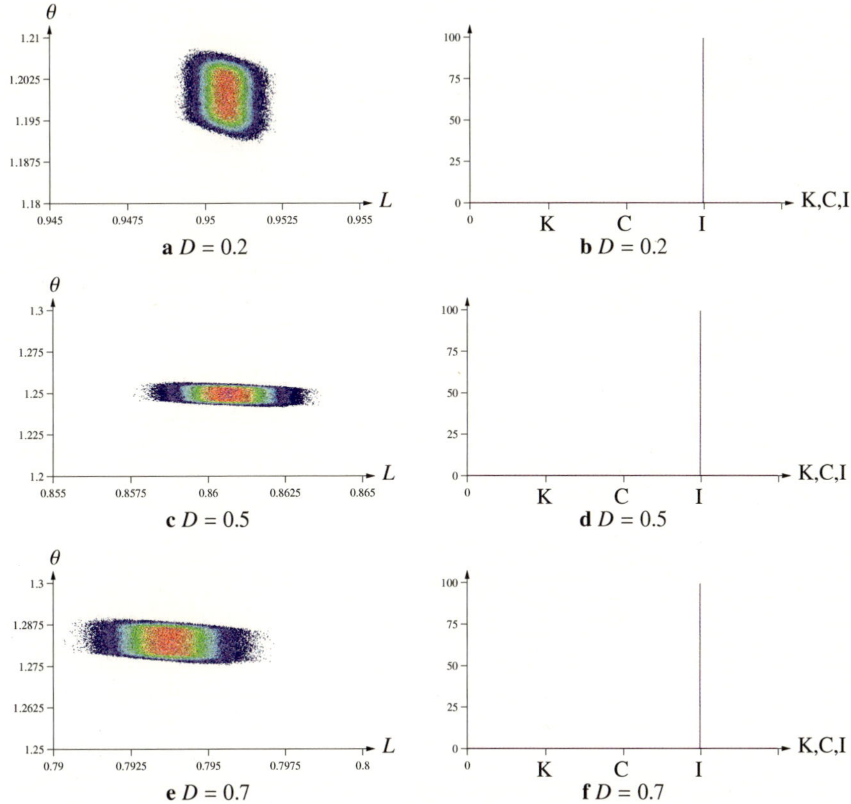

Fig. 8.29 Phillips curve under $B(\omega)$ uniform i.i.d. $\sim$ $[0.79, 0.81]$ for $D \in \{0.2, 0.5, 0.7\}$, $g = 0.5$, $\tau = 0.35$, $T = 10^6$

For $\mathbb{E}\{B(\omega)\} = 0.9$ the Phillips curve has a much wider spread of employment and the elasticity of labor supply partly induces bimodal distributions, Figure 8.30.

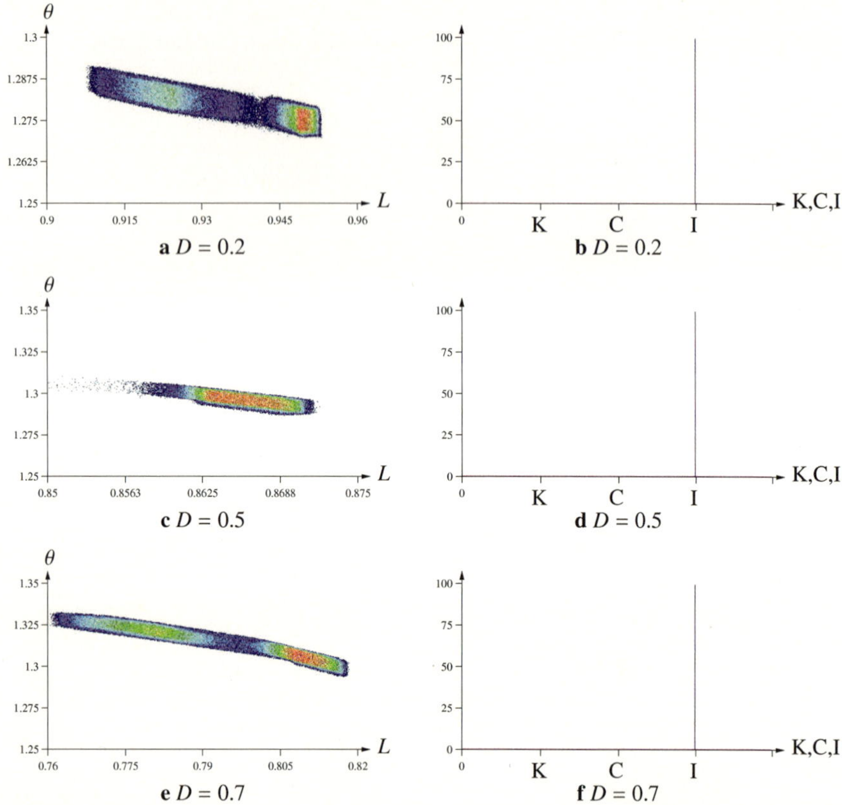

Fig. 8.30 Phillips curve under $B(\omega)$ uniform i.i.d. $\sim$ $[0.89, 0.91]$ for $D \in \{0.2, 0.5, 0.7\}$, $g = 0.5$, $\tau = 0.35$, $T = 10^6$

8.3.3 Employment and the Real-Wage Tradeoff with $A(\omega)$

In a final comparison of the productivity shocks of scale versus curvature, it is interesting to observe that the attracting stationary data of the labor market differ structurally between the two cases. The effects from endogenous labor supply on the tradeoff between the real wage and employment are much weaker under both productivity shocks with low values of the elasticity of production. They show an almost identical negative correlation for the values of D considered. However, for a random elasticity of production the sign of the correlation in the Keynesian case is reversed, negative under $A(\omega)$ (Figure 8.31) versus positive under $B(\omega)$ (Figure 8.35). In other words, while the two associated Phillips curves were indistinguishable, their labor market correlations under the Keynesian regime are opposite to each other. The impact from the elasticity of the supply of labor seems to be secondary.

Figures 8.31 to 8.34 show the results of the same Hicksian productivity shock as the production elasticity ranges between $B = 0.6$ to $B = 0.9$ for different values of D. At $B = 0.6$ the attractor is completely in the Keynesian region with a dominating negative correlation between employment and the real wage. Moreover, it displays properties comparable to an AR1 system with real roots without rotation. As B is increased beyond $B = 0.6$ regime switching away from a purely Keynesian attractor occurs.

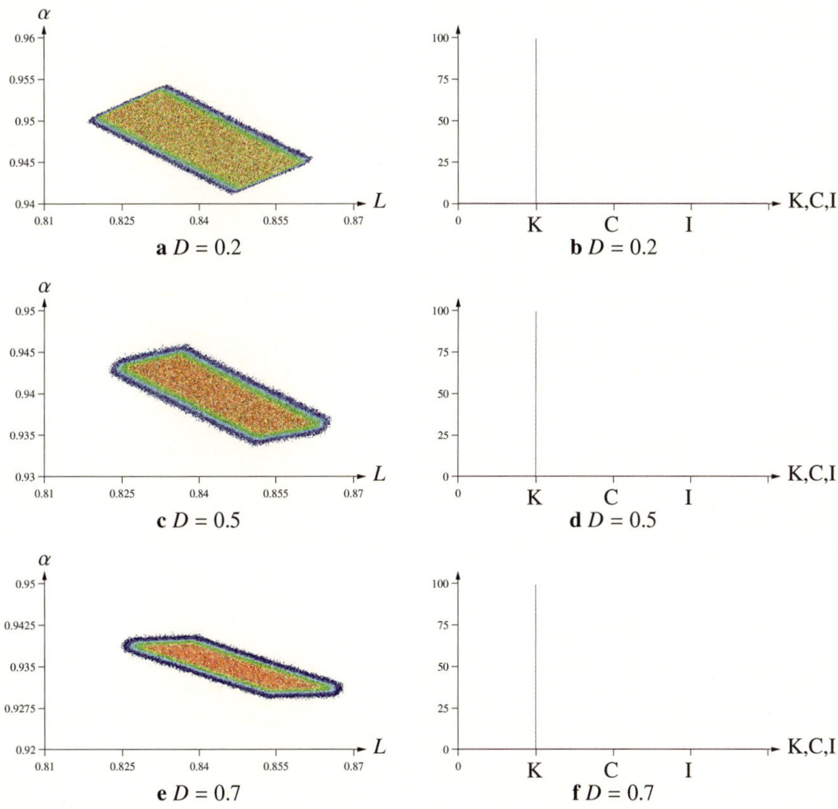

a $D = 0.2$ **b** $D = 0.2$ **c** $D = 0.5$ **d** $D = 0.5$ **e** $D = 0.7$ **f** $D = 0.7$

Fig. 8.31 Employment with $B = 0.6$ under $A(\omega)$ uniform i.i.d. $\sim [0.99, 1.01]$ for $D \in \{0.2, 0.5, 0.7\}$, $g = 0.5$, $\tau = 0.35$

At $B = 0.7$ switching away from Keynesian states to classical and inflationary states is observed influenced by the elasticity of labor supply. The higher elasticity decreases the frequency of the occurrence of classical states and increases the rotation in the inflationary region, Figure 8.32, **a**, **c**, and **e**, also in Figure 8.33. During the switching phase the correlation between the real wage and employment is reversed to a positive one, from **a** and **c** in Figure 8.32. Attractors and distributions

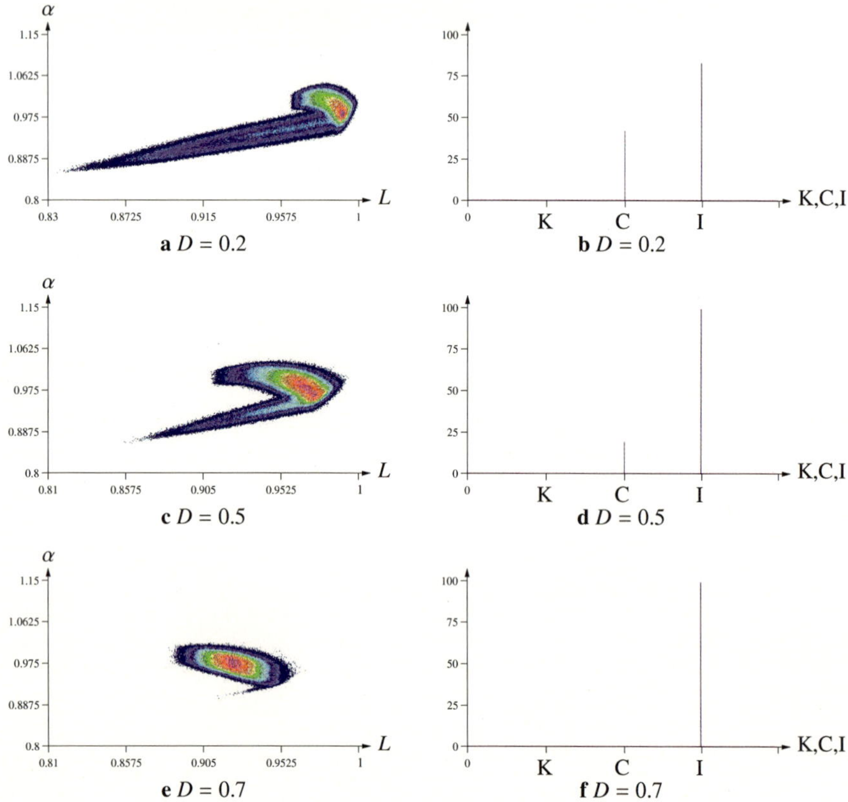

Fig. 8.32 Employment with $B = 0.7$ under $A(\omega)$ uniform i.i.d. ~ $[0.99, 1.01]$ for $D \in \{0.2, 0.5, 0.7\}$, $g = 0.5$, $\tau = 0.35$

appear to be similar for the two kinds of perturbations of scale or curvature. No clear distinct differences were detected in the numerical experiments.

As the elasticity of production increases to $B = 0.8$ stationary states are of the inflationary type only. The attractor displays a clear negative correlation with more rotation. The role of the endogeneity of labor supply appears to be secondary. Comparing the results for the two cases of noise - random scale versus random curvature - one observes some similarity of the attractors for the case with $A = 1$ and random elasticity with $\mathbb{E}\{B(\omega)\} = 0.8$, see Figure 8.37. However, under random elasticity there is no clear correlation between employment and the real wage, but a strong indication for rotation.

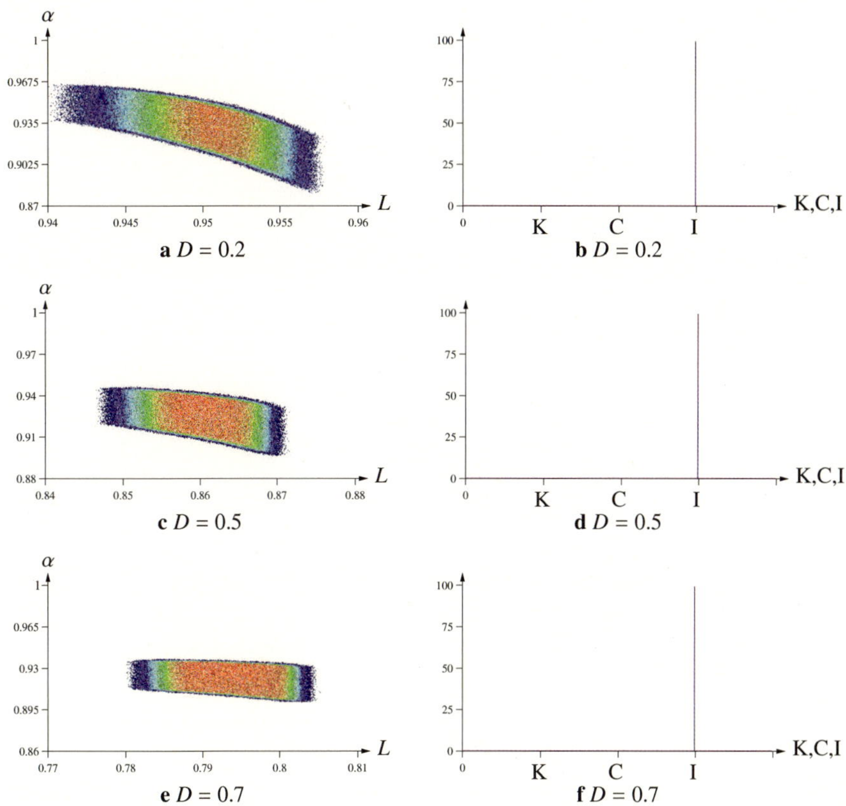

Fig. 8.33 Employment with $B = 0.8$ under $A(\omega)$ uniform i.i.d. ~ [0.99, 1.01] for $D \in \{0.2, 0.5, 0.7\}$, $g = 0.5$, $\tau = 0.35$

For $B = 0.9$ the attracting stationary tradeoff between employment and the real wage is no longer monotonic, see Figure 8.34 indicating a clear dependence on the elasticity of labor supply. When it is large ($D = 0.2$) then the attractor shows a bimodal distribution for the tradeoff which weakens for $D = 0.5$ and disappears for $D = 0.7$.At this level the second mode in the distribution, which existed at

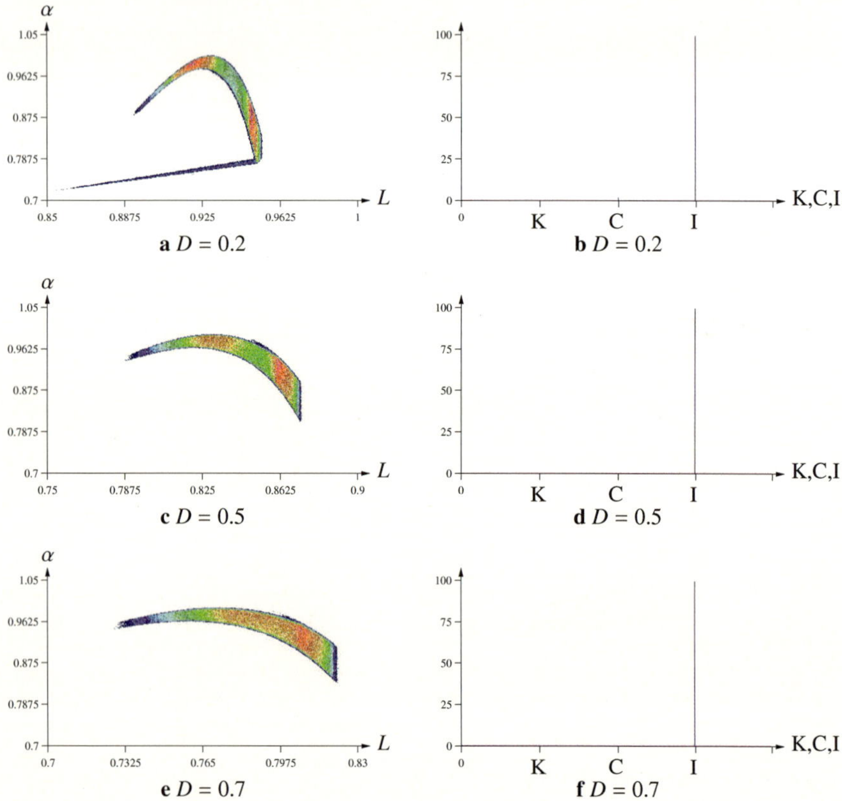

Fig. 8.34 Employment with $B = 0.9$ under $A(\omega)$ uniform i.i.d. ~ [0.99, 1.01] for $D \in \{0.2, 0.5, 0.7\}$, $g = 0.5$, $\tau = 0.35$

lower levels of employment, has disappeared and the attractor suggests a curved but negative correlation.

8.3.4 Employment and the Real-Wage Tradeoff with $B(\omega)$

The tradeoff between employment and the real wage under random curvature but constant production scale $A = 1$ shows a very high strong positive correlation with an attractor which is almost linear, see Figure 8.35. This contrasts with the reverse correlation under Hicks neutral production shocks (random scale) for the comparable set of parameters, see Figure 8.31.

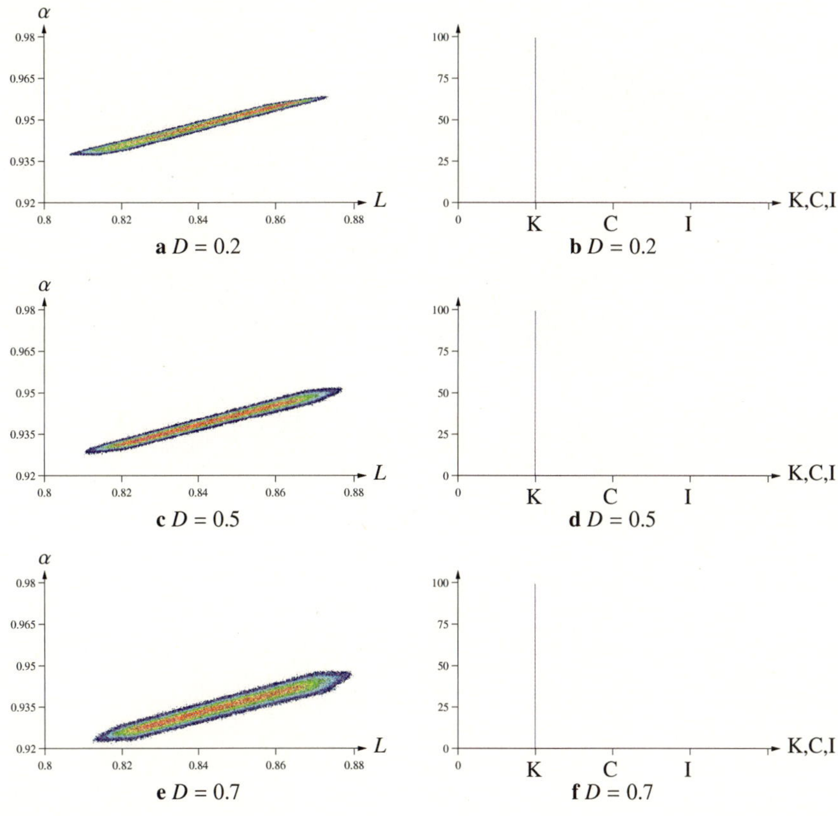

Fig. 8.35 Employment under $B(\omega)$ uniform i.i.d. $\sim$ $[0.59, 0.61]$ for $D \in \{0.2, 0.5, 0.7\}$, $g = 0.5$, $\tau = 0.35$

There seems to be no clear pattern which distinguishes the case with random curvature from random scale when the switching occurs, i.e. when $\mathbb{E}\{B(\omega)\} = 0.7$ with $A = 1$ compared to $\mathbb{E}\{A(\omega)\} = 1$ and $B = 0.7$. However, under both experiments a lower elasticity of labor supply seems to enforce the switching, the attractivity of the inflationary states, and a dominating rotation of attractors in the inflationary cases.

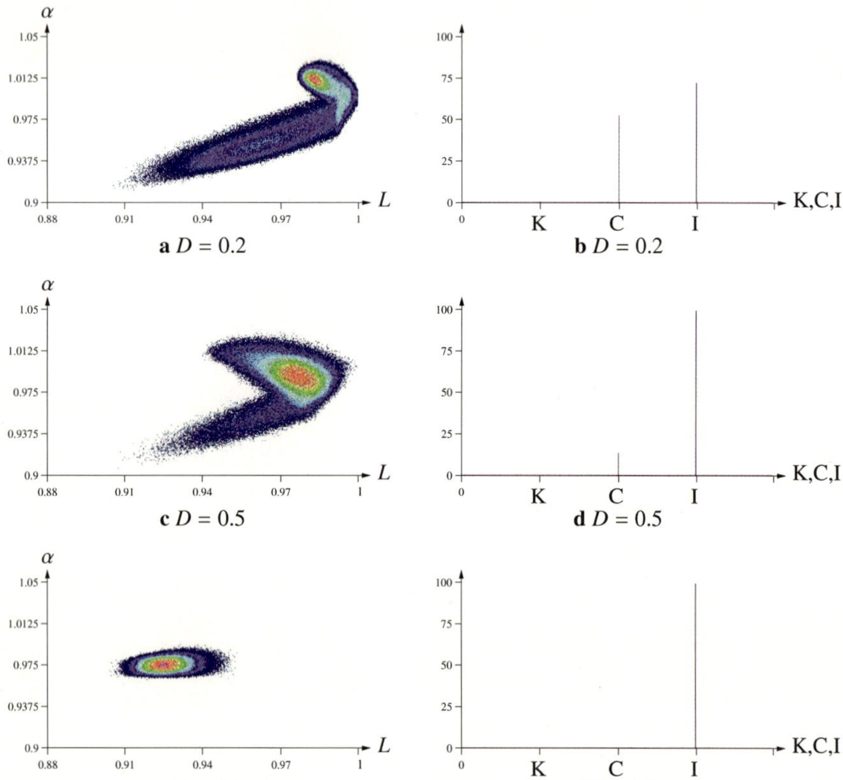

a $D = 0.2$ **b** $D = 0.2$ **c** $D = 0.5$ **d** $D = 0.5$ **e** $D = 0.7$ **f** $D = 0.7$

Fig. 8.36 Employment under $B(\omega)$ uniform i.i.d. $\sim [0.69, 0.71]$ for $D \in \{0.2, 0.5, 0.7\}$, $g = 0.5$, $\tau = 0.35$

For $\mathbb{E}\{B(\omega)\} = 0.8$ the attracting inflationary states show strong signs of rotation without clear correlation and almost no influence from the labor supply elasticity. This typical feature, however, disappears and changes to a narrow attractor without clear evidence of rotation but a completely different narrow form when the mean noise increases to $\mathbb{E}\{B(\omega)\} = 0.9$, Figure 8.38. Thus, the effect of the mean noise and the elasticity of labor supply play structurally different roles.

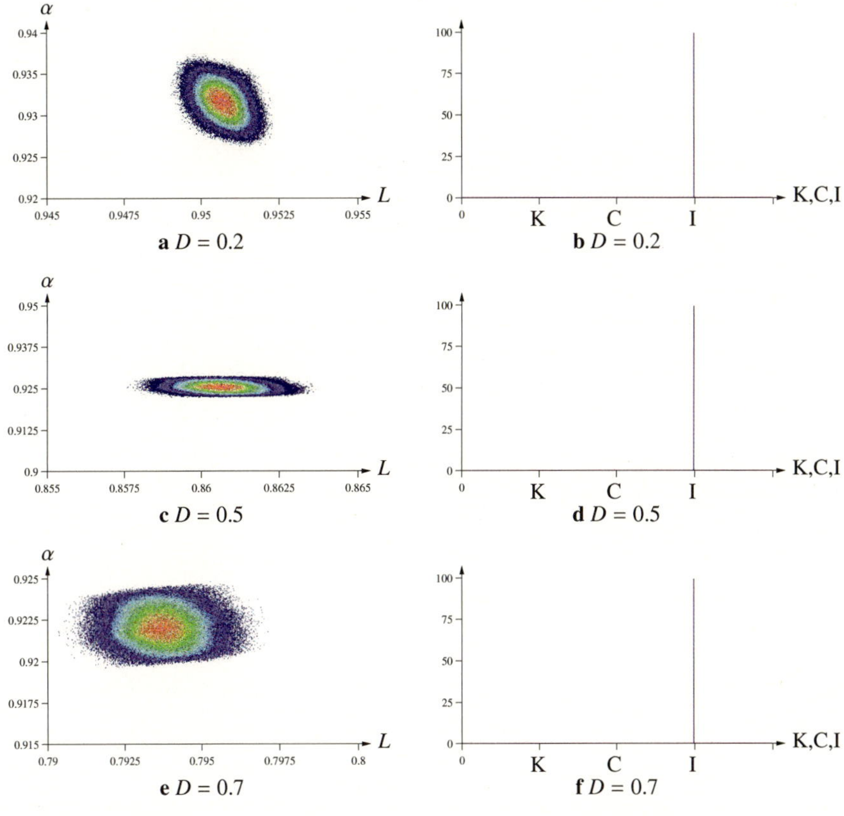

Fig. 8.37 Employment under $B(\omega)$ uniform i.i.d. $\sim [0.79, 0.81]$ for $D \in \{0.2, 0.5, 0.7\}$, $g = 0.5$, $\tau = 0.35$

With $\mathbb{E}\{B\} = 0.9$ the attracting sets under both experiments are very similar to each other. For high elasticity ($D = 0.2$) the tradeoff is not monotonic with a bimodal distribution (as in Figure 8.34). The attractor remains curved but with negative correlation for $D = 0.7$, keeping the bimodality in contrast to the comparable case under Hicksian productivity shocks.

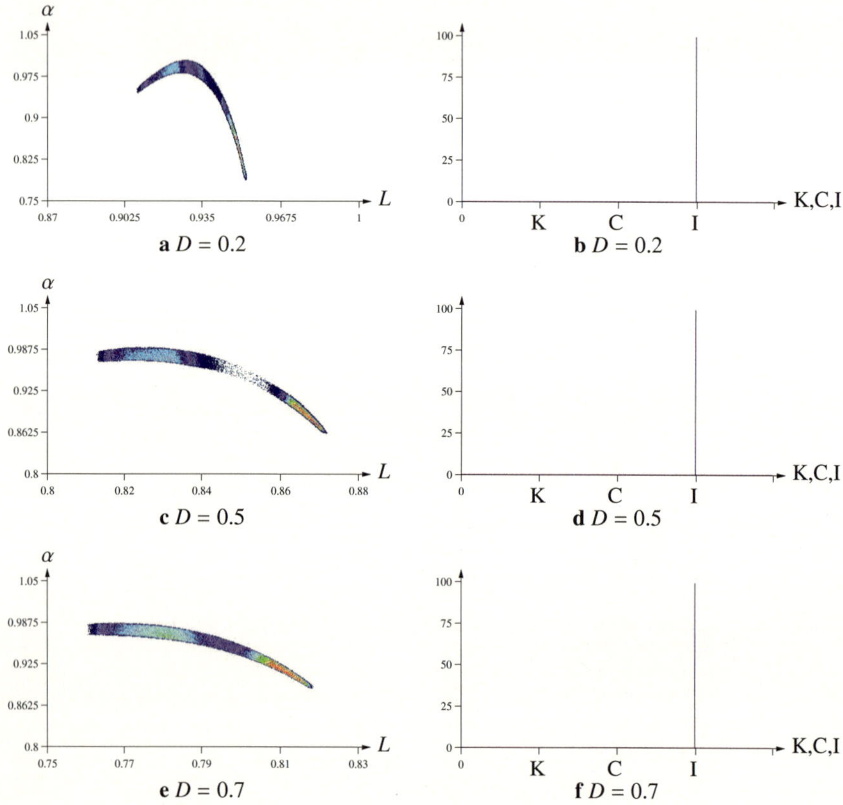

Fig. 8.38 Employment under $B(\omega)$ uniform i.i.d. $\sim [0.89, 0.91]$ for $D \in \{0.2, 0.5, 0.7\}$, $g = 0.5$, $\tau = 0.35$

8.4 Stochastic Disequilibrium Dynamics: An Appraisal

The analysis of this chapter was designed to exhibit major qualitative characteristics of the asymptotic (the long-run or stationary) behavior of stochastic real business cycles under disequilibrium in monetary macroeconomic systems of the type presented in Chapter 6. Different scenarios of the parameters were examined when some of them are subjected to an exogenous noise process. Formally, such stochastic difference equations correspond to the randomization over a family of difference equations (i.e. of deterministic time-one mappings) generating random sample paths as orbits of perturbed dynamical systems. The time series were generated using standard numerical iterative methods applied to a fully parametrized strictly forward recursive mathematical model of the macroeconomic system. These correspond to empirically observable time series of an economy in a stochastic environment which are the generic objects of a theoretical and an empirical investigation.

From the methodological point of view the insight to be gained from this observation consists of the fact that the **simultaneity** *of the dynamic forces and of the random perturbations* creates a complex interaction within a class of *nonautonomous* dynamical systems whose properties, in general, cannot be judged or derived using only partial information about the family of the dynamical systems, for example from its set of deterministic fixed points, their random mixing, or from the characterization of parametrized stationary solutions, approaches which are often used in dynamic macroeconomics. The long-run behavior of *orbits* of stochastic difference equations is neither a mixing of deterministic fixed points nor of deterministic orbits. In general, features of *stochastic* orbits are *poorly* describable by the orbits of linearized approximations of the nonlinear mappings or by the mixing of their deterministic fixed points, whose limiting properties often do not converge to the nonlinear solutions even under small noise. Therefore, a statistical or econometric analysis of such approximating solutions will necessarily generate biased estimates of the time series of the true system or make wrong forecasts since the wrong numerical data of the true system are used. There exists little knowledge about the convergence of statistics of approximating affine systems for nonlinear stochastic equations.

As a consequence, a useful and adequate time series analysis of such systems requires that the conceptual framework as well as the mathematical techniques should be chosen to take a correct account of these interactions. It seems that the methods chosen here from the theory of random dynamical systems in the sense of Arnold (1998) are best suited for the type of theoretical and empirical analysis of macroeconomic models in discrete time. This approach stipulates that the long-run behavior of stochastic nonlinear equations takes place on *random attractors* sometimes given by *random fixed points* often with complex invariant distributions, concepts which are well defined and for which geometric descriptions and numerical approximations have been introduced.

The strictly forward recursive modeling of the state variables allows the application of exact numerical techniques. They provide a method to compute forward iterations with all nonlinearities and most stationary random processes of the noise. These can be applied efficiently and successfully for random dynamical systems in discrete time. Judging from the results presented in this chapter these methods are particularly useful in the context of models described by piecewise differentiable nonlinear mappings implying typical regime switching and stationary attracting sets with complex limiting distributions for which analytical solutions or qualitative statistical properties like moments are typically unavailable. In such cases their limiting properties can be determined and investigated numerically from sufficiently long orbits invoking the implications from ergodic theory for the validity of the results.

The numerical results in Sections 8.1 to 8.3 reveal a large range of different qualitative properties and their dependence on the parameters. The numerical analysis concentrated first on the issue of whether linear approximations near the Walrasian equilibrium can be used to derive qualitative properties of the stationary features of the model under small and varying noise. The randomization *and* the dynamics interact strongly inducing simultaneous sources for regime switching. Specifically, the disequilibrium adjustment in markets and the occurrence of the shocks interact inducing frequent and recurrent regime switching and spillovers. These generate e.g. specific stationary tradeoffs between employment and inflation and other distinct cross correlations in the long run. Therefore, the results provide evidence that an approximation via linear techniques is not useful and conclusive due to the nonlinearities of the model and the occurrence of frequent switching near the Walrasian equilibria.

The second economic investigation examined mainly the role of two types of productivity shocks: the effects of random scale and of random curvature. Their impact on the long-run correlation/tradeoff of employment with respect to the real wage and to inflation, the so-called Phillips Curve, was examined. This led to a characterization of attracting stationary features of real business cycles which are specific either to the Keynesian, the inflationary regime, or whether regime switching dominates the stationary behavior. The results clearly show that there are decisive differences of scale effects and curvature effects from production which depend also on the elasticity of labor supply. Specifically, for both production shocks, i.e. under Hicks neutrality or random labor productivity, positive (almost linear), zero, or negative correlations between the real wage and the level of employment were shown to appear. The attracting set in employment-real wage space appears to be similar to the limiting set of an AR1 process if it is contained in the Keynesian region. It reveals nonlinear features in the inflationary region or when there is regime switching.

The results also show that the long-run behavior in typical generic cases are easily identifiable as random fixed points or random cycles featuring properties similar to those occurring under deterministic bifurcations. In particular, a systematic numerical analysis with respect to single parameters exhibits the qualitative properties of the stochastic and dynamic consequences under parametric changes. Thus, numerical analysis provides powerful tools to examine the existence of stochastic bifurcations, i.e. identifying the economic causes for structural changes of the attracting long-run behavior from random fixed points to cycles or higher order stationary solutions leading to multimodular invariant distributions.

Appendix A
Dynamical Systems in Discrete Time

A.1 Stability and Cycles in Deterministic Systems

This section presents some selected definitions, concepts, and results. For more details, the reader should consult any standard text on dynamical systems (for example Guckenheimer & Holmes, 1983; Hale & Koçak, 1991; Kuznetsov, 1995).

A.1.1 Basic Concepts

Definition A.1.1. A dynamical system in discrete time $\mathbb{N} = \{0, 1, 2, \dots\}$ on a set $\mathcal{X} \subset \mathbb{R}^n$ is given by the so-called *time-one map*

$$F : \mathcal{X} \to \mathcal{X}$$

describing the change of the state of the system by one time step. The set $\mathcal{X}$ is called the *state space* of the system.
The system is called *autonomous* if the time-one map is independent of time t. In this case the system is said to reach the state x_t, $t \geq 1$ in period t from an initial state $x \in \mathcal{X}$ after t successive applications of the mapping F, i.e.

$$x_t = \underbrace{F \circ \cdots \circ F(x)}_{t\text{-times}} \qquad \forall t \in \mathbb{N}. \tag{A.1.1}$$

Therefore, one defines the mapping $\varphi : \mathbb{N} \times \mathcal{X} \to \mathcal{X}$

$$\varphi(t, x) := \begin{cases} F \circ \cdots \circ F(x) \equiv F^t(x) & t \geq 1 \\ x & t = 0 \end{cases} \tag{A.1.2}$$

which describes the state of the system reached at time t from an arbitrary initial position. The mapping satisfies

$$\varphi(s+t,x) = \varphi(t,x) \circ \varphi(s,x) \quad \text{for all } s,t \in \mathbb{N}. \tag{A.1.3}$$

For any initial state $x_0 \in \mathcal{X}$ the iterated values $x_k := F^k(x_0)$ satisfy the *first order difference equation*

$$x_{k+1} = F(x_k) \quad k \in \mathbb{N}.$$

Definition A.1.2. The positive *orbit* $\gamma^+(x_0)$ of any $x_0 \in \mathcal{X}$ is defined as

$$\gamma^+(x_0) := \left\{F^k(x_0)\right\}_{k=0}^{\infty} \tag{A.1.4}$$

Definition A.1.3. $\bar{x} \in \mathcal{X}$ is called a *fixed point* or *steady state* of F if $F(\bar{x}) = \bar{x}$.

Definition A.1.4. $\bar{x} \in \mathcal{X}$ is called a *periodic point of F with period m* ≥ 1, if $F^m(\bar{x}) = \bar{x}$ and $F^k(\bar{x}) \neq \bar{x} \quad 0 < k < m$.

Any periodic point $\bar{x}$ with period $m > 1$ is a fixed point of the mapping $F^m(\bar{x}) = \bar{x}$. The periodic point $\bar{x}$ coexists with $m-1$ additional distinct periodic points $\bar{x}_j := F^j(\bar{x})$, $j = 1, \ldots, m-1$, satisfying $F^m(\bar{x}_j) = \bar{x}_j$. The list $(\bar{x}, \bar{x}_j, \ldots, \bar{x}_{m-1})$ is also-called a *cycle* of order m.

Definition A.1.5. A fixed point $\bar{x}$ of F is called *stable*, if for every $\epsilon > 0$ there exists a $\delta(\epsilon) > 0$ such that

$$\|F^k(x_0) - \bar{x}\| < \epsilon \quad \text{for all} \quad \|x_0 - \bar{x}\| < \delta(\epsilon)$$

holds for all k.

Definition A.1.6. A fixed point $\bar{x}$ of F is called *asymptotically stable*, if it is stable and there exists a $\delta > 0$, such that

$$\lim_{k\to\infty} F^k(x_0) = \bar{x} \quad \text{for all} \quad \|x_0 - \bar{x}\| < \delta.$$

Definition A.1.7. A fixed point of a differentiable dynamical system $F : \mathcal{X} \to \mathcal{X}$, $\mathcal{X} \subset \mathbb{R}^m$ is called *hyperbolic*, if all eigenvalues λ_j, $j = 1, \ldots, n$ of the Jacobian matrix $DF(\bar{x})$ have modulus different than one, i.e. $|\lambda_j| \neq 1$, $j = 1, \ldots, m$.

Corollary A.1.1. *Hyperbolic fixed points are either asymptotically stable or unstable.*

A.1.2 One Dimensional Systems

For one dimensional differentiable systems there is a simple and intuitive criterion for asymptotic stability of fixed points and periodic points.

Lemma A.1.1. *A hyperbolic fixed point $\bar{x}$ of F is asymptotically stable if and only if* $|DF(\bar{x})| < 1$.

Lemma A.1.2. *A hyperbolic periodic point $\bar{x}_m = F^m(\bar{x}_m)$ of order $m > 1$ is asymptotically stable if and only if*

$$\prod_{j=1}^{m} |(DF(\bar{x}_j))| < 1, \quad \bar{x}_j := F^j(\bar{x}_m) \quad j = 1, \ldots, m. \tag{A.1.5}$$

A.1.3 Linear Systems

Figure A.1 displays the graphs of two mappings F (red curve) and the induced time

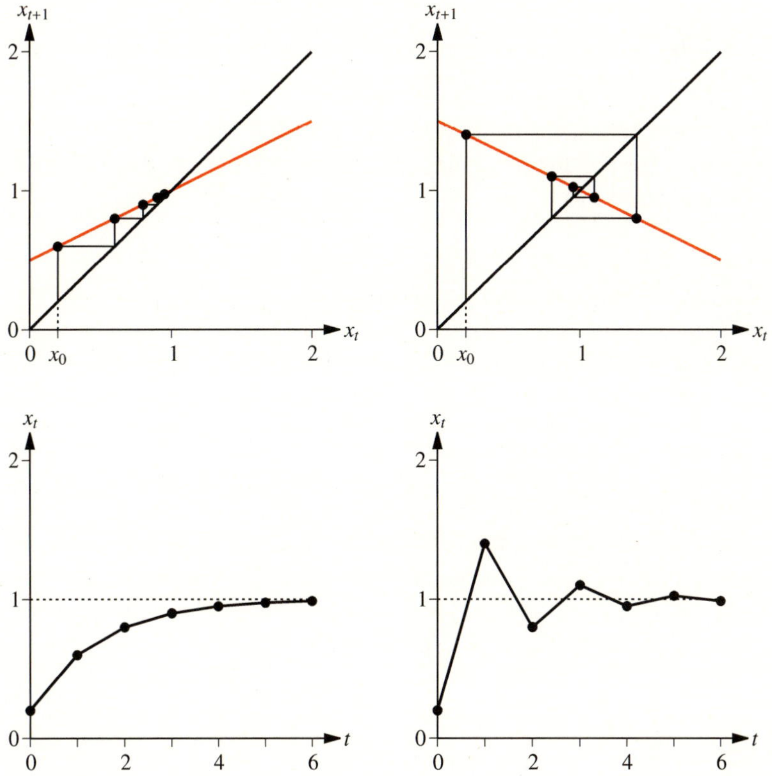

Fig. A.1 State space representations and time profiles of affine systems

series of part of an orbit. Affine systems are either stable or unstable, depending on whether the slope of the graph is less than or larger than one in absolute value. Orbits are monotonically converging when the slope is positive and less than one.

A.1.4 Examples of Nonlinear Systems

Continuous *monotonically increasing* mappings have no fixed points of order larger than one.

Example A.1.1 (A Third Order Polynomial). For $\mu > 0$ let $f : \mathbb{R} \to \mathbb{R}$ be defined by

$$f(x) = \frac{1}{\mu}(x^3 + x) \tag{A.1.6}$$

Example A.1.2 (A sigmoid function). For $0 \leq \mu < 1$, let

$$f(x) = \frac{\frac{\mu}{1-\mu}x}{\sqrt{1 + \left(\frac{\mu}{1-\mu}x\right)^2}} \tag{A.1.7}$$

Lemma A.1.3. *Let $f : \mathbb{R} \to \mathbb{R}$ be continuous and monotonically decreasing. Then, f cannot have cycles of order higher than two.*

Therefore, one-dimensional maps with cycles of order higher than two or with complex dynamics require discontinuities or sufficient variations in slope from positive to negative or vice versa.

Example A.1.3 (The Logistic Map). $F : [0, 1] \to [0, 1]$ defined by

$$F(x, \mu) := \mu x(1 - x), \quad 0 < \mu \leq 4 \tag{A.1.8}$$

Example A.1.4 (The Tent Map). For $0 < \mu < 1$, let $f : [0, 1] \to [0, 1]$ be defined by

$$f(x) = \begin{cases} \frac{x}{\mu} & 0 \leq x \leq \mu \\ \frac{1-x}{1-\mu} & \mu < x \leq 1 \end{cases} \tag{A.1.9}$$

Example A.1.5 (The Dyadic Map). For $0 < \mu < 1$, let $f : [0, 1] \to [0, 1]$ be defined by

$$f(x) = \begin{cases} \frac{x}{\mu} & 0 \leq x \leq \mu \\ \frac{x-\mu}{1-\mu} & \mu < x \leq 1 \end{cases} \tag{A.1.10}$$

Example A.1.6 (A Nonlinear Cobweb Model).

- Demand function: $D(p_t) = 10 - 4p_t$.
- Supply function: $S(p^e_{t-1,t}) = 8p^e_{t-1,t}(2 - p^e_{t-1,t}) + 2$.
- Static Expectations: $p^e_{t-1,t} = p_{t-1}$
- Equilibrium Price:

$$p_t = (D^{-1} \circ S)(p_{t-1}) = 2 - 2p_{t-1}(2 - p_{t-1}). \tag{A.1.11}$$

– $\Longrightarrow$ $F := D^{-1} \circ S : [0,2] \to [0,2]$

Figure A.2 **a** shows the graph of the mapping $F := D^{-1} \circ S$. F has two unstable fixed points. Figures **a** and **b** show part of a complex orbit starting at $p_0 \approx .2$. The first ten iterations induce a fluctuating path reaching $p_{11} \approx 1$, $p_{12} \approx 0$, and $p_{13} < 2$ below the unstable fixed point, then declining back to low values.

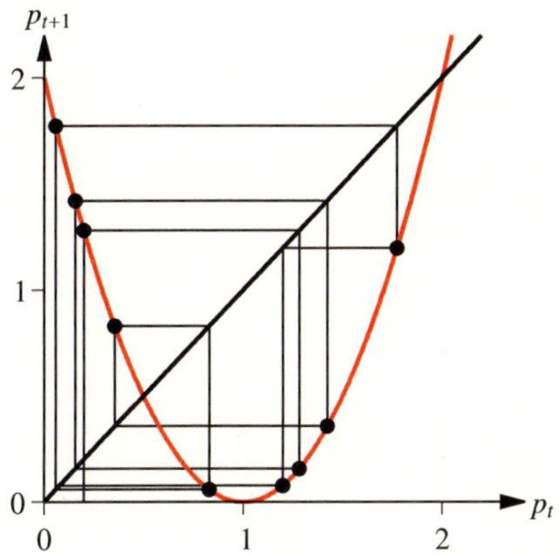

a Part of a complex orbit in (p_t, p_{t+1})-space

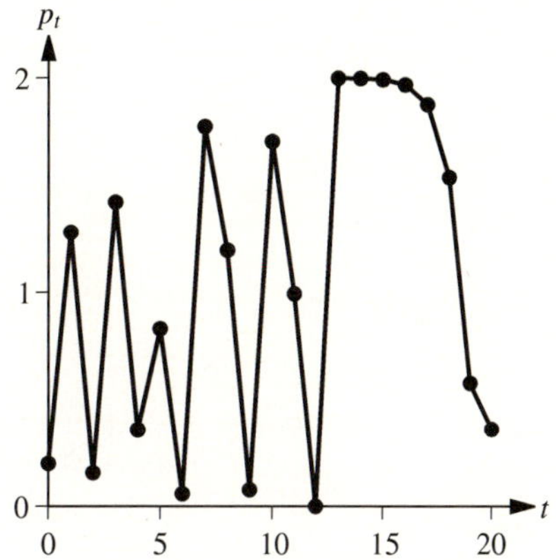

b Time profile of a complex orbit

Fig. A.2 Complex behavior in the nonlinear Cobweb model

A.1.5 Loss of Stability and Local Bifurcations

Consider the logistic map as given in Example A.1.3 (see Devaney, 1989, for more details and results of the logistic map)

$$F(x,\mu) := \mu x(1-x) \quad 0 < \mu \leq 4.$$

- $x = 0$ is a fixed point of $F(\cdot,\mu)$ for all μ. Since

$$\frac{\partial F}{\partial x}(x,\mu) = \mu(1-2x) \tag{A.1.12}$$

  Therefore, $x = 0$ is stable if and only if $0 < \mu < 1$ since

$$\frac{\partial F}{\partial x}(0,\mu) = \mu < 1 \quad \text{if and only if} \quad 0 < \mu < 1. \tag{A.1.13}$$

- F has an additional unique fixed point

$$x(\mu) = 1 - \frac{1}{\mu} \qquad \mu > 1 \tag{A.1.14}$$

- For their asymptotic stability one finds

$$\frac{\partial F}{\partial x}(x(\mu), \mu) = 2 - \mu. \tag{A.1.15}$$

$x(\mu)$ is stable if and only if $1 < \mu < 3$. $x(3)$ is a non hyperbolic fixed point with

$$\frac{\partial F}{\partial x}(x(3), 3) = -1$$

and

$$\frac{\partial}{\partial x}F^2(x(\mu), \mu) = \frac{\partial F}{\partial x}(x(\mu), \mu)\frac{\partial F}{\partial x}(x(\mu), \mu) = (2-\mu)^2. \tag{A.1.16}$$

$$\frac{\partial}{\partial x}F^2(x(3), 3) = 1 \tag{A.1.17}$$

Therefore,

$$\frac{\partial}{\partial \mu}\left(\frac{\partial}{\partial x}F^2(x(\mu), \mu)\right) = 2\mu - 4, \tag{A.1.18}$$

so that

$$\frac{\partial}{\partial \mu}\left(\frac{\partial}{\partial x}F^2(x(3), 3)\right) = 2 > 0.$$

Therefore, a period doubling bifurcation occurs at $\mu = 3$.

Figure A.3 displays the set of fixed points of the logistic map. Figure A.4 shows the

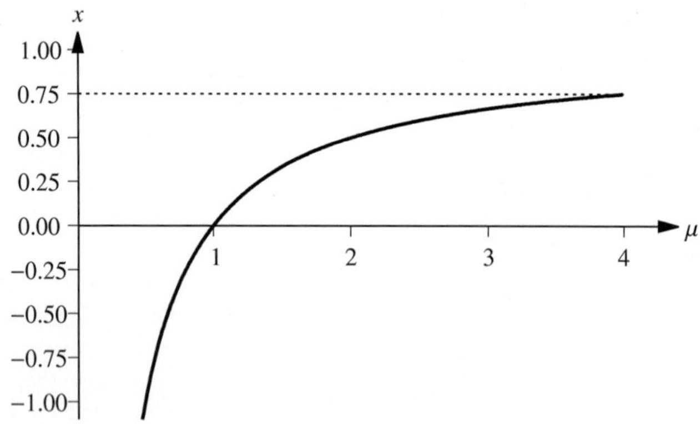

Fig. A.3 Fixed point manifold of the logistic map

graphs of F and of the second iterate F^2 for values of $\mu = 2.5, 3, 3.2, 3.5$, displaying the appearance of the two additional fixed points of period 2 for $\mu > 3$.

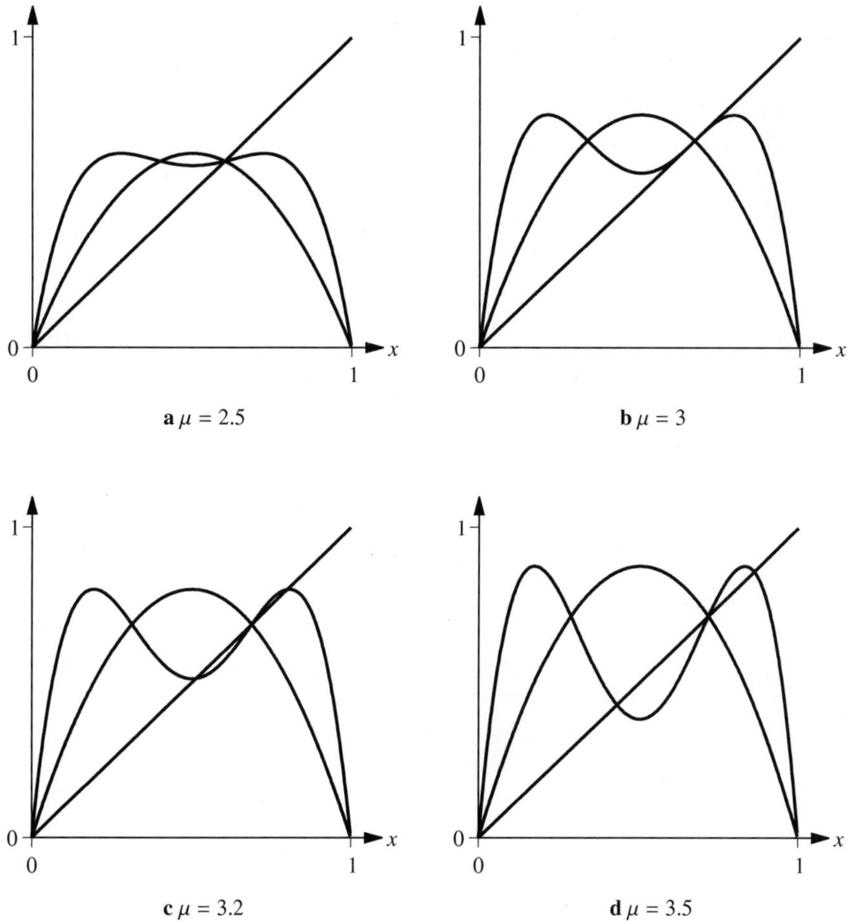

Fig. A.4 The role of the parameter μ for the logistic function F and F^2

Other Features of the Logistic Map

- The period doubling phenomenon continues as μ increases.
- There are corresponding bifurcation points for all 2^n, $n > 2$ periodic points, creating a cascade of period doubling scenarios.
- For $\mu = 1 + \sqrt{8}$, F has a periodic point of order three and a fold bifurcation inducing a stable and an unstable cycle of order three as μ increases.
- For $\mu > 1 + \sqrt{8}$, chaotic orbits occur.
- There is sensitive dependence of orbits on initial conditions.
- Figure A.6 shows the time profile of two orbits with close initial conditions.

Figure A.5 is the so-called bifurcation diagram of the logistic map, showing the limiting behavior of an orbit for each value of μ. Figure A.7 displays the histogram of a long orbit for $\mu = 4$ whose limiting density function is given by $f(x) = 1/(\pi\sqrt{x(1-x)})$ (see Lasota & Mackey, 1994).

Theorem A.1.1 (Li & Yorke (1975)). *Let $F : X \to X$, $X \subset \mathbb{R}$ an interval, be continuous and suppose there exists a point x such that either*

$$F^3(x) \le x < F(x) < F^2(x) \qquad or \qquad F^3(x) \ge x > F(x) > F^2(x). \tag{A.1.19}$$

Then:

(a) There exist periodic points for every period $m \ge 1$.
(b) There exists an uncountable subset $E \subset X$ containing no periodic points such that for all $x, y \in E$ one has:

$$\limsup |F^n(x) - F^n(y)| > 0 \qquad and \qquad \liminf |F^n(x) - F^n(y)| = 0. \tag{A.1.20}$$

(c) if y is a periodic point, then for all points $x \in E$:

$$\limsup |F^n(x) - F^n(y)| > 0$$

The four examples A.1.3 - A.1.6 represent mappings often chosen as applications of Theorem A.1.1 to demonstrate the existence of cycles of high order and of complex behavior. The tent map (A.1.9) satisfies the conditions of the Li and York Theorem for all $0 < \mu < 1$, while the logistic map (A.1.8) does so for μ sufficiently large, exhibiting *topological chaos* in the sense defined by the theorem. Because of condition (A.1.19) the theorem is referred to as demonstrating that "period three implies chaos". The example of the nonlinear Cobweb model (A.1.11) also satisfies the conditions of the theorem. Variations of the logistic function appear in many areas of

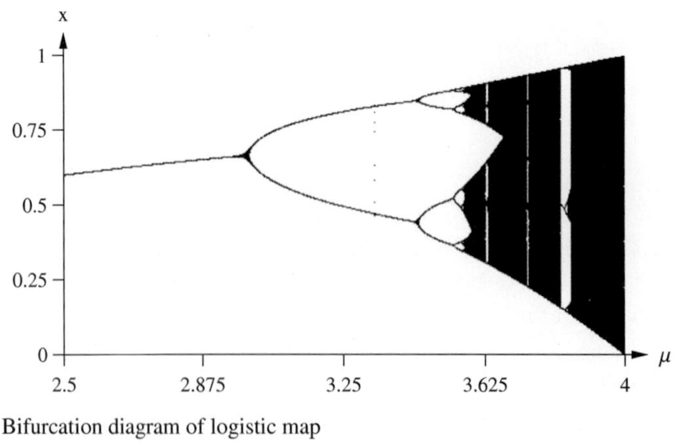

Fig. A.5 Bifurcation diagram of logistic map

economic theory with applications of the results by Li & Yorke (1975) and Devaney

(1989) (as in Hommes, 1991, 1994; Böhm & Kaas, 2000; Onozaki, Sieg & Yokoo, 2000) with numerical results confirming the bifurcation features.
Piecewise smooth maps (as Examples A.1.4 and A.1.5) were discussed as well at relatively early stages (Hicks, 1950; Day, 1984; Day & Shafer, 1987). Economically they are of special interest describing business cycles with regime switching (for example Keener, 1980; Day, 1982; Ishida & Yokoo, 2003; Tramontana, Gardini & Agliari, 2011). These show additional and different bifurcation scenarios beyond the smooth case of the logistic map with a rich mathematical structure (see Avrutin & Sushko, 2013, for a recent survey).

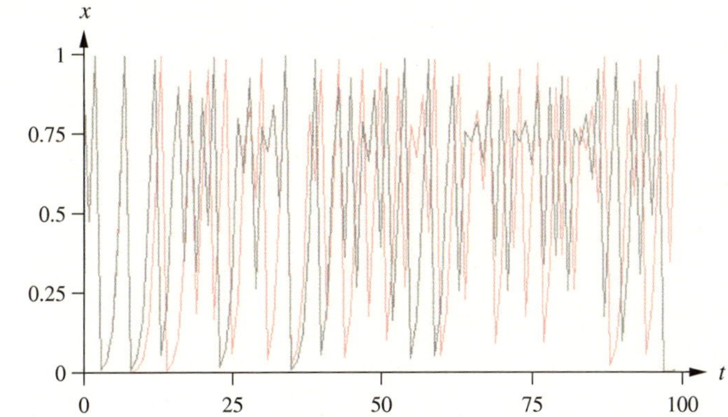

Fig. A.6 Two orbits with close initial conditions: gray: $x_0 = 0.31415$ and pink: $x'_0 = 0.31425$

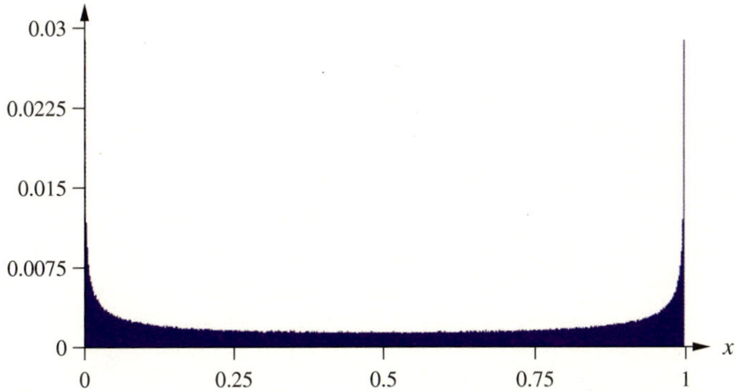

Fig. A.7 Histogram of the logistic map for $\mu = 4$ approximating the density $f(x) = \left(\pi\sqrt{x(1-x)}\right)^{-1}$

The characteristics of the limiting behavior of chaotic orbits is often described by statistical properties for $t \to \infty$. These are approximated numerically by calculating the associated histograms for long orbits. There exists an established mathematical methodology taking a statistical approach of chaotic dynamical systems: "studying chaos with densities" and relating them to stochastic models (see Lasota & Mackey, 1994, for the standard concepts and results). One of the basic results is the following theorem.

Theorem A.1.2 (Day & Pianigiani (1991)). *Let $X \subset \mathbb{R}$ denote an interval of the real line and consider a mapping $F : X \to X$ which is piecewise C^2. If $F'(x) \geq \lambda > 1$ for almost every $x \in X$, then*

(a) all periodic points are unstable
(b) there exists an absolutely continuous invariant measure

The tent map (A.1.9) and the dyadic map (A.1.10) satisfy the conditions of the Theorems A.1.1 and A.1.2. Therefore, for all $0 < \mu < 1$ they have an invariant distribution with associated density defined on the unit interval. For $\mu = 1/2$ this is the uniform distribution on [0, 1]. For the logistic map (A.1.8) with $\mu = 4$ the invariant measure has a continuous density function given by $f(x) := \left(\pi\sqrt{x(1-x)}\right)^{-1}$ and displayed in Figure A.7 (see Lasota & Mackey, 1994).

A.1.6 Two Dimensional Systems

Lemma A.1.4. *Let $F : \mathcal{X} \to \mathcal{X}$ be continuously differentiable with $\mathcal{X} \subset \mathbb{R}^2$. A hyperbolic fixed point $\bar{x} \in \mathcal{X}$ of F is asymptotically stable if and only if both eigenvalues λ_j, $j = 1, 2$ of the Jacobian matrix $DF(\bar{x})$ satisfy $|\lambda_j| < 1$ for $j = 1, 2$.*

To analyze the asymptotic stability of a hyperbolic fixed point according to Lemma A.1.4, one needs to determine the eigenvalues of a 2×2 matrix $DF(\bar{x}) \equiv A = \begin{vmatrix} a & b \\ c & d \end{vmatrix}$. Define the characteristic polynomial of A as

$$\begin{aligned} \chi_A(\lambda) :=& \det \begin{vmatrix} a-\lambda & b \\ c & d-\lambda \end{vmatrix} = \lambda^2 - (a+d)\lambda + (ad - bc) \\ =&\lambda^2 - (\mathrm{tr}A)\,\lambda + \det A. \end{aligned} \tag{A.1.21}$$

Then, the eigenvalues are given by the formula

$$\lambda_{1,2} = \frac{1}{2}\mathrm{tr}A \pm \sqrt{\Delta}, \qquad \Delta := \left(\frac{1}{2}\mathrm{tr}A\right)^2 - \det A \tag{A.1.22}$$

which are inside the unit circle of the complex plane if and only if the three conditions

$$\begin{aligned} &\det A < 1 \\ &\det A > \mathrm{tr}A - 1 \\ &\det A > -\mathrm{tr}A - 1 \end{aligned} \tag{A.1.23}$$

hold. The pairs (det, tr) satisfying the three conditions define the so-called stability triangle, i.e. the relationships between the determinant det and the trace tr of the matrix A under stability, see Figure A.8 (see Pampel, 2010).

a) if $\det A = 0$, then $\lambda_1 = 0$ and $\lambda_2 = \mathrm{tr}A$.
b) if $\det A = 1$ and $|\mathrm{tr}| < 2$, $\Longrightarrow$ the roots are complex with

$$|\lambda_{1,2}| = \left|\frac{\mathrm{tr}A}{2} \pm i\sqrt{1-\left(\frac{\mathrm{tr}A}{2}\right)^2}\right| = \sqrt{\left(\frac{\mathrm{tr}A}{2}\right)^2 + 1 - \left(\frac{\mathrm{tr}A}{2}\right)^2} = 1 \qquad \text{(A.1.24)}$$

c) if $\det A = \mathrm{tr}A - 1$, $\Longrightarrow$

$$\lambda_{1,2} = \frac{\mathrm{tr}A}{2} \pm \sqrt{\left(\frac{\mathrm{tr}A}{2}\right)^2 - \mathrm{tr}A + 1} = \frac{\mathrm{tr}A}{2} \pm \left(\frac{\mathrm{tr}A}{2} - 1\right) = \begin{cases} \mathrm{tr}A - 1 \\ 1 \end{cases} \qquad \text{(A.1.25)}$$

d) if $\det A = -\mathrm{tr}A - 1$, $\Longrightarrow$

$$\lambda_{1,2} = \frac{\mathrm{tr}A}{2} \pm \sqrt{\left(\frac{\mathrm{tr}A}{2}\right)^2 + \mathrm{tr}A + 1} = \frac{\mathrm{tr}A}{2} \pm \left(\frac{\mathrm{tr}A}{2} + 1\right) = \begin{cases} -1 \\ \mathrm{tr}A + 1 \end{cases} \qquad \text{(A.1.26)}$$

The conditions b) - d) for the boundaries identify the critical values for the occurrence of the corresponding bifurcation: complex unit roots for b), one positive unit root for c) and one negative unit root for d). The diagram has been augmented by

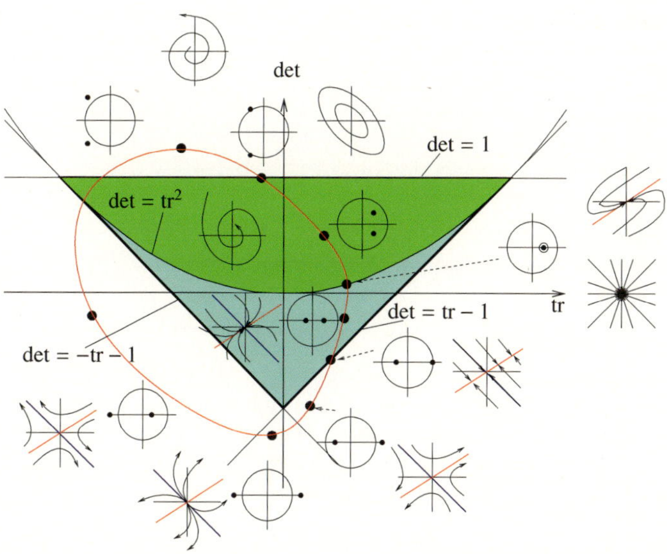

Fig. A.8 Stability triangle with characteristics of eigenvalues and associated phase portraits

the insertion of the unit circles indicating the location of the associated roots in the respective regions plus a graph of the associated local phase portrait showing convergence or divergence respectively.

A.1.7 Equivalent Dynamical Systems

Definition A.1.8. Two differentiable dynamical systems $f : X \to X$ and $F : Y \to Y$ are said to be *topologically equivalent* if there exists a diffeomorphism $T : X \to Y$ such that $F = T \circ f \circ T^{-1}$.

Equivalent dynamical systems (see Kuznetsov, 1995, p. 38) are also referred to as conjugate systems. They display the same dynamical characteristics, i.e. they have the same number of steady states with identical asymptotic behavior, even though they may be defined on spaces with different dimensions. In economic applications the dynamical features are often described and analyzed via equilibrium configurations in spaces with dimensions larger than the minimal dimension required for the state space, i.e. larger than the dimension of the so-called reduced form. If the latter is well defined it is often useful or informative to analyze the dynamics of the economy on subsets of a larger Euclidean space than the state space itself. These are typically equilibrium sets or manifolds defined by a temporary equilibrium mapping or configuration. The next result states that the dynamic equivalence holds if the subsets are defined by the graph of a function. In other words, the dynamics in the larger environment cannot have any other dynamic properties than those induced by the evolution in the state space.

Lemma A.1.5. *For $i = 1, \dots, n$, let $P_i : X \to \mathbb{R}$ denote a finite family of differentiable mappings describing equilibrium configurations in $Y := \mathbb{R}^n$ of an economy with state space X. Let $\mathcal{G}_P : X \to Y$, $\mathcal{G}_P(x) \mapsto (x, P_1(x), \dots, P_n(x))$ denote the function generating*

$$\mathtt{Graph}_P := \{(x,y) \in X \times Y \,|\, y = P(x)\} \subset X \times \mathbb{R}^n, \qquad \text{(A.1.27)}$$

the graph of P.
If $f : X \to X$ is the time-one map of the economic system, then the equation

$$(x_{t+1}, y_{t+1}) = (f(x_t), P(f(x_t))) \quad t = 0, 1, \dots, \qquad \text{(A.1.28)}$$

induces unique orbits $\{(x_t, y_t)\}_0^\infty$ generated by an equivalent dynamical system

$$F : \mathtt{Graph}_P \to \mathtt{Graph}_P \qquad \text{(A.1.29)}$$

defined by $F := \mathcal{G}_P \circ f \circ proj_X$, i.e. $(x,y) \mapsto F(x,y) := \mathcal{G}_P \circ f \circ proj_X(x,y)$

Proof. f and F are equivalent for P since $\mathcal{G}_P \circ proj_X = id_{\mathtt{Graph}_P}$. □

A.2 Balanced Expansion in Homogeneous Dynamical Systems

A dynamical system $F : \mathbb{R}^n_+ \to \mathbb{R}^n_+$ is called *homogeneous* if the mapping F is positively homogeneous of degree one, i.e. if $F(\lambda x) = \lambda F(x)$ holds for all $x \geq 0$ and all $\lambda > 0$. Homogeneous dynamical systems appear in economics in most closed flow economies, predominantly in monetary models and in models of economic growth. Fixed points of homogeneous systems are rare. If they exist, they are nonhyperbolic and, typically, there is a continuum. If $x = F(x) \neq 0$ is a solution, then λx is also a fixed point for all $\lambda > 0$, because $F(\lambda x) = \lambda F(x) = \lambda x$.

Definition A.2.1. An orbit $\{x_t\}_{t=0}^{\infty} = \{F^t(x_0)\}_{t=0}^{\infty}$ is called *balanced* if there exists a positive constant $\rho > 0$ such that $x_t = \rho^t x_0$ for all $t = 0, 1, \ldots,$. The number ρ is called the *growth factor* or the contraction factor, $\rho - 1$ the *growth rate* or contraction rate.

Let $\|\cdot\|$ denote a norm on $\mathbb{R}^n$ and define the unit circle as $S := \{y \in \mathbb{R}^n \mid \|y\| = 1\}$. Define the mapping $s : \mathbb{R}^n \to S,\ x \mapsto s(x) := \frac{1}{\|x\|}x$ and the dynamical system $f : S \to S$ associated with F as $f = s \circ F$, i.e.

$$f(y) := \frac{1}{\|F(y)\|} F(y). \tag{A.2.1}$$

If $\{x_t\}_{t=0}^{\infty}$ is a balanced orbit, then the positive halfline $L(x_0) := \{y \in \mathbb{R}^n_+ \mid \gamma x_0, \gamma \geq 0\} \subset \mathbb{R}^n_+$ associated with initial point x_0 is the union of all balanced orbits with the same growth factor, referred to as a balanced path.

Lemma A.2.1. *Let F be homogeneous and continuous. There exists a balanced orbit $\{x_t\} = \{F^t(x_0)\}_{t=0}^{\infty}$ if and only if there exists a fixed point $\bar{y} = f(\bar{y}) \neq 0$ and $\lambda > 0$ such that $x_0 = \lambda \bar{y}$ and $\rho(x_0) = \rho(\bar{y}) = \|F(\bar{y})\|$.*

Proof. Let $\gamma(x_0) = \{F^t(x_0)\}_{t=0}^{\infty}$ denote a balanced orbit with growth factor $\rho(x_0)$. For $F(x_0) = \rho(x_0)x_0$, choose $\lambda > 0$ such that $\|\bar{y}\| = \|\lambda x_0\| = 1$. Then,

$$f(\bar{y}) = \frac{1}{\|F(\bar{y})\|} F(\bar{y}) = \frac{1}{\|\lambda \rho(x_0)\bar{y}\|} \rho(x_0)\lambda \bar{y} = \bar{y}. \tag{A.2.2}$$

Conversely, if $\bar{y}$ is a fixed point of f, i.e.

$$\bar{y} = f(\bar{y}) = \frac{1}{\|F(\bar{y})\|} F(\bar{y}), \tag{A.2.3}$$

$x_0 = \lambda \bar{y}$ with $\lambda > 0$ induces a balanced orbit with growth factor $\|F(\bar{y})\|$. Therefore,

$$x_1 = F(x_0) = F(\lambda \bar{y}) = \lambda F(\bar{y}) = \lambda \|F(\bar{y})\| \bar{y} = \lambda \|F(\bar{y})\| \frac{1}{\lambda} x_0 = \|F(\bar{y})\| x_0. \tag{A.2.4}$$

By induction, $x_t = \|F(\bar{y})\|^t x_0$, i.e. x_0 expands with grow factor $\|F(\bar{y})\| = \rho(x_0)$. □

It is well known that convergence of an orbit on the unit sphere to a fixed point $\bar{y}$ induced by the system f is only a necessary condition for an orbit of F to converge to the balanced path $L(\bar{y})$ defined by $\bar{y}$. In order to analyze the convergence, let $\varDelta_1 := ||(x_1 - ||x_1||\,\bar{y})||$ denote the distance of an arbitrary point $x_1 \in \mathbb{R}^n_+$ from $L(\bar{y})$. Then,

$$\begin{aligned}\varDelta_1 : &= ||(x_1 - ||x_1||\,\bar{y})|| = \left\|\left(\frac{1}{||x_1||}x_1 - \bar{y}\right)\right\| \, ||x_1|| \\ &= \frac{||y_1 - \bar{y}||}{||y_0 - \bar{y}||}\,||y_0 - \bar{y}||\,||F(x_0)|| \\ &= \frac{||f(y_0) - \bar{y}||}{||y_0 - \bar{y}||}\,||F(y_0)||\,\varDelta_0 =: h(y_0, \varDelta_0)\end{aligned} \tag{A.2.5}$$

defines the time-one change of the distance for each $y_0 \in S$ induced by an orbit $\gamma(x_0)$ of F. The pair $(\bar{y}, 0) \in S \times \mathbb{R}$ is a fixed point of the augmented dynamical system $(f, h) : S \times \mathbb{R}_+ \to S \times \mathbb{R}_+$.

Definition A.2.2. An orbit $\gamma(x_0)$ of the homogeneous system F is said to *converge* to the balanced path $L(\bar{y})$ for $\bar{y} \in S$ if $\lim \varDelta_t \to 0$ as $y_t \to \bar{y}$ for $f : S \to S$ and $h : S \times \mathbb{R} \to \mathbb{R}$ as defined in (A.2.1) and (A.2.5).

Theorem A.2.1. *Let F be continuously differentiable and let $\bar{y} \neq 0$ be an asymptotically stable fixed point of the system $f : S \to S$ defined in* (A.2.1) *with $y_0 \in \mathcal{B}(\bar{y})$, the basin of attraction of $\bar{y}$. Let $\gamma(x_0)$ be an orbit of F with $y_0 := \frac{1}{||x_0||}x_0 \neq \bar{y}$, $\varDelta_0 \neq 0$, and let $\gamma(y_0, \varDelta_0)$ be the associated orbit of (f, h). Then:*

$$\begin{aligned}&\textit{If} \quad \lim_{t\to\infty} \frac{||f(y_t) - \bar{y}||}{||y_t - \bar{y}||}\,||F(\bar{y})|| > 1, \quad \textit{then} \quad \lim_{t\to\infty} |\varDelta_t| = \infty \\ &\textit{If} \quad \lim_{t\to\infty} \frac{||f(y_t) - \bar{y}||}{||y_t - \bar{y}||}\,||F(\bar{y})|| < 1, \quad \textit{then} \quad \lim_{t\to\infty} |\varDelta_t| = 0\end{aligned} \tag{A.2.6}$$

Proof. Let $y_0 \in S$ and $y_0 \neq \bar{y}$. Applying (A.2.5) implies

$$\frac{\varDelta_{t+1}}{\varDelta_t} = \frac{||f(y_t) - \bar{y}||}{||y_t - \bar{y}||}\,||F(y_t)||.$$

Since y_t converges to $\bar{y}$ and $y_0 \in \mathcal{B}(\bar{y})$, one has

$$\lim_{t\to\infty} \frac{\varDelta_{t+1}}{\varDelta_t} = \lim_{t\to\infty} \frac{||f(y_t) - \bar{y}||}{||y_t - \bar{y}||}\,||F(\bar{y})|| \quad \text{with} \quad \lim_{t\to\infty} \frac{||f(y_t) - \bar{y}||}{||y_t - \bar{y}||} < 1.$$

This implies

$$\left|\frac{\varDelta_{t+1}}{\varDelta_t} - \lim_{t\to\infty} \frac{||f(y_t) - \bar{y}||}{||y_t - \bar{y}||}\,||F(\bar{y})||\right| < \epsilon$$

for t larger than some t_0. Therefore,

$$\left[\lim_{t\to\infty} \frac{||f(y_t) - \bar{y}||}{||y_t - \bar{y}||}\,||F(\bar{y})|| - \epsilon\right] |\varDelta_t| < |\varDelta_{t+1}| < \left[\lim_{t\to\infty} \frac{||f(y_t) - \bar{y}||}{||y_t - \bar{y}||}\,||F(\bar{y})|| + \epsilon\right] |\varDelta_t|, \quad t \geq t_0,$$

and by induction, for $\tau > 0$,

$$\left[\lim_{t\to\infty}\frac{\|f(y_t)-\bar{y}\|}{\|y_t-\bar{y}\|}\|F(\bar{y})\|-\epsilon\right]^{\tau}|\Delta_{t_0}| < |\Delta_{\tau+t_0}| < \left[\lim_{t\to\infty}\frac{\|f(y_t)-\bar{y}\|}{\|y_t-\bar{y}\|}\|F(\bar{y})\|+\epsilon\right]^{\tau}|\Delta_{t_0}|.$$

Therefore,

$$\lim_{t\to\infty}\frac{\|f(y_t)-\bar{y}\|}{\|y_t-\bar{y}\|}\|F(\bar{y})\| < 1 \quad \text{implies} \quad \lim_{t\to\infty}\frac{\|f(y_t)-\bar{y}\|}{\|y_t-\bar{y}\|}\|F(\bar{y})\|+\epsilon < 1$$

for ϵ small enough so that $\lim_{t\to\infty}\Delta_t = 0$. Conversely,

$$\lim_{t\to\infty}\frac{\|f(y_t)-\bar{y}\|}{\|y_t-\bar{y}\|}\|F(\bar{y})\| > 1, \quad \text{implies} \quad \lim_{t\to\infty}\frac{\|f(y_t)-\bar{y}\|}{\|y_t-\bar{y}\|}\|F(\bar{y})\|-\epsilon > 1$$

for ϵ sufficiently small, so that $\lim_{t\to\infty}\Delta_t = \infty$. □

Since the mapping f is independent of Δ and h is linear in Δ, the Lipschitz condition (A.2.6) is a product of the contraction rate for the real system f and the growth factor $\|F(\bar{y})\|$ requiring

$$\lim_{t\to\infty}\frac{\|f(y_t)-\bar{y}\|}{\|y_t-\bar{y}\|} < \frac{1}{\|F(\bar{y})\|}, \tag{A.2.7}$$

i.e. the growth factor cannot be too large or the contractivity of the real system should be sufficiently strong. For $n = 2$ the condition (A.2.6) translates into a product of two eigenvalues of the Jacobian matrix, see Theorem 4.2.3.

A.3 Random Dynamical Systems in Discrete Time

The theory of stochastic processes and the theory of random dynamical systems are two areas of mathematics using closely related mathematical tools to analyze an evolving dynamical system subjected to regular and ongoing exogenous stochastic perturbations. The mathematical literature provides different approaches and the term random dynamical system is not used uniformly in the literature[1]. Depending on the perspective and the objective of the desired properties and results one may be more appropriate than the other. For many economic applications it seems most natural to use an approach which uses stochastic orbits as the primitive object of investigations, as proposed by Arnold (1998). These are the observable objects in empirical economics, rather than distributions or Markov kernels which are theoretical concepts and empirically unobservable.

In order to describe and emphasize the nature of an orbit oriented (or dynamical systems) approach to stochastic dynamics, consider a *parametrized* dynamical

[1] For example, Bhattacharya & Majumdar (2004) use the term random dynamical system in a different way than the one adopted here from Arnold (1998).

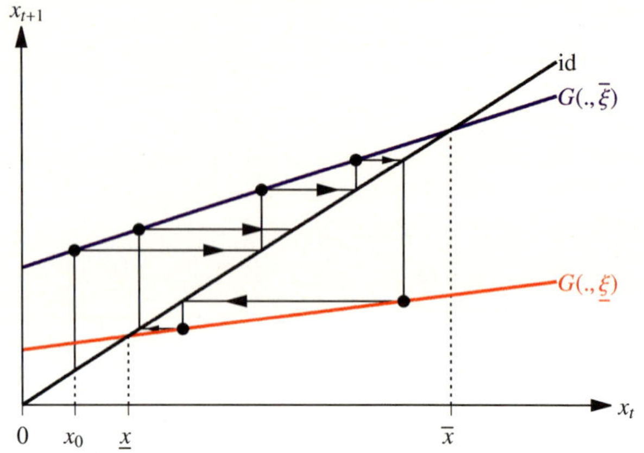

Fig. A.9 A random orbit of x for $\omega = (\ldots, \overline{\xi}, \overline{\xi}, \overline{\xi}, \underline{\xi}, \underline{\xi}, \overline{\xi}, \ldots)$

system $G : \mathbb{R}^n \times \mathbb{R}^m \to \mathbb{R}^n$ given by a family of continuous mappings

$$G(\cdot, \xi) : \mathcal{X} \subset \mathbb{R}^n \to \mathcal{X} \qquad x \mapsto G(k, \xi). \tag{A.3.1}$$

- $\xi \in \mathbb{R}^m$ is a vector of parameters and x is the vector of endogenous or state variables.
- The change of x for a *given* value of the parameter $\xi \in \mathbb{R}^m$ is described by

$$x_{t+1} = G_\xi(x_t) \qquad G_\xi \equiv G(\cdot, \xi). \tag{A.3.2}$$

- The *dynamics* follows the rules and the description of a deterministic dynamical system once the value of a *particular* ξ is given.
- Assume that ξ is following a stationary stochastic process generic random paths are described by $\omega := (\ldots, \xi_{s-2}, \xi_{s-1}, \xi_s, \xi_{s+1}, \ldots)$.
- The change of the value ξ along ω implies choosing at each iteration for x a *different mapping*.
- If G is a contraction mapping for each ξ with upper and lower bounds $[\underline{\xi}, \overline{\xi}]$, then for any random path ω the associated evolution of x will eventually be trapped in some compact interval $[\underline{x}, \overline{x}]$.
- For a one dimensional model with a discrete perturbation concentrated on two values $\{\underline{\xi}, \overline{\xi}\}$, the evolution of the state variable $\{x_t\}$ can be visualized as in Figure A.9 for any given initial condition x_0.

A random dynamical system in the sense of Arnold (1998) has two building blocks:

- a model describing a *dynamical system* perturbed by noise
- and an explicit *model of the noise*.

(1) The exogenous noise process is modeled as a so-called *metric dynamical system* known from ergodic theory satisfying the following assumptions:

a) Let $\vartheta : \Omega \to \Omega$ be a measurable invertible mapping on a probability space $(\Omega, \mathcal{F}, \mathbb{P})$ with measurable inverse ϑ^{-1};

b) ϑ is measure preserving with respect to $\mathbb{P}$: $\mathbb{P}(\vartheta^{-1}(E)) = \mathbb{P}(E)$ for all $E \in \mathcal{F}$, i.e. $\mathbb{P}$ is invariant under ϑ.

c) ϑ is ergodic: $\vartheta^{-1}(E) = E$ implies $\mathbb{P}(E) = 0$ or $\mathbb{P}(E) = 1$, i.e. invariant sets under ϑ have zero or full measure.

d) Let ϑ^t denote the t-th iterate of the map ϑ. The collection $(\Omega, \mathcal{F}, \mathbb{P}, \{\vartheta^t\}_{t\in\mathbb{Z}})$ is called an *ergodic metric dynamical system* (see Arnold, 1998, for details).

e) It is well known that under general assumptions a stationary ergodic process $\{u_t\}_{t\in\mathbb{N}}$, $u_t : \Omega \to \mathbb{R}^m$ can be represented by an ergodic dynamical system. This implies that there exists a measurable map $u : \Omega \to \mathbb{R}^m$ such that for each fixed $\omega \in \Omega$, a sample path of the noise process is given by $u_t(\omega) = u(\vartheta^t\omega)$, $t \in \mathbb{Z}$. Such a process is often referred to as a *real noise process*.

(2) The second ingredient is a parametrized family of time-one maps of a *topological dynamical systems* $F : X \times \mathbb{R}^m \to X$, $X \subset \mathbb{R}^K$ inducing

a) the *random difference equation* $F : X \times \Omega \to X$,

$$x_{t+1} = F(x_t, u(\vartheta^t\omega)) \equiv F(\vartheta^t\omega)x_t. \tag{A.3.3}$$

b) For any x_0, the iteration of the map F under the perturbation ω induces a measurable map $\phi : \mathbb{Z} \times \Omega \times X \to X$ defined by

$$\phi(t, \omega, x_0) := \begin{cases} F(\vartheta^{t-1}\omega) \circ \ldots \circ F(\omega)x_0 & \text{if} \quad t > 0 \\ x_0 & \text{if} \quad t = 0 \end{cases} \tag{A.3.4}$$

such that $x_t = \phi(t, \omega, x_0)$ is the state of the system at time t.

c) For any $x_0 \in X$ and any $\omega \in \Omega$, the sequence $\gamma(\omega, x_0) := \{x_t\}_{t\in\mathbb{Z}}$ with $x_t = \phi(t, \omega, x_0)$ is called an orbit of the random dynamical system ϕ.

d) For any t and s one has:

$$\begin{aligned} \phi(t + s, \omega, x_0) &= F(\vartheta^{t+s}\omega) \circ \ldots \circ F(\omega))x_0 \\ &= \phi(t, \vartheta^s\omega) \circ \phi(s, \omega, x_0) \end{aligned}$$

e) Most random economic processes can be described as metric dynamical systems, due to the representation result by Kolmogoroff of so-called canonical processes.

(3) For i.i.d. processes one obtains a standard representation. Let $\{\xi_t\}$ denote a family of independent and identically distributed random variables with values in $\mathbb{R}^m$, and with common distribution (measure) λ. Then:

a) $\Omega := W^{\mathbb{Z}} = \cdots\cdots W \times W \times W \times \cdots$

b) $\mathcal{F} = \mathcal{B}(\Omega)$ is the Borel σ-algebra

c) $\omega = (\ldots, \xi_{s-1}, \xi_s, \xi_{s+1}, \ldots)$ with $\omega(s) \equiv \xi_s$

d) $\vartheta : \Omega \to \Omega$, 'the shift map' with, $\omega \mapsto \vartheta\omega$ with $\vartheta\omega(s) = \omega(s+1) \equiv \xi_{s+1}$

e) $\xi : \Omega \to \mathbb{R}^m$ "the evaluation map" with $\xi(\omega) \equiv \omega(0)$

f) $\xi_t = \xi(\vartheta^t\omega)$

g) $\mathbb{P} = \lambda^{\mathbb{Z}}$

A.3.1 Random Fixed Points

The long-run behavior of a random dynamical system is described by random attractors, the random analogue of an attractor of a deterministic dynamical system, the random fixed point being a special case (see Arnold, 1998, p. 483), (also Schmalfuß, 1996, 1998).

Definition A.3.1. Consider an ergodic dynamical system $(\Omega, \mathcal{F}, \mathbb{P}, (\vartheta^t))$ as described in (A.3.4) and let $\phi : \mathbb{Z} \times \Omega \times X \to X$ be as defined by (A.3.4) derived from the random difference equation (A.3.3). A *random fixed point* of ϕ is a random variable $x_* : \Omega \to X$ on $(\Omega, \mathcal{F}, \mathbb{P})$ such that $\mathbb{P}$-almost surely[2]

$$x_*(\vartheta\omega) = \phi(1, \omega, x_*(\omega)) = F(x_*(\omega), u(\omega)) \quad \text{for all} \quad \omega \in \Omega', \tag{A.3.5}$$

where $\Omega' \subset \Omega$ is a ϑ-invariant set of full measure, $\mathbb{P}(\Omega') = 1$.

Therefore, a random fixed point is a stationary solution of the stochastic difference equation generated by the metric dynamical system. Some implications of the definition can be observed directly.

- If F is independent of the perturbation ω, then the condition A.3.5 coincides with the one of a deterministic fixed point.
- Definition A.3.1 implies that $x_*(\vartheta^{t+1}\omega) = F(x_*(\vartheta^t\omega), u(\vartheta^t\omega))$ for all times t. Therefore, the orbit $\{x_*(\vartheta^t\omega)\}_{t\in\mathbb{N}}$, $\omega \in \Omega$ generated by x_* solves the random difference equation
$$x_{t+1} = F(u_t(\omega), x_t).$$
Stationarity and ergodicity of ϑ imply that the stochastic process $\{x_*(\vartheta^t)\}_{t\in\mathbb{N}}$ is stationary and ergodic.
- The fixed point x_* induces an invariant distribution $x_*\mathbb{P}$ on $\mathbb{R}^K$ defined by
$$(x_*\mathbb{P})(B) := \mathbb{P} \circ (x_*)^{-1}(B) = \mathbb{P}\{\omega \in \Omega \,|\, x_*(\omega) \in B\} \tag{A.3.6}$$
The invariance of the measure $\mathbb{P}$ under the shift ϑ, i. e. $\mathbb{P} = \vartheta\mathbb{P} = \mathbb{P} \circ \vartheta^{-1}$ implies the stationarity of $x_*\mathbb{P}$, since
$$((x_*\vartheta)\mathbb{P})(B) = \mathbb{P} \circ (\vartheta^{-1} \circ (x_*)^{-1})(B) = \mathbb{P}((x_*)^{-1}(B)) = (x_*\mathbb{P})(B) \tag{A.3.7}$$

[2] The term *almost surely (a.s.)* is used here in a non-standard sense: a property holds a.s. if there exists a ϑ-invariant set $\Omega' \subset \Omega$ (i.e. $\vartheta\Omega' = \Omega'$) with full measure such that the property holds for all $\omega \in \Omega'$.

– If, in addition, $\mathbb{E}\|x_*\| < \infty$, then

$$\lim_{T\to\infty} \frac{1}{T} \sum_{t=0}^{T} 1_B(x_*(\vartheta^t\omega)) = x_*\mathbb{P}(B) := \mathbb{P}\{\omega \in \Omega | x_*(\omega) \in B\}$$

for every $B \in \mathcal{B}(X)$. In other words, the empirical law of an orbit is well defined and it is equal to the distribution $x_*\mathbb{P}$ of x_*.

– If the perturbation corresponds to an i.i.d. process the orbit of the fixed point x_* will be an ergodic Markov equilibrium in the sense of (cf. Duffie, Geanakoplos, Mas-Colell & McLennan, 1994).

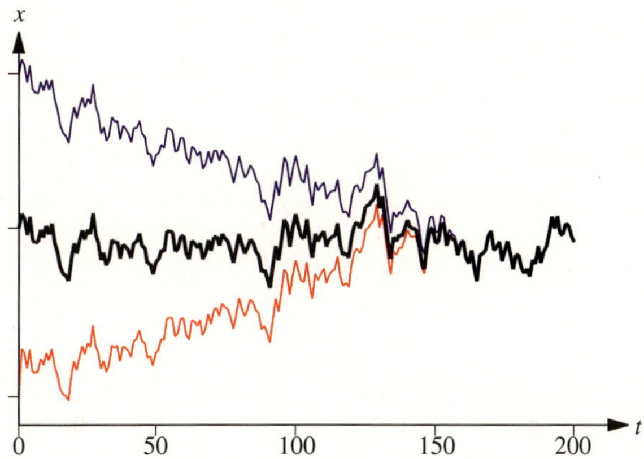

Fig. A.10 Asymptotic convergence of three orbits to a random fixed point

The definition of a stable random fixed point is due to Schmalfuß (1996, 1998). It includes the notion of stability given by Definition 7.4.6 in Arnold (1998).

Definition A.3.2. A random fixed point x_* is called *asymptotically stable* with respect to a norm $\|\cdot\|$, if there exists a random neighborhood $U(\omega) \subset X$, $\omega \in \Omega$ such that $\mathbb{P}$-almost surely (see (A.3.1))

$$\lim_{t\to\infty} \|\phi(t, \omega, x_0) - x_*(\vartheta^t\omega)\| = 0 \quad \text{for all } x_0(\omega) \in U(\omega).$$

Figure A.10 portrays the convergence property of three orbits with different initial conditions and the same noise path ω to an orbit of the random fixed point.

Existence and uniqueness of stable random fixed points require typical contraction properties for the random dynamical system. There are several results available in the literature, for example Arnold (1998); Schmalfuß (1996, 1998); Evstigneev & Pirogov (2007). Economic applications can be found in Böhm (1999, 2006); Schenk-Hoppé & Schmalfuß (2001); Böhm & Wenzelburger (2002); Böhm & Chiarella (2005); Schenk-Hoppé (2005).

Theorem A.3.1. *Consider a random dynamical system ϕ induced by the continuous mapping $F : X \times Y \to X$, $Y \subset \mathbb{R}^m$, with real noise process $u_t = u \circ \vartheta^t$, $u : \Omega \to Y$ measurable, over the ergodic dynamical system $(\Omega, \mathcal{F}, \mathbb{P}, (\vartheta^t))$.*
Let $F(\cdot, y) : X \to X$ be a contraction for all $y \in Y$ compact. Then, there exists a unique random fixed point $x_* : \Omega \to X$ which is asymptotically stable.

A.3.2 Affine Random Dynamical Systems

For affine stochastic difference equations of the form

$$x_{t+1} = A(\omega_t)x_t + b(\omega_t) \tag{A.3.8}$$

there exists an extensive literature with numerous results and applications. A useful general result is provided by the following theorem.

Theorem A.3.2 (Arnold (1998), Theorem 5.6.5/Corollary 5.6.6). *Consider $G : X \times \mathbb{R}^m \times \mathbb{R}^m \times \mathbb{R}^m \to X$ with real noise process $(A, b)_t := (A, b) \circ \vartheta^t$, $A : \Omega \to \mathbb{R}^m \times \mathbb{R}^m$ and $b : \Omega \to \mathbb{R}^m$ measurable, over the ergodic dynamical system $(\Omega, \mathcal{F}, \mathbb{P}, (\vartheta^t))$, where $G \equiv (A, b)$ is the family of hyperbolic affine difference equations*

$$x_{t+1} = A(\vartheta^t \omega)x_t + b(\vartheta^t \omega).$$

a) If

$$\log^+ \|A^{\pm 1}\| \in L^1(\mathbb{P}), \quad \log^+ \|b\| \in L^1(\mathbb{P}) \tag{A.3.9}$$

there exists a unique random fixed point.
b) If $A(\omega)$ is contractive, the fixed point is globally attracting satisfying

$$x_*(\omega) = \sum_{n=-\infty}^{-1} A(\vartheta^{n+1}\omega)^{-1} b(\vartheta^n \omega). \tag{A.3.10}$$

The relationship to Markov processes is discussed in Arnold (1998) Chapter 2 (compare Bhattacharya & Majumdar, 2004, for the notion of stability in distributions).

A.3.3 Iterated Function Systems

The mathematical properties of random dynamical systems with finite discrete i.i.d. noise have been derived and analyzed extensively in the literature of so-called *iterated function systems* (IFS) which appear in connection with image generation and encoding (see Barnsley, 1988; Hutchinson, 1981)[3].

[3] The summary of results here is taken from Arnold & Crauel (1992)

Definition A.3.3 (Iterated Function System (IFS)). Let (X, d) denote a compact metric space. An iterated function system is a finite collection of maps $\mathcal{T} = \{T_i : X \to X \,|\, i = 1, \dots, N\}$, such that each T_i is a contraction, i.e. there exists a constant $c < 1$ such that $d(T_i(x), T_i(y)) \le c\, d(x, y)$ for all $x, y \in X$ and $i = 1, \dots, N$.

Theorem A.3.3 (Hutchinson 1981). *Let $\mathcal{T} = \{T_i : X \to X \,|\, i = 1, \dots, N\}$ denote an iterated function system.*

a) There exists a unique compact set $A_{\mathcal{T}} \neq \emptyset$ satisfying $A_{\mathcal{T}} = \cup_1^N W_i(A_{\mathcal{T}})$.
b) $A_{\mathcal{T}} = closure\{fixed\ points\ of T_{i_k} \circ T_{i_{k-1}} \circ \dots \circ T_{i_2} \circ T_{i_1} \,|\, 1 \le i_j \le N, k \in \mathbb{N}\}$
c) $A_{\mathcal{T}}$ is an invariant set under the mapping $K \mapsto \cup_1^N T_i(K)$ of subsets K in X, which is a contraction on the set of compact subsets of X equipped with the Hausdorff metric.
d) Define $\Omega := \{1, \dots, N\}^{\mathbb{N}}$. Then, each $\omega \in \Omega$ has a representation $\omega = \omega_0\omega_1\omega_2\dots$.

$$\bar{x}(\omega) := \lim_{n\to\infty} T_{\omega_1} T_{\omega_2} T_{\omega_3} \cdots T_{\omega_n} x \tag{A.3.11}$$

exists and is independent of $x \in X$.
e) The map $\omega \mapsto \bar{x}(\omega)$ is continuous and surjective onto $A_{\mathcal{T}}$, (Ω equipped with the product topology).

Definition A.3.4 (Iterated Function System with Probabilities (IFSP)). A finite collection of contraction mappings $\mathcal{T}\{T_i : X \to X \,|\, i = 1, \cdots, N\}$ is called an *iterated function system with probabilities* (IFSP) if there are associated probabilities $p = (p_1, \dots, p_N)$, $p_i > 0$, $\sum_1^N p_i = 1$.

To describe the Markov process associated with an ISFP, using the apparatus suggested by Arnold, let $\Omega = \{1, \dots, N\}^{\mathbb{N}}$ be equipped with the product σ–algebra. For $p = (p_1, \dots, p_N)$, $p_1 > 0$, $\sum_1^N p_i = 1$, let $\mathbb{P} = p^{\mathbb{N}}$ denote the product measure on Ω. Then, $\mathbb{P}$ is invariant under the left shift ϑ, i.e. $\mathbb{P} \circ \vartheta^{-1} = \mathbb{P}$ for $\vartheta : \Omega \to \Omega$ defined by with $(\vartheta\omega)_i = \omega_{i+1}$. Define the random mapping

$$T : \Omega \to \{T_1, \dots, T_N\} \subset X^X \quad \text{by} \quad \omega \mapsto T_{\omega_0}.$$

Then, $\{T(\vartheta^n\omega) \,|\, n \in \mathbb{N})\}$ is an i.i.d. sequence of maps of X whose iteration $T(\vartheta^n\omega)$ induces a Markov chain on X. For each $x \in X$, the process

$$x_n(\omega) = T(\vartheta^{n-1}\omega) \circ \cdots \circ T(\vartheta\omega) \circ T(\omega)x, \quad n > 0, \quad x_0 = x \tag{A.3.12}$$

is a homogeneous Markov process with transition probabilities

$$Q(x, B) := \sum_1^N p_i 1_B(T_i(x))$$

Let Prob (X) denote the set of probability measures on X . Then, the transition probabilities induce a Markov operator Q on Prob (X) by

$$Q\rho := \sum_{1}^{N} p_i(T_i\rho) = \sum_{1}^{N} p_i(\rho \circ T_i^{-1}) \tag{A.3.13}$$

where $\rho \in \text{Prob}(X)$ and $T : X \to X$ measurable.
A measure ρ is invariant if $Q\rho = \rho$. Any attractive measure is invariant and it is the unique measure with this property. The following theorem collects further results from Hutchinson (1981) describing the properties of iterated function systems with probabilities.

Theorem A.3.4 (Hutchinson (1981)). *Let* $\{(T_1, \ldots, T_N); p = (p_1, \ldots, p_N)\}$ *be an IFS with probabilities.*

a) The Markov operator defined in (A.3.13) *is a contraction on Prob* (X) *with compact support, equipped with the metric*

$$d(\mu, \nu) = \sup\left\{\int_X f d\mu - \int_X f d\nu \,|\, Lip(f) \leq 1\right\} \tag{A.3.14}$$

where $Lip(f)$ *denotes the Lipschitz constant of* $f : X \to \mathbb{R}$.
b) There exists a unique stationary measure $\rho \in Prob(X)$ *with* $Q\rho = \rho$ *with* $A_{\mathcal{T}} = \text{supp}\ \rho$, *where* supp ρ *is the support of* ρ.
c) The measure ρ *is attractive for the Markov operator* Q, *i.e.* $Q^n\sigma$ *converges to* ρ *exponentially fast in the metric* (A.3.14) *for all* $\sigma \in Prob(X)$ *with compact support.*

Theorem 1 of Barnsley & Elton (1988) (p.20) provides a significant generalization of Hutchinson's result where none of the maps need to be contractions.

Theorem A.3.5 (Barnsley & Elton (1988)). *Let* $\mathcal{T} = (T_i : X \to |\, i = 1, \ldots, N)$ *be a list of maps with associated probabilities* $p = (1_1, \ldots, p_N)$, $p_i \geq 0$, $\sum_1^N p_i = 1$. *If for some* $c < 1$

$$\prod_{1}^{N} d(T_i x, T_i y)^{p_i} \leq c\, d(x, y) \tag{A.3.15}$$

for all $x, y \in X$,

a) there exists an attractive measure ρ *for the Markov operator* Q *as defined in* (A.3.13)*;*
b) for $\mathbb{P}$*-almost all* $\omega \in \Omega = (1, \ldots, N)^{\mathbb{N}}$ *the limit*

$$\bar{x}(\omega) = \lim_{n\to\infty} T_{\omega_1} T_{\omega_2} T_{\omega_3} \ldots T_{\omega_n} x \tag{A.3.16}$$

exists and is independent of $x \in X$*;*
c) the random variable $\bar{x}$ *is distributed like* ρ, *i.e.* $\mathbb{P}(\bar{x}(\omega) \in B) = \rho(B)$.

Property (A.3.15) is called average contractivity and is equivalent to

$$\mathbb{E}\left(\log \frac{d(T(\omega)x, T(\omega)y)}{d(x, y)}\right) < 0 \tag{A.3.17}$$

uniformly in $x, y \in X$. Arnold & Crauel (1992) also provide a simple affine example

$$T_1(x) = \begin{pmatrix} r & 0 \\ 0 & s \end{pmatrix} x + b_1 \quad \text{and} \quad T_2(x) = \begin{pmatrix} r & 0 \\ 0 & s \end{pmatrix} x + b_2 \tag{A.3.18}$$

which is average contractive for $r < s^{-1}$ and $p_1 = p_2 = \frac{1}{2}$. Thus, a collection of maps can be average contractive without any of the individual maps being contractions.

It follows from Theorem A.3.1 that a hyperbolic iterated function system has a unique asymptotically stable random fixed point. In addition, the literature provides many important results concerning the invariant distribution (or measure) of the random fixed point.

Theorem A.3.6 (Lasota & Mackey (1994), Barnsley (1988)). *Let $\{\mathcal{X}; w_1, \ldots, w_r; \pi_1, \ldots, \pi_r\}$ denote a hyperbolic iterated function system with probabilities and associated compact metric space $(\mathcal{X}, d)$. Let $M : \mathcal{P}(\mathcal{X}) \to \mathcal{P}(\mathcal{X})$ be the associated Markov operator. Then:*

a) M is a contraction with contractivity factor $s := \max\{s_i \mid i = 1, \ldots, r\}$
b) there exists a unique measure $\mu^* \in \mathcal{P}$ such that $M\mu^* = \mu^*$,
i.e. $\mu^*(W^{-1}(B)) = \mu^*(B)$ for all $B \subset \mathcal{X}$
c) the support of μ is equal to the attractor $A := \lim_{i=0}^{\infty} W(A_i)$

A.3.4 Affine Difference Equations with Discrete i.i.d. Noise

A special case of an iterated function system with probabilities occurs when each mapping $w_i, \quad i = 1, \cdots, r$ is affine. Let

$$G_i(x) := A_i x + b_i \qquad i \in I := \{1, \ldots, r\}$$

denote a finite family $\{G_i\}$ of affine maps $G_i : \mathbb{R}^K \to \mathbb{R}^K$ with associated probabilities $\{\pi_i\}$, $\pi_i > 0$, $i = 1, \ldots, r$, and $\sum \pi_i = 1$.
If all mappings G_i are contractions, then $\{(G_i); (\pi_i), i = 1, \ldots, r\}$ is called a *hyperbolic affine iterated function system with probabilities.*

Lemma A.3.1 (Barnsley (1988); Arnold (1998)). *Let $\{(G_i); (\pi_i), i = 1, \ldots, r\}$ be a hyperbolic affine iterated function system.*
a) $\{(G_i); (\pi), i = 1, \ldots, r\}$ has a unique compact attractor $A \subset \left[\bar{k}_1, \bar{k}_r\right]$,
b) A is independent of the probabilities $\{\pi_i\}$.
c) A is the limit of a decreasing sequence of finite unions of compact cubes,
d) there exists a unique invariant measure μ on A,
e) there exists a unique globally attracting random fixed point $x_\star$ of the associated random dynamical system, whose empirical distribution is μ

Often A is a 'fractal' set (Cantor set) and the measure μ^* is very complex without density.

A.3.5 Moments of Univariate Random Variables

Let X denote a random variable in $\mathbb{R}$ and let $\{x_t\}_{t=1}^{T}$ denote T realizations of the associated stochastic process. The so-called raw moments of order r, $r = 1, 2, 3, \ldots$ of X are defined as

$$\mu_r := \mathbb{E}[X^r],$$

while the rth *central moment* of X about $\mathbb{E}(X) = \mu_1$ is denoted $\mu_{(r)} := \mathbb{E}[(X - \mu_1)^r]$ (see Jungeilges, 2003). This yields in particular

$$\begin{aligned} \mu_{(2)} &= \mathbb{E}[X^2] - \mu^2 \\ \mu_{(3)} &= \mathbb{E}[X^3] - 3\mu\mathbb{E}[X^2] + 2\mu^3 \\ \mu_{(4)} &= \mathbb{E}[X^4] - 4\mu\mathbb{E}[X^3] + 6\mu^2\mathbb{E}[X^2] - 3\mu^4. \end{aligned} \tag{A.3.19}$$

Table A.1 lists the definitions of commonly used moments of univariate time series and their estimators from finite samples.

Table A.1 Moments of univariate random variables and their estimators from time series

Statistic	Definition	Estimator
Mean	$\mathbb{E}[X] = \mu_1 \equiv \mu$	$\hat{\mu} = \frac{1}{T}\sum_{t=1}^{T} x_t$
Variance	$\mathbb{V}[X] = \mu_{(2)} = \mathbb{E}\left[(X - \mu_1)^2\right]$	$\hat{\sigma}^2 = \frac{1}{T}\sum_{t=1}^{T}(x_t - \hat{\mu})^2$
Skewness	$\frac{\mu_{(3)}}{(\mu_{(2)})^{\frac{3}{2}}}$	$\hat{\gamma}_1 = \frac{\sqrt{T}\sum_1^T (x_t - \hat{\mu})^3}{\left[\sum_1^T (x_t - \hat{\mu})\right]^{\frac{3}{2}}}$
Kurtosis	$\frac{\mu_{(4)}}{\mu_{(2)}^2} - 3$	$\hat{\gamma}_2 = \frac{1}{T}\frac{\sum_1^T (x_t - \hat{\mu})^4}{\left[\sum_1^T (x_t - \hat{\mu})^2\right]^2} - 3$

Appendix B
Proofs and Further Results

B.1 Proofs for Chapter 3

B.1.1 Proof of Lemma B.1.1

Lemma B.1.1. *Let intertemporal preferences of the worker be given by a continuously differentiable utility function* $U_w : \mathbb{R}^3_+ \to \mathbb{R}$, $(c_1, c_2, \ell) \mapsto U_w(c_1, c_2, \ell)$ *with*

$$U_w(c_1, c_2, \ell) \equiv c_1 u(c_2/c_1) - v(\ell), \tag{B.1.1}$$

where $u : \mathbb{R}_+ \to \mathbb{R}_+$ *is strictly concave and increasing while* $v : [0, \ell_{max}] \to \mathbb{R}_+$ *is strictly convex and increasing. Let* u *and* v *satisfy the respective Inada conditions (see 3.1.1)*

$$\lim_{x\to 0} u'(x) = \infty \qquad \lim_{x\to\infty} u'(x) = 0 \tag{B.1.2}$$

$$\lim_{x\to 0} v'(x) = 0 \qquad \lim_{x\to\ell_{max}} v'(x) = \infty. \tag{B.1.3}$$

Define

$$V(\theta^e) := \max_{0\le x\le 1} x\,u\left(\frac{1-x}{\theta^e x}\right) \quad \text{and} \quad x(\theta^e) := \arg\max_{0\le x\le 1} x\,u\left(\frac{1-x}{\theta^e x}\right). \tag{B.1.4}$$

Then:

a) $V(\theta^e)$ *is a strictly decreasing, globally invertible, and differentiable function whose elasticity* $E_V(\theta^e)$ *satisfies*

$$0 < -E_V(\theta^e) = E_u\left(\frac{1-x(\theta^e)}{\theta^e x(\theta^e)}\right) < 1. \tag{B.1.5}$$

b) The labor supply function has the explicit form

$$\ell^\star(\alpha,\theta^e) := (v')^{-1}(\alpha V(\theta^e)), \qquad \alpha^e := \alpha V(\theta^e) \tag{B.1.6}$$

which is increasing in α, *decreasing in* θ^e, *and decreasing in* p

c) with elasticities

$$E_{\ell^\star}(w) = E_{(v')^{-1}}(\alpha V(\theta^e)) = \frac{1}{E_{v'}(\ell^\star(\alpha V(\theta^e)))} > 0, \tag{B.1.7}$$

and

$$E_{\ell^\star}(p) = -\frac{1}{E_{v'}(\ell^\star(\alpha V(\theta^e)))}(1 + E_V(\theta^e)) < 0. \tag{B.1.8}$$

d) If v *is isoelastic and of the from* $v(\ell) = C/(C+1)\ell^{(1+1/C)}$, *then*

$$\ell^\star(\alpha V(\theta^e)) = (\alpha V(\theta^e))^C. \tag{B.1.9}$$

e) Consumption demand is multiplicatively separable and of the form

$$c_1^\star(\alpha,\theta^e) = x(\theta^e)\,\alpha\,\ell^\star(\alpha V(\theta^e)); \tag{B.1.10}$$

if the function $u : \mathbb{R}_+ \to \mathbb{R}_+$ *is isoelastic with elasticity* $0 \le 1-\delta \le 1$, *then*

$$x(\theta^e) = \delta \qquad \text{and} \qquad V(\theta^e) = \left(\frac{1}{\theta^e}\right)^{1-\delta} \delta^\delta (1-\delta)^{1-\delta}. \tag{B.1.11}$$

Proof. Let $0 \le x \le 1$ denote the propensity to consume of a worker out of wage income. Then, using the notation $pc_1 = xw\ell$ and $p^e c_2 = (1-x)w\ell$, the consumer decision problem can be factorized into two successive maximization problems, i.e.

$$\max_{c_1,c_2,\ell}\left\{c_1 u\left(\frac{c_2}{c_1}\right) - v(\ell) \,|\, pc_1 + p^e c_2 \le w\ell\right\} = \max_{x,\ell}\left\{\frac{w}{p}\ell x u\left(\frac{1-x}{\theta^e x}\right) - v(\ell)\right\}$$
$$= \max_{\ell}\left\{\frac{w}{p}\ell\left(\max_x\, x\,u\left(\frac{1-x}{\theta^e x}\right)\right) - v(\ell)\right\}.$$

The first order condition for $V(\theta^e) := \max_x\, x\,u\left(\frac{1-x}{\theta^e x}\right)$ is

$$u\left(\frac{1-x}{\theta^e x}\right) = xu'\left(\frac{1-x}{\theta^e x}\right)\frac{1}{x^2\theta^e} \tag{B.1.12}$$

which implies that $x(\theta^e)$ solves

$$x(\theta^e) = 1 - E_u\left(\frac{1-x(\theta^e)}{\theta^e x(\theta^e)}\right) \tag{B.1.13}$$

As a consequence of $V(\theta^e) := x(\theta^e)u\left(\frac{1-x(\theta^e)}{\theta^e x(\theta^e)}\right)$ one obtains, using the Envelop Theorem,

$$V'(\theta^e) = -\frac{1-x(\theta^e)}{(\theta^e)^2} u'\left(\frac{1-x(\theta^e)}{\theta^e x(\theta^e)}\right) < 0 \tag{B.1.14}$$

which implies

$$0 < -E_V(\theta^e) = E_u\left(\frac{1-x(\theta^e)}{\theta^e x(\theta^e)}\right) < 1. \tag{B.1.15}$$

Therefore, it follows from (B.1.12) and the first order conditions that

$$\ell^\star(\alpha V(\theta^e) := \arg\max_{\ell}\{V(\theta^e)\alpha\ell - v(\ell)\} = (v')^{-1}(\alpha V(\theta^e)) \tag{B.1.16}$$

which is increasing in α, decreasing in θ^e, and decreasing in p since v is strictly convex and increasing.
The conditions for the elasticities and for the isoelastic situation follow by straightforward calculations from the first order conditions using the Implicit Function Theorem. □

B.1.2 Proof of Lemma 3.2.2

Proof. The real wage $\alpha > 0$ is a labor market clearing for $n_f h(\alpha) = n_w \phi_\ell(\alpha, \theta^e)$ if and only if $0 < \ell < \ell_{\max}$ is a solution of

$$\alpha = (1-\tau_w)V(\theta^e)F'\left(\frac{n_w \ell}{n_f}\right) = v'(\ell). \tag{B.1.17}$$

The Inada conditions imply existence, uniqueness, and interiority of such a solution $\ell(\theta^e)$ which is strictly decreasing and satisfying $\lim_{\theta^e\to 0}\ell(\theta^e) = \ell_{\max}$ and $\lim_{\theta^e\to\infty}\ell(\theta^e) = 0$. Therefore, $\ell(\theta^e)$ induces the aggregate supply function

$$y^s = AS(\theta^e, 1-\tau_w) := n_f F\left(\frac{n_w \ell(\theta^e)}{n_f}\right)$$

which is strictly decreasing satisfying (3.2.29).

For the real wage function

$$\alpha = W(\theta^e, 1-\tau_w) := F'((n_w\ell(\theta^e))/n_f)$$

one finds that $\lim_{\theta^e\to 0} W(\theta^e, 1-\tau_w) = F'(n_w \ell_{\max/n_f})$ and $\lim_{\theta^e\to\infty} W(\theta^e, 1-\tau_w) = \infty$ in addition to

$$\frac{\partial W}{\partial \theta^e} = \frac{n_w}{n_f}F''\cdot\frac{V'F'}{v'' - (n_w/n_f)VF''} < -E_V\frac{\alpha}{\theta^e} \leq \frac{\alpha}{\theta^e}$$

implying $0 < E_W(\theta^e) < 1$. □

B.2 Proofs for Chapter 4

Lemma B.2.1. *Let the aggregate supply function* $AS : \mathbb{R}_{++} \to \mathbb{R}_{++}$ *be continuous, strictly decreasing, and strictly convex satisfying* $\lim_{\theta^e \to \infty} AS(\theta^e) = 0$.

a) Its inverse AS^{-1} *is well defined and continuous on its range* $\mathbb{R}_+$ *or* $[0, AS(0)]$, *strictly decreasing, and strictly convex satisfying*

$$\lim_{y\to 0} AS^{-1}(y) = 0, \qquad \lim_{y\to AS(0)} AS^{-1}(y) = +\infty$$

b) The mapping

$$f(y) := (\tilde{c} - \tau)\frac{y}{AS^{-1}(y)} \tag{B.2.1}$$

is strictly increasing and strictly convex. It satisfies $f(0) = 0$ *and*

$$\begin{aligned} &\lim_{y\to 0} f(y)/y = 0 \\ &\lim_{y\to\infty} \frac{f(y)}{y} = \infty \qquad \textit{for} \qquad AS^{-1} : \mathbb{R}_+ \to \mathbb{R}_+ \\ &\lim_{y\to AS(0)} \frac{f(y)}{y} = \infty \qquad \textit{for} \qquad AS^{-1} : [0, AS(0)] \to \mathbb{R}_+ \end{aligned} \tag{B.2.2}$$

c) The function $h(y) := f(y) - y$ *attains a global minimum at a unique* $y^\star > 0$ *with* $f(y^\star) - y^\star < 0$, *i.e.* $f(y) - y > h(y^\star) = f(y^\star) - y^\star$, *for all* $y \neq y^\star$.
d) f *has two fixed points satisfying* $0 = y_1 < y^\star < y_2$.

Proof. a) and b) can be verified directly from the assumptions.
The mapping h is strictly convex, $h(0) = 0$ and decreasing near zero with $\lim_{y\to AS(0)} h(y) = +\infty$. Therefore $0 < y^\star = \arg\min h(y)$ exists and is unique which proves c).
d) follows from c) and strict convexity since there exists $y_2 > y^\star$ with $h(y^\star) < 0 = h(y_2)$ □

Remark: Without strict convexity, the limiting behavior of AS^{-1} stated in the lemma still holds true, but this may imply more than two steady states in Theorem 4.2.1.

B.2.1 Proof of Theorem 4.2.1

Theorem 4.2.1 Let the aggregate demand function be of the form

$$D(m, g) := \frac{m+g}{1 - c(1-\tau)} \equiv \frac{m+g}{\tilde{c}}, \quad 0 < c, \tau < 1, \quad 0 < \tilde{c} = 1 - c(1-\tau) < \tau,$$

and assume that the aggregate supply function $AS : \mathbb{R}_+ \to \mathbb{R}_+$ is globally invertible, strictly decreasing, and strictly convex. For all $g \geq 0, 0 < c, \tau < 1$:

a) The time one map of real balances under perfect foresight $\mathcal{F} : \mathbb{R}_+ \to \mathbb{R}_+$ is strictly monotonically increasing and strictly convex, implying that the dynamics is monotonic.
b) If $g = 0$, there are two steady states with $\underline{m} = 0$ being stable while the positive steady state $\overline{m}$ is unstable.
c) There exists a critical level $g^\star > 0$, such that there is no steady state for $g > g^\star$.
d) If $g > 0$, there are at most two positive steady states $0 < \underline{m}(g) < \overline{m}(g)$, one of which is always unstable.

Proof. The dynamics of money balances and prices under perfect foresight

$$\begin{aligned} M_{t+1} &= M_t + p_t(g - \tau D(m_t, g)) \\ p_{t+1} &= p_t AS^{-1}(D(m_t, g)) \end{aligned} \tag{B.2.3}$$

imply the time-one map of real money balances given by

$$\mathcal{F}(m) := \left(\frac{\tilde{c} - \tau}{\tilde{c}}\right) \frac{m + g}{AS^{-1}\left(\frac{m+g}{\tilde{c}}\right)} \tag{B.2.4}$$

where

$$\tilde{c} := 1 - c(1 - \tau) \quad \text{with} \quad 0 < \left(\frac{\tilde{c} - \tau}{\tilde{c}}\right) < 1.$$

a) Since $g \geq 0$ induces an additive shift of the map $\mathcal{F}$, it is best to consider the dynamics under the change of variable $x := m + g$ inducing an equivalent dynamical system $F : \mathbb{R} \to \mathbb{R}$

$$x_1 = m_1 + g = F(x, g) := g + f(x) := g + \left(\frac{\tilde{c} - \tau}{\tilde{c}}\right) \frac{x}{AS^{-1}\left(\frac{x}{\tilde{c}}\right)} \tag{B.2.5}$$

According to Lemma B.2.1 $f : \mathbb{R} \to \mathbb{R}$ is strictly increasing and strictly convex and satisfies

$$f(0) = 0; \quad \lim_{x \to 0} f(x)/x = 0; \quad \lim_{x \to \infty} f(x)/x = \infty.$$

Therefore, $\mathcal{F}$ is also strictly increasing and strictly convex for all $g \geq 0$.
b) For $g = 0$, it follows that $\underline{x} = 0$ and $\bar{x} = \tilde{c}AS\left(\left(\frac{\tilde{c}-\tau}{\tilde{c}}\right)\right)$ are the only two fixed points of $F(x, 0)$ which induce two fixed points $\underline{m}(0) = \underline{x} = 0$ and $\overline{m}(0) = \overline{x}(0) = \tilde{c}AS\left(\frac{\tilde{c}-\tau}{\tilde{c}}\right)$. Lemma B.2.1 implies that there are two positive fixed points $0 < \underline{x}(g) < \overline{x}(g)$ of $F(x, g)$ for small $g > 0$.
c) Let $x^\star = \arg\min_x f(x) - x$ and define $0 < g^\star := -\min_x f(x) - x$. Then, $F(x, g)$ has no fixed point for $g > g^\star$. Thus, $x^\star := x(g^\star) = \underline{x}(g^\star) = \overline{x}(g^\star) > 0$, such that $F(x^\star, g^\star) = x^\star$. From convexity and monotonicity it follows that $\underline{x}$ is increasing in g while $\overline{x}$ is decreasing in g.

The critical steady state $x^\star$ is independent of g. This yields the condition

$$\frac{\partial F}{\partial x}(x^\star, g) = f'(x^\star) = \frac{\frac{\tilde{c}-\tau}{\tilde{c}}}{(AS)^{-1}\left(\frac{x^\star}{\tilde{c}}\right)}\left(1 - E_{(AS)^{-1}}\left(\frac{x^\star}{\tilde{c}}\right)\right) \stackrel{!}{=} 1 \tag{B.2.6}$$

implying

$$x^\star = \tilde{c}AS\left(\frac{\tilde{c}-\tau}{\tilde{c}}\left(1 - E_{(AS)^{-1}}\left(\frac{x^\star}{\tilde{c}}\right)\right)\right). \tag{B.2.7}$$

For the isoelastic case the right hand side of (B.2.7) is a constant which can be either above or below one. Therefore, at the critical fixed point one may have inflation or deflation determined by

$$\theta^\star := \left(\frac{\tilde{c}-\tau}{\tilde{c}}\right)\left(1 - E_{(AS)^{-1}}\left(\frac{x^\star}{\tilde{c}}\right)\right). \tag{B.2.8}$$

Define stationary real money balances with the associated inflation factors by

$$\begin{aligned} \underline{m}(g) &:= \underline{x}(g) - g \quad \text{and} \quad \underline{\theta}(g) := (AS)^{-1}\left(\frac{\underline{x}(g)}{\tilde{c}}\right) \\ \overline{m}(g) &:= \overline{x}(g) - g \quad \text{and} \quad \overline{\theta}(g) := (AS)^{-1}\left(\frac{\overline{x}(g)}{\tilde{c}}\right), \end{aligned} \tag{B.2.9}$$

for which one has the following properties for $0 \le g < g^\star$:

- $\overline{m}(g) > \underline{m}(g)$ with $\overline{m}(g)$ decreasing and $\underline{m}(g)$ increasing in g
- $\underline{\theta}(g) > \overline{\theta}(g)$ with $\underline{\theta}(g)$ decreasing and $\overline{\theta}(g)$ increasing in g
- $\lim_{g\to 0} \underline{\theta}(g) = \infty$, and $\lim_{g\to 0} \overline{\theta}(g) > 0$.

d) To determine the stability of the two fixed points for $0 < g < g^\star$, one has

$$\frac{\partial F}{\partial x}(x, g) = f'(x), \tag{B.2.10}$$

which is strictly increasing in x due to strict convexity of f. Therefore,

$$\frac{\partial F}{\partial x}(\underline{x}(g), g) \quad \text{is increasing,} \quad \text{and} \quad \frac{\partial F}{\partial x}(\overline{x}(g), g) \quad \text{is decreasing}$$

in g, while $\frac{\partial F}{\partial x}(x(g^\star), g^\star) = 1$. Therefore, for all $0 < g < g^\star$, the smaller fixed point $\underline{x}(g)$ is asymptotically stable while the higher one $\overline{x}(g)$ is asymptotically unstable. Strict monotonicity of f implies that convergence and divergence are monotonic. □

Generically, a balanced path for $g = \tau AS(1)$, i.e. one with a balanced budget, can be either on the stable or the unstable branch. In other words, one may have either $\overline{\theta}(\tau AS(1)) = 1$ or $\underline{\theta}(\tau AS(1)) = 1$, depending on the parameters of the supply and the demand side. Thus, typically, the balanced budget path will coexist with an inflationary one or a deflationary one.

B.2.2 Proof of Theorem 4.2.5

Proof. The delay system (4.2.38) implies the Jacobian matrix

$$DS(\hat{q},\hat{q}) = \begin{pmatrix} 0 & 1 \\ -\frac{\mathcal{P}'}{1+g\mathcal{P}} & (2+g\mathcal{P})\frac{\mathcal{P}'}{1+g\mathcal{P}} \end{pmatrix} \tag{B.2.11}$$

implying

$$\begin{aligned} 0 < \operatorname{tr} DS(\hat{q},\hat{q}) &= (2+g\mathcal{P}(1,\hat{q}))\frac{\mathcal{P}'(1,\hat{q})}{1+g\mathcal{P}(1,\hat{q})}, \\ 0 < \det DS(\hat{q},\hat{q}) &= \frac{\mathcal{P}'(1,\hat{q})}{1+g\mathcal{P}(1,\hat{q})} < 1, \\ \operatorname{tr} DS(\hat{q},\hat{q}) &= (2+g\mathcal{P}(1,\hat{q}))\det DS(\hat{q},\hat{q}). \end{aligned} \tag{B.2.12}$$

Applying Lemma A.1.4 one finds

$$\begin{aligned} &\operatorname{tr} DS(\hat{q},\hat{q}) + \det DS(\hat{q},\hat{q}) > -1, \\ &\operatorname{tr} DS(\hat{q},\hat{q}) - \det DS(\hat{q},\hat{q}) = (2+g\mathcal{P}(1,\hat{q}))\det DS(\hat{q},\hat{q}) - \det DS(\hat{q},\hat{q}) \\ &\qquad = \mathcal{P}'(1,\hat{q}) = E_{\mathcal{P}}(1,\hat{q}) < 1. \end{aligned} \tag{B.2.13}$$

Therefore, the two roots of the matrix (B.2.11)

$$\lambda_{1,2} = \frac{1}{2}\left(\operatorname{tr} \pm \sqrt{\operatorname{tr}^2 - 4\det}\right) \tag{B.2.14}$$

have modulus less than one. If they are real they satisfy $0 < \lambda_1 < \lambda_2 < 1$. Therefore, $(\hat{q},\hat{q})$ is asymptotically stable under S.

In order to prove (b) define the distance from the balanced path as

$$\Delta_t = M_t(q_t^e - \hat{q}) = \frac{M_t}{M_{t-1}}\frac{q^e - \hat{q}}{q_{t-1}^e - \hat{q}}\Delta_{t-1} = \frac{\tilde{c}-\tau}{\tilde{c}}\left(1 + g\mathcal{P}(q_{t-1}^e)\right)\frac{S(q_{t-2}^e, q_{t-1}^e) - \hat{q}}{q_{t-1}^e - \hat{q}}\Delta_{t-1}. \tag{B.2.15}$$

Together with the delay system $S : \mathbb{R}_+^2 \to \mathbb{R}_+^2$ from (4.2.36) one obtains the auxiliary system $(\Delta, S) : \mathbb{R}\times\mathbb{R}_+^2 \to \mathbb{R}\times\mathbb{R}_+^2$. The balanced path $\hat{q}$ is said to be asymptotically unstable if the distance $|\Delta_t|$ of all orbits from the path associated with $\hat{q}$ diverges to $+\infty$ while $\lim_{t\to\infty} q_t^e = \hat{q}$. The Jacobian matrix of (Δ, S) is given by

$$\frac{\partial(\Delta, S)}{\partial(\Delta_{-1}, q^e_{-2}, q^e_{-1})}(0, \hat{q}, \hat{q}) = \begin{pmatrix} \frac{\tilde{c}-\tau}{\tilde{c}}(1+g\mathcal{P})\frac{S(q^e_{-2}, q^e_{-1}) - \hat{q}}{q^e_{-1} - \hat{q}} & 0 & 0 \\ 0 & 0 & 1 \\ 0 & -\frac{\mathcal{P}'}{1+g\mathcal{P}} & \frac{\mathcal{P}'(2+g\mathcal{P})}{1+g\mathcal{P}} \end{pmatrix}$$

$$= \begin{pmatrix} \frac{\tilde{c}-\tau}{\tilde{c}}\mathcal{P}'(2+g\mathcal{P}) & 0 & 0 \\ 0 & 0 & 1 \\ 0 & -\frac{\mathcal{P}'}{1+g\mathcal{P}} & \frac{\mathcal{P}'(2+g\mathcal{P})}{1+g\mathcal{P}} \end{pmatrix} \tag{B.2.16}$$

implying the roots λ_1, λ_2 of the matrix (B.2.11) and

$$\begin{aligned} \lambda_3 &= \frac{\tilde{c}-\tau}{\tilde{c}}\mathcal{P}'(1,\hat{q})\,(2 + g\mathcal{P}(1,\hat{q})) = \frac{\tilde{c}-\tau}{\tilde{c}}(1 + g\mathcal{P}(1,\hat{q}))\,[\det DS(\hat{q},\hat{q}) + 1] \\ &= \frac{\tilde{c}-\tau}{\tilde{c}}(1 + g\mathcal{P}(1,\hat{q}))\,[\lambda_1\lambda_2 + 1]\,. \end{aligned} \tag{B.2.17}$$

Therefore, $\lambda_3 > 1$ if the growth rate of money

$$\hat{M}(\hat{q}) := \frac{\tilde{c}-\tau}{\tilde{c}}(1 + g\mathcal{P}(1,\hat{q})) > 1.$$

Applying Theorem A.2.1 (or the proof of Theorem 4.2.3) implies the result. □

B.2.3 Proof of Lemma 4.3.1

Proof. Consider the labor market clearing condition with random productivity

$$n_f h\left(\frac{w}{pZ}\right) = N\left(\frac{w}{p}V(\theta^e)\right).$$

For every $(p, p^e, Z) \gg 0$ there is a unique positive solution $w > 0$ since h is decreasing and surjective and N is non decreasing. The solution is homogeneous of degree one in (p, p^e) so that the equilibrium wage function can be written as

$$w = pW\left(Z, \frac{p^e}{p}\right) \tag{B.2.18}$$

where W denotes the real wage function.

Properties of the real wage function:

1) Since h is invertible, one has $\alpha \equiv w/p = W(Z, \theta^e)$ if and only if

$$\frac{\alpha}{h^{-1}\left(\dfrac{N(\alpha V(\theta^e))}{n_f}\right)} = Z, \qquad \theta^e := p^e/p. \tag{B.2.19}$$

$h^{-1}(N(\alpha V(\theta^e))/n_f)$ is decreasing in α which makes the left hand side an increasing function of α, so that W must be increasing in Z for each θ^e.

2) Similarly, since N is strictly increasing and V strictly decreasing the equality $n_f h(\alpha) = N(\alpha V(\theta^e))$ implies that W must be increasing in θ^e for every Z.
3) If the mappings N and V are invertible, the real wage function W is invertible with respect to both variables yielding explicit inverses given by

$$\theta^e = V^{-1}\left[\frac{\alpha}{N^{-1}\left(n_f h\left(\frac{\alpha}{Z}\right)\right)}\right] \quad \text{and} \quad Z = \frac{\alpha}{h^{-1}\left[\frac{1}{n_f} N\left(\alpha V\left(\theta^e\right)\right)\right]} \tag{B.2.20}$$

4) If the functions F, N, and V are isoelastic, the real wage function is isoelastic as well and of the form $W(Z, \theta^e) = W(1, 1)Z^a(\theta^e)^b$.

Properties of the random aggregate supply function:

1) The identity

$$\frac{W(Z, \theta^e)}{h^{-1}\left(\frac{1}{n_f} N(W(Z, \theta^e)V(\theta^e))\right)} \equiv Z \tag{B.2.21}$$

implies that

$$\frac{W(Z, \theta^e)}{Z} \equiv h^{-1}\left(\frac{1}{n_f} N(W(Z, \theta^e)V(\theta^e))\right) \tag{B.2.22}$$

is a decreasing function of Z. As a consequence, the aggregate supply is given by

$$AS\,(Z, \theta^e) := Z\, n_f F\left(\frac{1}{n_f} N(W(Z, \theta^e)V(\theta^e))\right) \tag{B.2.23}$$

and satisfies $E_{AS}(Z) > 1$. $E_{AS}(\theta^e) + E_W(\theta^e) > 0$ follows.

2) If the functions F, N, and V are isoelastic the aggregate supply function is isoelastic as well and of the form $AS\,(Z, \theta^e) = AS\,(1, 1)Z^a(\theta^e)^b$. □

B.2.4 Proof of Proposition 4.3.2

Proof. It follows from the properties of the random price law given (see Proposition 4.3.1) that the mean price law is strictly monotonically increasing in p^e, satisfying $\lim_{p^e\to\infty}(\mathbb{E}\mathcal{P})(M, p^e) = \infty$, $\lim_{p^e\to 0}(\mathbb{E}\mathcal{P})(M, p^e) = 0$, and $(\mathbb{E}\mathcal{P})(M, p^e) \geq \mathcal{P}(M, p^e, Z_{\max})$ while $(\mathbb{E}\mathcal{P})(M, p^e)/p^e$ is strictly decreasing with

$$\lim_{p^e\to\infty}\frac{(\mathbb{E}\mathcal{P})(M,p^e)}{p^e}\geq\lim_{p^e\to\infty}\frac{\mathcal{P}(M,p^e,Z_{\max})}{p^e}=\frac{1}{AS^{-1}\left(Z_{\max},\frac{g}{\tilde{c}}\right)}. \tag{B.2.24}$$

Therefore, $\psi^*(M,p^e_{-1},\mu) := (\mathbb{E}\mathcal{P})^{-1}(M,p^e_{-1},\mu)$ exists for every p^e_{-1}. Moreover, $\psi^*(M,p^e_{-1},\mu)/p^e_{-1}$ is strictly increasing and therefore satisfies (4.3.23)

$$\lim_{p^e_{-1}\to 0}\frac{\psi^*(M,p^e_{-1},\mu)}{p^e_{-1}}=0,\quad \lim_{p^e_{-1}\to\infty}\frac{\psi^*(M,p^e_{-1},\mu)}{p^e_{-1}}\leq AS^{-1}\left(Z_{\max},\frac{g}{\tilde{c}}\right). \qquad \square$$

B.3 Proofs for Chapter 7

The parametric version of the two dimensional system of Chapter 7 used in Theorem 7.5.1 and for the numerical analysis in Section 7.5 is given by

$$\begin{aligned}\alpha_{t+1}&=\alpha_t\frac{1+w(s^\ell(\alpha_t,m_t))}{1+p(s^c(\alpha_t,m_t))}\\ m_{t+1}&=\frac{\min((1-\tau)y_t,m_t+g-\tau y_t)}{1+p(s^c(\alpha_t,m_t))}\end{aligned} \tag{B.3.1}$$

such that for $0<\gamma<1$, $0<\kappa<1$:

$$p(s^c(\alpha_t,m_t))=\begin{cases}\gamma\dfrac{y^D_t-y_t}{y^D_t} & y^D_t>y_t\\[2ex] \kappa\dfrac{y_t-y^*_t}{y^*_t} & \text{otherwise,}\end{cases}$$

and for $0<\mu<1$, $0<\lambda<1$:

$$w(s^\ell(\alpha_t,m_t))=\begin{cases}\lambda\dfrac{L_t-L_{\max}}{L_{\max}} & L_{\max}>L_t\\[2ex] \mu\dfrac{z^*_t-L_t}{z^*_t} & \text{otherwise}\end{cases}$$

where $y_t=\min\left(y^D_t,y^*_t,y_{\max}\right)$ and

$$y^D_t=\frac{m_t+g}{1-c(1-\tau)},\quad y_{\max}=\frac{A}{B}(L_{\max})^B,\quad y^*_t=\frac{A}{B}\left(\frac{\alpha_t}{A}\right)^{\frac{B}{B-1}},\quad z^*_t=\left(\frac{\alpha_t}{A}\right)^{\frac{1}{B-1}}.$$

To prove the statements of stability and the necessary conditions for period doubling of Theorem 7.5.1 it is useful and informative to collect and establish the findings in a different form than in Kaas (1995). Theorem 7.5.1 then becomes a corollary to the Auxiliary Theorem B.3.1.

Theorem B.3.1 (Auxiliary Theorem). *Let the conditions of Theorem 7.5.1 be satisfied and assume that* $A = L_{\max} = 1$ *holds.*
Let $Bg > \tau$ *and* $(\tilde{\alpha}, \tilde{m}, \tilde{\theta}) \in \mathbf{I}$ *denote an inflationary steady state.*

(1) The steady state values as functions of the parameters satisfy

$$\begin{aligned} 1 < \tilde{\theta}(g,\tau,\gamma,B) &< \min(1+\mu, 1+\gamma), \\ \tilde{m}(g,\tau,\gamma,B) &> \frac{1-c(1-\tau)-Bg}{B}, \\ \tilde{\alpha}(g,\tau,\gamma,B,\mu) &= \left(\frac{\mu+1-\tilde{\theta}(g,\tau,\gamma,B)}{\mu}\right)^{1-B} < 1. \end{aligned} \tag{B.3.2}$$

(2) The comparative statics effects satisfy

$$\begin{array}{lll} \left.\frac{\partial \tilde{\theta}}{\partial g}\right|^{\mathbf{I}} > 0, & \left.\frac{\partial \tilde{\theta}}{\partial \tau}\right|^{\mathbf{I}} < 0, & \left.\frac{\partial \tilde{\theta}}{\partial \gamma}\right|^{\mathbf{I}} > 0, \\ \left.\frac{\partial \tilde{m}}{\partial g}\right|^{\mathbf{I}} > 0, & \left.\frac{\partial \tilde{m}}{\partial \tau}\right|^{\mathbf{I}} < 0, & \left.\frac{\partial \tilde{m}}{\partial \gamma}\right|^{\mathbf{I}} > 0, \\ \left.\frac{\partial \tilde{\alpha}}{\partial g}\right|^{\mathbf{I}} > 0, & \left.\frac{\partial \tilde{\alpha}}{\partial \tau}\right|^{\mathbf{I}} < 0, & \left.\frac{\partial \tilde{\alpha}}{\partial \gamma}\right|^{\mathbf{I}} > 0. \end{array} \tag{B.3.3}$$

(3) $(\tilde{\alpha}, \tilde{m}, \tilde{\theta}) \in \mathbf{I}$ *is asymptotically stable if and only if*

$$2 > \frac{\mu+1-\tilde{\theta}}{\tilde{\theta}(1-B)}. \tag{B.3.4}$$

(4) There exist parameters (g,τ,γ,B,μ) *satisfying* (B.3.4) *with equality, i.e.*

$$\mu + 1 = [2(1-B)+1]\,\tilde{\theta}(g,\tau.\gamma,B).$$

There are two real eigenvalues $-1 = \nu_1 < 0 < \nu_2 < 1$ *such that the steady state undergoes a period doubling bifurcation at* (g,τ,γ,B,μ)*, i.e.*

$$\nu_1(g,\tau,\gamma,B,\mu) = -1 \iff \mu+1 = [2(1-B)+1]\,\tilde{\theta}(g,\tau,\gamma,B). \tag{B.3.5}$$

(5) Moreover,

$$\left.\frac{\partial \nu_1}{\partial \tau}\right|^{\mathbf{I}}_{\nu_1=-1} < 0, \quad \left.\frac{\partial \nu_1}{\partial g}\right|^{\mathbf{I}}_{\nu_1=-1} > 0 \Rightarrow \left.\frac{d\tau}{dg}\right|^{\mathbf{I}}_{\nu_1=-1} > 0; \quad \left.\frac{\partial \nu_1}{\partial B}\right|^{\mathbf{I}}_{\nu_1=-1} > 0 \tag{B.3.6}$$

holds for the bifurcation curves in parameter space (g,τ).

Let $Bg < \tau$ *and* $(\tilde{\alpha}, \tilde{m}, \tilde{\theta}) \in \mathbf{K}$ *denote a Keynesian steady state.*

(1) The steady state values as functions of the parameters satisfy

$$\max(C, 1-\lambda, 1-\kappa) < \tilde{\theta}(g,\tau,\lambda,B) < 1, \quad C := \frac{(1-c)(1-\tau)}{1-c(1-\tau)} < 1,$$

$$\tilde{m}(g,\tau,\lambda,B) < \frac{1-c(1-\tau)-Bg}{B}, \tag{B.3.7}$$

$$\tilde{\alpha}(g,\tau,\lambda,B,\kappa) = \left(\frac{\kappa-1+\tilde{\theta}(g,\tau,\lambda,B)}{\kappa B} \Big/ \frac{\tilde{m}(g,\tau,\lambda,B)+g}{1-c(1-\tau)}\right)^{\frac{1-B}{B}}.$$

(2) The comparative statics effects satisfy

$$\begin{aligned}
&\left.\frac{\partial\tilde{\theta}}{\partial g}\right|^{\mathbf{K}} > 0, && \left.\frac{\partial\tilde{\theta}}{\partial \tau}\right|^{\mathbf{K}} < 0, && \left.\frac{\partial\tilde{\theta}}{\partial \lambda}\right|^{\mathbf{K}} < 0,\\
&\left.\frac{\partial\tilde{m}}{\partial g}\right|^{\mathbf{K}} < 0, && \left.\frac{\partial\tilde{m}}{\partial \tau}\right|^{\mathbf{K}} > 0, && \left.\frac{\partial\tilde{m}}{\partial \lambda}\right|^{\mathbf{K}} > 0,\\
&\left.\frac{\partial\tilde{\alpha}}{\partial g}\right|^{\mathbf{K}} > 0, && \left.\frac{\partial\tilde{\alpha}}{\partial \tau}\right|^{\mathbf{K}} < 0, && \left.\frac{\partial\tilde{\alpha}}{\partial \lambda}\right|^{\mathbf{K}} < 0.
\end{aligned} \tag{B.3.8}$$

(3) $(\tilde{\alpha}, \tilde{m}, \tilde{\theta}) \in \mathbf{K}$ *is asymptotically stable if and only if*

$$2 > \frac{\tilde{\theta}-1+\kappa}{\tilde{\theta}(1-B)}\left(B + \frac{C(1-\lambda)}{\tilde{\theta}^2}\right) - \frac{C}{\tilde{\theta}^2}(1-\kappa+\tilde{\theta}) \tag{B.3.9}$$

(4) There exist parameters $(g,\tau,\lambda,B,\bar{\kappa})$ *such that* (B.3.9) *holds with equality. The two eigenvalues* ν_1, ν_2 *are real with* $-1 = \nu_1 < 0 < \nu_2 < 1$ *such that*

$$\nu_1(g,\tau,\lambda,B,\bar{\kappa}) = -1 \iff 2 = \frac{\tilde{\theta}-1+\bar{\kappa}}{\tilde{\theta}(1-B)}\left(B + \frac{C(1-\lambda)}{\tilde{\theta}^2}\right) - \frac{C}{\tilde{\theta}^2}(1-\bar{\kappa}+\tilde{\theta}). \tag{B.3.10}$$

The steady state undergoes a period doubling bifurcation at $(g,\tau,\lambda,B,\bar{\kappa})$*, i.e.* $1 \geq \kappa > \bar{\kappa}$ *implies* $\nu_1(g,\tau,\lambda,B,\kappa) < -1$.

(5) Moreover,

$$\left.\frac{\partial\nu_1}{\partial\tau}\right|^{\mathbf{K}}_{\nu_1=-1} > 0, \quad \left.\frac{\partial\nu_1}{\partial g}\right|^{\mathbf{K}}_{\nu_1=-1} < 0 \Rightarrow \left.\frac{d\tau}{dg}\right|^{\mathbf{K}}_{\nu_1=-1} > 0; \quad \left.\frac{\partial\nu_1}{\partial B}\right|^{\mathbf{K}}_{\nu_1=-1} > 0 \tag{B.3.11}$$

holds for the bifurcation curves in parameter space (g,τ).

B.3.1 Proof of Theorem B.3.1

Steady States in I

Assume $Bg > \tau$ and let the dynamics be given by (B.3.1) with

$$w(s^{\ell}(\alpha_t, m_t)) = \mu\left(1 - \frac{L_{\max}}{z_t^*(\alpha_t)}\right) \qquad \text{and} \qquad p(s^c(\alpha_t, m_t)) = \gamma\left(1 - \frac{y_{\max}}{y_t^D(m_t)}\right).$$

It is useful to write the conditions for a steady state as three equations using $L_{\max} = 1$, $y_{\max} = 1/B$, and making the level of inflation explicit, i.e.

$$\begin{aligned}
\theta &= 1 + \gamma\left(1 - \frac{1 - c(1-\tau)}{B(m+g)}\right) \\
\theta &= \frac{1}{Bm}(\min(1, B(m+g)) - \tau) \qquad\qquad \text{(B.3.12)} \\
\theta &= 1 + \mu\left(1 - \frac{1}{z^*(\alpha)}\right).
\end{aligned}$$

Let $(\tilde{\alpha}, \tilde{m}, \tilde{\theta})$ denote the solution with $\tilde{\theta} > 1$ and $(\tilde{\alpha}, \tilde{m})$ the steady state of (B.3.1).

(B.3.2):
The properties in (B.3.2) follow from the determination of inflation and real money balances by the first two equations in (B.3.12) which are independent of the real wage. Therefore, the levels of stationary inflation and stationary real money balances are independent of the adjustment parameter μ on the labor market which, however, determines the stationary real wage.

(B.3.3):
The role of the parameters in (B.3.3) are obtained by applying the Implicit Function Theorem to the system (B.3.12) for $Bg > \tau$ and using type uniqueness.

(B.3.4):
Since m_{t+1} is independent of α_t in (B.3.1) the Jacobian matrix has diagonal form and the eigenvalues are real and equal to the diagonal entries. Therefore, the conditions for asymptotic stability (B.3.4) are determined by

$$\begin{aligned}
\nu_1 := \frac{\partial \alpha_t}{\partial \alpha_{t+1}} &= \frac{1+w}{1+p} - \frac{\tilde{\alpha}}{1+p} \cdot \mu \cdot L_{\max} \cdot \frac{1}{1-B}\left(\frac{\tilde{\alpha}}{A}\right)^{\frac{B}{1-B}} \cdot \frac{1}{A} \\
&= 1 - \frac{\mu + 1 - \tilde{\theta}}{\tilde{\theta}(1-B)} \qquad\qquad \text{(B.3.13)} \\
&\in \left(1 - \frac{\mu}{1-B}, 1 - \frac{\mu - \gamma}{(1+\gamma)(1-B)}\right)
\end{aligned}$$

where $1 < \tilde{\theta} < \mu + 1$. Thus, $\nu_1 < 1$. For the second eigenvalue two cases may occur. If $y_{\max} > \tilde{m} + g$ one has

$$\begin{aligned}
\nu_2 := \frac{\partial m_{t+1}}{\partial m_t} &= \frac{1}{1+p} - \frac{\tilde{m} + g - \tau y_{\max}}{1+p)^2} \cdot \gamma \cdot y_{\max} \cdot \frac{1 - c(1-\tau)}{(\tilde{m}+g)^2} \\
&= \frac{1}{\tilde{\theta}}\left(1 - \frac{\tilde{m}}{\tilde{m}+g}(1 + \gamma - \tilde{\theta})\right) \in (0, 1).
\end{aligned}$$

If $y_{\max} < \tilde{m} + g$, one has

$$\begin{aligned} \nu_2 := \frac{\partial m_{t+1}}{\partial m_t} &= -\frac{y_{\max}(1-\tau)}{(1+p)^2} \cdot \gamma \cdot y_{\max} \cdot \frac{1-c(1-\tau)}{(\tilde{m}+g)^2} \\ &= -\frac{\tilde{m}}{\tilde{\theta}} \cdot \frac{1+\gamma-\tilde{\theta}}{\tilde{m}+g} \in (-1,0) \end{aligned}$$

Therefore, ν_2 has always modulus less than one so that (B.3.13) implies that an inflationary steady state is asymptotically stable if and only if

$$2 > \frac{\mu+1-\tilde{\theta}}{(1-B)\tilde{\theta}} \quad \Longleftrightarrow \quad \tilde{\theta}(2(1-B)+1) > \mu+1. \tag{B.3.14}$$

The right hand side of the first inequality is decreasing in $\tilde{\theta}$. With $1+\mu > \tilde{\theta} > 1$, the inequality implies that $2(1-B) > \mu$ is a sufficient condition for stability which is satisfied for $B \leq 1/2$ and all $0 < \mu, \lambda, \gamma, \kappa < 1$.
Condition (B.3.14) shows that stability is violated if and only if

$$\mu + 1 \geq \tilde{\theta}(2(1-B)+1) > (2(1-B)+1)$$

implying that $\mu > 2(1-B)$ must hold if cycles occur, i.e. the condition is necessary for instability and proves part **I** in the Theorem.

(B.3.5):
Some further arguments indicate sufficient conditions for the occurrence of a period doubling bifurcation. The inflation rate $\tilde{\theta}$ is a nonlinear bounded function of (g, τ, B, γ) independent of μ if $\mu \geq \gamma$. Therefore, increasing $\tilde{\theta}$ decreases ν_1 which may cross the value -1 for some $B > 1/2$ inducing a period doubling bifurcation as the only possibility. From

$$[2(1-B)+1]\,\tilde{\theta}(g,\tau,\gamma,B) < [2(1-B)+1]\,(1+\gamma)$$

one obtains

$$\lim_{B\to 1}\,[2(1-B)+1]\,\tilde{\theta}(g,\tau,\gamma,B) \leq (1+\gamma) \leq 1+\mu,$$

which implies the existence of $1 > B > 1/2$ such that

$$[2(1-B)+1]\,\tilde{\theta}(g,\tau,\gamma,B) = (1+\mu)$$

for $\mu \geq \gamma$. Therefore, $\nu_1(g,\tau,\gamma,B,\mu) \leq -1$ with $2(1-B) < \mu$. This implies that $2(1-B) < \mu$ after the occurrence of the period-doubling bifurcation.

(B.3.6):
The properties in (B.3.6) follow from (B.3.3) and (B.3.5).

Steady States in K
In the Keynesian regime with $Bg < \tau$ the two dimensional system (B.3.1) $(\alpha, m) \mapsto \mathcal{F}(\alpha, m)$ takes the form

$$\alpha_{t+1} = \alpha_t \frac{1 + w(s^\ell(\alpha_t, m_t))}{1 + p(s^c(\alpha_t, m_t))}$$

$$m_{t+1} = \frac{C}{1 + p(s^c(\alpha_t, m_t))}(m_t + g), \qquad C = \frac{(1-c)(1-\tau)}{1-c(1-\tau)} < 1$$

with

$$w(s^\ell(\alpha_t, m_t)) = \lambda\left(\frac{F^{-1}(y^D(m_t))}{L_{\max}} - 1\right) \quad \text{and} \quad p(s^c(\alpha_t, m_t)) = \kappa\left(\frac{y^D(m_t)}{y^*(\alpha_t)} - 1\right).$$

The same procedure as above will be used to prove part **K**. Let $Bg < \tau$ and $(\tilde{\alpha}, \tilde{m}) \in$ **K** denote a steady state of the system with $0 < \tilde{\theta} < 1$ where $(\tilde{\alpha}, \tilde{m}, \tilde{\theta})$ is the unique solution of

$$\begin{aligned}
\theta =& C\left(1 + \frac{g}{m}\right) = \frac{C(m+g)}{m} \\
\theta =& 1 + \lambda\left(\frac{F^{-1}(m+g)/(1-c(1-\tau))}{L_{\max}} - 1\right) = 1 - \lambda + \lambda\left(\frac{B(m+g)}{1-c(1-\tau)}\right)^{1/B} \qquad \text{(B.3.15)} \\
\theta =& 1 + \kappa\left(\frac{m+g}{(1-c(1-\tau))y^\star(\alpha)} - 1\right) = 1 - \kappa + \frac{\kappa(m+g)}{(1-c(1-\tau))y^\star(\alpha)}.
\end{aligned}$$

(B.3.7):
$\tilde{\theta}(g, \tau, \lambda, B)$ and $\tilde{m}(g, \tau, \lambda, B)$ are uniquely determined by the first two equations of (B.3.15) which are independent of κ. The real wage α follows from the third equation where κ becomes decisive. This yields (B.3.7).

(B.3.8):
Using the Implicit Function Theorem on the system (B.3.15) one obtains the comparative statics properties given in (B.3.8).

(B.3.9):
Denote by

$$D\mathcal{F}(\tilde{\alpha}, \tilde{m}) = \begin{pmatrix} a & b \\ c & d \end{pmatrix}$$

the Jacobian matrix of $\mathcal{F}$, one finds

$$\begin{aligned}
a &:= \frac{\partial \alpha_{t+1}}{\partial \alpha_t} = 1 - \frac{1}{\tilde{\theta}} \cdot \frac{B}{1-B}(\tilde{\theta} - 1 + \kappa) \\
b &:= \frac{\partial \alpha_{t+1}}{\partial m_t} = \frac{C\tilde{\alpha}}{\tilde{m}\tilde{\theta}^2} \cdot \left(\frac{1}{B}(\tilde{\theta} - 1 + \lambda) - (\tilde{\theta} - 1 + \kappa)\right) \\
c &:= \frac{\partial m_{t+1}}{\partial \alpha_t} = -\frac{\tilde{m}}{\tilde{\theta}\tilde{\alpha}} \cdot \frac{B}{1-B}(\tilde{\theta} - 1 + \kappa) \\
d &:= \frac{\partial m_{t+1}}{\partial m_t} = \frac{C}{\tilde{\theta}^2} \cdot (1 - \kappa)
\end{aligned}$$

This implies as the value of the determinant

$$\begin{aligned} ab - cd &= \frac{C}{\tilde{\theta}^2}(1-\kappa) + \frac{C}{\tilde{\theta}^3} \cdot \frac{B}{1-B}(\tilde{\theta} - 1 + \kappa)\left(\frac{1}{B}(\tilde{\theta} - 1 + \lambda) - \tilde{\theta}\right) \\ &< \frac{C}{\tilde{\theta}^2}(1-\kappa) + \frac{C}{\tilde{\theta}^3} \cdot \frac{B}{1-B}(\tilde{\theta} - 1 + \kappa)\left((\tilde{\theta} - 1 + \lambda) - \tilde{\theta}\right) \\ &< \frac{C}{\tilde{\theta}^2}(1-\kappa) + \frac{C}{\tilde{\theta}^3} \cdot (\tilde{\theta} - 1 + \kappa)(\tilde{\theta} - 1 + \lambda) \\ &< \frac{C}{\tilde{\theta}^2}(1-\kappa) + \frac{C}{\tilde{\theta}^2} \cdot (\tilde{\theta} - 1 + \kappa) = \frac{C}{\tilde{\theta}} \leq 1 \end{aligned}$$

since $0 < C < 1$ and $\tilde{m}\tilde{\theta} = C(\tilde{m} + g)$. Therefore, complex eigenvalues have modulus less than one.
Let the eigenvalues be real. Using Lemma A.1.4 the larger one of the two becomes

$$\nu_2 = \frac{a+d}{2} + \sqrt{\left(\frac{a+d}{2}\right)^2 - (ad - bc)}$$

which is less than one if $\det \mathcal{F} > \operatorname{tr} \mathcal{F} - 1$, i.e. if and only if $-(ad - bc) < 1 - a - d$. This is equivalent to

$$a + d < 1 + \frac{C}{\tilde{\theta}^2}(1-\kappa) < 1 + \frac{1-\kappa}{\tilde{\theta}} < 2.$$

Therefore, $1 > \tilde{\theta} > C$ and $\tilde{\theta} > 1 - \lambda$ imply

$$1 - \frac{C}{\tilde{\theta}} > -\frac{C}{\tilde{\theta}^2} \cdot \frac{1}{B}(\tilde{\theta} - 1 + \lambda)$$

so that $\nu_2 < 1$. As a consequence, a Keynesian steady state $(\tilde{\alpha}, \tilde{m}, \tilde{\theta})$ is asymptotically stable if and only if

$$\nu_1 = \frac{a+d}{2} - \sqrt{\left(\frac{a+d}{2}\right)^2 - (ad - bc)} \tag{B.3.16}$$

which is equivalent to $a + d + 1 > -ad + bc$. In parameter form this translates into the condition

$$\begin{aligned} 2 &> \frac{B}{1-B}\left(1 - \frac{1-\kappa}{\tilde{\theta}}\right) - \frac{C(1-\kappa)}{\tilde{\theta}^2} - \frac{C}{\tilde{\theta}} + \frac{C(1-\lambda)(\tilde{\theta} - 1 + \kappa)}{\tilde{\theta}^3(1-B)} \\ &= \frac{\tilde{\theta} - 1 + \kappa}{\tilde{\theta}(1-B)}\left(B + \frac{C(1-\lambda)}{\tilde{\theta}^2}\right) - \frac{C}{\tilde{\theta}^2}(1 - \kappa + \tilde{\theta}) \end{aligned} \tag{B.3.17}$$

which proves (B.3.9)

(B.3.10):
Applying the same procedure as in the inflationary case, let $\kappa > \lambda$ one finds

$$\nu_1 \leq -1 \quad \iff \quad 2 \leq \frac{\tilde{\theta} - 1 + \kappa}{\tilde{\theta}(1 - B)} \left(B + \frac{C(1 - \lambda)}{\tilde{\theta}^2} \right) - \frac{C}{\tilde{\theta}^2}(1 - \kappa + \tilde{\theta}).$$

For $\kappa > \lambda$, the right hand side of the inequality is linear and increasing in κ while $\tilde{\theta}$ is independent of κ. Therefore,

$$\begin{aligned} &\lim_{\kappa \to 1} \frac{\tilde{\theta} - 1 + \kappa}{\tilde{\theta}(1 - B)} \left(B + \frac{C(1 - \lambda)}{\tilde{\theta}^2} \right) - \frac{C}{\tilde{\theta}^2}(1 - \kappa + \tilde{\theta}) \\ &= \frac{1}{1 - B} \left(B + \frac{C(1 - \lambda)}{\tilde{\theta}^2} \right) - \frac{C}{\tilde{\theta}} > \frac{1}{1 - B}(2B - 1) \end{aligned}$$

which is larger than 2 for $B > 3/4$. Continuity and monotonicity of the right hand side of (B.3.17) imply that there exists a unique $1 > \bar{\kappa} \geq \lambda > 0$ with $\nu_1(g, \tau, \lambda, B, \bar{\kappa}) = -1$, such that $\kappa \geq \bar{\kappa} \Rightarrow \nu_1(g, \tau, \lambda, B, \kappa) < -1$, proving the necessary condition for cycles for $B > 1/2$.

(B.3.11):
The properties in (B.3.11) follow from (B.3.8) and (B.3.10). □

B.3.2 Proof of Theorem 7.5.1

a) With constant labor supply and no expectations feed back in consumption type uniqueness follows from Theorem 7.4.1. The remaining properties (b) to (d) are proved using the results of Theorem B.3.1.
b) Global stability for $0 < B \leq 1/2$ is proved in (B.3.14) and (B.3.17) of Theorem B.3.1.
c) Condition (B.3.14) shows that stability is violated if and only if

$$\mu + 1 \geq \tilde{\theta}(2(1 - B) + 1) > (2(1 - B) + 1).$$

Therefore,

$$\mu > \bar{\mu} := \tilde{\theta}(2(1 - B) + 1) - 1 > (2(1 - B) + 1) - 1 = 2(1 - B)$$

implies that $\mu > 2(1 - B)$ must hold if cycles occur in **I**.
d) If condition (B.3.9) is violated for $(\tilde{\alpha}, \tilde{m}, \tilde{\theta}) \in \mathbf{K}$, (B.3.10) shows that there exists $(g, \tau, \lambda, B, \bar{\kappa})$ with $B > 1/2$, $1 > \bar{\kappa} \geq \lambda >$ and $\nu_1(g, \tau, \lambda, B, \bar{\kappa}) = -1$ such that $\kappa > \bar{\kappa}$ implies $\nu_1(g, \tau, \lambda, B, \kappa) < -1$, proving the occurrence of a period doubling bifurcation and the necessary condition for cycles. □

References

ABEL, A. B. & B. S. BERNANKE (2005): *Macroeconomics*. Addison-Wesley, New York.

AKERLOF, G. & J. YELLEN (1986): *Efficiency Wage Models of the Labor Market*. Cambridge (UK): Cambridge University Press.

AKERLOF, G. & J. L. YELLEN (1985): "A Near-Rational Model of the Business Cycle, with Wage and Price Inertia", *Quarterly Journal of Economics*, 100, Supplement, 823–838.

ALLEN, R. (1967): *Macro-Economic Theory: A Mathematical Treatment*. MacMillan, London a.o.

ARNOLD, L. (1998): *Random Dynamical Systems*. Springer-Verlag, Berlin a.o.

ARNOLD, L., G. BLECKERT & K. R. SCHENK-HOPPÉ (1999): *The Stochastic Brusselator: Parametric Noise Destroys Hopf Bifurcation*pp. 71–92. Springer New York, New York, NY.

ARNOLD, L. & P. BOXLER (1991): "Additive noise turns a hyperbolic fixed point into a stationary solution", in *Lyapunov Exponents*, ed. by L. Arnold, H. Crauel & J. Eckmann, Vol. 1486 of *Lecture Notes on Mathematics*, pp. 159–164. Springer Verlag, Berlin.

——— (1992): "Stochastic bifurcation: Instructive examples in dimension one", in *Stochastic Flows*, ed. by M. A. Pinsky & V. Wistutz, Vol. II of *Diffusion Processes and Related Problems in Analysis*, pp. 241–255. Birkhäuser, Boston.

ARNOLD, L. & H. CRAUEL (1992): "Iterated Function Systems and Multiplicative Ergodic Theory", in *Diffusion Processes and Related Problems in Analysis*, ed. by M. Pinsky & V. Wihstutz, Vol. II, pp. 283–305. Birkhäuser.

ARNOLD, L. & V. WIHSTUTZ (1982): "Stationary solutions of linear systems with additive and multiplicative noise", *Stochastics*, 7, 133–155.

ARROW, K. J. (1951): "An Extension of the Basic Theorems of Classical Welfare Economics", in *Proceedings of the Second Berkeley Symposium on Mathematical Statistics and Probability*, ed. by J. Neyman. University of California Press, Berkeley, Cali.

ARROW, K. J. & F. H. HAHN (1971): *General Competitive Analysis*. Holden Day, San Francisco, Cali.

AVRUTIN, V. & I. SUSHKO (2013): "A Gallery of Bifurcation Scenarios in Piecewise Smooth 1D Maps", in *Global Analysis of Dynamic Models in Economics and Finance*, ed. by G.-I. Bischi, C. Chiarella & I. Sushko. Springer.

AZARIADIS, C. (1993): *Intertemporal Macroeconomics*. Blackwell Publishers, Oxford a.o.

BARNSLEY, M. (1988): *Fractals Everywhere*. Academic Press, New York a.o.

BARNSLEY, M. F. & J. ELTON (1988): "A new class of Markov processes for image encoding", *Advances in Applied Probability*, 20, 14–32.

BARRO, R. J. (1994): "The Aggregate-Supply/Aggregate-Demand Model", *Eastern Economic Journal*, 20(1), 1–6.

BARRO, R. J. & H. I. GROSSMAN (1971): "A General Disequilibrium Model of Income and Employment", *American Economic Review*, 61, 82–93.

——— (1976): *Money, Employment and Inflation*. Cambridge University Press.

BELLMAN, R. (1957): *Dynamic Programming*. Princeton University Press.

BENASSY, J.-P. (1975a): "Disequilibrium Exchange in Barter and in Monetary Economies", *Economic Inquiry*, 13, 131–156.

BENASSY, J.-P. (1975b): "Neo-Keynesian Disequilibrium Theory in a Monetary Economy", *Review of Economic Studies*, 42, 503–523.

——— (1982): *The Economics of Market Disequilibrium*, Economic Theory, Econometrics, and Mathematical Economics. Academic Press, New York a.o.

——— (1983): "The three regimes of the IS-LM model", *European Economic Review*, 23, 1–17.

BENASSY, J.-P. (1984): "A Non-Walrasian Model of the Business Cycle", *Journal of Economic Behavior & Organization*, 5, 77–89.

BENASSY, J.-P. (1986): *Macroeconomics: An Introduction to the Non-Walrasian Approach*. Academic Press, New York a.o.

——— (1987): "Imperfect Competition, Unemployment and Policy", *European Economic Review*, 31, 417–426.

——— (1988): "The Objective Demand Curve in General Equilibrium with Price Makers", *The Economic Journal*, 98(390 Supplement), 37–49.

BHATTACHARYA, R. & M. MAJUMDAR (2004): "Random Dynamical Systems", *Economic Theory*, 23, 13–38.

BINMORE, K., A. RUBINSTEIN & A. WOLINSKY (1986): "The Nash Bargaining Solution in Economic Modelling", *Rand Journal*, 17, 176–188.

BLANCHARD, O. (2003): *Macroeconomics*. Prentice Hall, 3rd edition.

BLANCHARD, O. J. (1979): "Backward and Forward Solutions for Economies with Rational Expectations", *American Economic Review – Papers and Proceedings*, 69(2), 114–118.

BLANCHARD, O. J. & S. FISCHER (1989): *Lectures on Macroeconomics*. MIT Press, Cambridge (Mass.) a.o.

BLANCHARD, O. J. & N. KIYOTAKI (1987): "Monopolistic Competition and the Effects of Aggregate Demand", *American Economic Review*, 77, 647–666.

BLANCHARD, O. J. & D. QUAH (1989): "The Dynamic Effects of Aggregate Demand and Supply Disturbances", *American Economic Review*, 79 No. 4, 655–673.

BLINDER, A. S. & R. M. SOLOW (1973): "Does fiscal policy matter", *Journal of Public Economics*, 2, 319–370.

BLUME, L., M. BRAY & D. EASLEY (1982): "Introduction to the Stability of Rational Expectations Equilibrium", *Journal of Economic Theory*, 26, 313–317.

BÖHM, V. (1978a): "Disequilibrium dynamics in a simple macroeconomic model", *Journal of Economic Theory*, 17, 179–199.

——— (1978b): "Zur Dynamik temporärer Gleichgewichtsmodelle mit Mengenrationierung", *Schriften des Vereins für Socialpolitik*, 98, 255–274.

——— (1980): *Preise, Löhne und Beschäftigung*. J.C.B. Mohr (Paul Siebeck), Tübingen.

——— (1989): *Disequilibrium and Macroeconomics*. Blackwell Publishers, Oxford a.o.

——— (1993): "Recurrence in Keynesian Macroeconomic Models", in *Nonlinear Dynamics in Economics and Social Sciences*, ed. by F. Gori, L. Geronazzo & M. Galeotti. Springer-Verlag, Heidelberg a.o.

——— (1994): "The Foundation of the Theory of Monopolistic Competition Revisited", *Journal of Economic Theory*, 63, 208–218.

——— (1999): "Stochastische Wachstumszyklen aus dynamischer Sicht", in *Trend und Zyklus: Zyklisches Wachstum aus der Sicht moderner Konjunktur- und Wachstumstheorie*, ed. by W. Franz, H. Hesse, H. J. Ramser & M. Stadler, Vol. 28, pp. 83–115.

——— (2001): "The Keynesian Business Cycle with Random Productivity", in *Beiträge zur Mikro- und zur Makroökonomik*, ed. by S. Berninghaus & M. Braulke. Springer-Verlag, Heidelberg a.o.

——— (2003): "MACRODYN – The Handbook", Discussion Paper 498, Department of Economics, Bielefeld University.

——— (2006): "Multiplier–Accelerator Models with Random Perturbations", in *Business Cycle Dynamics – Models and Tools*, ed. by T. Puu & I. Sushko, Vol. VII, Chap. 4, pp. 113–142. Springer-Verlag, Heidelberg a.o.

——— (2009): "Lectures on the Theory of Economic Growth", Monograph, Department of Economics, Bielefeld University, Bielefeld.

Böhm, V. & C. Chiarella (2005): "Mean Variance Preferences, Expectations Formations, and the Dynamics of Random Asset Prices", *Mathematical Finance*, 15(1), 61–97.

Böhm, V., C. Chiarella, X.-Z. He & T. Hüls (2013): "A Homoclinic Route to Volatility: Dynamics of Asset Prices Under Autoregressive Forecasting", in *Global Analysis of Dynamic Models in Economics and Finance*, ed. by G. Bischi, C. Chiarella & I. Sushko, pp. 287–314. Springer-Verlag, Berlin Heidelberg.

Böhm, V. & O. Claas (2017): "Efficient Wage Bargaining in a Dynamic Macroeconomic Model", Discussion Paper 465 (revised), Institute of Mathematical Economics, Bielefeld University.

Böhm, V. & J. Eichberger (2006): "Production Efficiency and Imperfect Competition", in *Institutions in Perspective*, ed. by U. Bindseil, J. Haucap & C. Wey, pp. 155–173. Mohr Siebeck, Tübingen.

Böhm, V. & L. Kaas (2000): "Differential Savings, Factor Shares, and Endogenous Growth Cycles", *Journal of Economic Dynamics and Control*, 24, 965–980.

Böhm, V., M. Lohmann & H.-W. Lorenz (1994): "Dynamic Complexity in a Keynesian Macroeconomic Model", Discussion Paper 288, Department of Economics, Bielefeld University.

Böhm, V., T. Pampel & J. Wenzelburger (2005): "On the Stability of Balanced Growth", Discussion paper no. 548, Department of Economics, Bielefeld University, Bielefeld.

Böhm, V. & K.-R. Schenk-Hoppé (1998): "MACRODYN A User's Guide", Discussion Paper 400, Department of Economics, Bielefeld University.

Böhm, V. & J. Wenzelburger (1996): "Cycles and Local Stability in Discrete Dynamical Systems with Delay Structure", Discussion Paper 320, Department of Economics, Bielefeld University.

——— (1999): "Expectations, Forecasting and Perfect Foresight – A Dynamical Systems Approach", *Macroeconomic Dynamics*, 3(2), 167–186.

——— (2002): "Perfect Predictions in Economic Dynamical Systems with Random Perturbations", *Macroeconomic Dynamics*, 6(5), 687–712.

——— (2004): "Expectational Leads in Economic Dynamical Systems", in *Economic Complexity: Non-linear dynamics, multi-agents economies, and learning*, ed. by W. Barnett, C. Deissenberg & G. Feichtinger, Vol. 14 of *International Symposia in Economic Theory and Econometrics*, Chap. 13, pp. 333–361. Elsevier, Amsterdam.

——— (2005): "On the Performance of Efficient Portfolios", *Journal of Economic Dynamics and Control*, 29(4), 721–740.

Bonanno, G. (1990): "General Equilibrium Theory with Imperfect Competition", *Journal of Economic Surveys*, 4(4), 297–328.

Booth, A. L. (1995): *The economics of the trade union*. Cambridge University Press, Cambridge, New York a.o.

Bray, M. (1982): "Learning, Estimation, and the Stability of Rational Expectations Equilibria", *Journal of Economic Theory*, 26, 318–339.

Bray, M. & N. E. Savin (1986): "Rational Expectations Equilibria, Learning, and Model Specification", *Econometrica*, 54, 1129–1160.

Brock, W. A. & C. H. Hommes (1997): "A Rational Route to Randomness", *Econometrica*, 65(5), 1059–1095.

Calvo, G. A. (1983): "Staggered Prices in a Utility-Maximizing Framework", *Journal of Monetary Economics*, 12(3), 383–398.

Chen, X. & H. White (1998): "Nonparametric Adaptive Learning with Feedback", *Journal of Economic Theory*, 82(1), 199–222.

Chiarella, C., R. Dieci & X.-Z. He (2009): "Heterogeneity, Market Mechanisms, and Asset Pricing", in *Handbook of Financial Markets: Dynamics and Evolution*, ed. by T. Hens & K. R. Schenk-Hoppé, Chap. 5. North-Holland, Elsevier, Oxford a.o.

Claas, O. (2017): "Essays on Wage Bargaining in Dynamic Macroeconomics", Ph.D. thesis, Department of Business Administration and Economics, Bielefeld University.

Clarida, R., J. Galí & M. Gertler (1999): "The Science of Monetary Policy: A New Keynesian Perspective", *Journal of Economic Literature*, 37(4), 1661–1707.

Clower, R. W. (1965): "The Keynesian Counterrevolution: A Theoretical Appraisal", in *The Theory of Interest Rates*, ed. by F. H. Hahn & F. Brechling, London. Macmillan.

——— (1967): "A Reconsideration of the Microfoundation of Monetary Theory", *Western Economic Journal*, 6, 1–9.

Crauel, H. & F. Flandoli (1998): "Additive Noise Destroys a Pitchfork Bifurcation", *Journal of Dynamics and Differential Equations*, 10(2), 259–274.

Day, R. & W. Shafer (1987): "Ergodic fluctuations in deterministic economic models", *Journal of Economic Bahaviour and Organization*, 8, 339–361.

Day, R. H. (1982): "Irregular Growth Cycles", *American Economic Review*, 72, 406–414.

Day, R. H. (1984): "Disequilibrium Economic Dynamics", *Journal of Economic Behavior & Organization*, 5, 57–76.

Day, R. H. & G. Pianigiani (1991): "Statistical dynamics and economics", *Journal of Economic Behavior & Organization*, 16, 37–83.

Deardorff, A. V. (1970): "Growth Paths in the Solow Neoclassical Growth Model", *Quarterly Journal of Economics*, 84, 134–139.

Debreu, G. (1959): *Theory of Value*. John Wiley & Sons, Inc., New York a.o.

Devaney, R. L. (1989): *An Introduction to Chaotic Dynamical Systems*. Addison-Wesley, Reading, 2nd edition.

Dierker, H. & B. Grodal (1986): "Non-Existence of Cournot-Walras Equilibrium in a General Equilibrium Model with Two Oligopolists", in *Contributions to Mathematical Economics. In honour of Gérard Debreu*, ed. by W. Hildenbrand & A. Mas-Colell, Chap. 10. North-Holland Publishing Company, Amsterdam a.o.

Dixon, H. D. (2008): "New Keynesian Macroeconomics", in *The New Palgrave Dictionary of Economics*, ed. by S. N. Durlauf & L. E. Blume. Palgrave Macmillan, Basingstoke.

Dornbusch, R. & S. Fischer (1991): *Lecture on Macroeconomics*. MIT Press, Cambridge (Mass.).

Drèze, J. H. (1975): "Existence of Exchange Equilibrium under Price Rigidity", *International Economic Review*, 16, 301–320.

——— (1991): *Underemployment Equilibria – Essays in Theory, Econometrics and Policy*. Cambridge University Press.

Drèze, J. H. (2001): "Indeterminateness of Equilibria and Macroeconomics", in *Equilibrium, Markets and Dynamics – Essays in Honour of Claus Weddepohl*, ed. by C. H. Hommes, R. Ramer & C. A. Withagen. Springer.

Duffie, D., J. Geanakoplos, A. Mas-Colell & A. McLennan (1994): "Stationary Markov Equilibria", *Econometrica*, 62(4), 745–781.

Evans, G. W. & S. Honkapohja (1999): "Learning Dynamics", in *Handbook of Macroeconomics*, ed. by J. Taylor & M. Woodford, Vol. 1A, Chap. 7, pp. 449–533. Elsevier Science B.V., Amsterdam a.o.

——— (2001): *Learning and Expectations in Macroeconomics*. Princeton University Press.

Evstigneev, I. V. & S. A. Pirogov (2007): "A stochastic contraction principle", *Random Operators and Stochastic Equations*, 15(2), 155–162.

Ezekiel, M. (1938): "The cobweb theorem", *Quarterly Journal of Economics*, 52, 255–280.

Farmer, R. E. A., D. F. Waggoner & T. Zha (2009a): "Indeterminacy in a forward looking regime switching model", *International Journal of Economic Theory*, 5, 69–84.

——— (2009b): "Understanding Markov-switching rational expectations models", *Journal of Economic Theory*, 144(5), 1849–1867.

Felderer, B. & S. Homburg (1987): *Macroeconomics and New Macroeconomics*. Springer-Verlag, Berlin a.o.

Friedman, M. (ed.) (1965): *Studies in the Quantity Theory of Money*. University of Chicago Press.

Frisch, R. (1933): "Propagation Problems and Impulse Problems in Dynamic Economics", in *Economic Essays in Honour of Gustav Cassel*, pp. 171–205. Allen & Unwin, London.

GAUBE, T. (1997): "Numéraire und Second-Best-Gleichgewicht: Eine Untersuchung allokativer Preisnormierungseffekte in Ramsey-Boiteux-Modellen", Inauguraldissertation, Universität Mannheim, Fakultät für Volkswirtschaftslehre.

GORDON, R. J. (1990): "What is New-Keynesian Economics?", *Journal of Economic Literature*, 28(3), 1115–1171.

GRANDMONT, J.-M. (1974): "On the Short-Run Equilibrium in a Monetary Economy", in *Allocation under Uncertainty: Equilibrium and ptimality*, ed. by J. H. Drèze. Macmillan, London.

GRANDMONT, J.-M. (1983): *Money and Value – A Reconsideration of Classical and Neoclassical Theories*. Cambridge University Press, Cambridge (Mass.) a.o.

——— (1985): "On Endogenous Competitive Business Cycles", *Econometrica*, 53(5), 995–1045.

——— (1988): "Introduction", in *Temporary Equilibrium — Selected Readings*, ed. by J.-M. Grandmont, pp. xiii–xxiv. Academic Press.

GRANDMONT, J.-M. & G. LAROQUE (1986): "Stability of Cycles and Expectations", *Journal of Economic Theory*, 40, 138–151.

——— (1991): "Economic Dynamics with Learning: Some Instability Examples", in *Equilibrium Theory and Applications*, ed. by W. A. Barnett, B. Cornet, J. D'Aspremont, J. Gabszewics & A. Mas-Collel, pp. 247–273. Cambridge University Press, Cambridge (Mass.) a.o.

GUCKENHEIMER, J. & P. HOLMES (1983): *Nonlinear Oscillations, Dynamical Systems, and Bifurcations of Vector Fields*. Springer, New York.

GÜTH, W. & R. SELTEN (1982): "Game Theoretical Analysis of Wage Bargaining in a Simple Business Cycle Model", *Journal of Mathematical Economics*, 10, 177–195.

HAHN, F. (1982): *Money and Inflation*. Basil Blackwell, Oxford.

HAHN, F. H. (1965): "On Some Problems of Proving the Existence of an Equilibrium of a Monetary Economy", in *The Theory of Interest Rates*, ed. by F. H. Hahn & F. P. R. Brechling. MacMillan Press.

HALE, J. & H. KOÇAK (1991): *Dynamics and Bifurcation*. Springer, New York.

HICKS, J. (1950): *A Contribution to the Theory of the Trade Cycle*. Oxford University Press, Oxford.

HICKS, J. R. (1939): *Value and Capital*. Clarendon Press Oxford, London, 1st edition.

HOMMES, C. H. (1991): *Chaotic Dynamics in Economic Models – Some Simple Case–Studies*. Wolters–Noordhoff, Groningen.

——— (1994): "Dynamics of the Cobweb Model with Adaptive Expectations and Nonlinear Supply and Demand", *Journal of Economic Behavior & Organization*, 24, 315–335.

——— (1998): "On The Consistency of Backward Looking Expectations: The Case of The Cobweb", *Journal of Economic Behavior & Organization*, 33, 333–362.

HUTCHINSON, J. E. (1981): "Fractals and Self Similarity", *Indiana Univ. Mathematical Journal*, 30, 713–747.

ISHIDA, J. & M. YOKOO (2003): "Heterogeneity-Induced Chaos: An Example", Discussion paper, Shinshu University and Okayama University.

JENSEN, B. S. (1994): *The dynamic systems of basic economic growth models*, No. 302 in Mathematics and its applications. Kluwer Acad. Publ., Dordrecht a.o.

JENSEN, R. V. & R. URBAN (1984): "Chaotic price behavior in a nonlinear cobweb model", *Economic Letters*, 15, 235–240.

JUNGEILGES, J. (2003): "Sequential Computation of Sample Moments and Sample Quantiles – A Tool for MACRODYN –", Discussion Paper 499, Bielefeld University, Department of Economics, Bielefeld, Germany.

KAAS, L. (1995): "Steady States and Local Bifurcations in a Dynamic Disequilibrium Model", Discussion Paper 300, Department of Economics, Bielefeld University.

KEENER, J. P. (1980): "Chaotic behavior in piecewise continuous difference equations", *Transactions of the American Mathematical Society*, 261, 589–604.

KEMPF, A. (1999): *Wertpapierliquidität und Wertpapierpreise*. Deutscher Universitäts-Verlag, Mannheim.

KEYNES, J. M. (1936): *The General Theory of Employment, Interest and Money*. Macmillan Press Ltd., London.

Kuznetsov, Y. A. (1995): *Elements of Applied Bifurcation Theory*, Applied Mathematical Sciences 112. Springer Verlag, New York.

Landmann, O. & J. Jerger (1999): *Beschäftigungstheorie*. Springer, Berlin a.o.

Lasota, A. & M. Mackey (1994): *Chaos, Fractals, and Noise: Stochastic Aspects of Dynamics*, Vol. 97 of *Applied Mathematical Science*. Springer-Verlag, New York a.o.

Leijonhufvud, A. (1968): *On Keynesian Economics and the Economics of Keynes*. Oxford University Press, Oxford.

Lettau, M. & H. Uhlig (2002): "The Sharpe Ratio and Preferences: A Parametric Approach", *Macroeconomic Dynamics*, 6, 242–265.

Li, T. & J. A. Yorke (1975): "Period Three Implies Chaos", *American Mathematical Monthly*, 82, 985–992.

Lindahl, E. (1939): *Studies in the Theory of Money and Capital*. Allen & Unwin, London.

Lohmann, M. & J. Wenzelburger (1996): "A Statistical Method for Detecting Cycles in Discrete Dynamical Systems", *International Journal of Bifurcation and Chaos in Applied Science and Engineering*, 6(12A), 2375–2388.

Lorenz, H.-W. & M. Lohmann (1996): "On the Role of Expectations in a Dynamic Keynesian Macroeconomic Model", *Chaos, Solitons & Fractals*, 7(12), 2135–2155.

Lucas, R. E. (1972): "Expectations and the Neutrality of Money", *Journal of Economic Theory*, 4, 103–124.

Lucas, R. E. (1973): "Some International Evidence on Output-Inflation Tradeoffs", *American Economic Review, Papers and Proceedings*, 63 (2), 326–334.

Lucas, R. E. & L. A. Rapping (1969a): "Price Expectations and the Phillips Curve", *American Economic Review*, 59(3), 345–350.

——— (1969b): "Real wages, employment and inflation", *Journal of Political Economy*, 77, 721–754.

Malinvaud, E. (1977): *The Theory of Unemployment Reconsidered*. Blackwell Publishers, Oxford.

——— (1980): *Profitability and Unemployment*. Cambridge University Press, Cambridge (Mass.).

Marcet, A. & T. J. Sargent (1989): "Convergence of Least Squares Learning Mechanisms in Self-Referential Linear Stochastic Models", *Journal of Economic Theory*, 48, 337–368.

Marschak, T. & R. Selten (1974): *General Equilibrium with Price-Making Firms*. Springer-Verlag, Heidelberg a.o.

McDonald, I. M. & R. M. Solow (1981): "Wage Bargaining and Employment", *The American Economic Review*, 71, 896–908.

Negishi, T. (1961): "Monopolistic Competition and General Equilibrium", *Review of Economic Studies*, 28, 196–201.

——— (1987): "Monopolistic Competition and General Equilibrium", in *The New Palgrave: A Dictionary of Economics*, ed. by J. Eatwell, M. Murray & P. Newman, Vol. 3. Macmillan Press Ltd., London a.o.

Nerlove, M. (1958): "Adaptive expectations and cobweb phenomena", *Quarterly Journal of Economics*, 72, 227–240.

Onozaki, T., G. Sieg & M. Yokoo (2000): "Complex Dynamics in a Cobweb Model with Adaptive Production Adjustment", *Journal of Economic Behavior & Organization*, 41(2), 101–115.

Pampel, T. (2009): "On the Dynamics of Basic Growth Models: Ratio Stability versus Convergence", *German Economic Review*, 10(4), 384–400.

——— (2010): *Mathematik für Wirtschaftswissenschaftler*. Springer.

Pashigian, P. B. (1987): "Cobweb Theorem", in *The New Palgrave*, ed. by J. Eatwell, M. Milgate & P. Newman, Vol. I. Macmillan Press, London a.o.

Patinkin, D. (1965): *Money, Interest and Prices: An Integration of Monetary and Value Theory*. Harper & Row, New York.

Pencavel, J. (1985): "Wages and Employment under Trade Unionism: Microeconomic Models and Macroeconomic Applications", *Scandinavian Journal of Economics*, 87, 197–225.

Pigou, A. C. (1933): *The Theory of Unemployment*. Macmillan Press Ltd., Routledge 2013 edition.

Roberts, D. J. & H. Sonnenschein (1976): "On the Existence of Cournot Equilibrium without Concave Profit Functions", *Journal of Economic Theory*, 8, 112–117.

Romer, D. (2000): "Keynesian Macroeconomics without the LM Curve", *The Journal of Economic Perspectives*, 14(2), 149–169.

——— (2005): *Advanced Macroeconomics*. McGraw–Hill, New York, third edition.

Rosenmüller, J. (2000): *Game theory : stochastics, information, strategies and cooperation*, No. 25 in Theory and decision library : Series C, Game theory, mathematical programming and operations research. Kluwer Academic Publ., Boston.

Samuelson, P. A. (1974): "Remembrances of Frisch", *European Economic Review*, 5, 7–23.

Samuelson, P. A. (1976): "Optimality of sluggish predictors under ergodic probabilities", *International Economic Review*, 17, 1–7.

Sargent, T. & N. Wallace (1975): "'Rational' Expectations, the Optimal Monetary Instrument, and the Optimal Money Supply Rule", *Journal of Political Economics*, 83(2), 241–254.

Sargent, T. J. (1993): *Bounded Rationality in Macroeconomics*. Clarendon Press, Oxford.

Schenk-Hoppé, K. R. (2005): "Poverty traps and business cycles in a stochastic overlapping generations economy with S-shaped law of motion", *Journal of Macroeconomics*, 27(2), 275–288.

Schenk-Hoppé, K.-R. & B. Schmalfuss (2001): "Random Fixed Points in a Stochastic Solow Growth Model", *Journal of Mathematical Economics*, 36(1), 19–30.

Schmalfuss, B. (1996): "A Random Fixed Point Theorem Based on Lyapunov Exponents", *Random and Computational Dynamics*, 4(4), 257–268.

Schmalfuss, B. (1998): "A Random Fixed Point Theorem and the Random Graph Transformation", *Journal of Mathematical Analysis and Application*, 225(1), 91–113.

Schulz, N. (1985): "Existence of Equilibria Based on Continuity and Boundary Behavior", *Economics Letters*, 19, 101–103.

Shapiro, C. & J. Stiglitz (1984): "Equilibrium Unemployment as a Worker Discipline Device", *American Economic Review*, 74, 433–44.

Silvestre, J. (1977): "A model of General Equilibrium with Monopolistic Behavior", *Journal of Economic Theory*, 16, 425–442.

Solow, R. M. (1979): "Another Possible Source of Wage Stickiness", *Journal of Macroeconomics*, 1, 79–82.

Stiefenhofer, A. (1999): "Dynamik in Cobweb-Modellen mit rekursiven Erwartungen", Inauguraldissertation, Universität Mannheim, Fakultät für Volkswirtschaftslehre, Mannheim.

Taylor, J. B. (1977): "Conditions for Unique Solutions in Stochastic Macroeconomic Models with Rational Expectations", *Econometrica*, 45(6), 1377–1386.

——— (1993): "Discretion Versus Policy Rules in Practice", in *Carnegie-Rochester Conference Series on Public Policy*, pp. 195–214.

Tobin, J. & W. Buiter (1976): "Longrun effects of fiscal and monetary policy on aggregate demand", in *Monetarism*, ed. by J. Stein, pp. 273–309. Amsterdam: North Holland.

Tramontana, F., L. Gardini & A. Agliari (2011): "Endogenous cycles in discontinuous growth models", *Mathematics and Computers in Simulations*, 81, 1625–1639.

Turnovsky, S. J. (1977): *Macroeconomic Analysis and Stabilization Policies*. Cambridge University Press, Cambridge (Mass.) a.o.

——— (1995): *Methods of Macroeconomic Dynamics*. MIT Press, Cambridge (Mass.) a.o.

Wenzelburger, J. (1996): "The Construction of Cell-to-Cell of Mappings from Frobenius-Perron Operators", in *Third Meeting on Current Ideas in Mechanics and Related Fields*, ed. by M. De León, M. Epstein & M. Elzanowski, Vol. 11 of *Extracta Mathematicae*, pp. 243–254. Universidad de Extremadura.

Woodford, M. (1984): "Indeterminacy of Equilibrium in the Overlapping Generations Model: A Survey", Mimeo, Columbia University, New York.

Younès, Y. (1975): "On the Role of Money in the Process of Exchange and the Existence of a Non-Walrasiansian Equilibrium", *Review of Economic Studies*.

Zenner, M. (1996): *Learning to Become Rational*, Lecture Notes in Economics and Mathematical Systems Vol. 439. Springer-Verlag, Berlin a.o.

Printed by Printforce, the Netherlands